New Developments in NMR

Titles in the series:

1: Contemporary Computer-Assisted Approaches to Molecular Structure Elucidation
2: New Applications of NMR in Drug Discovery and Development
3: Advances in Biological Solid-State NMR
4: Hyperpolarized Xenon-129 Magnetic Resonance: Concepts, Production, Techniques and Applications
5: Mobile NMR and MRI: Developments and Applications
6: Gas Phase NMR
7: Magnetic Resonance Technology: Hardware and System Component Design
8: Biophysics and Biochemistry of Cartilage by NMR and MRI
9: Diffusion NMR of Confined Systems: Fluid Transport in Porous Solids and Heterogeneous Materials
10: NMR in Glycoscience and Glycotechnology
11: Fast NMR Data Acquisition: Beyond the Fourier Transform
12: Cross-relaxation and Cross-correlation Parameters in NMR: Molecular Approaches
13: Contrast Agents for MRI: Experimental Methods
14: NMR-based Metabolomics
15: Modern Methods in Solid-state NMR: A Practitioner's Guide
16: Paramagnetism in Experimental Biomolecular NMR
17: Optimizing NMR Methods for Structure Elucidation: Characterizing Natural Products and Other Organic Compounds
18: Field-cycling NMR Relaxometry: Instrumentation, Model Theories and Applications
19: Hybrid MR-PET Imaging of the Brain: Systems, Methods and Applications
20: NMR Methods for Characterization of Synthetic and Natural Polymers
21: In-cell NMR Spectroscopy: From Molecular Sciences to Cell Biology

22: Long-lived Nuclear Spin Order: Theory and Applications
23: NMR and MRI of Gels
24: Advanced Diffusion Encoding Methods in MRI
25: NMR and MRI of Electrochemical Energy Storage Materials and Devices
26: NMR Spectroscopy for Probing Functional Dynamics at Biological Interfaces
27: Fast 2D Solution-state NMR: Concepts and Applications
28: Practical NMR for Oil and Gas Exploration
29: Biophysics of Molecular Chaperones: Function, Mechanisms and Client Protein Interactions

How to obtain future titles on publication:
A standing order plan is available for this series. A standing order will bring delivery of each new volume immediately on publication.

For further information please contact:
Book Sales Department, Royal Society of Chemistry, Thomas Graham House, Science Park, Milton Road, Cambridge, CB4 0WF, UK
Telephone: +44 (0)1223 420066, Fax: +44 (0)1223 420247
Email: booksales@rsc.org
Visit our website at www.rsc.org/books

Biophysics of Molecular Chaperones
Function, Mechanisms and Client Protein Interactions

Edited by

Sebastian Hiller
Biozentrum, University of Basel, Switzerland
Email: sebastian.hiller@unibas.ch

Maili Liu
Chinese Academy of Sciences, Wuhan, China
Email: ml.liu@wipm.ac.cn

and

Lichun He
Chinese Academy of Sciences, Wuhan, China
Email: helichun@wipm.ac.cn

ROYAL SOCIETY OF CHEMISTRY

New Developments in NMR No. 29

Print ISBN: 978-1-83916-282-4
PDF ISBN: 978-1-83916-598-6
EPUB ISBN: 978-1-83916-599-3
Print ISSN: 2044-253X
Electronic ISSN: 2044-2548

A catalogue record for this book is available from the British Library

The Royal Society of Chemistry is a charity, registered in England and Wales, Number 207890, and a company incorporated in England by Royal Charter (Registered No. RC000524), registered office: Burlington House, Piccadilly, London W1J 0BA, UK, Telephone: +44 (0)20 7437 8656.

For further information see our web site at www.rsc.org

Printed in the United Kingdom by CPI Group (UK) Ltd, Croydon, CR0 4YY, UK

Preface

The complex and dynamic interactions between chaperones and their client proteins are critical for cellular homeostasis. Despite its essentiality, the process of protein folding is prone to errors that can lead to misfolding, aggregation, and consequently disease. By preventing and correcting these errors, the chaperone machinery plays a crucial role in all kingdoms of life.

Remarkably, certain fundamental principles of molecular chaperones and their interactions with client proteins have only recently begun to emerge, as biophysical methods have sufficiently matured to resolve dynamic processes and transient interactions down to the atomic level. The resulting detailed insights into mechanisms and structural details have proved to be highly valuable.

This book presents an overview of modern biophysical methods used to study chaperones and chaperone–client interactions, as well as key findings emerging from these works. Experts in the respective fields have contributed the individual chapters, highlighting how recent technical advances have sharpened our understanding of chaperone-mediated protein folding and quality control. The strengths and limitations of each technique and how they can be used in combination to provide a more complete picture of chaperone–client interactions are discussed.

Chapter 1 starts with a general overview of molecular chaperones. Chapters 2–6 discuss different aspects of solution nuclear magnetic resonance (NMR) spectroscopy methods and their applications, with Chapter 5 focusing specifically on the biochemical preparation of suitable samples. Chapters 7 and 11 describe the use of single-molecule biophysics, while Chapter 8 reviews applications of X-ray crystallography to chaperone–client interactions. Chapters 9 and 10 describe electron paramagnetic resonance

New Developments in NMR No. 29
Biophysics of Molecular Chaperones: Function, Mechanisms and Client Protein Interactions
Edited by Sebastian Hiller, Maili Liu and Lichun He

Published by the Royal Society of Chemistry, www.rsc.org

(EPR) methods and their applications to chaperones. Finally, Chapters 12 and 13 discuss two emerging aspects of chaperone biophysics, biological phase separation and in-cell NMR methods, respectively.

Synergizing from these contributions emerges the functional diversity of chaperone systems, their roles in protein folding, misfolding, and degradation, as well as common principles of chaperone–client interactions. The book thus emphasizes the importance of interdisciplinary approaches to bridge the gap between biophysics, biochemistry, and cell biology.

We hope that this overview will inspire researchers and students interested in molecular chaperones and stimulate new insights and collaborations, ultimately leading to a deeper understanding of chaperone–client interactions and their roles in health and disease.

Sebastian Hiller
Basel

Acknowledgement

The editors deeply thank all authors for their exciting contributions and Alexandra Meng and Matt Caffrey for their excellent assistance.

New Developments in NMR No. 29
Biophysics of Molecular Chaperones: Function, Mechanisms and Client Protein Interactions
Edited by Sebastian Hiller, Maili Liu and Lichun He

Published by the Royal Society of Chemistry, www.rsc.org

Contents

New Developments in NMR No. 29
Biophysics of Molecular Chaperones: Function, Mechanisms and Client Protein Interactions
Edited by Sebastian Hiller, Maili Liu and Lichun He

Published by the Royal Society of Chemistry, www.rsc.org

CHAPTER 1

Introduction: Molecular Chaperones and Protein Quality Control

D. BALCHIN,*[a] M. AGUILAR RANGEL[b] AND R. S. SAMANT*[c]

[a] Protein Biogenesis Laboratory, The Francis Crick Institute, London, UK; [b] Department of Structural Biology, Stanford University, Stanford, California, USA; [c] Signalling Programme, The Babraham Institute, Cambridge, UK
*Emails: david.balchin@crick.ac.uk; rahul.samant@babraham.ac.uk

1.1 Lifecycle of a Protein

The lifecycle of a typical protein—from synthesis through to degradation—is regulated at almost every stage by molecular chaperones (see Figure 1.1).[1,2] The initial journey of a newly synthesised nascent chain is especially fraught with pitfalls, with a linear polypeptide chain growing from the peptidyl transferase centre of a ribosome requiring folding into its native 3D structure, translocation to the correct subcellular location(s), and incorporation into one or more protein complexes at the appropriate stoichiometry. Even once mature, some proteins are inherently 'metastable' and require constant chaperoning to maintain their function. Others are prone to misfolding in response to any of a range of environmental stresses a cell encounters during its lifespan. Finally, once a protein has fulfilled its physiological function (*e.g.*, transition through a specific phase of the cell cycle; transduction of a signal from the cell surface to the nucleus), it must be degraded in a timely manner so as not to interfere with other processes. Shepherding the proteome through all of these processes is mediated by a specialised set of

New Developments in NMR No. 29
Biophysics of Molecular Chaperones: Function, Mechanisms and Client Protein Interactions
Edited by Sebastian Hiller, Maili Liu and Lichun He

Published by the Royal Society of Chemistry, www.rsc.org

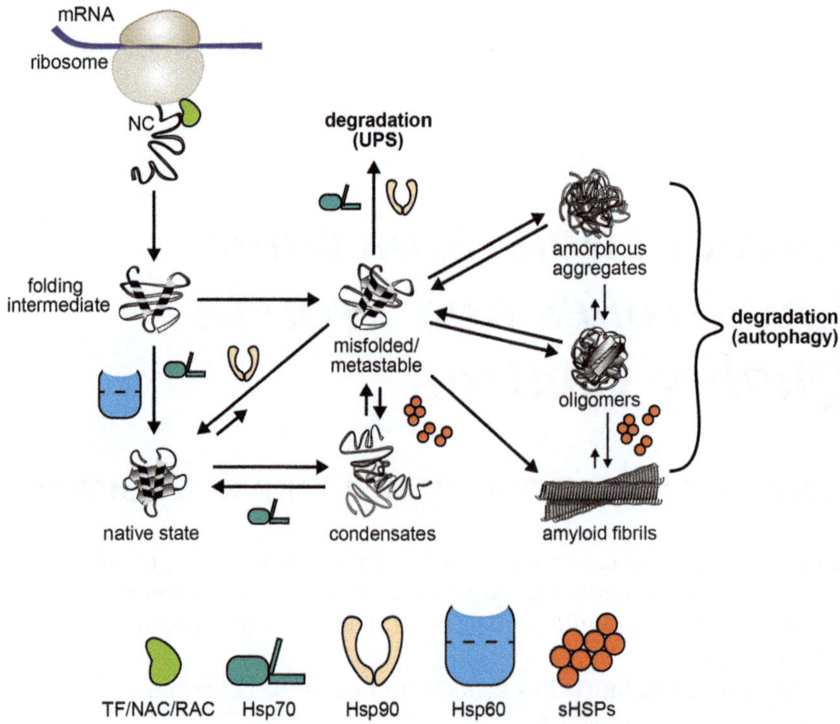

Figure 1.1 Molecular chaperones assist protein folding to the native state and protect against toxic misfolded protein conformations. A nascent chain (NC) emerging from the ribosome is folded cotranslationally by ribosome-associated chaperones (TF in prokaryotes; NAC/RAC in eukaryotes) and may be further assisted post-translationally by the ATP-dependent chaperones Hsp60 (GroEL/GroES in prokaryotes and TRiC/CCT in eukaryotes), Hsp70, and/or Hsp90. Off-pathway misfolded or metastable proteins are recognised *via* the "holdase" function of chaperones and triaged for re-folding or degradation. Misfolded proteins may be temporarily sequestered in biomolecular condensates (*e.g.*, Q-bodies, stress foci) by ATP-independent sHSPs if the ATP-dependent chaperones are overwhelmed (*e.g.*, during stress). sHSPs also play a role in the control of higher-order intermolecular interactions between misfolded proteins, protecting against the accumulation of toxic oligomers and/or amorphous aggregates.

machineries—of which chaperones are a central component—that together make up the protein homeostasis, or "proteostasis", network.

1.1.1 *De Novo* Folding and Assembly

De novo protein synthesis and folding (termed protein biogenesis) is a highly regulated process required to establish a functional proteome. Folding starts at the ribosome and is further directed by a diverse array of molecular chaperones. Together, the protein biogenesis network promotes and accelerates folding, avoids and reverses misfolding, and inhibits toxic aggregation.

1.1.1.1 Fundamental Concepts in Protein Refolding

The basic physical principles that underlie the refolding of isolated proteins are now quite well understood, at least for single-domain proteins with simple folding pathways.[3] The conformational search for the native state ensemble is thought to be driven by the burial of hydrophobic residues and the relative stability of native-like interactions that nucleate the folding reaction. This process is often described using the analogy of a "folding funnel", which places the native state at a free energy minimum with low configurational entropy (see Figure 1.2). Proteins are envisioned to traverse the funnel during folding, moving energetically "downhill" towards the native state. Although a funnel-shaped energy landscape implies the existence of multiple folding pathways, it has also been argued that proteins

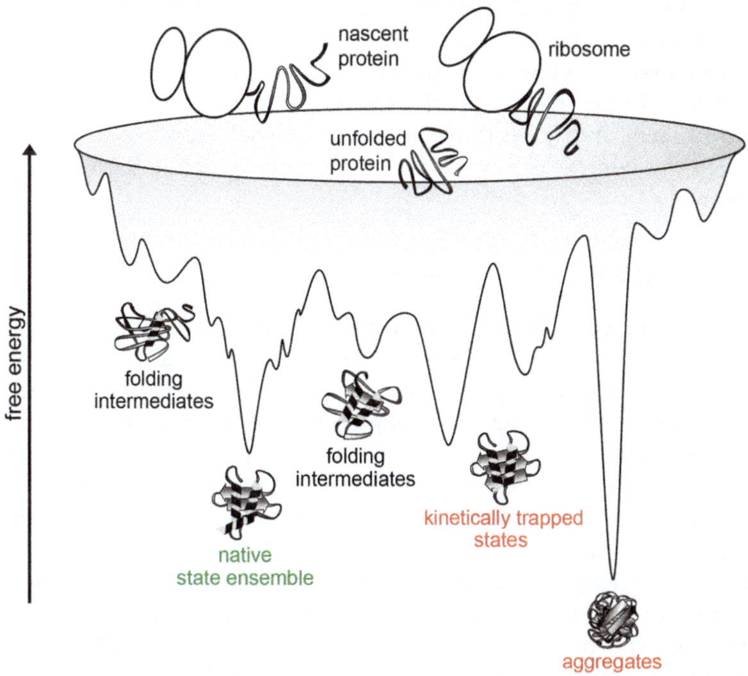

Figure 1.2 Proteins navigate complex energy landscapes during folding. The conformational search for the native state can be envisioned as occurring on a funnel-shaped energy landscape, with the native state ensemble at a local free energy minimum. On-pathway intermediates are common, while off-pathway kinetically-trapped states may give rise to stable aggregates. Proteins at the top of the folding funnel may be in high-energy unfolded states or nascent chains emerging from the ribosome.

instead fold *via* an obligate sequence of specific substructures (termed "foldons").[4] These interpretations are not necessarily mutually exclusive: heterogeneity in folding at a microscopic scale may give rise to discrete foldons at the macroscopic level.[5]

Several factors complicate the folding process. The folding free energy landscape is typically rugged such that protein chains must overcome substantial energy barriers *en route* to the native state and consequently populate folding intermediates (see Figure 1.2). Off-pathway intermediates and kinetic traps can slow folding, and non-native interactions may lead to stable misfolding. Moreover, misfolded and intermediate states often expose hydrophobic surfaces that can engage in non-functional intermolecular interactions or aggregation. The folding problem is significantly exacerbated for multidomain proteins, typically >100 residues in length, that comprise the vast majority of proteomes in both bacteria and human cells.[6] Larger proteins have a substantially increased tendency to populate folding intermediates and are at risk of inter-domain misfolding.[7]

Consistent with these observations, most proteins do not refold efficiently (rapidly and to a high yield) in the absence of additional factors *in vitro*.[8] On the other hand, cells routinely support the folding of a huge diversity of large, topologically complex proteins, even in the crowded cytosol. These observations imply that authentic folding reactions differ profoundly from *in vitro* models. Indeed, folding *in vivo* is a closely supervised process that begins in the context of translation and employs a cooperative network of chaperones.

1.1.1.2 Cotranslational Folding

During protein synthesis, once the ribosome has successfully completed the initiation phase, it will proceed to decode the mRNA at a mean rate of ∼4–6 codons per second in eukaryotes and ∼10–20 in prokaryotes. As a consequence, the synthesis of an average eukaryotic protein of 400 amino acids in length could take ∼1 minute. This time-scale can be explained in terms of the synergistic effect of several mechanisms aiding the proper folding of the nascent polypeptide in a cotranslational manner: mRNA codon composition, tRNA pools, ribosome–nascent chain interactions, and chaperone recruitment.

Translation elongation does not occur at a constant rate.[9] It is directly influenced, in the first instance, by the codon composition of the mRNA. The genetic code can be split into codons that are abundant and/or translated faster (optimal codons) and those whose rate of decoding is slow and/or might be rare in the translatome (non-optimal codons).[10] Most transcripts have a globally unbalanced composition of optimal and non-optimal codons and can also present local codon optimality biases (*e.g.*, stretches of non-optimal codons within a globally optimal transcript).[11–13] Therefore, for any transcript, the translation elongation rate is not uniform, causing the vectorial emergence of the nascent chain to happen at a differential rate. A major consequence of this rate-regulated N- to C-terminus vectorial synthesis of a nascent chain is its controlled exposure to the ribosome exit tunnel. Spanning ~100 Å in length, the exit tunnel can accommodate ~30–40 amino acids in sub-regions of wide (~20 Å) and narrow (<10 Å) diameters.[14,15] Given its size and negatively charged electrostatic nature, the tunnel reduces the conformational sampling space of the nascent chain, which can promote the formation of secondary structure elements[16,17] and even facilitate the folding of small domains.[18] It is here where the cotranslational folding of the protein can start.

As the nascent chain approaches the exit vestibule of the ribosome and begins to be exposed to the cytosolic environment, the accessible conformational space for the emerging protein increases dramatically as a result of the removal of the tunnel constraints. An immediate response to this scenario is the interaction between the nascent chain and the surface of the ribosome, a phenomenon that has been reported for various proteins.[19–23] Through interactions with positive and aromatic residues in the nascent chain, the ribosome surface has been proposed to act as a holdase, and although its role is not fully understood, it possibly prevents hydrophobic regions from forming unfavourable interactions and/or regulates the interactions between the nascent chain and molecular chaperones.[22]

The crowded nature of the cytosol, with macromolecular concentrations of up to 400 mg mL^{-1},[24] imposes a challenge for the correct folding of the nascent chains, as non-specific interactions can bias the conformational sampling space that the emerging protein can explore. Fortunately, as the nascent chain is exposed to the cytosol, a diverse array of molecular chaperones are ready to engage it, preventing its aggregation and promoting the correct folding of the protein into its native state.[25] Chaperones acting cotranslationally can bind directly to the ribosome,[26] thus forming the first line of action, or can engage with the nascent chain in a ribosome-independent manner.[25] A more detailed overview of the constituents of both classes is given in Section 1.3.

1.1.1.3 Cotranslational Assembly of Oligomeric Complexes

The vast majority of the proteome is oligomeric, with dimers predominating.[27] Thus, newly made proteins must not only fold efficiently, but also assemble into functional complexes. Oligomeric assembly *in vivo* has classically been assumed to occur *via* random diffusion of subunits. However, strictly spontaneous assembly is difficult to reconcile with the fact that many proteins are present at very low abundance in cells,[28] the highly crowded environment of the cytosol, and the requirement that promiscuous subunits avoid misassembly.[29] One solution is for cells to coordinate protein synthesis with oligomeric assembly. Although cotranslational assembly was observed as early as the 1960s,[30] recent work has revived this concept, revealed that it is a widespread phenomenon, and begun to hint at some of the underlying mechanisms.[31–33] Specific examples of complexes that assemble during translation include beta-galactosidase,[30] bacterial luciferase,[34] coiled coil proteins,[33] certain proteasome subunits,[35] several metabolic enzymes,[31] nuclear transcription complexes,[36] and nucleosome sub-complexes.[37,38] Several observations point to a role for chaperones in cotranslational assembly, especially of heteromeric complexes,[31,34] although how exactly the proteostasis network co-ordinates assembly at the ribosome remains to be determined.

1.1.2 Conformational Maintenance

The native state of a protein is not a single rigid structure, but rather an ensemble of energetically similar conformations. Indeed, the functional conformation of a protein may not even lie at the global energy minimum for that amino acid sequence. Such proteins are termed "metastable" and present a particular challenge to the cell. Some metastable proteins need constant interaction with chaperones to be maintained in a functional conformation, as is the case for constitutive Hsp90 clients,[39] discussed later. Other metastable proteins do not require constitutive chaperoning under physiological conditions but are especially vulnerable to misfolding under proteotoxic stress exposure. In these cases, Hsp70 chaperones can mitigate the damage by converting the energy of ATP hydrolysis to stabilising the native state out-of-equilibrium.[40,41] Stress-induced misfolding is a constant pressure for the chaperome and may present unique challenges not encountered during *de novo* misfolding, in which nascent chains most likely sample fewer off-pathway conformations due to the vectorial nature of translation and the accessory chaperoning function of the ribosome exit tunnel and surface (discussed in Section 1.1.1.2). That is not to say that misfolded nascent chains represent a simpler problem for chaperones to handle; in contrast, actively-translating chains may be the most immediately vulnerable to stress-induced misfolding,[42] as they are in a non-native conformation at a time when the cell has not yet had a chance to activate its cytoprotective stress response programmes. Cotranslational off-pathway folding intermediates can even persist and influence the post-translational folding route, as shown for disease-associated misfolded alpha-1-antitrypsin.[43]

1.1.3 Degradation

Although some proteins are maintained for years and decades at a time,[44] most of the proteome is turned over in the range of minutes to hours, with an average half-life of roughly 24 hours in humans.[45,46] Broadly speaking, the reasons for degradation can be split into two categories: regulatory (because the protein has fulfilled its physiological function) and quality-controlled (because the protein is in some way damaged or dysfunctional). Much of our early mechanistic understanding of protein degradation systems came from the former category.[47] For example, the tightly-choreographed turnover of different cyclins and cyclin-dependent kinases plays a critical role in timing progression through each phase of the cell cycle.[48] Similar timer-based turnover mechanisms are now well-established for gene transcription,[49–51] membrane trafficking,[52] and apoptosis.[53]

The majority of the proteome is turned over through either the ubiquitin–proteasome system (UPS) or the autophagy–lysosome system,[54] although other mechanisms exist for both dividing (*e.g.*, dilution, asymmetric inheritance) and non-dividing (*e.g.*, exocytosis, terminal sequestration) cells.[55] In general, the UPS is considered the preferred pathway for short-lived and globular proteins, whereas the autophagy–lysosome system exists for longer-lived (*e.g.*, cytoskeletal) or transmembrane (*e.g.*, cell surface receptor) proteins[56]—although this division of labour likely represents trends rather than absolute rules.[57–59]

1.1.3.1 Ubiquitin–Proteasome System

Degradation *via* the UPS generally relies on the protein in question being 'tagged' on an internal lysine or N-terminal methionine with one or more molecules of the highly-conserved, 76-amino-acid protein ubiquitin.[54] 'Protein ubiquitylation' involves an enzymatic cascade for sequential activation, conjugation, and ligation of ubiquitin onto the client. Whereas there are only two E1 and 30–50 E2 enzymes in humans for steps 1 and 2, respectively, the final step—through which client specificity is conferred—can be mediated by one of over 600 E3 ubiquitin ligases.[56] Furthermore, ubiquitin can be removed from a client by deubiquitylases ('DUBs'; >100 in humans). Added together with the expanding list of identified ubiquitin-binding receptors and the multi-subunit 26S proteasome, a human cell dedicates around 1000 proteins, *i.e.*, well over 5% of protein-coding genes, to the UPS.

The degradative machine of the UPS is the 26S proteasome: a 2.5 MDa holoenzyme consisting of a 20S core particle capped at one or both ends by 19S regulatory particles.[60] Substrate degradation proceeds through a highly ordered series of steps, involving binding of a ubiquitylated substrate to ubiquitin receptors in the 19S lid and insertion of an unstructured region within the substrate into the pore of the 19S base, followed by mechanistically coupled substrate translocation and deubiquitylation steps that allow the substrate to be threaded through to the proteolytic subunits within the inner chamber of the 20S core particle.

Initially considered purely to be a signal for proteasomal degradation, it is now appreciated that protein ubiquitylation signals diverse proteolytic and non-proteolytic fates.[61] Much of this diversity comes from the ability of ubiquitin to form chains of different topologies, due to covalent conjugation of additional ubiquitin moieties onto any of seven internal lysines of ubiquitin or its N-terminal methionine. The canonical signal for targeting the proteasome is a chain of four or more ubiquitins linked *via* the lysine at position 48 in each moiety ('tetra-K48-ubiquitin'), although other chain topologies appear to be involved in different contexts. In the case of misfolded proteins, for example, branched K11/K48-ubiquitin chains play a role in chaperone-assisted degradation pathways to the proteasome.[62–64]

1.1.3.2 Autophagy–Lysosome System

Unlike proteasomes, which require substrates to be threaded through one polypeptide chain at a time, lysosomes are acidic compartments that can degrade substrates in bulk, consisting of proteases and peptidases, as well as digestive enzymes for other molecules (*e.g.*, nucleic acids, carbohydrates, lipids, and metabolites). The most well-characterised pathway that ends up with lysosomal degradation is macro-autophagy.[65] In response to a variety of stress-related signals (with the best characterised being starvation), intracellular resources such as nutrients and macromolecules can be recycled into their constituent building blocks by engulfing regions of the cellular milieu in bulk in double-membrane structures called autophagosomes and fusing them with lysosomes—thereby releasing the lysosomal digestive enzymes to degrade the autophagosomal contents. Whereas this 'bulk' macro-autophagy will result in indiscriminate degradation of proteins that happen to be in the subcellular volume engulfed by autophagosomes, specific proteins (*e.g.*, those that misfolded in response to the same stress that necessitated macro-autophagy induction) can also be selectively targeted to autophagosomes *via* K63-linked ubiquitin chain attachment. The K63-ubiquitylated protein cargo is recognised by one of several selective autophagy receptors (*e.g.*, p62/SQSTM1, Optineurin, or Tollip), which target the cargo to the ubiquitin-like protein Atg8/LC3 that studs autophagosome membranes.[65]

Outside of bulk and selective macro-autophagy, proteins can end up in the lysosome through various other pathways. Intrinsic membrane receptors (*e.g.*, receptor tyrosine kinases) are commonly internalised upon ligand binding into endosomes, which are subsequently either recycled to the cell surface for another round of ligand-induced signalling or sent to the lysosome *via* endosomal fusion (analogous to autophagosome–lysosome fusion). Chaperone-mediated autophagy employs the Hsp70 chaperone machinery for direct insertion of a subset of misfolded proteins into the lysosome *via* the intrinsic lysosomal receptor LAMP2A.[66] Several alternative routes exist for direct protein insertion into lysosomes—collectively referred to as micro-autophagy[67]—as well as non-canonical autophagy pathways.[68]

1.2 Protein Aggregation

1.2.1 Biophysical Perspective

The folding of a protein can be understood as a free diffusion process on an energy landscape (see Figure 1.2). In solution, proteins exist in an equilibrium between their native or folded state and a small fraction of unfolding intermediates.[69] The latter are the result of a protein overcoming the energy barriers separating the energy minimum of its native state and the neighbouring local minima in the free energy landscape of the protein. Depending on the environment, these intermediate states have the ability to transition to the unfolded state and/or to start self-associating into aggregates.[70] The feasibility of protein transition to the aggregated state depends on a combination of the aggregated state's kinetic accessibility and thermodynamic stability. It is also dependent on protein concentration since this factor determines the frequency of intermolecular contacts and can tilt the balance in favour of the aggregated state at high concentrations, when the protein reaches supersaturation.[71] Therefore, it is crucial for proteins to exist at permissible levels of solubility to avoid a high frequency of detrimental interactions.

Unfolding intermediates are prone to exposing hydrophobic patches that are otherwise inaccessible in the native state.[72] These regions can promote the formation of unfavourable protein–protein interactions and the formation of aggregates. Several studies have proposed and analysed the existence of aggregation-prone regions (APRs),[73–75] cataloguing them as short regions (5–10 residues) with a high content of hydrophobic residues and a low net charge, which have a tendency to form β-sheet structures. APRs can be classified as buried (the hydrophobic core of a protein) or exposed (*e.g.*, protein–protein dimerisation interfaces).[76] Flanking these regions with charged amino acids that could induce electrostatic repulsion or disrupt the β-sheet propensity can attenuate their aggregation.[77] In the cell, this strategy is present by placing Arg, Lys, Glu, Asp, and/or Pro residues near APRs, which has gained them the name of aggregation gatekeepers.[78]

In instances where aggregation does occur, the products can be classified based on their ultrastructural conformation: either as an amorphous aggregate or as an amyloid.[69] The former used to be described as a mostly disordered conformation, but it has recently been shown that amorphous aggregates can contain certain degrees of beta strands.[72] On the other hand, amyloids are highly ordered structures, rich in β-strands assembling in a cross-β pattern. Fibrillar in nature, amyloids possess outstanding mechanical properties, such as high Young's modulus and tensile strength.[70] Thermodynamically, they are very stable structures, mostly observed in short proteins. It is suggested that long proteins will have difficulties assembling into amyloid states due to the topological constraints imposed by the packaging of the polypeptide into the fibrillar core.[70] The mechanism by which amyloids are formed can be described using a heterogeneous polymerisation system, where the aggregation is dependent not only on the interactions of soluble monomers (primary nucleation-derived aggregates) but also on monomer-independent processes, such as fibril fragmentation, and on monomer-dependent secondary pathways involving fibril surface-catalysed nucleation.[79] This model highlights the importance of protein concentration during aggregation for both the monomeric and the aggregated species, as the presence of the latter, in the form of seeds, can catalyse the aggregation reaction. Recent work has focused on understanding the instances where aggregation-prone proteins are found in environments that push them over their solubility limits, promoting their aggregation.[80–82] Therefore, liquid droplets formed by phase separation of these proteins can provide an amenable environment for fibrillar formation, as has been observed for some amyloidogenic proteins.[80] It is still unclear how this liquid–liquid phase separation-driven aggregation might relate to the pathogenicity associated with protein aggregation or the sequestration and clearance of aggregates in the cell.

1.2.2 Physiological Roles of Protein Aggregation *In Vivo*

Protein aggregation—especially the formation of beta-amyloid fibrils—is generally considered to be pathologic and is best known for its association with a wide-range of degenerative human and animal diseases (see Section 1.5). However, research over the past two decades on "functional amyloids" has revealed the widespread and diverse roles of beta-amyloids throughout cellular life, including in biofilm stabilisation, cell cycle regulation, host–pathogen interactions, spider-silk production, and melanin pigment biosynthesis.[83,84] The physiological prevalence of functional amyloids would appear incongruent with the idea of these same structures being intrinsically causative for pathology.[85]

Conceptual insight into this dilemma comes from the work in the neurodegeneration field, with evidence that the most toxic conformation of disease-associated amyloidogenic misfolded proteins such as amyloid-beta and tau was actually the short oligomeric form and not the long amyloid fibrils characteristic of pathology.[85–87] Therefore, in addition to preventing the initial oligomer formation and dissociating or degrading oligomers, sequestration into structured aggregates would serve as an alternative route for mitigating gain-of-function oligomer toxicity. Sequestration into thermodynamically stable beta-fibrils might actually be the safest option for pre-existing oligomers, as (i) dissolution into monomers would not prevent future re-oligomerisation and potentially risk further oligomer seeding and propagation[87,88] and (ii) oligomers can "clog-up" 20S proteasomes and are proposed to contribute to the toxicity of these neurodegenerative diseases.[89] Further research into how cells avoid oligomer toxicity during the production of functional amyloids could therefore be translated directly to inform our understanding of protein misfolding disease pathology and drive the development of therapeutic interventions.[90]

It is also important to distinguish between biophysical protein aggregation, *i.e.*, due to the intrinsic properties of the protein itself, and cytoprotective protein sequestration by chaperones and other machineries of the proteostasis network (see Section 1.4). Historically these have been conflated *in vivo*, as traditional methods for visualising aggregates (*e.g.*, GFP-tagged misfolded proteins) did not differentiate between the two. Recent advances in single-molecule imaging techniques and tools are already paying dividends for deconvoluting cytoprotective *vs.* biophysical aggregation processes *in situ*.[91,92]

1.3 Major Molecular Chaperone Families

1.3.1 Generic Ribosome-associated Chaperones

As the nascent chain starts its vectorial emergence from the ribosomal exit tunnel, ribosome-associated molecular chaperones are ready to engage with it, constraining the high-energy state of the yet-to-be-folded polypeptide intermediates. Several ribosomal proteins, conserved across the tree of life, surround the exit of the ribosomal tunnel and serve as binding sites for this type of chaperone.[93–95] However, it is becoming clearer that not only riboproteins are involved in the docking of nascent chain processing factors; rRNA-mediated contacts can also be involved in this process.

Despite being present in all branches of the tree of life, ribosome-associated chaperones are not conserved across them. Whereas prokaryotic cells possess a single system, named trigger factor (TF),[96] eukaryotic organisms have two independent systems: the nascent polypeptide-associated complex (NAC) and the ribosome-associated complex (RAC).[26]

1.3.1.1 Trigger Factor (TF)

Present in bacteria and chloroplasts,[97] TF is the only ribosome-associated prokaryotic chaperone. Monomeric TF binds in a 1 : 1 ratio to bacterial ribosomes; but it also exists in a dimeric ribosome-unbound form in the cytosol.[98] Nuclear magnetic resonance-derived structures of TF revealed four substrate binding sites, consisting of hydrophobic patches with the capacity for accommodating ~40–50 amino acids in total.[99] This explains why long polypeptides had been shown to recruit multiple TF units. When associated with the ribosome, TF can bind to hydrophobic regions upon their emergence from the tunnel, keeping them in a folding-competent state until they are handed over to other chaperones or until other regions of the protein, which may promote favourable interactions, emerge.[96,100] The hydrophobic nature of the binding sites of this chaperone, in turn decorated with polar residues suitable for hydrogen bond formation with the substrate, can also destabilise and unfold transient protein intermediates, helping the substrate to re-enter the appropriate folding pathway.[99] Therefore, TF possesses a dual holdase/unfoldase character.

Selective ribosome profiling of TF-bound translating ribosomes catalogued most of the bacterial translatome as a TF substrate, except for the nascent chains recognised by the signal recognition particle (SRP).[101] This study also pointed out that TF engages its substrates only after the first 100 amino acids of the protein have been translated, presumably allocating time for the pre-processing of the N-terminus. Among the identified substrates, the β-barrel outer membrane proteins (Omps) were the strongest interactors. Later studies established the mechanism by which TF binds to a model Omp, proOmpA, aiding its recruitment to the translocase system.[102] However, the binding between TF and its substrate is too stable for the latter to be released and thus begin translocation; therefore, the assistance of another chaperone, SecB, is required to facilitate the release of the Omp protein.[102] This example highlights the power of selective ribosome profiling for studying chaperone–client interactions during cotranslational folding.

1.3.1.2 The Nascent Polypeptide-associated Complex (NAC)

The NAC is a heterodimeric complex, formed by one α and one β subunit, that adopts a β-barrel-like structure and is present in archaea and eukaryotes.[103,104] However, while a knockout of the *nac* genes has little phenotypic effect in yeasts, it is essential in higher eukaryotes.[105] The NAC interacts with the ribosome *via* electrostatic interactions mediated by conserved stretches of positively charged residues located in both subunits.[95] In higher eukaryotes, this interaction is negatively regulated by a cluster of negatively charged residues located in the N-terminus of the α-subunit. Ablation of this regulator results in tighter binding of the NAC to the ribosomal surface and causes protein biogenesis defects both *in vitro* and *in vivo*.[95] Therefore, the NAC relies on a self-regulated, transient interaction mechanism to play its chaperone role during translation.

Earlier biochemical studies evidenced how the NAC is able to engage nascent chains even before they start to emerge from the ribosomal tunnel, crosslinking with emerging chains as short as 17 residues.[106] Not being a ribosomal protein, this posed NAC as the first molecular chaperone acting during the cotranslational folding of proteins. A recent cryo-EM structure of the NAC in complex with the ribosome revealed an exquisite mode of action for this early nascent chain–NAC interaction.[95] Through insertion of its flexible β-subunit N-terminus, the NAC is able to reach up to the constriction site of the ribosomal tunnel, near the protrusion of uL22, and interact with the nascent chain just after the first rounds of elongation.[95] As translation progresses, the nascent chain pushes out the inserted β-NAC N-terminus, which can in turn be seen as a sensing probe for translating ribosomes. Given that the NAC is the only molecular chaperone that exists in a 1 : 1 ratio with respect to ribosome concentration[107] and that its binding to the ribosome can sterically outcompete other ribosome-associated factors such as SRP or RAC,[108,109] it is possible that it functions as a mechanism to discern between translating and non-translating ribosomes. Indeed, the NAC was found to regulate the engagement of active ribosomes to the translocon system Sec61.[110,111] Additionally, experiments carried out in yeasts show how some nascent chains bind less effectively to SRP in the absence of the NAC.[112] Therefore, it is possible that the NAC serves as a sorting mechanism for nascent chains to engage with their appropriate cotranslational folding pathways.

1.3.1.3 The Ribosome-associated Complex (RAC)

In addition to the NAC system, eukaryotes possess a second ribosome-associated chaperone system, RAC. It is a heterodimer formed by the stable association of a heat shock protein 40 (Hsp40; Zuo1 in yeasts and MPP11 in mammals) and a non-canonical heat shock protein 70 (Hsp70; Ssz1 in yeasts and Hsp70L1 in mammals) with impaired catalytic activity.[113,114] This complex serves as the J-domain co-chaperone partner of the ribosome-associated Hsp70 Ssb in yeasts or a cytosolic Hsp70 partner in higher eukaryotes.[115] Association of the RAC to the ribosome is mediated by its Hsp40 unit. The structure of the yeast RAC–ribosome complex shows that Zuo1 forms direct contacts with both ribosomal subunits, binding to the 60S subunit *via* its zuotin homology domain and to the 40S subunit through its 4-helix-bundle region.[116]

Although the RAC is present in mammals, most of our knowledge on its mechanism and function in protein folding comes from the yeast system. There, Ssz1 is recruited to the ribosome *via* its interaction with Zuo1.[117] Ssz1 can bind ATP at its nucleotide binding domain (NBD), but lacks the ability to hydrolyse it. Compared to canonical Hsp70s, Ssz1 is missing the α-subdomain of the substrate binding domain (SBDα), but retains the SBDβ subdomain.[116,118] For a long time, this atypical Hsp70 character posed the question of whether Ssz1 had an active role in the RAC–Ssb triad or instead served as a passive scaffold for Ssb recruitment. Recent studies showed that Ssz1 can interact with nascent chains in a low-affinity manner, which in turn is regulated through the binding between an LP-motif located in Zuo1 and the SBDβ of Ssz1, in a manner akin to how SBDα functions as a lid in canonical Hsp70s.[118]

Ssb is a canonical Hsp70 protein that binds to the ribosome *via* electrostatic interactions.[119] It interacts directly with nascent chains, displaying preferential binding for stretches rich in positive and hydrophobic residues[120]—common features of aggregation-prone regions. Advances in understanding the mechanism by which Ssb interacts with the RAC and the nascent chain came largely from cross-linking experiments. When Ssb is in the ATP-bound conformation, it crosslinks to the NBD of Ssz1, meaning that it is proximal to the J-domain of Zuo1 (required for promoting ATP hydrolysis) and the ribosome exit tunnel.[121] After ATP is hydrolysed and Ssb engages with the nascent chain, Ssb crosslinks map to the ribosome, suggesting that it has freed the Ssz1 binding interface, leaving it ready for the coupling of another Ssb protein.[121] This mode of action is consistent with observations of Ssz1 acting transiently with the nascent chains before Ssb engages them.[118] It has yet to be established what drives the transition from the Ssb–Ssz1-bound state to the Ssb–ribosome-bound state, as well as the role of nascent chain-relaying from Ssz1 to Ssb instead of their direct recruitment to Ssb. However, it is now clear that all components of the RAC–Ssb triad have active roles during cotranslational folding, as opposed to the RAC serving as a mere scaffold for Ssb-assisted protein folding.

1.3.2 Hsp70

Hsp70s are among the most abundant and versatile molecular chaperones.[122,123] They play critical roles in *de novo* folding,[124,125] unfolding misfolded states,[41,126] refolding of stress-denatured proteins,[127] disaggregation,[128,129] disassembly of biomolecular condensates,[130] protein trafficking,[131] and the import of proteins into organelles.[132]

Hsp70s consist of two highly conserved domains: a nucleotide binding domain (NBD) with an actin-type fold and a substrate binding domain (SBD) containing a groove that accommodates peptides. Through the SBD, the chaperone recognises short linear motifs of ~7 amino acids in length in client proteins, preferentially consisting of a stretch of hydrophobic residues flanked by positive charges.[133,134] These sequence motifs are common in the proteome and are typically buried in the core of properly folded proteins, consistent with the idea that Hsp70s are a "hub" of the chaperone network.[135] Central to the mechanism of Hsp70s is their nucleotide-driven conformational cycle, which dictates the affinity of the chaperone for substrates. In the apo or ATP-bound states, Hsp70 is structurally biased towards the "domain-docked" state, where the SBD is in an open conformation and substrates bind with high on/off rates. ATP hydrolysis stabilises the "domain-undocked" state, characterised by an occluded SBD and stable client engagement.

Regulation of the Hsp70 conformational cycle is "outsourced" to co-chaperones. Most prominent among these are nucleotide exchange factors (NEFs) and J-domain containing Hsp40s (*e.g.*, Zuo1 in the RAC, described in Section 1.3.1.3). NEFs trigger ADP:ATP exchange in the Hsp70 NBD and in some cases directly interfere with the binding of substrates by interaction with the SBD, thereby controlling the residence time of clients on the chaperone.[136] Some NEFs also act as adaptors that bridge different components of the proteostasis network. Bag3, for example, connects Hsp70 to an sHSP to facilitate selective autophagy.[137]

Hsp40s interface with the Hsp70 conformational cycle in two ways. First, they deliver client proteins to Hsp70, presumably *via* a transient ternary complex.[138] Second, acting synergistically with the substrate peptide, they stimulate ATP hydrolysis by Hsp70, a non-equilibrium process that enables high-affinity client binding.[139] Together, these two features underlie the ability of Hsp40s to act as client-specific adaptors for Hsp70s.[140] Different Hsp40s direct the core Hsp70 activity to diverse targets, leveraging the functional diversification of J-domains.[141]

A largely unresolved question is how the conformational cycle of Hsp70, driving repeated cycles of peptide engagement and release, enacts its various cellular functions. Is there a universal underlying mechanism? A compelling explanation, in at least some instances, is "entropic pulling".[142] In this model, binding of Hsp70 entropically favours translocation, disaggregation, or unfolding *via* an excluded volume effect. There is now strong experimental evidence that entropic pulling contributes to Hsp70-mediated disaggregation of fibrils.[129] However, whether all Hsp70 functions are based on entropic pulling, in particular those related to *de novo* folding, is unclear. Additional layers of complexity are suggested by studies that have probed the conformation of client proteins or model domains during association with Hsp70.[41,143–146] Stochastic engagement and release of substrates by multiple Hsp70 molecules may modulate the folding pathway of the client, in some cases resulting in increased folding rates.[41,147]

1.3.3 Hsp90

Hsp90s are another essential, highly abundant, and evolutionarily conserved family of chaperones. While the Hsp90 machinery also uses ATP binding and hydrolysis cycles to drive client protein maturation, it is structurally and mechanistically distinct from the Hsp70 machinery and employs its own suite of co-chaperones.[148] Hsp90 functions as a constitutive homodimer, with each ~90 kDa chain consisting of three domains: N-terminal domain (NTD), middle domain (MD), and C-terminal domain (CTD). Dimerisation occurs *via* CTDs to create a V-shaped molecular clamp that binds immature client proteins. ATP-binding in a pocket within the NTD triggers a series of conformational changes throughout the Hsp90 structure, including NTD dimerisation and closing of the molecular clamp into a "tense" or "twisted" state.[149] ATP hydrolysis relaxes this conformation, opens the clamp, and enables nucleotide exchange to allow another round of cycling. The ordered, sequential binding and dissociation of co-chaperones plays a critical role in coupling this cycle to client maturation[150]—although it is important to note that co-chaperone requirements are also dependent on the client in question. For example, the co-chaperone Cdc37 is generally accepted as a protein kinase-specific factor.[151,152] Surprisingly, however, even clients from the same class can have different co-chaperone requirements, as demonstrated for five steroid hormone receptor clients.[153]

Although initial mechanistic work on the Hsp90 conformational cycle was performed in the absence of client protein substrates—generally due to the difficulty in working with recombinant unfolded proteins—a single-particle electron microscopy structure of Hsp90 with the protein kinase client Cdk4 and the kinase-specific co-chaperone Cdc37[154] provided a template for considerable progress in our understanding of the structure–function relationship of the Hsp90 machinery and client protein maturation.[155–162] Unlike Hsp70—which is required for most nascent polypeptides to reach their correctly-folded, native conformation—chaperones and co-chaperones of the Hsp90 machinery have a much more exclusive client list (see https://www.picard.ch/downloads/Hsp90interactors.pdf for an updated list of experimentally-verified clients and co-chaperones). Hsp90 is considered to function downstream of Hsp70, with clients that are more mature yet still require "finishing" to become and/or remain active. Indeed, most Hsp90 clients are transferred to the chaperone directly from Hsp70 *via* the bridging co-chaperone Sti1/Hop ('Hsp70–Hsp90 organising protein'), although Hsp70 and Hsp90 can also interact directly—at least in budding yeasts.[163] More recent work has revealed unexpected roles for Hsp90 in breaking unproductive Hsp70 cycling.[164]

Why ~300 structurally and functionally unrelated clients need this extra chaperone machinery remains an open question. It has been proposed that Hsp90 is a "capacitor for evolution",[165] effectively buffering novel, metastable protein variants that would otherwise be degraded and enabling them to persist for long enough to confer selectable advantages to the cell. Several studies have built upon this hypothesis to identify key roles for Hsp90 in shaping RNA and protein evolution,[166,167] as well as in the emergence of new genotypic and phenotypic traits.[168–172] A recent protein–protein interaction network analysis across two different yeast species found that chaperone clients diverged evolutionarily more quickly than non-clients, with both Hsp70 and Hsp90 machineries playing a major role in the rates of protein evolution.[173]

In support of the evolutionary capacitor model, a common thread linking Hsp90's otherwise diverse clientele is their metastability. Indeed, a landmark system-wide quantification of Hsp90–client interactions found that thermal stability was a major determinant of binding strength[174]—at least for the human protein kinase complement. For example, oncogenic mutations in the protein kinases v-Src, $BRAF^{V600E}$, and $EGFR^{T790M}$ considerably increase their reliance on Hsp90, as compared with the wild-type c-Src, BRAF, and EGFR.[175] All of these mutants are intrinsically less stable than their wild-type counterparts but are key driver oncoproteins across different cancer indications. Inhibitors of Hsp90 have therefore been pursued for over a decade as a means of specifically targeting cancer cells driven by oncoprotein client mutations.[176] Note that, while the above specific cases confer client status by mutation, most Hsp90 clients are encoded in the native genome (*e.g.*, wild-type ERBB2—belonging to the same family as EGFR—is a strong Hsp90 interactor[177]).

A further development building upon Hsp90's classification system of 'strong', 'weak', and 'non'-interactors was provided by proteome-wide mass spectrometric pulse-chase analysis that measured proteome-wide synthesis and degradation dynamics in response to Hsp90 inhibition.[39] This study categorised clients as either 'synthesis' or 'constitutive', based on whether their Hsp90 requirement was limited to their initial folding from the nascent polypeptide or whether they needed Hsp90 throughout their life cycle. As expected, constitutive clients had lower thermal stability than synthesis clients.

1.3.4 Hsp60

Hsp60s (usually referred to as chaperonins) are paradigmatic protein folding machines. They specialise in the *de novo* folding of particularly challenging protein structures and are essential in all domains of life.[178] The characteristic feature of chaperonins is their cage-like architecture, creating a cavity in which single protein molecules can fold while being isolated from aberrant interactions in the bulk cytosol. The "folding cage" is formed by the assembly of 14–18 subunits into near-megadalton complexes that are typically arranged as two identical rings stacked back-to-back.[179,180] Energy from ATP hydrolysis drives the transition between the closed and open states of the chaperonins, controlling the transient encapsulation of substrates in the central cavity.[178] Each subunit has a conserved 3-domain structure. The equatorial domains form the base of each ring and mediate inter-ring contacts. These domains also bind and hydrolyse ATP, thereby coordinating the dramatic conformational rearrangements that drive the functional cycle of chaperonins. Connected to the equatorial domains are the intermediate domains, which act as hinges to the apical domains at the tip of each subunit. The apical domains are chiefly responsible for substrate capture and are highly responsive to allosteric signals originating from the ATP-binding pocket in the equatorial domains, transmitted *via* the intermediate domains.

Despite retaining the same overall architecture and apparently similar conformational cycles, archaeal and eukaryotic cytosolic chaperonins (TRiC or CCT) differ from their bacterial counterparts (GroEL/ES) in several important ways.[181,182] First, while bacterial GroEL cooperates with a co-chaperone (GroES) to encapsulate substrates, eukaryotic TRiC incorporates a subdomain that acts as a built-in lid. Triggered by ATP hydrolysis, dynamic extensions at the tip of each apical domain form helical structures that assemble with their neighbours to close the folding chamber.[183,184] Second, unlike GroEL which consists of identical subunits, eukaryotic and some archaeal chaperonins are heterooligomeric; TRiC is a heterooligomer with eight unique subunits per ring. The subunits of TRiC are arranged in a defined order,[185,186] with important (albeit incompletely understood) implications for substrate recognition and folding.[187–189]

While the interactome of chaperonins is relatively broad (>10% of cytosolic proteins[190]), it has been challenging to determine which of these are obligate clients. In part, this is a result of different definitions of "obligate" in regard to chaperone substrates. By the most stringent criteria, proteins are obligatorily chaperone-dependent when they cannot fold in the absence of the chaperone, including when aggregation is avoided.[147] However, this is difficult to establish unambiguously, even when single clients are studied in isolation. Regardless, multiple studies have converged on the idea that chaperonin interactors fall into different classes, depending on the extent to which they rely on the chaperonin to fold.[191–193] Chaperonins can promiscuously interact with non-native proteins, but it has emerged that only a minority tangibly benefit from specifically engaging this class of proteins. That said, this minority includes several essential proteins, explaining the universal requirement for chaperonins.

The folding mechanism of chaperonins has been the subject of considerable debate, especially in the case of the better-studied bacterial system.[178] By encapsulating single protein molecules, chaperonins support folding at "infinite dilution". The prevention of off-pathway intermolecular interactions (aggregation) is therefore a core function of chaperonins that applies to almost all clients. Some substrates that are too large to fit in the cavity may benefit from alternative interaction modes.[194] Beyond simply preventing aggregation, a number of studies have demonstrated an exciting additional role for chaperonins in reshaping the conformational search for the native state to catalyse client folding.[147,189,195–199] In the "active cage model", not all GroEL interactors are expected to experience accelerated folding. Rather, only those proteins that populate intermediates stabilised by high configurational entropy during folding are likely to benefit from confinement in the chaperonin cavity. In this regard, it is notable that endogenous clients often fold very efficiently *via* GroEL or TRiC, presumably because they have co-evolved with their cognate chaperonin.

1.3.5 Small HSPs (sHSPs)

Unlike the highly-conserved Hsp70, Hsp90, and Hsp60 systems—which use ATP-coupled cycling to facilitate client folding—small heat shock proteins (sHSPs) are a much more heterogenous family of molecular chaperones with no ATP-binding domains.[200,201] With the molecular mass ranging from 12 to 43 kDa, the only common structural element conserved between all sHSPs is an alpha-crystallin domain—named after the alpha-crystallin protein, an eye lens-specific sHSP tasked with maintaining this highly-crowded and post-mitotic environment free from misfolding, in the absence of turnover mechanisms, throughout the lifespan of an organism.[202] Consistent with this specialised role of alpha-crystallin in the eye lens, sHSPs as a group do not exhibit the "foldase" activity, but rather act as "holdases", *i.e.*, chaperones that bind misfolded proteins and prevent them from interfering with the functional cellular (or extracellular) milieu. This is not to say that sHSP binding represents permanent sequestration of a misfolded protein client: in many cases, sHSPs act as "first-responders" for stress-induced misfolded proteins, with subsequent handover to the ATP-dependent machineries described above allowing re-folding attempts to be made.[203] In fact, a key proposed function of sHSPs is to shepherd misfolded proteins into aggregates in such a way as to enable eventual recovery to their native, folded state.[204–206] Consistent with this hypothesis, exclusion of sHSPs from aggregates and inclusion bodies makes their resident misfolded proteins much more refractory to disaggregation and recovery.[207–210]

Outside of the conserved alpha-crystallin domain, sHSPs contain highly-variable and flexible flanking N-terminal and C-terminal regions. These flanking regions, which have a high level of intrinsic disorder, are critical for the structural plasticity of sHSPs. A characteristic IX(I/V) motif in the C-terminal region guides homo- and hetero-oligomerisation, with the N-terminal region also playing a role in oligomerisation, as well as in client binding. At steady state, sHSPs are found to be large, dynamic oligomers (typically 12- to 32-mers). These larger assemblies could be seen as stores of sHSP dimers, which appear to be the functional units for client recognition.[211] Upon proteotoxic stress, the distribution of these polydisperse oligomers skews towards smaller complexes and the release of dimers that can bind client proteins.[201,212] Although the structure–function relationship between the sHSP homo-/hetero-oligomerisation state and client protein binding, aggregation, and disaggregation has yet to be fully revealed—in large part due to the technical challenges posed by intrinsically disordered regions—the relatively recent revolution in our understanding of liquid–liquid phase separation and intrinsically disordered proteins[213] holds promise for translating through to mechanistic insights in this field.[214–216]

1.3.6 AAA+ Family Chaperones

"ATPases associated with diverse cellular activities" (AAA+), as their name suggests, comprise a functionally and evolutionarily diverse super-family of proteins.[217] They are characterised by a ∼230 amino acid domain containing two ATPase motifs (Walker A and B) and typically form hexameric ring-shaped complexes with a central pore through which its macromolecular substrates can be threaded, as is the case for the DNA helicase, cytoskeletal motor protein, and protein quality control (PQC) members of the family. As one would expect from this description, AAA+ modules are present in several protein degradation machineries as 'unfoldases' that thread the polypeptide substrate into the protease core.[218] Examples include all 5 major ATP-dependent proteases in *Escherichia coli* (ClpAP, ClpXP, HslUV, FtsH, and Lon) and the 19S proteasome and p97/VCP in eukaryotes, which extract and/or unfold ubiquitylated proteins for degradation in the 20S proteasome core particle.[219] Note that not all AAA+ machines important for PQC necessarily unfold their substrates purely in preparation for proteolysis, as exemplified by bacterial ClpB and yeast Hsp104. Rather than being coupled to downstream proteases, these two homologous Hsp100 family proteins function as disaggregases for intracellular misfolded protein deposits and thus serve a crucial function in PQC networks.[220] Metazoans do not possess ClpB or Hsp104 homologues for cytoplasmic and nuclear PQC but appear to have devolved at least some of the same functions to various co-chaperones of the Hsp70 machinery.[221]

1.4 Protein Quality Control

Based on broader definitions of molecular chaperones, it would be easy to perceive chaperones as benevolent, pro-folding agents for troubled proteoforms. Given the potentially catastrophic cellular and organismal consequences of misfolded protein accumulation, however, molecular chaperones have an equally important role in shepherding proteins that either cannot be re-folded or are too "expensive" to be re-folded for degradation. In this way, chaperones are seen as central mediators of protein triage—the decision of whether a misfolded protein is sent down a pro-folding or pro-degradation route.[222] Defining factors governing this triage decision represents an active field of research, with both protein- and context-specific determinants identified.

1.4.1 Folding *vs.* Degradation

Two mechanistic determinants of triage have consistently emerged across disparate PQC systems and species: the chaperone residence time and co-factor competition. In almost all cases, the two are fundamentally intertwined. The proposal that the residence time of a client on a chaperone could determine whether it is folded or degraded is not new[222] and stems from kinetic partitioning models of chaperone function.[223–225] Although specifics vary, these models posit that a chaperone binds non-native clients and partitions them away from aggregated states through repeated and stochastic binding-and-release cycles, until either the substrate is sufficiently mature so as to not be recognised by a chaperone or a competing interaction takes over. Extending this model for the folding-*vs.*-degradation triage decision, too many binding-and-release cycles for the same client polypeptide necessitate some sort of off-ramp for degradation so as to prevent fruitless titration of cellular chaperone pools. The competition between downstream factors for the same binding site on the chaperone–client complex would create such an off-ramp, with degradation-targeting factors taking over if physiologic binding factors that only recognise the mature client structure cannot bind.

An elegant example of both residence time and co-factor competition determining triage has been established for tail-anchored ER membrane proteins.[226] A newly synthesised tail-anchored protein is recognised by the co-chaperone SGTA, which acts to keep the client in a partitioned and uncommitted state whilst being bound. Transfer of the client to TRC40—an ER membrane targeting factor—is a rapid and selective process for mature clients and leads to their insertion into the ER membrane for its physiological function. If the client fails to mature and lingers on SGTA for a prolonged period of time, it is released and instead detected by the PQC co-chaperone BAG6, which targets the client for ubiquitylation and proteasomal degradation. Importantly, binding of either TRC40 or BAG6 commits the client for ER targeting or degradation, respectively, as they release clients much less readily than SGTA despite having comparable binding affinities.[226] Similar PQC mechanisms are likely to exist for triage of proteins that need targeting to other membranes or subcellular compartments, as well as obligate protein complex subunits.[227]

Perhaps, the most widely studied triage process is that of the co-chaperone and E3 ubiquitin ligase CHIP/STUB1.[228] CHIP (the 'C-terminal of Hsp70-interacting protein') is one of several co-chaperones that bind the C-terminal tails of Hsp70 and/or Hsp90. Whereas most co-chaperones aid in client maturation in some way, CHIP can ubiquitylate clients and send them to the proteasome, autophagy, or aggresome.[228,229] Importantly, although CHIP can ubiquitylate clients directly, the reaction is more efficient on the chaperone machinery.[230] Therefore, the relative competition between CHIP and other co-chaperones for the C-terminal of Hsp70/Hsp90 is a central determinant of triage.[231–233] For example, a recent study found that the J-domain co-chaperone DNAJB4 limits the ubiquitylation activity of Hsp70–CHIP complexes, and CHIP suppresses the client folding activity of Hsp70–DNAJB4 complexes.[234]

Note that other E3 ligases have also been shown to be involved in client degradation, including by triggering dissociation of pro-folding co-chaperones on non-functional Hsp90–client complexes.[235,236] Furthermore, competition between different factors can influence which degradation pathway is employed, as is the case for BAG1 *vs.* BAG3 to target Hsp70-bound clients to the UPS *vs.* autophagy, respectively.[237] Importantly, triage governed by co-chaperone competition would provide the cell a means of fine-tuning the proteostasis balance in response to a variety of physiological, patho-physiological, and stressed states (*e.g.*, by modulating the relative expression levels of co-chaperones).[238,239]

1.4.2 Spatial Sequestration

The relationship between molecular chaperones and intracellular misfolded protein aggregates is complicated. Given that a defining characteristic of molecular chaperones is to prevent the aggregation of non-native proteins, any physiological role they might have in promoting protein aggregation was initially difficult to reconcile. However, in an analogous manner to protective sequestration of oligomers into higher-order amyloid plaques in neurodegeneration, spatial sequestration of misfolded proteins into defined subcellular sites has emerged as a widespread cytoprotective mechanism for PQC that is not just limited to disease-associated proteins.[240]

A key discovery in the spatial PQC field was the description of the aggresome: a juxta-nuclear and membrane-free inclusion that stores misfolded proteins when the proteasome capacity is insufficient to keep up with misfolded protein production.[241] It has since been established that juxtanuclear sequestration at the aggresome and functionally analogous sites such as the JUNQ (juxtanuclear QC compartment)[242] or INQ (intranuclear QC compartment)[243] is conserved across eukaryotic life and occurs in response to a wide range of proteotoxic stresses.[55,240] Furthermore, spatial sequestration is not merely employed in conditions where preferred degradation pathways fail or are overwhelmed, but rather occurs early in stressed conditions, *e.g.*, formation of "Q-bodies" and other stress foci.[244] If PQC machineries cannot successfully resolve the misfolded proteins within these foci in sufficient time, they eventually coalesce at the juxta-nuclear site. Therefore, spatial sequestration—whether it be in Q-bodies, the aggresome, JUNQ, or INQ—serves as a rapid response to concentrate potentially aggregation-prone misfolded proteins away from the functional cellular milieu until triage decisions can be made.[245]

As one would expect, spatial sequestration at almost every stage requires molecular chaperones. Misfolded proteins that are recognised by the holdase function of chaperones are sequestered into Q-bodies through the action of sHSPs, whose intrinsic oligomerisation propensity might facilitate the process. Q-bodies can be resolved through Hsp70-dependent folding-*vs.*-degradation triage as described in Section 1.4.1. However, if the misfolded protein load exceeds the available triage capacity of the cell (*e.g.*, during environmental stress, genetic perturbations, or ageing), Q-bodies are targeted to the juxta-nuclear site by Hsp70–BAG3 and/or p62–HDAC6 complexes.[55]

Importantly, misfolded proteins in this juxta-nuclear site are maintained—presumably by holdases—in a state that allows them to be triaged once the PQC capacity is available again. The juxtanuclear location is convenient for targeting to ER-budded autophagosomes and lysosomes, which cluster around the site. Although proteasomes are enriched here, it has been assumed that these are present in a proteolytically-inactive state and either are themselves passive substrates for autophagy ('proteaphagy') by CHIP-mediated ubiquitylation[246] or provide a functional source of K63-ubiquitin chains for recruitment of selective autophagy cargo receptors, in a process mediated by Hsp90.[247] However, much of the research on aggresome biology was performed in conditions with inactivated proteasomes (*e.g.*, by pharmacologic inhibition) or proteasome-refractory PQC substrates, where autophagy is the predominant degradation route available. It is possible that proteasomal clearance plays a larger role in physiological settings. For example, clearance of aggresomes during recovery from heat shock (a stress that does not directly inactivate proteasomes) involves the UPS in an Hsp70-mediated pathway.[248]

In contrast to this reversible sequestration system that exists for most of the proteome, a certain subset of proteins appear to pose such a threat to cellular function that they are immediately and permanently sequestered by sHSPs at the periphery of the cell.[242] This insoluble protein deposit, or 'IPOD', contains terminal aggregates of disease-associated prion and amyloidogenic proteins that persist throughout the lifespan of the cell. Importantly, although IPOD clients are not typically processed through Q-bodies, forcing them to the juxta-nuclear site (*via* ubiquitylation) impairs the dynamics of the entire inclusion and promotes cell death.[242,249] Therefore, keeping these spatial sequestration pathways distinct appears to be critical for maintaining proteome health.

1.5 Chaperone Dysfunction in Diseases

Given the central role of molecular chaperones in maintaining proteome health, it is unsurprising that chaperone dysregulation is associated with a wide range of diseases.[250] Most prominently, neurodegenerative diseases characterised by the accumulation of large amyloid deposits rich in misfolded proteins and ubiquitin—*e.g.*, Alzheimer's, Parkinson's, and Huntington's disease, and prionopathies such as the Creutzfeldt–Jakob disease—have long been attributed to failures in PQC, with chaperones being neuroprotective across several models.[251–253] Amyloid deposits are also characteristic of diverse pathologies outside the nervous system (collectively termed 'amyloidoses'). Non-amyloid misfolded protein deposits seem to be no less toxic, as illustrated by such deposits in amyotrophic lateral sclerosis, fronto-temporal dementia, and spino-cerebellar ataxia-1.[250] While the major misfolded protein in these deposits appears to be disease-specific, several other molecular and structural elements are shared across conditions. These similarities in diverse sets of diseases—regardless of whether they are systemic *vs.* localised or sporadic *vs.* hereditary—suggest at least some mechanistic commonality in their aetiology. The similarities also suggest that the pathology of most protein misfolding diseases is largely due to the gain-of-toxic-function caused by misfolded protein accumulation rather than by the loss-of-physiologic-function of the aggregating protein in question.

As discussed in earlier sections, the large deposits observed in these diseases may represent protective, inert end-points to pacify the toxic effects of oligomeric misfolded species. It is possible, therefore, that chaperones are neuroprotective due to their ability to act at both early and late stages of protein aggregation, *e.g.*, by triaging the misfolded protein monomers before they have a chance to form toxic oligomers and by aggregating into inert fibrils any oligomers that do form. Chaperone capacity declines during the "normal" ageing process;[254–256] therefore, oligomeric forms would begin to accumulate with age and further reduce functional chaperone pools. In addition to resulting in higher levels of toxic oligomers that persist for longer, titrating away chaperones would initiate misfolding and aggregation of other metastable proteins now neglected by PQC systems. Such an age-induced positive-feedback loop is consistent with age being the major risk factor in most protein misfolding diseases.[257] Boosting chaperone function by various methods is therefore an area of active pursuit for therapeutic interventions in such diseases.[258–260]

That is not to say that higher chaperone levels are always protective. Genetic and sporadic changes to chaperones themselves (*e.g.*, mutations, post-translational modifications) contribute to another diverse set of diseases known as chaperonopathies.[261] Even in the case of the neurodegenerative diseases described above, chaperones potentially contribute to spreading of misfolded species from neuron to neuron by disaggregating fibrils and/or other intracellular deposits, thus propagating "seeds" in a similar way to prions (which also require chaperones).[262] Chaperones are additionally overexpressed in several cancers and are proposed to contribute to oncogenesis and malignant progression by enabling mutated or amplified oncoproteins to remain stable without causing proteostasis collapse and cell death[176,263,264]—although more direct roles in inflammation and extracellular matrix remodelling are also likely.[265] In fact, both cancers and neurodegeneration—which could be seen as "opposing" pathologies with respect to the direction of chaperone dysregulation—appear to possess rigid "epichaperomes", *i.e.*, multi-chaperone and co-chaperone assemblies that are chronically adapted to the proteostatic disease landscape.[266,267] Therefore, several mechanistic questions need to be addressed in different healthy and diseased settings before rational therapeutic interventions can realise their full potential.

1.6 Outstanding Questions and Future Directions

Although the structures and conformational cycles of most major chaperones are by now thoroughly characterised, far less is understood about how exactly client protein folding is modified by interaction with chaperones. Advances in NMR spectroscopy, structural mass spectrometry, and single-molecule techniques hold substantial promise in this regard.[268,269] Furthermore, it is now clear that chaperone function is determined not just by the characteristics of the chaperone, but by the features of the client.[270] A future challenge will therefore be to establish new model clients, especially endogenous aggregation-prone substrates, that encompass a broad range of topologies and folding obstacles. For example, a recent study in yeast showed that some of the molecular chaperone requirements defined for model misfolded protein aggregates are dispensable for endogenous misfolded protein condensates.[130] It is also important to consider that the substrate of a chaperone will in many cases not be a complete protein chain, but rather a nascent polypeptide on the ribosome. Future work will need to address chaperone function in the context of translation dynamics, proximity to the ribosome surface, cotranslational folding, and competition with other chaperones with access to the nascent chain.

The majority of the chaperone field to date has focused on the "core" machineries discussed in this chapter. However, several lines of recent research reveal additional layers of complexity and heterogeneity in multicellular organisms. For example, global surveys in both worms and humans have demonstrated tissue-specific chaperone subsets, which is unsurprising given the diverse proteostatic environments throughout an organism (*e.g.*, stem cells *vs.* cardiac muscle *vs.* eye lens).[271–273] Tissue-specific differences—which also hold true for stress programs such as heat-shock response—appear to be dictated by the dominant client classes for the tissue in question, as is the case for actin and SRP in the worm muscle and intestine, respectively.[274] Important progress is also being made in cell non-autonomous mechanisms of proteostasis, *e.g.*, how misfolded proteins and proteotoxic stress are communicated throughout an organism.[272,275,276] Recently discovered chaperone binding and clearance mechanisms for misfolded proteins in ATP-poor extracellular environments[277,278] are important for diseases characterised by extracellular aggregation or systemic amyloidosis. Finally, editing of the chaperone structure and function by post-translational modifications is a rapidly-moving area of exploration;[279–282] cracking this so-called "chaperone code" could not only help resolve several unanswered mechanistic aspects of the physiologic chaperone function, but also provide approaches for rescuing aberrant chaperone function in diseases such as chaperonopathies.[261] Integrating all of the additional layers of regulation into the wealth of existing research on the core chaperone machineries will likely prove a fruitful avenue for further fundamental and translational discoveries in the coming years.

Acknowledgements

We thank Professors Sebastian Hiller, Maili Liu, and Lichun He for inviting this chapter and apologise to all the studies and groups we did not reference due to space constraints. The work of DB was supported by the Francis Crick Institute, which receives its core funding from Cancer Research UK (FC001985), the UK Medical Research Council (FC001985), and the Wellcome Trust (FC001985). The work of RSS was supported by an Institute Strategic Programme Grant (BB/P013384/1) from the UK Biotechnology and Biological Sciences Research Council.

References

1. H. R. Saibil, *J. Biol. Chem.*, 2021, **296**, 100744.
2. C. L. Klaips, G. G. Jayaraj and F. U. Hartl, *J. Cell Biol.*, 2018, **217**, 51–63.
3. K. A. Dill and J. L. MacCallum, *Science*, 2012, **338**, 1042–1046.
4. S. W. Englander and L. Mayne, *Proc. Natl. Acad. Sci. U. S. A.*, 2017, **114**, 8253–8258.
5. G. C. Rollins and K. A. Dill, *J. Am. Chem. Soc.*, 2014, **136**, 11420–11427.
6. J.-H. Han, S. Batey, A. A. Nickson, S. A. Teichmann and J. Clarke, *Nat. Rev. Mol. Cell Biol.*, 2007, **8**, 319–330.
7. A. Borgia, K. R. Kemplen, M. B. Borgia, A. Soranno, S. Shammas, B. Wunderlich, D. Nettels, R. B. Best, J. Clarke and B. Schuler, *Nat. Commun.*, 2015, **6**, 8861.
8. P. Clark, *Trends Biochem. Sci.*, 2004, **29**, 527–534.
9. M. V. Rodnina, *Protein Sci.*, 2016, **25**, 1390–1406.
10. M. dos Reis, R. Savva and L. Wernisch, *Nucleic Acids Res.*, 2004, **32**, 5036–5044.
11. G. Hanson and J. Coller, *Nat. Rev. Mol. Cell Biol.*, 2018, **19**, 20–30.
12. V. Presnyak, N. Alhusaini, Y.-H. Chen, S. Martin, N. Morris, N. Kline, S. Olson, D. Weinberg, K. E. Baker, B. R. Graveley and J. Coller, *Cell*, 2015, **160**, 1111–1124.
13. F. Zhao, Z. Zhou, Y. Dang, H. Na, C. Adam, A. Lipzen, V. Ng, I. V. Grigoriev and Y. Liu, *Proc. Natl. Acad. Sci. U. S. A.*, 2021, **118**, e2022590118.
14. N. Ban, P. Nissen, J. Hansen, P. B. Moore and T. A. Steitz, *Science*, 2000, **289**, 905–920.
15. R. Beckmann, C. M. T. Spahn, N. Eswar, J. Helmers, P. A. Penczek, A. Sali, J. Frank and G. Blobel, *Cell*, 2001, **107**, 361–372.
16. J. Lu and C. Deutsch, *Nat. Struct. Mol. Biol.*, 2005, **12**, 1123–1129.
17. S. Bhushan, M. Gartmann, M. Halic, J.-P. Armache, A. Jarasch, T. Mielke, O. Berninghausen, D. N. Wilson and R. Beckmann, *Nat. Struct. Mol. Biol.*, 2010, **17**, 313–317.
18. J. Marino, G. von Heijne and R. Beckmann, *FEBS Lett.*, 2016, **590**, 655–660.
19. A. J. Samelson, M. K. Jensen, R. A. Soto, J. H. D. Cate and S. Marqusee, *Proc. Natl. Acad. Sci. U. S. A.*, 2016, **113**, 13402–13407.
20. M. K. Jensen, A. J. Samelson, A. Steward, J. Clarke and S. Marqusee, *J. Biol. Chem.*, 2020, **295**, 11410–11417.
21. K. Liu, K. Maciuba and C. M. Kaiser, *Mol. Cell*, 2019, **74**, 310–319.e7.

22. A. M. E. Cassaignau, T. Włodarski, S. H. S. Chan, L. F. Woodburn, I. V. Bukvin, J. O. Streit, L. D. Cabrita, C. A. Waudby and J. Christodoulou, *Nat. Chem.*, 2021, **13**, 1214–1220.
23. K. Liu, J. E. Rehfus, E. Mattson and C. M. Kaiser, *Protein Sci.*, 2017, **26**, 1439–1451.
24. K. S. Hingorani and L. M. Gierasch, *Curr. Opin. Struct. Biol.*, 2014, 81–90.
25. D. Balchin, M. Hayer-Hartl and F. U. Hartl, *Science*, 2016, **353**, aac4354.
26. E. Deuerling, M. Gamerdinger and S. G. Kreft, *Cold Spring Harbor Perspect. Biol.*, 2019, **11**, a033977.
27. L. Danielli, X. Li, T. Tuller and R. Daniel, *Sci. Rep.*, 2020, **10**, 17689.
28. M. Y. Hein, N. C. Hubner, I. Poser, J. Cox, N. Nagaraj, Y. Toyoda, I. A. Gak, I. Weisswange, J. Mansfeld, F. Buchholz, A. A. Hyman and M. Mann, *Cell*, 2015, **163**, 712–723.
29. A. Schwarz and M. Beck, *Trends Cell Biol.*, 2019, **29**, 791–803.
30. D. Zipser, *J. Mol. Biol.*, 1963, **7**, 739–751.
31. A. Shiber, K. Döring, U. Friedrich, K. Klann, D. Merker, M. Zedan, F. Tippmann, G. Kramer and B. Bukau, *Nature*, 2018, **561**, 268–272.
32. G. Kramer, A. Shiber and B. Bukau, *Annu. Rev. Biochem.*, 2019, **88**, 337–364.
33. M. Bertolini, K. Fenzl, I. Kats, F. Wruck, F. Tippmann, J. Schmitt, J. J. Auburger, S. Tans, B. Bukau and G. Kramer, *Science*, 2021, **371**, 57–64.
34. Y.-W. Shieh, P. Minguez, P. Bork, J. J. Auburger, D. L. Guilbride, G. Kramer and B. Bukau, *Science*, 2015, **350**, 678–680.
35. O. O. Panasenko, S. P. Somasekharan, Z. Villanyi, M. Zagatti, F. Bezrukov, R. Rashpa, J. Cornut, J. Iqbal, M. Longis, S. H. Carl, C. Peña, V. G. Panse and M. A. Collart, *Nat. Struct. Mol. Biol.*, 2019, **26**, 110–120.
36. I. Kamenova, P. Mukherjee, S. Conic, F. Mueller, F. El-Saafin, P. Bardot, J.-M. Garnier, D. Dembele, S. Capponi, H. T. M. Timmers, S. D. Vincent and L. Tora, *Nat. Commun.*, 2019, **10**, 1740.
37. O. Lautier, A. Penzo, J. O. Rouvière, G. Chevreux, L. Collet, I. Loïodice, A. Taddei, F. Devaux, M. A. Collart and B. Palancade, *Mol. Cell*, 2021, **81**, 2417–2427.e5.
38. M. Seidel, A. Becker, F. Pereira, J. J. M. Landry, N. T. D. de Azevedo, C. M. Fusco, E. Kaindl, N. Romanov, J. Baumbach, J. D. Langer, E. M. Schuman, K. R. Patil, G. Hummer, V. Benes and M. Beck, *Nat. Commun.*, 2022, **13**, 1224.
39. M. M. Savitski, N. Zinn, M. Faelth-Savitski, D. Poeckel, S. Gade, I. Becher, M. Muelbaier, A. J. Wagner, K. Strohmer, T. Werner, S. Melchert, M. Petretich, A. Rutkowska, J. Vappiani, H. Franken, M. Steidel, G. M. Sweetman, O. Gilan, E. Y. N. Lam, M. A. Dawson, R. K. Prinjha, P. Grandi, G. Bergamini and M. Bantscheff, *Cell*, 2018, **173**, 260–274.e25.
40. P. Goloubinoff, A. S. Sassi, B. Fauvet, A. Barducci and P. De Los Rios, *Nat. Chem. Biol.*, 2018, **14**, 388–395.
41. R. Imamoglu, D. Balchin, M. Hayer-Hartl and F. U. Hartl, *Nat. Commun.*, 2020, **11**, 365.

42. B. Medicherla and A. L. Goldberg, *J. Cell Biol.*, 2008, **182**, 663–673.
43. E. Plessa, L. P. Chu, S. H. S. Chan, O. L. Thomas, A. M. E. Cassaignau, C. A. Waudby, J. Christodoulou and L. D. Cabrita, *Nat. Commun.*, 2021, **12**, 6447.
44. B. H. Toyama, J. N. Savas, S. K. Park, M. S. Harris, N. T. Ingolia, J. R. Yates and M. W. Hetzer, *Cell*, 2013, **154**, 971–982.
45. E. Eden, N. Geva-Zatorsky, I. Issaeva, A. Cohen, E. Dekel, T. Danon, L. Cohen, A. Mayo and U. Alon, *Science*, 2011, **331**, 764–768.
46. F.-M. Boisvert, Y. Ahmad, M. Gierliński, F. Charrière, D. Lamont, M. Scott, G. Barton and A. I. Lamond, *Mol. Cell. Proteomics*, 2012, **11**, M111.011429.
47. A. Varshavsky, *Protein Sci.*, 2006, **15**, 647–654.
48. S. Bansal and S. Tiwari, *Cell Div.*, 2019, **14**, 14.
49. R.-C. Wu, Q. Feng, D. M. Lonard and B. W. O'Malley, *Cell*, 2007, **129**, 1125–1140.
50. G. Ee and N. Lehming, *Transcription*, 2012, **3**, 235–239.
51. M. Muratani, C. Kung, K. M. Shokat and W. P. Tansey, *Cell*, 2005, **120**, 887–899.
52. M. J. Clague and S. Urbé, *FEBS J.*, 2017, **284**, 1753–1766.
53. F. Gonzalvez, D. Lawrence, B. Yang, S. Yee, R. Pitti, S. Marsters, V. C. Pham, J.-P. Stephan, J. Lill and A. Ashkenazi, *Mol. Cell*, 2012, **48**, 888–899.
54. I. Dikic, *Annu. Rev. Biochem.*, 2017, **86**, 193–224.
55. H. E. Johnston and R. S. Samant, *FEBS J.*, 2021, **288**, 4464–4487.
56. A. Hershko and A. Ciechanover, *Annu. Rev. Biochem.*, 1998, **67**, 425–479.
57. R. M. Gronostajski, A. B. Pardee and A. L. Goldberg, *J. Biol. Chem.*, 1985, **260**, 3344–3349.
58. J. Ahlberg, A. Berkenstam, F. Henell and H. Glaumann, *J. Biol. Chem.*, 1985, **260**, 5847–5854.
59. T. Zhang, S. Shen, J. Qu and S. Ghaemmaghami, *Cell Rep.*, 2016, **14**, 2426–2439.
60. J. A. M. Bard, E. A. Goodall, E. R. Greene, E. Jonsson, K. C. Dong and A. Martin, *Annu. Rev. Biochem.*, 2018, **87**, 697–724.
61. M. Akutsu, I. Dikic and A. Bremm, *J. Cell Sci.*, 2016, **129**, 875–880.
62. R. G. Yau, K. Doerner, E. R. Castellanos, D. L. Haakonsen, A. Werner, N. Wang, X. W. Yang, N. Martinez-Martin, M. L. Matsumoto, V. M. Dixit and M. Rape, *Cell*, 2017, **171**, 918–933.e20.
63. R. S. Samant, C. M. Livingston, E. M. Sontag and J. Frydman, *Nature*, 2018, **563**, 407–411.
64. D. E. Leto, D. W. Morgens, L. Zhang, C. P. Walczak, J. E. Elias, M. C. Bassik and R. R. Kopito, *Mol. Cell*, 2019, **73**, 377–389.e11.
65. Y. Aman, T. Schmauck-Medina, M. Hansen, R. I. Morimoto, A. K. Simon, I. Bjedov, K. Palikaras, A. Simonsen, T. Johansen, N. Tavernarakis, D. C. Rubinsztein, L. Partridge, G. Kroemer, J. Labbadia and E. F. Fang, *Nat. Aging*, 2021, **1**, 634–650.
66. S. Kaushik and A. M. Cuervo, *Nat. Rev. Mol. Cell Biol.*, 2018, **19**, 365–381.

67. S. Schuck, *J. Cell Sci.*, 2020, **133**, jcs246322.
68. J. Durgan, A. H. Lystad, K. Sloan, S. R. Carlsson, M. I. Wilson, E. Marcassa, R. Ulferts, J. Webster, A. F. Lopez-Clavijo, M. J. Wakelam, R. Beale, A. Simonsen, D. Oxley and O. Florey, *Mol. Cell*, 2021, **81**, 2031–2040.e8.
69. G. Invernizzi, E. Papaleo, R. Sabate and S. Ventura, *Int. J. Biochem. Cell Biol.*, 2012, **44**, 1541–1554.
70. T. P. J. Knowles, M. Vendruscolo and C. M. Dobson, *Nat. Rev. Mol. Cell Biol.*, 2014, **15**, 384–396.
71. G. Vecchi, P. Sormanni, B. Mannini, A. Vandelli, G. G. Tartaglia, C. M. Dobson, F. U. Hartl and M. Vendruscolo, *Proc. Natl. Acad. Sci. U. S. A.*, 2020, **117**, 1015–1020.
72. J. Tyedmers, A. Mogk and B. Bukau, *Nat. Rev. Mol. Cell Biol.*, 2010, **11**, 777–788.
73. S. Ventura, J. Zurdo, S. Narayanan, M. Parreño, R. Mangues, B. Reif, F. Chiti, E. Giannoni, C. M. Dobson, F. X. Aviles and L. Serrano, *Proc. Natl. Acad. Sci. U. S. A.*, 2004, **101**, 7258–7263.
74. F. Chiti, M. Calamai, N. Taddei, M. Stefani, G. Ramponi and C. M. Dobson, *Proc. Natl. Acad. Sci. U. S. A.*, 2002, **99**, 16419–16426.
75. D. R. Booth, M. Sunde, V. Bellotti, C. V. Robinson, W. L. Hutchinson, P. E. Fraser, P. N. Hawkins, C. M. Dobson, S. E. Radford, C. C. F. Blake and M. B. Pepys, *Nature*, 1997, **385**, 787–793.
76. P. M. Buck, S. Kumar and S. K. Singh, *PLoS Comput. Biol.*, 2013, **9**, e1003291.
77. F. Rousseau, L. Serrano and J. W. H. Schymkowitz, *J. Mol. Biol.*, 2006, **355**, 1037–1047.
78. J. Beerten, W. Jonckheere, S. Rudyak, J. Xu, H. Wilkinson, F. De Smet, J. Schymkowitz and F. Rousseau, *Protein Eng., Des. Sel.*, 2012, **25**, 357–366.
79. S. I. A. Cohen, M. Vendruscolo, C. M. Dobson and T. P. J. Knowles, *J. Mol. Biol.*, 2012, **421**, 160–171.
80. W. M. Babinchak and W. K. Surewicz, *J. Mol. Biol.*, 2020, **432**, 1910–1925.
81. S. Boeynaems, S. Alberti, N. L. Fawzi, T. Mittag, M. Polymenidou, F. Rousseau, J. Schymkowitz, J. Shorter, B. Wolozin, L. Van Den Bosch, P. Tompa and M. Fuxreiter, *Trends Cell Biol.*, 2018, **28**, 420–435.
82. F. Frottin, F. Schueder, S. Tiwary, R. Gupta, R. Körner, T. Schlichthaerle, J. Cox, R. Jungmann, F. U. Hartl and M. S. Hipp, *Science*, 2019, **365**, 342–347.
83. S. A. Levkovich, E. Gazit and D. Laor Bar-Yosef, *Trends Microbiol.*, 2021, **29**, 251–265.
84. D. M. Fowler, A. V. Koulov, W. E. Balch and J. W. Kelly, *Trends Biochem. Sci.*, 2007, **32**, 217–224.
85. E. I. Yakupova, L. G. Bobyleva, S. A. Shumeyko, I. M. Vikhlyantsev and A. G. Bobylev, *Biology*, 2021, **10**, 394.
86. I. Benilova, E. Karran and B. De Strooper, *Nat. Neurosci.*, 2012, **15**, 349–357.

87. J. Gerson and R. Kayed, *Front. Neurol.*, 2013, **4**, 93.
88. H. Mirbaha, D. Chen, O. A. Morazova, K. M. Ruff, A. M. Sharma, X. Liu, M. Goodarzi, R. V. Pappu, D. W. Colby, H. Mirzaei, L. A. Joachimiak and M. I. Diamond, *eLife*, 2018, 7, e36584.
89. T. A. Thibaudeau, R. T. Anderson and D. M. Smith, *Nat. Commun.*, 2018, **9**, 1097.
90. M. P. Jackson and E. W. Hewitt, *Biomolecules*, 2017, 7, 71.
91. M. Lu, C. F. Kaminski and G. S. K. Schierle, *Phys. Biol.*, 2020, **17**, 021001.
92. J. A. J. Housmans, G. Wu, J. Schymkowitz and F. Rousseau, *FEBS J.*, 2023, **290**, 554–583.
93. A. Jomaa, Y.-H. H. Fu, D. Boehringer, M. Leibundgut, S. Shan and N. Ban, *Nat. Commun.*, 2017, **8**, 15470.
94. K. Peisker, D. Braun, T. Wölfle, J. Hentschel, U. Fünfschilling, G. Fischer, A. Sickmann and S. Rospert, *Mol. Biol. Cell*, 2008, **19**, 5279–5288.
95. M. Gamerdinger, K. Kobayashi, A. Wallisch, S. G. Kreft, C. Sailer, R. Schlömer, N. Sachs, A. Jomaa, F. Stengel, N. Ban and E. Deuerling, *Mol. Cell*, 2019, **75**, 996–1006.e8.
96. L. Ferbitz, T. Maier, H. Patzelt, B. Bukau, E. Deuerling and N. Ban, *Nature*, 2004, **431**, 590–596.
97. M. Rohr, F. Ries, C. Herkt, V. L. Gotsmann, L. D. Westrich, K. Gries, R. Trösch, J. Christmann, F. Chaux-Jukic, M. Jung, D. Zimmer, T. Mühlhaus, F. Sommer, M. Schroda, S. Keller, T. Möhlmann and F. Willmund, *Plant Physiol.*, 2019, **179**, 1093–1110.
98. T. Saio, S. Kawagoe, K. Ishimori and C. G. Kalodimos, *eLife*, 2018, 7, e35731.
99. T. Saio, X. Guan, P. Rossi, A. Economou and C. G. Kalodimos, *Science*, 2014, **344**, 1250494.
100. A. Hoffmann, B. Bukau and G. Kramer, *Biochim. Biophys. Acta, Mol. Cell Res.*, 2010, **1803**, 650–661.
101. E. Oh, A. H. Becker, A. Sandikci, D. Huber, R. Chaba, F. Gloge, R. J. Nichols, A. Typas, C. A. Gross, G. Kramer, J. S. Weissman and B. Bukau, *Cell*, 2011, **147**, 1295–1308.
102. J. De Geyter, A. G. Portaliou, B. Srinivasu, S. Krishnamurthy, A. Economou and S. Karamanou, *EMBO Rep.*, 2020, **21**, e49054.
103. B. Wiedmann, H. Sakai, T. A. Davis and M. Wiedmann, *Nature*, 1994, **370**, 434–440.
104. S. Rospert, Y. Dubaquié and M. Gautschi, *Cell. Mol. Life Sci.*, 2002, **59**, 1632–1639.
105. A. Koplin, S. Preissler, Y. Ilina, M. Koch, A. Scior, M. Erhardt and E. Deuerling, *J. Cell Biol.*, 2010, **189**, 57–68.
106. S. Wang, H. Sakai and M. Wiedmann, *J. Cell Biol.*, 1995, **130**, 519–528.
107. U. Raue, S. Oellerer and S. Rospert, *J. Biol. Chem.*, 2007, **282**, 7809–7816.
108. M. Halic, T. Becker, M. R. Pool, C. M. T. Spahn, R. A. Grassucci, J. Frank and R. Beckmann, *Nature*, 2004, **427**, 808–814.

109. R. M. Voorhees and R. S. Hegde, *eLife*, 2015, **4**, e07975.
110. M. Gamerdinger, M. A. Hanebuth, T. Frickey and E. Deuerling, *Science*, 2015, **348**, 201–207.
111. I. Möller, M. Jung, B. Beatrix, R. Levy, G. Kreibich, R. Zimmermann, M. Wiedmann and B. Lauring, *Proc. Natl. Acad. Sci. U. S. A.*, 1998, **95**, 13425–13430.
112. M. del Alamo, D. J. Hogan, S. Pechmann, V. Albanese, P. O. Brown and J. Frydman, *PLoS Biol.*, 2011, **9**, e1001100.
113. Y. Zhang, I. Sinning and S. Rospert, *Nat. Struct. Mol. Biol.*, 2017, **24**, 611–619.
114. S. Preissler and E. Deuerling, *Trends Biochem. Sci.*, 2012, **37**, 274–283.
115. W. Yan, B. Schilke, C. Pfund, W. Walter, S. Kim and E. A. Craig, *EMBO J.*, 1998, **17**, 4809–4817.
116. Y. Zhang, C. Ma, Y. Yuan, J. Zhu, N. Li, C. Chen, S. Wu, L. Yu, J. Lei and N. Gao, *Nat. Struct. Mol. Biol.*, 2014, **21**, 1042–1046.
117. M. Gautschi, H. Lilie, U. Fünfschilling, A. Mun, S. Ross, T. Lithgow, P. Rücknagel and S. Rospert, *Proc. Natl. Acad. Sci. U. S. A.*, 2001, **98**, 3762–3767.
118. Y. Zhang, G. Valentín Gesé, C. Conz, K. Lapouge, J. Kopp, T. Wölfle, S. Rospert and I. Sinning, *Nat. Commun.*, 2020, **11**, 1504.
119. A. Gumiero, C. Conz, G. V. Gesé, Y. Zhang, F. A. Weyer, K. Lapouge, J. Kappes, U. von Plehwe, G. Schermann, E. Fitzke, T. Wölfle, T. Fischer, S. Rospert and I. Sinning, *Nat. Commun.*, 2016, 7, 13563.
120. K. Döring, N. Ahmed, T. Riemer, H. G. Suresh, Y. Vainshtein, M. Habich, J. Riemer, M. P. Mayer, E. P. O'Brien, G. Kramer and B. Bukau, *Cell*, 2017, **170**, 298–311.e20.
121. K. Lee, T. Ziegelhoffer, W. Delewski, S. E. Berger, G. Sabat and E. A. Craig, *Nat. Commun.*, 2021, **12**, 5666.
122. R. Rosenzweig, N. B. Nillegoda, M. P. Mayer and B. Bukau, *Nat. Rev. Mol. Cell Biol.*, 2019, **20**, 665–680.
123. A. Finka, S. K. Sharma and P. Goloubinoff, *Front. Mol. Biosci.*, 2015, **2**, 29.
124. E. Deuerling, A. Schulze-Specking, T. Tomoyasu, A. Mogk and B. Bukau, *Nature*, 1999, **400**, 693–696.
125. S. A. Teter, W. A. Houry, D. Ang, T. Tradler, D. Rockabrand, G. Fischer, P. Blum, C. Georgopoulos and F. U. Hartl, *Cell*, 1999, **97**, 755–765.
126. S. K. Sharma, P. De los Rios, P. Christen, A. Lustig and P. Goloubinoff, *Nat. Chem. Biol.*, 2010, **6**, 914–920.
127. L. Zhao, G. Vecchi, M. Vendruscolo, R. Körner, M. Hayer-Hartl and F. U. Hartl, *Cell Rep.*, 2019, **28**, 1335–1345.e6.
128. N. B. Nillegoda, J. Kirstein, A. Szlachcic, M. Berynskyy, A. Stank, F. Stengel, K. Arnsburg, X. Gao, A. Scior, R. Aebersold, D. L. Guilbride, R. C. Wade, R. I. Morimoto, M. P. Mayer and B. Bukau, *Nature*, 2015, **524**, 247–251.
129. A. S. Wentink, N. B. Nillegoda, J. Feufel, G. Ubartaitė, C. P. Schneider, P. De Los Rios, J. Hennig, A. Barducci and B. Bukau, *Nature*, 2020, **587**, 483–488.

130. H. Yoo, J. A. M. Bard, E. V. Pilipenko and D. A. Drummond, *Mol. Cell*, 2022, **82**, 741–755.e11.
131. R. Sousa and E. M. Lafer, *Traffic*, 2006, 7, 1596–1603.
132. E. A. Craig, *BMC Biol.*, 2018, **16**, 11.
133. S. Rüdiger, L. Germeroth, J. Schneider-Mergener and B. Bukau, *EMBO J.*, 1997, **16**, 1501–1507.
134. E. M. Clerico, A. K. Pozhidaeva, R. M. Jansen, C. Özden, J. M. Tilitsky and L. M. Gierasch, *Proc. Natl. Acad. Sci. U. S. A.*, 2021, **118**, e2016962118.
135. G. Calloni, T. Chen, S. M. Schermann, H.-C. Chang, P. Genevaux, F. Agostini, G. G. Tartaglia, M. Hayer-Hartl and F. U. Hartl, *Cell Rep.*, 2012, **1**, 251–264.
136. A. Bracher and J. Verghese, *Front. Mol. Biosci.*, 2015, **2**, 10.
137. V. Arndt, N. Dick, R. Tawo, M. Dreiseidler, D. Wenzel, M. Hesse, D. O. Fürst, P. Saftig, R. Saint, B. K. Fleischmann, M. Hoch and J. Höhfeld, *Curr. Biol.*, 2010, **20**, 143–148.
138. Y. Jiang, P. Rossi and C. G. Kalodimos, *Science*, 2019, **365**, 1313–1319.
139. P. De Los Rios and A. Barducci, *eLife*, 2014, **3**, e02218.
140. H. H. Kampinga and E. A. Craig, *Nat. Rev. Mol. Cell Biol.*, 2010, **11**, 579–592.
141. B. L. Piette, N. Alerasool, Z.-Y. Lin, J. Lacoste, M. H. Y. Lam, W. W. Qian, S. Tran, B. Larsen, E. Campos, J. Peng, A.-C. Gingras and M. Taipale, *Mol. Cell*, 2021, **81**, 2549–2565.e8.
142. P. De Los Rios, A. Ben-Zvi, O. Slutsky, A. Azem and P. Goloubinoff, *Proc. Natl. Acad. Sci. U. S. A.*, 2006, **103**, 6166–6171.
143. R. Rosenzweig, A. Sekhar, J. Nagesh and L. E. Kay, *eLife*, 2017, **6**, e28030.
144. A. Sekhar, R. Rosenzweig, G. Bouvignies and L. E. Kay, *Proc. Natl. Acad. Sci. U. S. A.*, 2016, **113**, E2794–E2801.
145. A. Sekhar, R. Rosenzweig, G. Bouvignies and L. E. Kay, *Proc. Natl. Acad. Sci. U. S. A.*, 2015, **112**, 10395–10400.
146. R. Kellner, H. Hofmann, A. Barducci, B. Wunderlich, D. Nettels and B. Schuler, *Proc. Natl. Acad. Sci. U. S. A.*, 2014, **111**, 13355–13360.
147. D. Balchin, M. Hayer-Hartl and F. U. Hartl, *FEBS Lett.*, 2020, **594**, 2770–2781.
148. F. H. Schopf, M. M. Biebl and J. Buchner, *Nat. Rev. Mol. Cell Biol.*, 2017, **18**, 345–360.
149. L. H. Pearl, *Biopolymers*, 2016, **105**, 594–607.
150. B. K. Zierer, M. Rübbelke, F. Tippel, T. Madl, F. H. Schopf, D. A. Rutz, K. Richter, M. Sattler and J. Buchner, *Nat. Struct. Mol. Biol.*, 2016, **23**, 1020–1028.
151. L. Stepanova, X. Leng, S. B. Parker and J. W. Harper, *Genes Dev.*, 1996, **10**, 1491–1502.
152. Y. Kimura, S. L. Rutherford, Y. Miyata, I. Yahara, B. C. Freeman, L. Yue, R. I. Morimoto and S. Lindquist, *Genes Dev.*, 1997, **11**, 1775–1785.
153. P. Sahasrabudhe, J. Rohrberg, M. M. Biebl, D. A. Rutz and J. Buchner, *Mol. Cell*, 2017, **67**, 947–961.e5.

154. C. K. Vaughan, U. Göhlke, F. Sobott, V. M. Good, M. M. U. Ali, C. Prodromou, C. V. Robinson, H. R. Saibil and L. H. Pearl, *Mol. Cell*, 2006, **23**, 697–707.
155. A. Giannoulis, A. Feintuch, Y. Barak, H. Mazal, S. Albeck, T. Unger, F. Yang, X.-C. Su and D. Goldfarb, *Proc. Natl. Acad. Sci. U. S. A.*, 2020, **117**, 395–404.
156. K. Lee, A. C. Thwin, C. M. Nadel, E. Tse, S. N. Gates, J. E. Gestwicki and D. R. Southworth, *Mol. Cell*, 2021, **81**, 3496–3508.e5.
157. F. Martino, M. Pal, H. Muñoz-Hernández, C. F. Rodríguez, R. Núñez-Ramírez, D. Gil-Carton, G. Degliesposti, J. M. Skehel, S. M. Roe, C. Prodromou, L. H. Pearl and O. Llorca, *Nat. Commun.*, 2018, **9**, 1501.
158. Y. Liu, M. Sun, A. G. Myasnikov, D. Elnatan, N. Delaeter, M. Nguyenquang and D. A. Agard, bioRxiv, 2020, DOI: 10.1101/2020.06.30.180695.
159. S. Wolf, B. Sohmen, B. Hellenkamp, J. Thurn, G. Stock and T. Hugel, *Chem. Sci.*, 2021, **12**, 3350–3359.
160. A. Rivera-Calzada, M. Pal, H. Muñoz-Hernández, J. R. Luque-Ortega, D. Gil-Carton, G. Degliesposti, J. M. Skehel, C. Prodromou, L. H. Pearl and O. Llorca, *Structure*, 2017, **25**, 1145–1152.e4.
161. J. Oroz, B. J. Chang, P. Wysoczanski, C.-T. Lee, Á. Pérez-Lara, P. Chakraborty, R. V. Hofele, J. D. Baker, L. J. Blair, J. Biernat, H. Urlaub, E. Mandelkow, C. A. Dickey and M. Zweckstetter, *Nat. Commun.*, 2018, **9**, 4532.
162. K. A. Verba, R. Y.-R. Wang, A. Arakawa, Y. Liu, M. Shirouzu, S. Yokoyama and D. A. Agard, *Science*, 2016, **352**, 1542–1547.
163. A. N. Kravats, J. R. Hoskins, M. Reidy, J. L. Johnson, S. M. Doyle, O. Genest, D. C. Masison and S. Wickner, *Proc. Natl. Acad. Sci. U. S. A.*, 2018, **115**, E2210–E2219.
164. T. Morán Luengo, R. Kityk, M. P. Mayer and S. G. D. Rüdiger, *Mol. Cell*, 2018, **70**, 545–552.e9.
165. S. L. Rutherford and S. Lindquist, *Nature*, 1998, **396**, 336–342.
166. R. Geller, S. Pechmann, A. Acevedo, R. Andino and J. Frydman, *Nat. Commun.*, 2018, **9**, 1781.
167. J. Lachowiec, T. Lemus, E. Borenstein and C. Queitsch, *Mol. Biol. Evol.*, 2015, **32**, 91–99.
168. C. Queitsch, T. A. Sangster and S. Lindquist, *Nature*, 2002, **417**, 618–624.
169. N. Rohner, D. F. Jarosz, J. E. Kowalko, M. Yoshizawa, W. R. Jeffery, R. L. Borowsky, S. Lindquist and C. J. Tabin, *Science*, 2013, **342**, 1372–1375.
170. B. Hummel, E. C. Hansen, A. Yoveva, F. Aprile-Garcia, R. Hussong and R. Sawarkar, *Nat. Struct. Mol. Biol.*, 2017, **24**, 234–242.
171. P. L. Yeyati, R. M. Bancewicz, J. Maule and V. van Heyningen, *PLoS Genet.*, 2007, **3**, e43.
172. D. F. Jarosz and S. Lindquist, *Science*, 2010, **330**, 1820–1824.

173. D. Alvarez-Ponce, J. Aguilar-Rodríguez and M. A. Fares, *Genome Biol. Evol.*, 2019, **11**, 2360–2375.
174. M. Taipale, I. Krykbaeva, M. Koeva, C. Kayatekin, K. D. Westover, G. I. Karras and S. Lindquist, *Cell*, 2012, **150**, 987–1001.
175. E. E. Boczek, L. G. Reefschläger, M. Dehling, T. J. Struller, E. Häusler, A. Seidl, V. R. I. Kaila and J. Buchner, *Proc. Natl. Acad. Sci. U. S. A.*, 2015, **112**, E3189–E3198.
176. A. M. Jaeger and L. Whitesell, *Annu. Rev. Cancer Biol.*, 2019, **3**, 275–297.
177. W. Xu, X. Yuan, Z. Xiang, E. Mimnaugh, M. Marcu and L. Neckers, *Nat. Struct. Mol. Biol.*, 2005, **12**, 120–126.
178. M. Hayer-Hartl, A. Bracher and F. U. Hartl, *Trends Biochem. Sci.*, 2016, **41**, 62–76.
179. P. B. Sigler, Z. Xu, H. S. Rye, S. G. Burston, W. A. Fenton and A. L. Horwich, *Annu. Rev. Biochem.*, 1998, **67**, 581–608.
180. Z. Xu, A. L. Horwich and P. B. Sigler, *Nature*, 1997, **388**, 741–750.
181. T. Lopez, K. Dalton and J. Frydman, *J. Mol. Biol.*, 2015, **427**, 2919–2930.
182. D. Gestaut, A. Limatola, L. Joachimiak and J. Frydman, *Curr. Opin. Struct. Biol.*, 2019, **55**, 50–58.
183. C. R. Booth, A. S. Meyer, Y. Cong, M. Topf, A. Sali, S. J. Ludtke, W. Chiu and J. Frydman, *Nat. Struct. Mol. Biol.*, 2008, **15**, 746–753.
184. A. S. Meyer, J. R. Gillespie, D. Walther, I. S. Millet, S. Doniach and J. Frydman, *Cell*, 2003, **113**, 369–381.
185. A. Leitner, L. A. Joachimiak, A. Bracher, L. Mönkemeyer, T. Walzthoeni, B. Chen, S. Pechmann, S. Holmes, Y. Cong, B. Ma, S. Ludtke, W. Chiu, F. U. Hartl, R. Aebersold and J. Frydman, *Structure*, 2012, **20**, 814–825.
186. N. Kalisman, C. M. Adams and M. Levitt, *Proc. Natl. Acad. Sci. U. S. A.*, 2012, **109**, 2884–2889.
187. L. A. Joachimiak, T. Walzthoeni, C. W. Liu, R. Aebersold and J. Frydman, *Cell*, 2014, **159**, 1042–1055.
188. S. Reissmann, L. A. Joachimiak, B. Chen, A. S. Meyer, A. Nguyen and J. Frydman, *Cell Rep.*, 2012, **2**, 866–877.
189. D. Balchin, G. Miličić, M. Strauss, M. Hayer-Hartl and F. U. Hartl, *Cell*, 2018, **174**, 1507–1521.e16.
190. D. Balchin, M. Hayer-Hartl and F. U. Hartl, *Science*, 2016, **353**, aac4354.
191. T. Niwa, T. Kanamori, T. Ueda and H. Taguchi, *Proc. Natl. Acad. Sci. U. S. A.*, 2012, **109**, 8937–8942.
192. M. J. Kerner, D. J. Naylor, Y. Ishihama, T. Maier, H.-C. Chang, A. P. Stines, C. Georgopoulos, D. Frishman, M. Hayer-Hartl, M. Mann and F. U. Hartl, *Cell*, 2005, **122**, 209–220.
193. K. Fujiwara, Y. Ishihama, K. Nakahigashi, T. Soga and H. Taguchi, *EMBO J.*, 2010, **29**, 1552–1564.
194. T. K. Chaudhuri, G. W. Farr, W. A. Fenton, S. Rospert and A. L. Horwich, *Cell*, 2001, **107**, 235–246.
195. A. K. Singh, D. Balchin, R. Imamoglu, M. Hayer-Hartl and F. U. Hartl, *J. Mol. Biol.*, 2020, **432**, 2304–2318.

196. A. J. Gupta, S. Haldar, G. Miličić, F. U. Hartl and M. Hayer-Hartl, *J. Mol. Biol.*, 2014, **426**, 2739–2754.
197. K. Chakraborty, M. Chatila, J. Sinha, Q. Shi, B. C. Poschner, M. Sikor, G. Jiang, D. C. Lamb, F. U. Hartl and M. Hayer-Hartl, *Cell*, 2010, **142**, 112–122.
198. J. Weaver, M. Jiang, A. Roth, J. Puchalla, J. Zhang and H. S. Rye, *Nat. Commun.*, 2017, **8**, 15934.
199. F. Georgescauld, K. Popova, A. J. Gupta, A. Bracher, J. R. Engen, M. Hayer-Hartl and F. U. Hartl, *Cell*, 2014, **157**, 922–934.
200. J. M. Webster, A. L. Darling, V. N. Uversky and L. J. Blair, *Front. Pharmacol.*, 2019, **10**, 1047.
201. M. Haslbeck, S. Weinkauf and J. Buchner, *J. Biol. Chem.*, 2019, **294**, 2121–2132.
202. M. A. Sprague-Piercy, M. A. Rocha, A. O. Kwok and R. W. Martin, *Annu. Rev. Phys. Chem.*, 2021, **72**, 143–163.
203. S. Żwirowski, A. Kłosowska, I. Obuchowski, N. B. Nillegoda, A. Piróg, S. Ziętkiewicz, B. Bukau, A. Mogk and K. Liberek, *EMBO J.*, 2017, **36**, 783–796.
204. M. Haslbeck, S. Walke, T. Stromer, M. Ehrnsperger, H. E. White, S. Chen, H. R. Saibil and J. Buchner, *EMBO J.*, 1999, **18**, 6744–6751.
205. D. M. Walther, P. Kasturi, M. Zheng, S. Pinkert, G. Vecchi, P. Ciryam, R. I. Morimoto, C. M. Dobson, M. Vendruscolo, M. Mann and F. U. Hartl, *Cell*, 2015, **161**, 919–932.
206. A. Strauch and M. Haslbeck, *Essays Biochem.*, 2016, **60**, 163–172.
207. H. H. Kampinga, J. F. Brunsting, G. J. Stege, P. W. Burgman and A. W. Konings, *Exp. Cell Res.*, 1995, **219**, 536–546.
208. G. J. Lee, A. M. Roseman, H. R. Saibil and E. Vierling, *EMBO J.*, 1997, **16**, 659–671.
209. A. G. Cashikar, M. Duennwald and S. L. Lindquist, *J. Biol. Chem.*, 2005, **280**, 23869–23875.
210. S. Ungelenk, F. Moayed, C.-T. Ho, T. Grousl, A. Scharf, A. Mashaghi, S. Tans, M. P. Mayer, A. Mogk and B. Bukau, *Nat. Commun.*, 2016, 7, 13673.
211. I. Santhanagopalan, M. T. Degiacomi, D. A. Shepherd, G. K. A. Hochberg, J. L. P. Benesch and E. Vierling, *J. Biol. Chem.*, 2018, **293**, 19511–19521.
212. E. V. Mymrikov, M. Riedl, C. Peters, S. Weinkauf, M. Haslbeck and J. Buchner, *J. Biol. Chem.*, 2020, **295**, 158–169.
213. S. Alberti, A. Gladfelter and T. Mittag, *Cell*, 2019, **176**, 419–434.
214. A. F. Clouser, H. E. Baughman, B. Basanta, M. Guttman, A. Nath and R. E. Klevit, *eLife*, 2019, **8**, e50259.
215. C. L. Johnston, N. R. Marzano, B. P. Paudel, G. Wright, J. L. P. Benesch, A. M. van Oijen and H. Ecroyd, *J. Biol. Chem.*, 2021, **296**, 100161.
216. C. Yu, S. K. P. Leung, W. Zhang, L. T. F. Lai, Y. K. Chan, M. C. Wong, S. Benlekbir, Y. Cui, L. Jiang and W. C. Y. Lau, *Nat. Commun.*, 2021, **12**, 3007.
217. Y. A. Khan, K. I. White and A. T. Brunger, *Crit. Rev. Biochem. Mol. Biol.*, 2022, **57**, 156–187.

218. J. Shorter and W. A. Houry, *Front. Mol. Biosci.*, 2018, **5**, 85.
219. S. Zhang and Y. Mao, *Biomolecules*, 2020, **10**, E629.
220. S. M. Doyle and S. Wickner, *Trends Biochem. Sci.*, 2009, **34**, 40–48.
221. N. B. Nillegoda and B. Bukau, *Front. Mol. Biosci.*, 2015, **2**, 57.
222. S. Gottesman, S. Wickner and M. R. Maurizi, *Genes Dev.*, 1997, **11**, 815–823.
223. S. J. Hardy and L. L. Randall, *Science*, 1991, **251**, 439–443.
224. J. F. Sinclair, M. M. Ziegler and T. O. Baldwin, *Nat. Struct. Biol.*, 1994, **1**, 320–326.
225. D. Thirumalai, D. K. Klimov and S. A. Woodson, *Theor. Chem. Acc.*, 1997, **96**, 14–22.
226. S. Shao, M. C. Rodrigo-Brenni, M. H. Kivlen and R. S. Hegde, *Science*, 2017, **355**, 298–302.
227. S. Juszkiewicz and R. S. Hegde, *Mol. Cell*, 2018, **71**, 443–457.
228. V. Joshi, A. Amanullah, A. Upadhyay, R. Mishra, A. Kumar and A. Mishra, *Front. Mol. Neurosci.*, 2016, **9**, 93.
229. Y. Sha, L. Pandit, S. Zeng and N. T. Eissa, *Mol. Cell. Biol.*, 2009, **29**, 116–128.
230. L. Quintana-Gallardo, J. Martín-Benito, M. Marcilla, G. Espadas, E. Sabidó and J. M. Valpuesta, *Sci. Rep.*, 2019, **9**, 5102.
231. C. Marques, W. Guo, P. Pereira, A. Taylor, C. Patterson, P. C. Evans and F. Shang, *FASEB J.*, 2006, **20**, 741–743.
232. P. Connell, C. A. Ballinger, J. Jiang, Y. Wu, L. J. Thompson, J. Höhfeld and C. Patterson, *Nat. Cell Biol.*, 2001, **3**, 93–96.
233. M. Stankiewicz, R. Nikolay, V. Rybin and M. P. Mayer, *FEBS J.*, 2010, **277**, 3353–3367.
234. O. T. Johnson, C. M. Nadel, E. C. Carroll, T. Arhar and J. E. Gestwicki, *J. Biol. Chem.*, 2022, **298**, 101697.
235. R. S. Samant, P. A. Clarke and P. Workman, *Proc. Natl. Acad. Sci. U. S. A.*, 2014, **111**, 6834–6839.
236. Z. Li, L. Zhou, C. Prodromou, V. Savic and L. H. Pearl, *Cell Rep.*, 2017, **19**, 2515–2528.
237. E. Stürner and C. Behl, *Front. Mol. Neurosci.*, 2017, **10**, 177.
238. L. B. Shelton, J. Koren and L. J. Blair, *Front. Neurosci.*, 2017, **11**, 724.
239. S. Altinok, R. Sanchez-Hodge, M. Stewart, K. Smith and J. C. Schisler, *Cells*, 2021, **10**, 3121.
240. E. M. Sontag, R. S. Samant and J. Frydman, *Annu. Rev. Biochem.*, 2017, **86**, 97–122.
241. J. A. Johnston, C. L. Ward and R. R. Kopito, *J. Cell Biol.*, 1998, **143**, 1883–1898.
242. D. Kaganovich, R. Kopito and J. Frydman, *Nature*, 2008, **454**, 1088–1095.
243. S. B. M. Miller, C.-T. Ho, J. Winkler, M. Khokhrina, A. Neuner, M. Y. H. Mohamed, D. L. Guilbride, K. Richter, M. Lisby, E. Schiebel, A. Mogk and B. Bukau, *EMBO J.*, 2015, **34**, 778–797.
244. S. Escusa-Toret, W. I. M. Vonk and J. Frydman, *Nat. Cell Biol.*, 2013, **15**, 1231–1243.

245. S. M. Hill, S. Hanzén and T. Nyström, *EMBO Rep.*, 2017, **18**, 377–391.
246. W. H. Choi, Y. Yun, S. Park, J. H. Jeon, J. Lee, J. H. Lee, S.-A. Yang, N.-K. Kim, C. H. Jung, Y. T. Kwon, D. Han, S. M. Lim and M. J. Lee, *Proc. Natl. Acad. Sci. U. S. A.*, 2020, **117**, 19190–19200.
247. P. Nanduri, R. Hao, T. Fitzpatrick and T.-P. Yao, *J. Biol. Chem.*, 2015, **290**, 9455–9464.
248. R. Hjerpe, J. S. Bett, M. J. Keuss, A. Solovyova, T. G. McWilliams, C. Johnson, I. Sahu, J. Varghese, N. Wood, M. Wightman, G. Osborne, G. P. Bates, M. H. Glickman, M. Trost, A. Knebel, F. Marchesi and T. Kurz, *Cell*, 2016, **166**, 935–949.
249. D. Öling, F. Eisele, K. Kvint and T. Nyström, *EMBO J.*, 2014, **33**, 747–761.
250. F. Chiti and C. M. Dobson, *Annu. Rev. Biochem.*, 2017, **86**, 27–68.
251. A. Ciechanover and Y. T. Kwon, *Front. Neurosci.*, 2017, **11**, 185.
252. P. J. Muchowski and J. L. Wacker, *Nat. Rev. Neurosci.*, 2005, **6**, 11–22.
253. M. Brehme and C. Voisine, *Dis. Models Mech.*, 2016, **9**, 823–838.
254. M. Brehme, C. Voisine, T. Rolland, S. Wachi, J. H. Soper, Y. Zhu, K. Orton, A. Villella, D. Garza, M. Vidal, H. Ge and R. I. Morimoto, *Cell Rep.*, 2014, **9**, 1135–1150.
255. M. S. Hipp, P. Kasturi and F. U. Hartl, *Nat. Rev. Mol. Cell Biol.*, 2019, **20**, 421–435.
256. C. Soti and P. Csermely, *Exp. Gerontol.*, 2003, **38**, 1037–1040.
257. K. Cuanalo-Contreras, A. Mukherjee and C. Soto, *Int. J. Cell Biol.*, 2013, **2013**, 638083.
258. W. E. Balch, R. I. Morimoto, A. Dillin and J. W. Kelly, *Science*, 2008, **319**, 916–919.
259. M. Brehme, A. Sverchkova and C. Voisine, *Curr. Opin. Syst. Biol.*, 2019, **15**, 74–81.
260. T. K. Chaudhuri and S. Paul, *FEBS J.*, 2006, **273**, 1331–1349.
261. A. J. L. Macario and E. Conway de Macario, in *Stress: Physiology, Biochemistry, and Pathology*, ed. G. Fink, Academic Press, 2019, pp. 135–152.
262. J. Tittelmeier, E. Nachman and C. Nussbaum-Krammer, *Front. Aging Neurosci.*, 2020, **12**, 581374.
263. A. H. Esfahani, A. Sverchkova, J. Saez-Rodriguez, A. A. Schuppert and M. Brehme, *PLoS Comput. Biol.*, 2018, **14**, e1005890.
264. C. Boudiaf-Benmammar, T. Cresteil and R. Melki, *PLoS One*, 2013, **8**, e60895.
265. S. K. Calderwood, *Philos. Trans. R. Soc., B*, 2018, **373**, 20160524.
266. M. C. Inda, S. Joshi, T. Wang, A. Bolaender, S. Gandu, J. KorenIII, A. Y. Che, T. Taldone, P. Yan, W. Sun, M. Uddin, P. Panchal, M. Riolo, S. Shah, A. Barlas, K. Xu, L. Y. L. Chan, A. Gruzinova, S. Kishinevsky, L. Studer, V. Fossati, S. A. Noggle, J. R. White, E. de Stanchina, S. Sequeira, K. H. Anthoney, J. W. Steele, K. Manova-Todorova, S. Patil, M. P. Dunphy, N. Pillarsetty, A. C. Pereira, H. Erdjument-Bromage, T. A. Neubert, A. Rodina, S. D. Ginsberg, N. De Marco Garcia, W. Luo and G. Chiosis, *Nat. Commun.*, 2020, **11**, 319.

267. A. Rodina, T. Wang, P. Yan, E. D. Gomes, M. P. Dunphy, N. Pillarsetty, J. Koren, J. F. Gerecitano, T. Taldone, H. Zong, E. Caldas-Lopes, M. Alpaugh, A. Corben, M. Riolo, B. Beattie, C. Pressl, R. I. Peter, C. Xu, R. Trondl, H. J. Patel, F. Shimizu, A. Bolaender, C. Yang, P. Panchal, M. F. Farooq, S. Kishinevsky, S. Modi, O. Lin, F. Chu, S. Patil, H. Erdjument-Bromage, P. Zanzonico, C. Hudis, L. Studer, G. J. Roboz, E. Cesarman, L. Cerchietti, R. Levine, A. Melnick, S. M. Larson, J. S. Lewis, M. L. Guzman and G. Chiosis, *Nature*, 2016, **538**, 397–401.
268. I. Sučec, B. Bersch and P. Schanda, *Front. Mol. Biosci.*, 2021, **8**, 762005.
269. A. Mashaghi, G. Kramer, D. C. Lamb, M. P. Mayer and S. J. Tans, *Chem. Rev.*, 2014, **114**, 660–676.
270. R. Mitra, V. V. Gadkari, B. A. Meinen, C. P. M. van Mierlo, B. T. Ruotolo and J. C. A. Bardwell, *Nat. Commun.*, 2021, **12**, 851.
271. N. Shemesh, J. Jubran, S. Dror, E. Simonovsky, O. Basha, C. Argov, I. Hekselman, M. Abu-Qarn, E. Vinogradov, O. Mauer, T. Tiago, S. Carra, A. Ben-Zvi and E. Yeger-Lotem, *Nat. Commun.*, 2021, **12**, 2180.
272. A. J. Sala, L. C. Bott and R. I. Morimoto, *J. Cell Biol.*, 2017, **216**, 1231–1241.
273. K. Nisaa and A. Ben-Zvi, *Trends Cell Biol.*, 2022, **32**, 470–474.
274. J. Ma, C. E. Grant, R. N. Plagens, L. N. Barrett, K. S. K. Guisbert and E. Guisbert, *G3: Genes, Genomes, Genet.*, 2017, **7**, 1011–1018.
275. J. Miles, R. Scherz-Shouval and P. van Oosten-Hawle, *Trends Biochem. Sci.*, 2019, **44**, 927–942.
276. S. Wolff, J. S. Weissman and A. Dillin, *Cell*, 2014, **157**, 52–64.
277. I. Gallotta, A. Sandhu, M. Peters, M. Haslbeck, R. Jung, S. Agilkaya, J. L. Blersch, C. Rödelsperger, W. Röseler, C. Huang, R. J. Sommer and D. C. David, *Nature*, 2020, **584**, 410–414.
278. E. Itakura, M. Chiba, T. Murata and A. Matsuura, *J. Cell Biol.*, 2020, **219**, e201911126.
279. Nitika, C. M. Porter, A. W. Truman and M. C. Truttmann, *J. Biol. Chem.*, 2020, **295**, 10689–10708.
280. S. J. Backe, R. A. Sager, M. R. Woodford, A. M. Makedon and M. Mollapour, *J. Biol. Chem.*, 2020, **295**, 11099–11117.
281. C. Caruso Bavisotto, G. Alberti, A. M. Vitale, L. Paladino, C. Campanella, F. Rappa, M. Gorska, E. Conway de Macario, F. Cappello, A. J. L. Macario and A. Marino Gammazza, *Front. Mol. Biosci.*, 2020, 7, 95.
282. M. P. Collier, K. B. Moreira, K. H. Li, Y.-C. Chen, D. Itzhak, R. Samant, A. Leitner, A. Burlingame and J. Frydman, *Sci. Rep.*, 2021, **11**, 13084.

CHAPTER 2

Structural Disorder in Chaperone Functions Probed by NMR

Y. HU*[a,b] AND M. LIU*[a,b]

[a] Key Laboratory of Magnetic Resonance in Biological Systems, State Key Laboratory of Magnetic Resonance and Atomic and Molecular Physics, National Center for Magnetic Resonance in Wuhan, Wuhan Institute of Physics and Mathematics, Innovation Academy for Precision Measurement Science and Technology, Chinese Academy of Sciences, Wuhan, 430071, China; [b] University of Chinese Academy of Sciences, Beijing, 10049, China
*Emails: huyunfei@wipm.ac.cn; ml.liu@wipm.ac.cn

2.1 Introduction

Molecular chaperones play essential roles in governing protein quality control and homeostasis in cells.[1,2] The general definition of chaperones includes a highly diverse collection of proteins that differentially affect various aspects of protein folding and unfolding processes. The main functions of chaperones include, but are not limited to, assisting protein folding, binding to a non-natively folded intermediate to avoid misfolding, unfolding proteins that are misfolded, binding proteins and preventing them from aggregation, catalysing peptide bond isomerization to facilitate folding, translocating proteins across membranes or inserting proteins into membranes and so on.[1–4] The sizes and shapes of chaperones also vary, ranging from small proteins with less than 100 amino acid residues to huge protein complexes with molecular weights of up to mega-Daltons. Nonetheless, all chaperones share some common characteristics, particularly

New Developments in NMR No. 29
Biophysics of Molecular Chaperones: Function, Mechanisms and Client Protein Interactions
Edited by Sebastian Hiller, Maili Liu and Lichun He

Published by the Royal Society of Chemistry, www.rsc.org

regarding their interactions with client proteins. Chaperones generally show little substrate specificity and are able to interact with a wide range of client proteins that are unfolded, partially folded or misfolded. These promiscuous chaperone–client interactions are also dynamic, forming the so-called "fuzzy" complexes, which have been extensively reviewed previously.[4–6]

In most cases, chaperones are well-folded proteins. They utilize an enclosed chamber (*e.g.* the GroEL/ES chaperonin),[7] a cavity or a cradle-like structure (*e.g.* the ATP-independent chaperones Skp[8] and trigger factor[9]) or an exposed surface with hydrophobic grooves (*e.g.* the unfoldase SecB[10]) to interact with unfolded polypeptides. However, there are also a number of special cases in which the chaperone does not adopt a well-folded structure when it actively associates with its client. These unique classes of chaperones may become partially or fully unfolded under stress conditions so as to perform their physiological functions, such as the stress-responsive Hsp33 and HdeA chaperones,[11–13] or are intrinsically disordered constitutively, such as the chaperone prothymosin-α (ProTα) of the linker histone H1.[14] A central question therefore arises: how do these disordered chaperones bind and protect client proteins if they themselves lack a well-defined structural scaffold? Because both interacting proteins show a high content of structural disorder which substantially adds to the sample heterogeneity, attempts to elucidate the molecular mechanisms of such chaperone–client interacting systems are met with exceptional challenges. Herein, we will focus on these specific classes of chaperones and summarize recent advances in understanding their functional mechanisms by applying solution NMR spectroscopy methods.

2.2 NMR Methods for Studying Dynamic Protein Interactions

The technological development of solution NMR over the past decades has established a variety of tools to detect protein–protein interactions at atomic resolution.[6,15,16] Among these various methods, some are particularly useful in investigating dynamic and heterogeneous protein interactions associated with protein disorder and are thus widely used in the studies of chaperone–client complexes. In Sections 2.2.1–2.2.4, we summarize some of the key techniques commonly used for studies of chaperones that show conditional or constitutive structural disorder.

2.2.1 2D HSQC Spectra

The two-dimensional (2D) heteronuclear single quantum coherence (HSQC) experiment is the most classical and commonly used method for protein characterization. Despite it being one of the simplest NMR experiments, HSQC spectra contain a wealth of information regarding protein conformations and interactions both globally and site-specifically. In particular, the ^{1}H–^{15}N HSQC spectrum functions as a fingerprint spectrum, in which each peak represents a single amino acid residue in the protein polypeptide.

First of all, the overall dispersion of the peaks in the HSQC spectrum offers a direct assessment of whether or not the protein is well-structured or contains large fractions of structural disorder, which is extremely practical in studies of conditionally unfolded chaperones. Additionally, the number of peaks (*e.g.* whether there are multiple sets of peaks) and the homogeneity of the peak intensities can provide hints and clues about the protein conformational heterogeneity. Second, protein interactions are commonly monitored by comparing the HSQC spectra in the absence and in the presence of the binding partner or by tracing the change of peak positions or intensities when different amounts of the binding partner are titrated into the sample. Under favorable conditions, chemical shift perturbation (CSP) analysis offers a most straightforward way of identifying binding interfaces at the residue level. However, chaperone–client interactions are intrinsically less specific and their complexes often show highly dynamic and heterogeneous properties, which may result in peak broadening or disappearance in the spectrum of the complex. In such cases, peak intensity changes and line-width analysis are commonly exploited to extract the binding site information. In some extreme cases, the majority of the peaks in the HSQC spectra may become severely broadened, making data interpretation impossible. Therefore, additional methods are needed to provide more detailed information. Taken together, simple as it is, the 2D HSQC spectrum plays an essential role in chaperone structural analysis as it guides the researcher to choose experimental approaches most suitable for the target protein in hand.

2.2.2 Relaxation Measurements

The relaxation of NMR signals is affected by microscopic dynamic processes and therefore contains ample information on protein motions. Measuring protein backbone amide ^{15}N relaxation parameters, including the longitudinal relaxation rate R_1, the transverse relaxation rate R_2 and the heteronuclear $\{^{1}H\}$–^{15}N NOE effect, was one of the earliest developed NMR methods to probe protein dynamics.[17,18] Combined with either the spectral density mapping or the Lipari–Szabo model-free method for analysis of the macroscopic relaxation rates, microscopic parameters reflecting protein motions at the residue level could be extracted.[19–21] It has been widely used in the past few decades to obtain both overall tumbling and local dynamics, particularly on the pico-to-nanosecond timescales.

Furthermore, many important physiological processes, including ligand binding, enzyme catalysis, allosteric regulation, *etc.*, occur on the micro-to-millisecond timescales, and the associated protein conformational fluctuation, more commonly referred to as conformational exchange, can significantly contribute to the apparent transverse relaxation rate. This effect is largely suppressed in the Carr–Purcell–Meiboom–Gill (CPMG) pulse train, and thus the μs–ms timescale motions cannot be adequately detected using routine R_2 experiments. To recapture protein dynamics on this time-scale, CPMG relaxation dispersion (RD)[22–24] and related R1ρ RD[24] methods were developed. Briefly, the CPMG RD experiment measures the apparent R_2 values as a function of τ_{cp}, the delay between two chemical shift-refocusing π-pulses in the CPMG pulse train, or ν_{CPMG}, the frequency of CPMG pulses. By fitting the data under specific exchange models, parameters describing the exchange rates, relative populations and chemical shift differences of major and minor conformational species could be obtained. The RD methods proved to be highly efficient in illuminating sparsely-populated "excited" state conformations that were previously invisible to other techniques.

2.2.3 Chemical Exchange Saturation Transfer (CEST)

The CEST experiment is another important NMR method that can be applied to probe sparsely-populated conformational states.[25] Compared to the CPMG RD method, which is most sensitive to exchanges on the time scale of approximately 0.25–5 ms, the CEST method is more sensitive to slower exchange processes, typically in the range of 2–25 ms.[15] The CEST method exploits the energy transfer between two conformational states (*e.g.* a major state A and a minor state B) in exchange with each other, such that saturation on the minor state by a weak radiofrequency field B_1 would be transferred to the major state and thus amplified. In practice, a series of HSQC spectra are recorded with a weak B_1 field scanning through different frequencies. When the B_1 frequency matches that of the minor state, the signal saturation effect transfers to the major state, resulting in signal reduction of the resonance corresponding to the major state that can be detected in the HSQC spectra. Therefore, by plotting the signal intensity ratio I/I_0 of major state resonances (where I and I_0 correspond to the signal intensities with and without the application of the B_1 field, respectively) against the B_1 frequency, the existence of minor conformational states together with their chemical shift information can be visualized in a very straightforward manner. The CEST method can be applied in studies of chaperone–client systems to detect sparsely-populated states in exchange equilibrium between folded and unfolded or bound and unbound states.

2.2.4 ^{19}F NMR Spectroscopy

^{19}F NMR is an extremely advantageous technique for obtaining protein conformational and interactional information,[26–29] particularly under circumstances when traditional ^{15}N- or ^{13}C-based NMR methods fail. These include "fuzzy" complexes formed between chaperones and clients that exhibit high structural heterogeneity, which often result in severe line-broadening and signal disappearance when performing ^{15}N-edited HSQC experiments.

^{19}F is a highly-sensitive nucleus, with a gyromagnetic ratio of ~83% that of ^{1}H. Unlike hydrogen, carbon, nitrogen or phosphorus atoms, fluorine is absent in proteins or nucleic acids and therefore exhibits very low background signals. The incorporation of ^{19}F into biomolecules can be achieved in different ways. Chemical ligation is a commonly used approach to attach ^{19}F-probes onto reactive moieties in a protein, such as thiol groups of cysteine residues.[29] Alternatively, ^{19}F labelling can be achieved during protein synthesis, either by using fluorinated amino acids or amino acid precursors[30,31] or by using the unnatural amino acid approach.[32–35]

The chemical shift of ^{19}F is widely dispersed, covering ~1000 ppm.[36] It is also highly susceptible to changes in the surroundings and thus is an ideal probe for reporting local conformational changes or interactions. Many different parameters, including chemical shifts, linewidths, solvent-induced isotope shifts and paramagnetic quenching (used in combination with the spin-labelling method), which were previously reviewed in detail,[27,29] can be obtained through simple 1D ^{19}F NMR spectra. Despite the high chemical shift anisotropy (CSA) effect of the ^{19}F nucleus and thus relatively fast relaxation at a high magnetic field, it has been successfully applied to obtain site-specific information on dynamic chaperone–client interacting systems.

2.3 Structural Disorder in Chaperone Functions

The involvement of structural disorder in chaperone functions is not uncommon. Herein, we summarize a few representative chaperone systems in which structural disorder plays a fundamental role, with special emphasis on how NMR methods have been employed to unveil the underlying molecular mechanisms.

2.3.1 Redox-regulated Heat Shock Protein Hsp33

The bacterial redox-responsive chaperone Hsp33 is one of the earliest recognized examples of conditional disordered chaperones.[37–39] It is a cytosolic ATP-independent chaperone that senses oxidizing stresses and acts as a holdase to prevent client proteins from aggregation. The domain architecture of Hsp33 comprises an N-terminal domain (NTD), a C-terminal zinc finger-containing redox switch domain and a 40-amino-acid linker region. Under normal reducing conditions, Hsp33 is an inactive, well-folded monomer. The redox switch domain coordinates a single zinc ion *via* four conserved cysteines, and the linker region folds into three helices and an extended stretch. Upon exposure to oxidative stress, Hsp33 undergoes an order-to-disorder transition, in which the zinc-coordinating cysteines become disulphide-bonded, the zinc ion is released and both the redox switch domain and the linker region become unfolded. This drastic structural change turns on the chaperone activity of Hsp33, enabling it to bind partially unfolded client proteins with high affinity. Extensive biochemical and biophysical efforts have elegantly established the correlation between "conditional disorder'' and the activation of Hsp33 chaperone activity, as well as a functional cycle that interconverts Hsp33 between its inactive and active states *via* local unfolding and folding.[37,39–45] Although the high-resolution crystal structure of the inactive reduced form of Hsp33 was available, the partially disordered nature of the active state prevents crystallization and hinders structural characterization.[38,46–48] Therefore, elucidating the molecular mechanisms of how partial unfolding transforms the chaperone structure and how the partially disordered chaperone binds and protects client proteins is technically challenging. The solution NMR method has been applied to address these difficulties in combination with other approaches.

Graf and co-workers employed a combination of biochemical and NMR methods to investigate the conformational changes of Hsp33 associated with oxidation-induced activation.[39] Both the CD and NMR spectroscopy results indicated extensive conformational disordering upon zinc depletion or oxidization. By comparing the two-dimensional (2D) HSQC spectra of both the zinc-binding redox switch domain and the full-length Hsp33 protein under reducing or oxidizing conditions, they demonstrated that the oxidation-induced structural unfolding occurs mainly within the redox switch domain.[39] The structure of the redox switch domain in its inactive, reduced form was determined by solution NMR[49] and was also visualized in an inactive dimeric crystal structure of full-length Hsp33.[47] However, investigation of the active state structure and interaction was less straightforward.

In order to directly detect client binding, Groitl and co-workers employed the ^{19}F NMR spectroscopy method by labelling individual sites in Hsp33[50] with the fluorine probe 4-(trifluoromethoxy)-phenylalanine (tFPA) *via* unnatural amino acid incorporation.[32–34] This strategy takes advantage of the high sensitivity of the ^{19}F probe both as an NMR nucleus and as an indicator of chemical environment changes. Among the five candidate sites, three sites were found to show the co-existence of two ^{19}F signals. Since the ^{19}F chemical shift itself is a sensitive reporter on the chemical environment, the authors were able to conclude that one signal corresponds to a more exposed conformation and probably represents the unfolded, active state, whereas the other signal corresponds to a relatively buried conformation representing the inactive state. A further experiment combining ^{19}F NMR and paramagnetic relaxation enhancement (PRE) methods (see Figure 2.1), in which Hsp33 was selectively ^{19}F labelled and a substrate peptide was spin labelled, provided direct evidence of client binding occurring at both the linker region and the linker-region docking surface of the NTD in Hsp33.[50] Furthermore, Lee *et al.* examined the various conformational states of Hsp33 by solution NMR.[51] Their results unveiled a dynamic interaction between the N-terminal segments and the linker region, which probably plays a role in the regulation of the chaperone activity.

2.3.2 pH-regulated Bacterial Periplasmic Chaperone HdeA

The bacterial acid-resistance chaperone HdeA is another unique conditional disordered chaperone. It locates in the periplasmic space and functions in

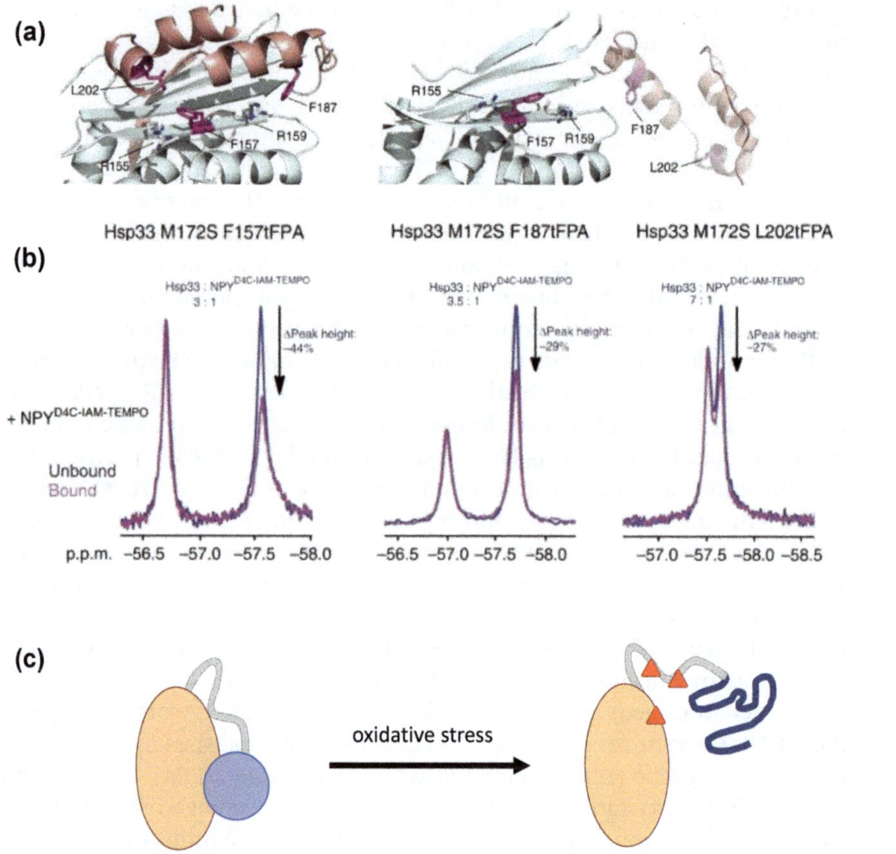

Figure 2.1 Probing Hsp33–client interactions by a combined usage of site-specific ^{19}F-labelling and PRE methods. (a and b) Selective ^{19}F-labelling sites mapped onto the linker region (pink) and the linker-docking surface (cyan) of *Escherichia coli* Hsp33 (a) and representative ^{19}F-NMR spectra perturbed by spin-labelling in the client peptide.[50] Reproduced from ref. 50 with permission from Springer Nature, Copyright 2016. (c) A schematic illustration of the Hsp33 activation mechanism in which partial unfolding enables exposure of the client binding sites (red triangles).

protecting client proteins from denaturation and precipitation under acidic environments, for example, in the gastric system of hosts.[13,52] Under neutral conditions, HdeA adopts a well-folded dimeric structure that lacks chaperone activity, whereas it undergoes dimer dissociation together with global unfolding when the pH decreases and becomes activated.[52,53] Compared to Hsp33, the HdeA chaperone is smaller and structurally simpler, containing a single domain of ~90 residues that dimerizes by burying a hydrophobic interface. Its activation process is accompanied by a more drastic and global unfolding event, which results in an active state of HdeA at pH 2–3 that is almost completely disordered.[52,54–56] Therefore, in the HdeA–client complexes, both the active chaperone and the acid-denatured client proteins are disordered or at least partially disordered. This feature adds to the sample heterogeneity and presents a critical challenge to elucidating the HdeA functional mechanism at the atomic level.

pH titration monitored by 2D $^{1}H–^{15}N$ HSQC NMR spectroscopy is a straightforward experiment to trace the acid-induced unfolding process of HdeA. Garrison and Crowhurst used this approach to examine the conformational changes of HdeA in the pH range between 6 and 3 and provided evidence for structural loosening accompanying a pH decrease.[57] However, the active state structure of HdeA at a pH lower than 3 was difficult to analyse, due to both its highly disordered nature and the conformational heterogeneity, which led to multiple sets of NMR peaks. Despite the extensive experimental and computational efforts to unravel the HdeA functional mechanism,[55,56,58–66] structural details of HdeA under low pH conditions are largely unclear.

Similar to the case of Hsp33, fluorine NMR stands out as an advantageous approach in the investigation of the HdeA active state structure. By adding fluorine-labelled amino acid precursors to a bacterial growth medium,[30,31] Zhai and co-workers selectively labelled phenylalanine or tryptophan residues in HdeA for ^{19}F NMR spectroscopy analysis.[67] They identified the potential active species among the multiple conformational states in slow exchange with each other (see Figure 2.2a), which is the first atomic-level characterization of the structural transition and client interaction in the low pH range, demonstrating the capacity of ^{19}F NMR in studying chaperone systems that are disordered and heterogeneous. Due to the limited number of available labelling sites in wild-type HdeA, a comprehensive understanding of the overall structural characteristics and client binding activity of the active state HdeA was lacking. To overcome this difficulty, Yu and co-workers carried out a more extensive ^{19}F NMR analysis *via* systematic incorporation of ^{19}F-labelled tryptophan using site-directed mutagenesis, which is accompanied by chaperone activity assays.[68] Using this approach, they succeeded in identifying two hydrophobic patches of residues that bind unfolded client proteins at a low pH (see Figure 2.2b). Furthermore, they found that the HdeA-F28W mutant shows a single set of peaks at pH 1.5, which is an ideal mimic of the monomeric species of wild-type HdeA under activated conditions. This finding enables structural characterization of the active, disordered state of HdeA using the conventional $^{15}N/^{13}C$ labelling method and triple-resonance NMR spectroscopy,[69,70] which offers the advantage of observing all residues in the protein simultaneously. Using ^{15}N-labelled samples, they performed pH-dependent ^{15}N CEST experiments to probe the partially unfolded intermediate states (see Figure 2.2c and d), unravelling a step-wise unfolding mechanism as well as the participation of the N-terminus in regulating activation.[68] Further NMR analyses provided additional insights into the roles of the N-terminal and loop regions in chaperone activation.[71,72]

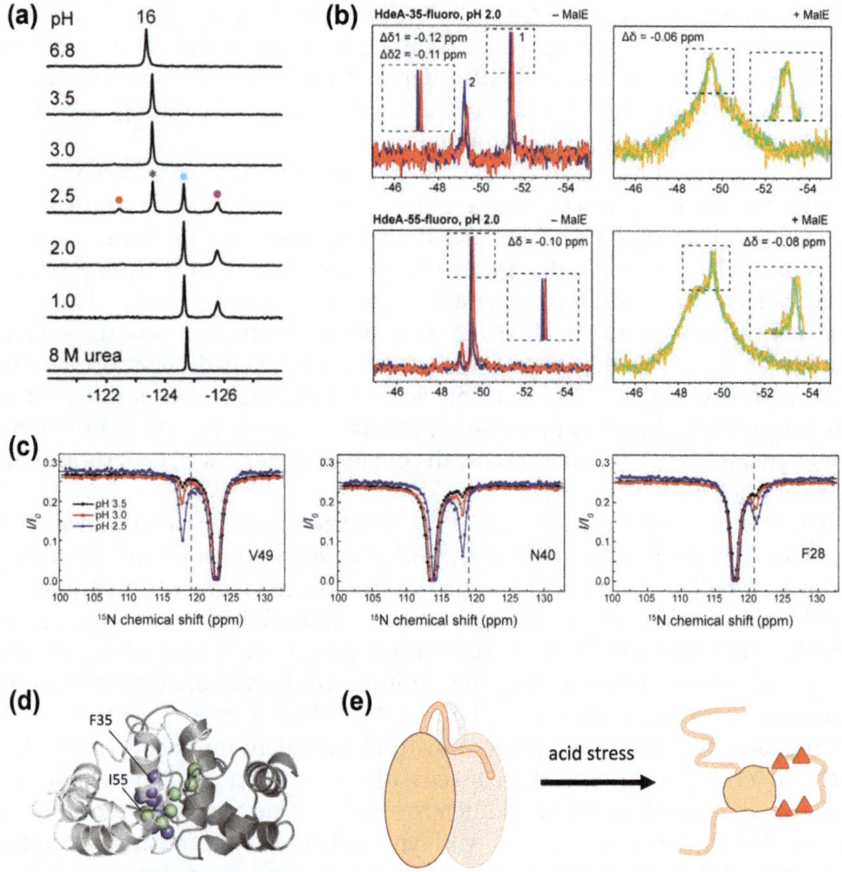

Figure 2.2 Probing the conformational dynamics,[67] the client-binding and activation process of HdeA using ^{19}F NMR and ^{15}N CEST methods.[68] (a) ^{19}F-NMR spectra showing the pH-dependent conformational heterogeneity of the Trp16 residue of HdeA. Reproduced from ref. 67 with permission from the Royal Society of Chemistry. (b and c) ^{19}F NMR spectra illustrating two representative client binding sites (b) and ^{15}N CEST data showing the pH-dependent local unfolding of HdeA (c). Reproduced from ref. 68, https://doi.org/10.1074/jbc.RA118.006398, under the terms of the CC BY 4.0 license, http://creativecommons.org/licenses/by/4.0/. (d) Three-dimensional structure of the inactive HdeA homodimer showing two client-binding segments (purple and green spheres) and the locations of two representative residues. (e) A schematic illustration of the HdeA activation mechanism in which dimer dissociation accompanied by global unfolding enables exposure of the client binding sites (red triangles).

2.3.3 Small Heat Shock Proteins

Small heat shock proteins (sHSPs) are a large family of ubiquitous and ATP-independent molecular chaperones that protect misfolded or partially unfolded client proteins from irreversible aggregation under stress conditions and are linked to various human diseases.[73,74] A unique feature of sHSPs is their ability to form large oligomers with different sizes that exhibit dynamic subunit exchanges.[73,75–77] The sHSP sequence contains a central compactly folded α-crystallin domain (ACD), flanked by a flexible N-terminal region (NTR) and a short C-terminal region (CTR). The ACD adopts a β-sandwich structure that dimerizes due to an antiparallel packing between the extended β6 + 7 strands and acts as the basic building block for a higher-order assembly. The CTR contains a three-residue IXI motif that can interact with the ACD and contribute to oligomer formation. The NTR is intrinsically disordered and of variable lengths, sometimes occupying over 50% of the full protein sequence. Complex roles such as regulating oligomerization and binding client proteins have been implicated in this region. Due to the heterogenous and dynamic nature of sHSPs, obtaining residue-resolution structural information outside the well-folded ACD region is extremely challenging. Hybrid methods in which NMR plays a critical role have been applied and have shed invaluable light on the functional mechanism of this chaperone family.

α-Crystallins are classical sHSP members that are found mainly in the eye lens (αA-crystallin) and in a variety of tissues (αB-crystallin).[76–78] Taking αB-crystallin as an example, despite extensive progress in unravelling the ACD structures, dynamics and oligomer formation,[79–87] the detailed mechanism of how it binds and protects client proteins remains elusive. Mainz and co-workers utilized a combined solid-state and solution NMR method to pursue the client-bound state of αB-crystallin, and they chose two distinct types of client proteins for investigation, the model client lysozyme that forms amorphous aggregates upon reduction and the $A\beta_{1-40}$ peptide that forms amyloid fibrils.[88] Intriguingly, they observed that the partially disordered NTR plays a major role in binding the amorphously aggregating lysozyme, whereas a hydrophobic edge of the β-sandwich-shaped ACD appears to form the interface for binding to the fibril-forming $A\beta_{1-40}$. These results, supplemented with further mutagenesis and functional assays, suggested a mechanism by which αB-crystallin uses different binding interfaces to adapt to different types of clients, highlighting the relationship between its structural flexibility and functional plasticity.

The human HSPB1, also known as Hsp27, is one of the most abundantly expressed sHSPs and has been the target of extensive investigations.[89,90] While structural insights have been gained for the ACD[91–93] and the CTR[94] *via* NMR and crystallographic methods, information regarding the NTR, which represents nearly half of the Hsp27 sequence, is largely lacking. By combining solution NMR with hydrogen–deuterium exchange mass spectrometry (HDXMS), Clouser and co-workers presented the first residue-level structural characterization of full-length Hsp27, particularly of the disordered NTR.[95] This study took advantage of a mutated version of Hsp27 that mimics the stress-activated dimeric state, which shows monodisperse properties and a low molecular weight suitable for NMR analysis. Classical CSP analysis of the 2D ^{1}H–^{15}N HSQC spectra of different constructs or sample conditions suggested wide-scale interaction between the NTR and the ACD, which coincides with the HDXMS results that revealed a transient order for regions in the NTR. Further NMR approaches, namely titration experiments using peptides corresponding to different NTR sub-regions and PRE experiments in which spin labels were attached to different NTR sub-regions, provided more detailed insights into the interaction specificities. Taken together, the authors suggested a 'quasi-ordered' model in which the NTR interacts with multiple sites on the ACD, providing the structural basis for Hsp27 heterogeneity and plasticity associated with the intrinsically disordered NTR. Interestingly, another study of Hsp27 by Alderson and co-workers focusing on the ACD suggested that the monomeric species is highly chaperone-active while at the same time prone to self-aggregation.[96] More importantly, by using the NMR relaxation dispersion method and high-pressure solution NMR spectroscopy, they observed local structural unfolding in the ACD that corresponds to the dimer interface (see Figure 2.3). This work, together with a follow-up study employing multiple NMR methods,[97] including pH-dependent backbone ^{15}N relaxation and residual dipolar coupling (RDC) analysis, highlighted the contribution of conditional disorder in the functional mechanism of Hsp27 and probably in other sHSPs as well. Taken together, the recent applications of NMR methods shed new light on the complex roles of structural disorder in the sHSP functional mechanisms.

2.3.4 Tight Complex Between the Linker Histone H1 and Prothymosin-α (ProTα)

In most typical cases, protein–protein interactions mediated by intrinsically disordered proteins (IDPs) are actually associated with structural orders. Many IDPs or disordered protein regions gain local structures upon binding or wrapping around the folded structure of the binding partner. Recently, tight binding between two extremely disordered proteins has been reported, revealing a completely new mode of interaction that has not been realized before.[14,98] Of these two proteins, one is the linker histone H1.0 (H1) and the other is its chaperone, a nuclear protein prothymosin-α (ProTα). H1 is largely disordered besides a small globular domain, and ProTα is fully disordered. The most striking characteristic of this protein–protein complex is that both parties exhibit very high net charges, H1 is positively charged and ProTα is negatively charged, and very low hydrophobicity.

By using solution NMR together with circular dichroism spectrometry and single-molecule Förster resonance energy transfer (smFRET), Borgia and co-workers suggested that both proteins remain unstructured in the complex and the binding is mainly mediated by the attraction of opposite charges.[14] Further molecular dynamics simulation indicated that the complex is a highly dynamic ensemble of various configurations without evident site-specific binding and was supported by smFRET and NMR relaxation measurements. Although there were debates regarding the exact binding affinity between the

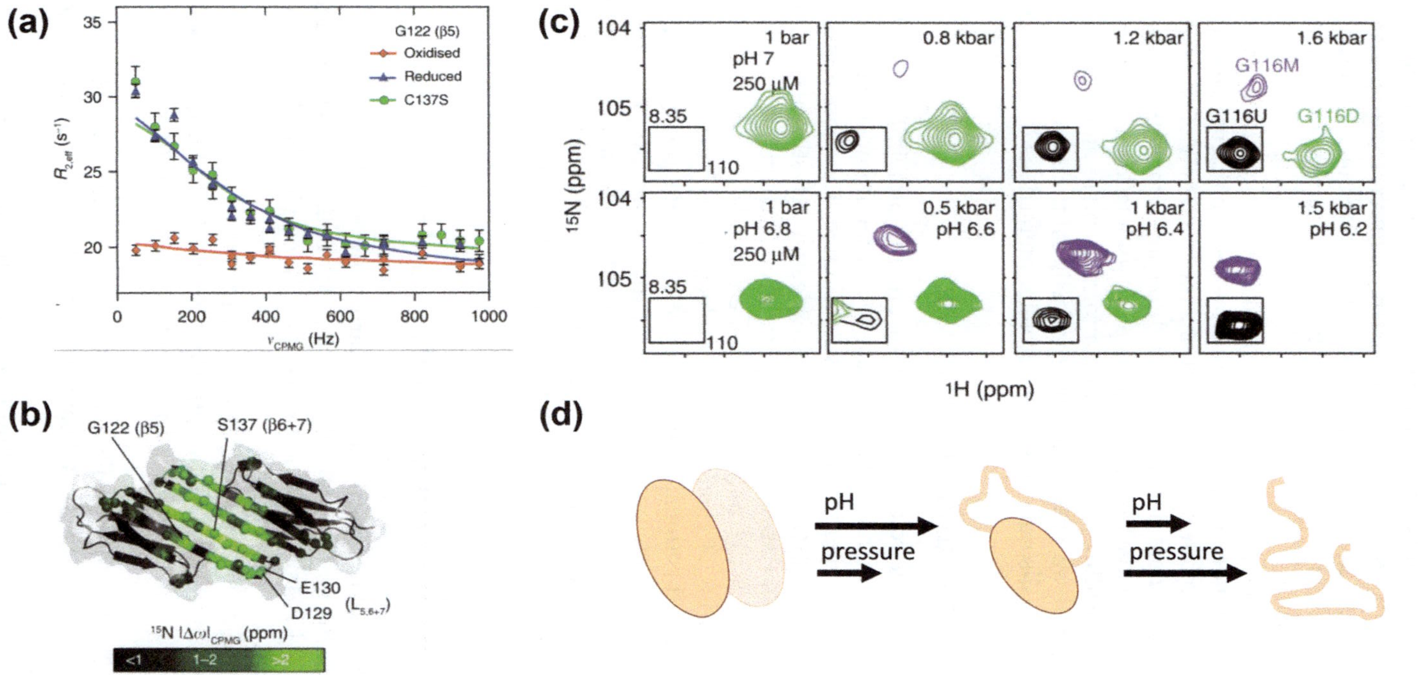

Figure 2.3 Probing local unfolding in Hsp27 during activation using CPMG RD and high-pressure NMR experiments.[96] (a and b) Representative CPMG RD profiles of Hsp27 under different sample conditions (a) and mapping of the conformational dynamics onto the dimer interface (b). (c) Representative spectral changes of Hsp27 under varying pH and pressure conditions. The panels (a–c) are reproduced from ref. 96, https://doi.org/10.1038/s41467-019-08557-8, under the terms of the CC BY 4.0 license, http://creativecommons.org/licenses/by/4.0/. (d) A schematic illustration of dimer dissociation and unfolding of Hsp27, modified based on Figure 6 of ref. 96.

two proteins,[14,99,100] it appears clear that a tight binding can be achieved between two highly disordered proteins *via* charge complementarity without obvious induced folding, which expands our conventional understanding of both protein interactions and the role of structural disorder in chaperone functional mechanisms.

It is worth noting that, in this highly dynamic and heterogenous system, neither solution NMR nor smFRET methods alone were sufficient to provide a comprehensive description of the interaction. Interpretation of the experimental data was also more difficult compared to conventional protein complex systems and thus largely relies on the help of computational simulations. Hybrid methods providing complementary information are more and more commonly used, and we anticipate that they will continue to provide invaluable insights into disorder-associated protein interactions in the future.

2.4 Conclusions and Perspectives

To summarize, while the role of structural disorder in protein function has gained increasing understanding in the past decades, our understanding of how it contributes to chaperone activities is still limited. The above examples suggest that unlike the more extensively investigated well-structured chaperones, such as Hsp70, Hsp90 and the supramolecular chaperonin systems, some chaperones are active only when they are partially or completely disordered. In the process of elucidating the molecular mechanism for chaperone activation and functionality, solution NMR methods play an essential and irreplaceable role. We anticipate with further application of NMR in these extremely dynamic and heterogeneous chaperone–client interacting systems, and with a combined usage of other biophysical and computational methods, deeper insights could be gained in the near future.

References

1. Y. E. Kim, M. S. Hipp, A. Bracher, M. Hayer-Hartl and F. U. Hartl, *Annu. Rev. Biochem.*, 2013, **82**, 323–355.
2. D. Balchin, M. Hayer-Hartl and F. U. Hartl, *Science*, 2016, **353**, aac4354.
3. A. Mogk, B. Bukau and H. H. Kampinga, *Mol. Cell*, 2018, **69**, 214–226.
4. A. Finka, R. U. Mattoo and P. Goloubinoff, *Annu. Rev. Biochem.*, 2016, **85**, 715–742.
5. S. Hiller, *Trends Biochem. Sci.*, 2019, **44**, 517–527.
6. H. J. Dyson and P. E. Wright, *Curr. Opin. Struct. Biol.*, 2021, **70**, 44–52.
7. M. Hayer-Hartl, A. Bracher and F. U. Hartl, *Trends Biochem. Sci.*, 2016, **41**, 62–76.
8. G. Mas, J. Thoma and S. Hiller, *Subcell. Biochem.*, 2019, **92**, 169–186.
9. T. Saio, X. Guan, P. Rossi, A. Economou and C. G. Kalodimos, *Science*, 2014, **344**, 1250494.
10. C. Huang, P. Rossi, T. Saio and C. G. Kalodimos, *Nature*, 2016, **537**, 202–206.

11. R. Fassler, L. Zuily, N. Lahrach, M. Ilbert and D. Reichmann, *Front. Mol. Biosci.*, 2021, **8**, 706039.
12. H. Kim, K. Wu and C. Lee, *Front. Mol. Biosci.*, 2021, **8**, 678697.
13. W. Hong, Y. E. Wu, X. Fu and Z. Chang, *Trends Microbiol.*, 2012, **20**, 328–335.
14. A. Borgia, M. B. Borgia, K. Bugge, V. M. Kissling, P. O. Heidarsson, C. B. Fernandes, A. Sottini, A. Soranno, K. J. Buholzer, D. Nettels, B. B. Kragelund, R. B. Best and B. Schuler, *Nature*, 2018, **555**, 61–66.
15. A. Sekhar and L. E. Kay, *Annu. Rev. Biophys.*, 2019, **48**, 297–319.
16. T. R. Alderson and L. E. Kay, *Cell*, 2021, **184**, 577–595.
17. L. E. Kay, D. A. Torchia and A. Bax, *Biochemistry*, 1989, **28**, 8972–8979.
18. N. A. Farrow, R. Muhandiram, A. U. Singer, S. M. Pascal, C. M. Kay, G. Gish, S. E. Shoelson, T. Pawson, J. D. Forman-Kay and L. E. Kay, *Biochemistry*, 1994, **33**, 5984–6003.
19. J. W. Peng and G. Wagner, *Biochemistry*, 1992, **31**, 8571–8586.
20. G. Lipari and A. Szabo, *J. Am. Chem. Soc.*, 1982, **104**, 4546–4559.
21. G. Lipari and A. Szabo, *J. Am. Chem. Soc.*, 1982, **104**, 4559–4570.
22. J. P. Loria, M. Rance and A. G. Palmer, *J. Am. Chem. Soc.*, 1999, **121**, 2331–2332.
23. F. A. Mulder, N. R. Skrynnikov, B. Hon, F. W. Dahlquist and L. E. Kay, *J. Am. Chem. Soc.*, 2001, **123**, 967–975.
24. A. G. Palmer III, C. D. Kroenke and J. P. Loria, *Methods Enzymol.*, 2001, **339**, 204–238.
25. P. Vallurupalli, G. Bouvignies and L. E. Kay, *J. Am. Chem. Soc.*, 2012, **134**, 8148–8161.
26. M. A. Danielson and J. J. Falke, *Annu. Rev. Biophys. Biomol. Struct.*, 1996, **25**, 163–195.
27. J. L. Kitevski-LeBlanc, F. Evanics and R. S. Prosser, *J. Biomol. NMR*, 2009, **45**, 255–264.
28. D. Rose-Sperling, M. A. Tran, L. M. Lauth, B. Goretzki and U. A. Hellmich, *Biol. Chem.*, 2019, **400**, 1277–1288.
29. J. L. Kitevski-LeBlanc and R. S. Prosser, *Prog. Nucl. Magn. Reson. Spectrosc.*, 2012, **62**, 1–33.
30. P. B. Crowley, C. Kyne and W. B. Monteith, *Chem. Commun.*, 2012, **48**, 10681–10683.
31. C. Li, G. F. Wang, Y. Wang, R. Creager-Allen, E. A. Lutz, H. Scronce, K. M. Slade, R. A. Ruf, R. A. Mehl and G. J. Pielak, *J. Am. Chem. Soc.*, 2010, **132**, 321–327.
32. J. C. Jackson, J. T. Hammill and R. A. Mehl, *J. Am. Chem. Soc.*, 2007, **129**, 1160–1166.
33. J. T. Hammill, S. Miyake-Stoner, J. L. Hazen, J. C. Jackson and R. A. Mehl, *Nat. Protoc.*, 2007, **2**, 2601–2607.
34. T. S. Young, I. Ahmad, J. A. Yin and P. G. Schultz, *J. Mol. Biol.*, 2010, **395**, 361–374.
35. S. E. Cellitti, D. H. Jones, L. Lagpacan, X. Hao, Q. Zhang, H. Hu, S. M. Brittain, A. Brinker, J. Caldwell, B. Bursulaya, G. Spraggon,

A. Brock, Y. Ryu, T. Uno, P. G. Schultz and B. H. Geierstanger, *J. Am. Chem. Soc.*, 2008, **130**, 9268–9281.
36. C. Dalvit, S. Y. Ko and A. Vulpetti, *J. Fluorine Chem.*, 2013, **152**, 129–135.
37. U. Jakob, W. Muse, M. Eser and J. C. Bardwell, *Cell*, 1999, **96**, 341–352.
38. J. Graumann, H. Lilie, X. Tang, K. A. Tucker, J. H. Hoffmann, J. Vijayalakshmi, M. Saper, J. C. Bardwell and U. Jakob, *Structure*, 2001, **9**, 377–387.
39. P. C. Graf, M. Martinez-Yamout, S. VanHaerents, H. Lilie, H. J. Dyson and U. Jakob, *J. Biol. Chem.*, 2004, **279**, 20529–20538.
40. J. Winter, M. Ilbert, P. C. Graf, D. Ozcelik and U. Jakob, *Cell*, 2008, **135**, 691–701.
41. D. Reichmann, Y. Xu, C. M. Cremers, M. Ilbert, R. Mittelman, M. C. Fitzgerald and U. Jakob, *Cell*, 2012, **148**, 947–957.
42. U. Jakob, M. Eser and J. C. Bardwell, *J. Biol. Chem.*, 2000, **275**, 38302–38310.
43. J. Winter, K. Linke, A. Jatzek and U. Jakob, *Mol. Cell*, 2005, **17**, 381–392.
44. M. Ilbert, J. Horst, S. Ahrens, J. Winter, P. C. Graf, H. Lilie and U. Jakob, *Nat. Struct. Mol. Biol.*, 2007, **14**, 556–563.
45. C. M. Cremers, D. Reichmann, J. Hausmann, M. Ilbert and U. Jakob, *J. Biol. Chem.*, 2010, **285**, 11243–11251.
46. J. Vijayalakshmi, M. K. Mukhergee, J. Graumann, U. Jakob and M. A. Saper, *Structure*, 2001, **9**, 367–375.
47. I. Janda, Y. Devedjiev, U. Derewenda, Z. Dauter, J. Bielnicki, D. R. Cooper, P. C. Graf, A. Joachimiak, U. Jakob and Z. S. Derewenda, *Structure*, 2004, **12**, 1901–1907.
48. S. W. Chi, D. G. Jeong, J. R. Woo, H. S. Lee, B. C. Park, B. Y. Kim, R. L. Erikson, S. E. Ryu and S. J. Kim, *FEBS Lett.*, 2011, **585**, 664–670.
49. H. S. Won, L. Y. Low, R. D. Guzman, M. Martinez-Yamout, U. Jakob and H. J. Dyson, *J. Mol. Biol.*, 2004, **341**, 893–899.
50. B. Groitl, S. Horowitz, K. A. T. Makepeace, E. V. Petrotchenko, C. H. Borchers, D. Reichmann, J. C. A. Bardwell and U. Jakob, *Nat. Commun.*, 2016, 7, 10357.
51. Y. S. Lee, J. Lee, K. S. Ryu, Y. Lee, T. G. Jung, J. H. Jang, D. W. Sim, E. H. Kim, M. D. Seo, K. W. Lee and H. S. Won, *J. Mol. Biol.*, 2015, **427**, 3850–3861.
52. K. S. Gajiwala and S. K. Burley, *J. Mol. Biol.*, 2000, **295**, 605–612.
53. F. Yang, K. R. Gustafson, M. R. Boyd and A. Wlodawer, *Nat. Struct. Biol.*, 1998, **5**, 763–764.
54. W. Hong, W. Jiao, J. Hu, J. Zhang, C. Liu, X. Fu, D. Shen, B. Xia and Z. Chang, *J. Biol. Chem.*, 2005, **280**, 27029–27034.
55. Y. E. Wu, W. Hong, C. Liu, L. Zhang and Z. Chang, *Biochem. J.*, 2008, **412**, 389–397.
56. T. L. Tapley, J. L. Körner, M. T. Barge, J. Hupfeld, J. A. Schauerte, A. Gafni, U. Jakob and J. C. Bardwell, *Proc. Natl. Acad. Sci. U. S. A.*, 2009, **106**, 5557–5562.

57. M. A. Garrison and K. A. Crowhurst, *Protein Sci.*, 2014, **23**, 167–178.
58. T. L. Tapley, T. M. Franzmann, S. Chakraborty, U. Jakob and J. C. Bardwell, *Proc. Natl. Acad. Sci. U. S. A.*, 2010, **107**, 1071–1076.
59. B. W. Zhang, L. Brunetti and C. L. Brooks, 3rd, *J. Am. Chem. Soc.*, 2011, **133**, 19393–19398.
60. M. Zhang, S. Lin, X. Song, J. Liu, Y. Fu, X. Ge, X. Fu, Z. Chang and P. R. Chen, *Nat. Chem. Biol.*, 2011, **7**, 671–677.
61. L. Foit, J. S. George, B. W. Zhang, C. L. Brooks, 3rd and J. C. Bardwell, *Proc. Natl. Acad. Sci. U. S. A.*, 2013, **110**, E1254–E1262.
62. L. S. Ahlstrom, S. M. Law, A. Dickson and C. L. Brooks, 3rd, *J. Mol. Biol.*, 2015, **427**, 1670–1680.
63. S. Zhang, D. He, Y. Yang, S. Lin, M. Zhang, S. Dai and P. R. Chen, *Proc. Natl. Acad. Sci. U. S. A.*, 2016, **113**, 10872–10877.
64. A. Dickson, L. S. Ahlstrom and C. L. Brooks III, *J. Comput. Chem.*, 2016, **37**, 587–594.
65. L. Salmon, F. Stull, S. Sayle, C. Cato, Ş. Akgül, L. Foit, L. S. Ahlstrom, E. Z. Eisenmesser, H. M. Al-Hashimi, J. C. A. Bardwell and S. Horowitz, *J. Mol. Biol.*, 2018, **430**, 33–40.
66. X. C. Yu, C. Yang, J. Ding, X. Niu, Y. Hu and C. Jin, *Biochemistry*, 2017, **56**, 5748–5757.
67. Z. Zhai, Q. Wu, W. Zheng, M. Liu, G. J. Pielak and C. Li, *Chem. Sci.*, 2016, **7**, 2222–2228.
68. X. C. Yu, Y. Hu, J. Ding, H. Li and C. Jin, *J. Biol. Chem.*, 2019, **294**, 3192–3206.
69. L. E. Kay, *Prog. Biophys. Mol. Biol.*, 1995, **63**, 277–299.
70. M. Sattler, J. Schleucher and C. Griesinger, *Prog. Nucl. Magn. Reson. Spectrosc.*, 1999, **34**, 93–158.
71. S. Pacheco, M. A. Widjaja, J. S. Gomez, K. A. Crowhurst and R. Abrol, *Biophys. Chem.*, 2020, **264**, 106406.
72. M. A. Widjaja, J. S. Gomez, J. M. Benson and K. A. Crowhurst, *Biochim. Biophys. Acta, Proteins Proteomics*, 2021, **1869**, 140576.
73. M. Haslbeck, S. Weinkauf and J. Buchner, *J. Biol. Chem.*, 2019, **294**, 2121–2132.
74. M. K. Janowska, H. E. R. Baughman, C. N. Woods and R. E. Klevit, *Cold Spring Harbor Perspect. Biol.*, 2019, **11**, a034025.
75. S. P. Delbecq and R. E. Klevit, *FEBS Lett.*, 2013, **587**, 1073–1080.
76. M. Haslbeck, J. Peschek, J. Buchner and S. Weinkauf, *Biochim. Biophys. Acta*, 2016, **1860**, 149–166.
77. M. A. Sprague-Piercy, M. A. Rocha, A. O. Kwok and R. W. Martin, *Annu. Rev. Phys. Chem.*, 2021, **72**, 143–163.
78. P. Chen, W. Ji, F. Y. Liu, H. Z. Tang, S. Fu, X. Zhang, M. Liu, L. Gong, M. Deng, W. F. Hu, X. H. Hu, X. W. Chen, Z. L. Li, X. Li, J. Liu and D. W. Li, *Curr. Mol. Med.*, 2012, **12**, 1164–1173.
79. S. Jehle, B. van Rossum, J. R. Stout, S. M. Noguchi, K. Falber, K. Rehbein, H. Oschkinat, R. E. Klevit and P. Rajagopal, *J. Mol. Biol.*, 2009, **385**, 1481–1497.

80. A. Mainz, B. Bardiaux, F. Kuppler, G. Multhaup, I. C. Felli, R. Pierattelli and B. Reif, *J. Biol. Chem.*, 2012, **287**, 1128–1138.
81. N. Braun, M. Zacharias, J. Peschek, A. Kastenmüller, J. Zou, M. Hanzlik, M. Haslbeck, J. Rappsilber, J. Buchner and S. Weinkauf, *Proc. Natl. Acad. Sci. U. S. A.*, 2011, **108**, 20491–20496.
82. J. A. Aquilina, J. L. Benesch, O. A. Bateman, C. Slingsby and C. V. Robinson, *Proc. Natl. Acad. Sci. U. S. A.*, 2003, **100**, 10611–10616.
83. S. Jehle, P. Rajagopal, B. Bardiaux, S. Markovic, R. Kühne, J. R. Stout, V. A. Higman, R. E. Klevit, B. J. van Rossum and H. Oschkinat, *Nat. Struct. Mol. Biol.*, 2010, **17**, 1037–1042.
84. A. J. Baldwin, G. R. Hilton, H. Lioe, C. Bagnéris, J. L. Benesch and L. E. Kay, *J. Mol. Biol.*, 2011, **413**, 310–320.
85. A. J. Baldwin, H. Lioe, C. V. Robinson, L. E. Kay and J. L. Benesch, *J. Mol. Biol.*, 2011, **413**, 297–309.
86. S. Jehle, B. S. Vollmar, B. Bardiaux, K. K. Dove, P. Rajagopal, T. Gonen, H. Oschkinat and R. E. Klevit, *Proc. Natl. Acad. Sci. U. S. A.*, 2011, **108**, 6409–6414.
87. A. J. Baldwin, P. Walsh, D. F. Hansen, G. R. Hilton, J. L. Benesch, S. Sharpe and L. E. Kay, *J. Am. Chem. Soc.*, 2012, **134**, 15343–15350.
88. A. Mainz, J. Peschek, M. Stavropoulou, K. C. Back, B. Bardiaux, S. Asami, E. Prade, C. Peters, S. Weinkauf, J. Buchner and B. Reif, *Nat. Struct. Mol. Biol.*, 2015, **22**, 898–905.
89. M. K. Singh, B. Sharma and P. K. Tiwari, *J. Therm. Biol.*, 2017, **69**, 149–154.
90. R. Shan, N. Liu, Y. Yan and B. Liu, *Pharmacol. Res.*, 2021, **166**, 105169.
91. E. V. Baranova, S. D. Weeks, S. Beelen, O. V. Bukach, N. B. Gusev and S. V. Strelkov, *J. Mol. Biol.*, 2011, **411**, 110–122.
92. G. K. Hochberg, H. Ecroyd, C. Liu, D. Cox, D. Cascio, M. R. Sawaya, M. P. Collier, J. Stroud, J. A. Carver, A. J. Baldwin, C. V. Robinson, D. S. Eisenberg, J. L. Benesch and A. Laganowsky, *Proc. Natl. Acad. Sci. U. S. A.*, 2014, **111**, E1562–E1570.
93. P. Rajagopal, Y. Liu, L. Shi, A. F. Clouser and R. E. Klevit, *J. Biomol. NMR*, 2015, **63**, 223–228.
94. T. R. Alderson, J. L. P. Benesch and A. J. Baldwin, *Cell Stress Chaperones*, 2017, **22**, 639–651.
95. A. F. Clouser, H. E. Baughman, B. Basanta, M. Guttman, A. Nath and R. E. Klevit, *eLife*, 2019, **8**, e50259.
96. T. R. Alderson, J. Roche, H. Y. Gastall, D. M. Dias, I. Pritišanac, J. Ying, A. Bax, J. L. P. Benesch and A. J. Baldwin, *Nat. Commun.*, 2019, **10**, 1068.
97. T. R. Alderson, J. Ying, A. Bax, J. L. P. Benesch and A. J. Baldwin, *J. Mol. Biol.*, 2020, **432**, 3033–3049.
98. R. B. Berlow and P. E. Wright, *Nature*, 2018, **555**, 37–38.
99. H. Feng, B. R. Zhou and Y. Bai, *Biochemistry*, 2018, **57**, 6645–6648.
100. A. Sottini, A. Borgia, M. B. Borgia, K. Bugge, D. Nettels, A. Chowdhury, P. O. Heidarsson, F. Zosel, R. B. Best, B. B. Kragelund and B. Schuler, *Nat. Commun.*, 2020, **11**, 5736.

CHAPTER 3

Solution NMR Approaches for Studying Molecular Chaperones

E. E. SELIG AND D. S. LIBICH*

Greehey Children's Cancer Research Institute and Department of Biochemistry and Structural Biology, University of Texas Health Science Center at San Antonio, San Antonio, TX, 78229, USA
*Email: libich@uthscsa.edu

3.1 Introduction

Molecular chaperones, also known as heat-shock proteins (HSPs), constitute one arm of the protein homeostasis (proteostasis) network. Chaperones are defined as any factor that interacts with and promotes the folding, oligomeric assembly, transport or conformational switching of a substrate without becoming incorporated into its final native structure.[1] As evidence for their importance, chaperones are highly conserved across all forms of life, and many are constitutively expressed and upregulated in response to cellular stress. Dysregulation of molecular chaperones is strongly associated with a diverse range of debilitating and life-threatening diseases, collectively termed protein conformational diseases which include neurodegenerative diseases (*i.e.*, dementia), systemic or organ-specific amyloidosis, cataracts, cystic fibrosis and type II diabetes. Although some of these diseases have well-characterized genetic underpinnings, others are age-related or can manifest sporadically. The social and economic burden of these diseases is considerable. The World Health Organization estimates that between 2015 and 2050, the proportion of the world's population over the age of 60 will increase from 12% to 22%. Currently, 5–8% of all people over the age of 60 are estimated to be living with some form of dementia. The economic impact of dementia was estimated to be

New Developments in NMR No. 29
Biophysics of Molecular Chaperones: Function, Mechanisms and Client Protein Interactions
Edited by Sebastian Hiller, Maili Liu and Lichun He

Published by the Royal Society of Chemistry, www.rsc.org

USD 1.3 trillion in 2019. Currently, dementia is the seventh leading cause of death in the United States. The Center for Disease Control and Prevention estimates that more than 37 million Americans have diabetes, 90–95% of which is type II. Type II diabetes is the eighth leading cause of death in the US and the associated medical costs and lost wages amount to approximately USD 327 billion annually. 20–30 million Americans are estimated to be living with cataracts or to have had a lens removed operatively due to cataracts. Cystic fibrosis is estimated to affect approximately 35 000 Americans. The social burden of systemic/organ-specific amyloidosis, myopathies and neuropathies is harder to quantify because these diseases manifest in numerous forms. Nevertheless, a thorough understanding of chaperone mechanisms is required to develop better strategies for treating or preventing protein conformational diseases.

Over the course of evolution, molecular chaperones were optimized for forming interactions with a multitude of substrate proteins and often interact with different conformations of a given substrate. To enable such broad substrate specificity, chaperones must be able to form protein–protein interactions with affinities ranging from high (low nM) to extremely transient (high μM); however, transient interactions between chaperones and substrates are more common. High-affinity interactions typically arise when both binding partners adopt a single, well-defined conformation. Conversely, transient chaperone–substrate interactions are usually due to conformational flexibility in the bound state. High-affinity interactions can be and have been elucidated with atomic resolution structural approaches including nuclear magnetic resonance (NMR) spectroscopy, X-ray crystallography and cryogenic electron microscopy (cryo-EM). However, transient chaperone–substrate complexes are particularly challenging to detect and characterize using X-ray crystallography and cryo-EM or biophysical mass-transport techniques including analytical ultracentrifugation, immunoprecipitation or size-exclusion chromatography because they are often lowly populated, conformationally heterogenous and exhibit rapid rates of complex dissociation. Though NMR is a suitable technique for investigating high-affinity chaperone–substrate interactions, it is sometimes the only viable method for characterizing transient chaperone–substrate interactions at atomic resolution. Therefore, NMR is emerging as an extremely important tool for studying the dynamic interactions of chaperones with their substrates.[2]

This chapter will introduce the key principles that underpin solution-based NMR spectroscopy, outline the general approach used by NMR spectroscopists to elucidate the structures of chaperone proteins and discuss recent developments in solution NMR spectroscopy that have been employed for investigating the interactions of chaperone proteins with their substrates.

3.2 Key Principles of Solution-based NMR Spectroscopy

3.2.1 NMR Observables

Atomic nuclei that are magnetically active have two (or more) quantum spin states and in the presence of an external magnetic field, the spin states have different energies.[3,4] NMR active nuclei in biological systems are ^{1}H, ^{15}N, ^{13}C, ^{31}P and ^{19}F; however, ^{15}N and ^{13}C isotopes have a low natural abundance and must be enriched in a sample of interest using isotopic labelling strategies.[5–9] Under equilibrium conditions in the presence of an external magnetic field, the lower energy nuclear spin state is marginally more populated than the higher energy state.[3] Because energy is proportional to frequency ($E = h\nu$), the difference in energy between the spin states corresponds to a particular frequency of microwave radiation.[10] Radio-frequency pulses can therefore be used to excite nuclei from the lower to the higher energy state. The excited nuclei subsequently relax back to the ground state, emitting microwave radiation in the process. This phenomenon is referred to as nuclear magnetic resonance.[11–13]

There are three major observables in NMR experiments: chemical shift, signal intensity and linewidth. Chemical shift, the resonance frequency of a nucleus of interest relative to a standard, is exquisitely sensitive to the surrounding chemical environment and therefore provides information about the local magnetic field experienced by the nucleus under investigation. The signal intensity is proportional to the population of nuclei in a specific chemical environment. The linewidth is governed by the rate of decay (relaxation) of the nuclear spin from the excited to the ground state and provides information about dynamic processes such as molecular motion and chemical and conformational exchange.[3]

3.2.2 Mechanisms of Magnetization Decay

As nuclear magnetic resonance is a bulk phase technique, individual nuclear spins are undetectable; however, the sum of all the individual nuclear spin states in the sample gives rise to a net magnetization vector. This vector is aligned in the direction of the applied magnetic field (the *z*-axis in a Cartesian co-ordinate system) and non-zero at equilibrium because the lower energy spin state is marginally more populated.[3] Radio-frequency pulses are applied to rotate the net magnetization vector away from its equilibrium position on the *z*-axis. Typically, a 90°_{y} pulse is used to align the magnetization vector along the *x*-axis, orthogonal to the *z*-axis (Figure 3.1). It is the net magnetization vector in the *x*–*y* plane that gives rise to signals in NMR spectroscopy. Once aligned along the *x*-axis, the magnetization of each nucleus precesses about the *z*-axis with a frequency dependent on its local chemical (magnetic) environment and the external magnetic field (B_0). Over time, the net magnetization vector decays in the *x*–*y* plane due to two distinct processes. The first is called spin–lattice decay, which is also referred to as T_1 relaxation or by its rate constant R_1 ($R_1 = 1/T_1$), and it occurs as nuclear spin states gain or lose magnetization along the *z*-axis (Figure 3.1). The second is termed spin–spin relaxation or T_2 relaxation and is interchangeably referred to by the rate constant R_2 ($R_2 = 1/T_2$). Spin–spin relaxation involves the loss of nuclear spin coherence (also called dephasing) in the *x*–*y* plane and therefore is related to the gain or loss of magnetization in this plane (Figure 3.1).

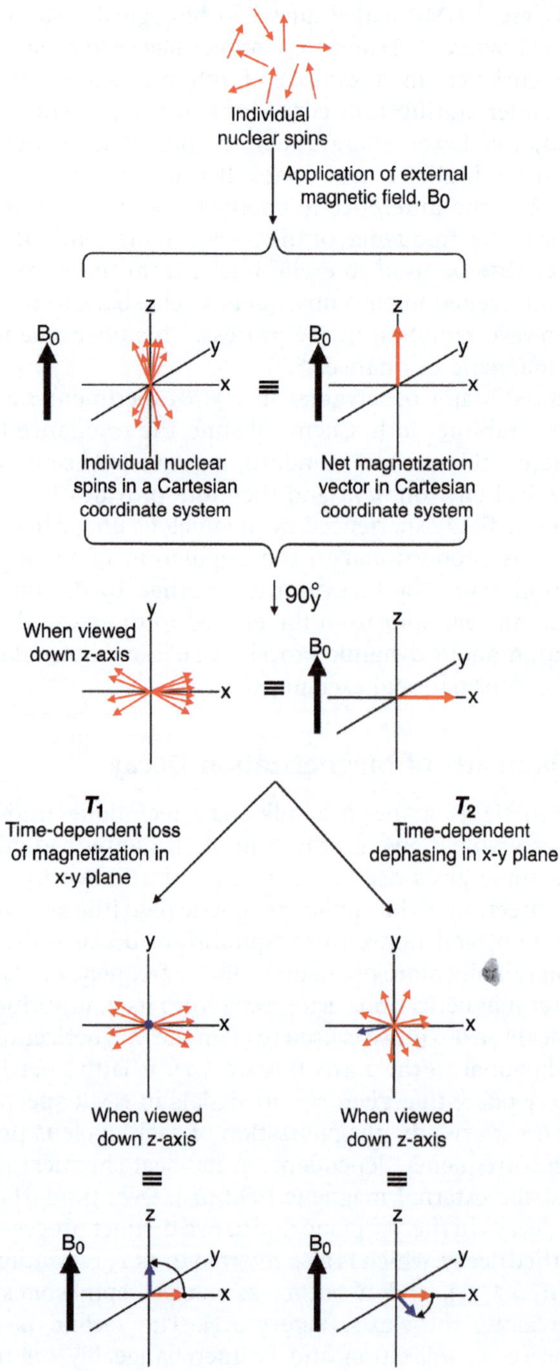
Individual
nuclear spins
Application of external
magnetic field, B_0
B_0
z
y
x
≡
Individual nuclear
spins in a Cartesian
coordinate system
Net magnetization
vector in Cartesian
coordinate system
90°_y
When viewed
down z-axis
T_1
Time-dependent loss
of magnetization in
x-y plane
T_2
Time-dependent
dephasing in x-y plane
When viewed
down z-axis
When viewed
down z-axis

Molecular motion, reorientation of molecules in solution (tumbling) and chemical or conformational exchange between two or more states create transient local magnetic fields that can stimulate both types of nuclear spin relaxation. However, T_2 relaxation processes are generally more efficient for proteins than T_1 processes. Both R_1 and R_2 relaxation rates govern the lifetimes of magnetic states and, consequently, linewidths. This can be rationalized by considering the Heisenberg uncertainty principle, which states that the precision with which the energy of a given spin state can be defined is inversely proportional to the lifetime of the state.[14] In other words, the shorter the lifetime of the magnetic state, the less precise the energy of the state can be defined and the broader the lines will appear in the NMR spectra.

3.2.3 Interactions Between Nuclear Spins and Higher Dimensional NMR Spectroscopy

Traditional one-dimensional (1D) NMR spectroscopy is a powerful analytical technique for the study of small molecules; however, for larger and more complex molecules such as proteins, a significant overlap of resonances makes the interpretation of the spectra difficult. These limitations can be overcome with 2D, 3D, or *n*D experiments.[15–19] Multi-dimensional experiments exploit the interactions (correlations) between two or more magnetically active nuclear spins due to their connectivity through chemical bonds (*J*-coupling) or close proximity in the three-dimensional space (dipolar coupling).[15,20–24] Additional dimensions provide enhanced resolution of the overlapping signals by spreading the correlations across greater frequency space. 2D experiments can be either homonuclear (in which the coupled spins are of the same type, *e.g.* ^{1}H and ^{1}H) or heteronuclear (in which the coupled spins arise from different nuclei, *e.g.* ^{1}H and ^{15}N). In a typical homonuclear 2D correlation experiment, the magnetization of the first nucleus in the spin-coupled system is carried out by an initial radiofrequency pulse, allowed to evolve on the first nucleus for a given amount of time, transferred from the first nucleus to the second nucleus *via* either *J*-coupling or dipolar coupling using a second radiofrequency pulse during a variable mixing time and finally detected.[25] Fourier transforms (FT) of both dimensions give rise to 2D spectra with two frequency axes where the peaks

Figure 3.1 Relaxation mechanisms in NMR spectroscopy. In the presence of an external magnetic field (B_0), individual nuclear spins will either align with or against the external magnetic field; however, the population of spins aligned with the magnetic field will be slightly greater. The sum of all the individual nuclear spin states in a sample then gives rise to a net magnetization vector that is aligned in the direction of the applied field. A 90°_y pulse rotates the net magnetization vector from the $+z$ axis to the x axis. The magnetization vector aligned with the x axis can then decay either due to spins losing magnetization in the x–y plane (T_1 relaxation) or due to a loss of phase coherence in the x–y plane (T_2 relaxation). Created with BioRender.com.

along the diagonal have identical chemical shifts in both dimensions and arise from nuclei whose magnetization was not transferred from the first nucleus to the second during the mixing time, while the cross peaks do not have the same chemical shift in both dimensions and result from nuclei that exchanged magnetization during the mixing time. The cross peaks indicate connectivity of the nuclei located on the diagonal *via* either *J*-coupling or dipolar coupling, depending on the nature of the experiment.

One of the most widely used heteronuclear 2D correlation experiments for proteins is the ^{1}H, ^{15}N heteronuclear single quantum coherence (HSQC) sequence, which exploits spin-coupling of ^{1}H and ^{15}N of the amide bonds of proteins. It is also known as a "fingerprint" experiment because each cross peak arises from the NH bond vector of the peptide plane and thus represents each amino acid except proline. For these experiments, magnetization is transferred from hydrogen to nitrogen with an insensitive nuclei enhanced by polarization transfer (INEPT) sequence,[26] allowed to evolve on nitrogen for a period of time t_1 and then transferred back to hydrogen with a reverse INEPT[27] for detection during the t_2 time period.[25] A 1D FT yields ^{1}H resonance frequencies. By incrementally increasing the t_1 evolution time, the resonance frequency of the nitrogen in the N–H spin system can be derived using a 2D FT. These spectra are depicted as 2D plots where the x axis usually represents the chemical shift of the hydrogen, and the y axis represents the chemical shift of the nitrogen. Higher dimensional experiments for biomolecules add further experimental parameters (and resolution) by combining two or more 2D experiments that select different nuclei such as ^{13}C.[15–19]

3.3 NMR as a Tool for Structural Elucidation

The key advantage of using NMR rather than X-ray crystallography or cryo-EM is that protein structures can be elucidated at a high resolution in their natural state in solution. Furthermore, it is increasingly apparent that high-resolution structural models obtained *via* X-ray crystallography or cryo-EM are sometimes insufficient for understanding the structure–function relationship of a protein of interest. Although these methodologies are exceedingly useful for characterizing highly populated, well-folded conformations of proteins, transiently populated conformers and proteins containing highly dynamic regions are extremely difficult to analyze using these structural approaches. Recent advances in solution-based NMR experiments uncover important structural and thermodynamic information of protein ensembles and protein–protein interactions.[28] This section will describe the typical steps used by NMR spectroscopists to assign spectra and subsequently build three-dimensional structural models.

3.3.1 NMR Experiments Used to Assign the Spectra and Elucidate the Structure

The first step in the structural elucidation using NMR is a spectral assignment. Prior to the early 1990s and the development of stable ^{15}N and ^{13}C isotope labelling of recombinant proteins, spectral assignment and structural elucidation were mainly achieved using a set of 2D homonuclear experiments such as correlation spectroscopy (COSY),[15] total correlation spectroscopy (TOCSY)[22] and nuclear Overhauser effect spectroscopy (NOESY).[29] COSY experiments elucidate the through-bond connectivity of nuclei of the same type up to a maximum of approximately three chemical bonds. TOCSY experiments are similar to COSY experiments but correlate all the *J*-coupled spins of an entire spin system (*i.e.*, an amino acid residue) within a protein. TOCSY experiments therefore give rise to characteristic patterns of signals for each amino acid and are useful for identifying residue types. NOESY experiments are the most widely used tool for protein structural elucidation and are essential for *de novo* structure determination. NOESY makes use of the nuclear Overhauser effect (NOE), the transfer of magnetization between nuclei that are correlated in 3-dimensional space but not *J*-coupled. Cross-peaks in the NOESY spectra are proportional to r^{-6}, where r is the internuclear distance. Nevertheless, these homonuclear experiments are generally only useful for the structure determination of proteins of less than 100 amino acids. For larger proteins, the significant signal overlap observed using 2D homonuclear experiments can sometimes be resolved using ^{15}N-edited TOCSY and NOESY methods. These methods improve the spectral resolution by spreading out the correlated ^{1}H–^{1}H peaks into a third dimension corresponding to the ^{15}N chemical shift of the backbone amide. ^{15}N-edited experiments provide the same information as standard TOCSY/NOESY experiments but require stable ^{15}N isotope labelling and prior assignment of backbone amides using the triple resonance approach, which will be discussed below.

As stable isotope labelling of recombinant proteins with ^{15}N and ^{13}C became more widespread and relatively inexpensive, a more sophisticated methodology based on a series of 3D experiments was developed for resonance assignment and structural elucidation.[30] The fundamental advancement this methodology employs over the homonuclear approach is the resolution of spin–spin couplings between adjacent residues. Various triple-resonance experiments exist for making assignments of ^{15}N, ^{13}C-labelled proteins and the exact set of experiments employed typically depends on the size of the protein/complex under investigation. A good approach for most proteins except for very large proteins >250 residues is a set of four experiments: HNCO, HN(CA)CO, HNCACB (or HNCA) and HN(CO)CACB (or HN(CO)CA).[31–33] A B_0 field strength of at least 600 MHz is recommended for adequate spectral resolution. 3D experiments used for resonance assignment are constructed from 2D experiments, typically the ^{1}H, ^{15}N HSQC. The third dimension arises from the correlation between the nuclear spins of the ^{13}C at the carbonyl, C_α or C_β position either in the same residue (i) as the N–H group or in the preceding ($i - 1$) residue, depending on the experiment. Consequently, the resulting spectra have an additional z-dimension corresponding to the ^{13}C chemical shift as well as the x and y dimensions corresponding to hydrogen and nitrogen, respectively. For example, the HNCO experiment correlates the nuclear spin of the carbonyl group of the $i-1$ residue with the N–H spin system of the i residue. The HN(CA)CO experiment correlates the nuclear spin of the carbonyl group of both the $i-1$ residue and the i residue with the N–H spin system of the i residue. Using both HNCO and HN(CA)CO allows sequential assignments to be made by "walking" through the sequence if there is sufficient resolution of the carbonyl chemical shifts. Other pairs of experiments are employed in similar ways to assist with sequential assignments.[33–38]

The *de novo* structure determination of *Escherichia coli* DnaJ, a close homolog of the human HSP40 family of ATP-dependent molecular chaperones, is a prominent example of a protein structure determination using NMR spectroscopy. The HSP40 family is the largest chaperone family in humans with approximately 50 members. They function in tandem with HSP70 and HSP110 family members in both substrate folding and substrate disaggregation and are therefore often referred to as co-chaperones.[39–43] All HSP40 family members have a conserved, approximately 75-residue J-domain (JD) and a C-terminal substrate binding domain.[44] Nearly complete backbone resonance assignment of the JD and a Gly/Phe-rich region (residues 2–108) of DnaJ was obtained from a combination of the ^{15}N-edited NOESY experiment, ^{15}N-edited TOCSY experiment and several of the triple resonance experiments mentioned above.[45,46] Subsequently, the overall fold of the JD was calculated based on NOE-derived distance constraints and *J*-coupling constants.[45,46] At the time, this fold was a novel topology not yet identified in any existing 3D protein structure. The flexibility of the C-terminal Gly/Phe-rich region likely precluded structure determination of this construct by X-ray crystallography, demonstrating the use of NMR for structural and functional studies of proteins with highly flexible regions.

3.3.2 Hydrogen–Deuterium Exchange

Hydrogen–deuterium exchange (H/DX) is a chemical reaction in which an exchangeable hydrogen atom in the molecule of interest is replaced with a deuterium atom from the solvent or *vice versa*.[47,48] While the ^{1}H atom is magnetically active and readily observed in solution NMR, ^{2}H has different magnetic properties and is undetectable using standard proton-detection NMR experiments. Since H/DX occurs readily for amide protons at a neutral pH without the need for a catalyst, it is a commonly used NMR experiment that reports on the exchange rates of protein amide groups.[49] Therefore, H/DX is interpreted as a protective factor and is useful for assisting in the structural elucidation and for identifying sites of interactions with binding partners.[50]

Goldberg and co-workers used H/DX NMR spectroscopy to investigate the structure of a GroEL-bound substrate, dihydrofolate reductase (DHFR).[51] GroEL is an *E. coli* chaperonin, a subfamily of heat-shock proteins that encapsulate their substrates in a folding chamber. The structure of GroEL comprises two rings of seven identical 60 kDa subunits per ring.[52]

Heptamer rings formed by the much smaller (10 kDa) GroES form a lid that encloses the substrate within the chamber in an ATP-dependent manner for ~15 seconds to allow for folding.[53,54] Subsequently, ATP hydrolysis promotes the release of the substrate from the GroEL chamber. To elucidate the structure of DHFR bound to GroEL, ^{15}N-labelled DHFR was unfolded in guanidine hydrochloride and then diluted into a solution containing GroEL, allowing unfolded DHFR to bind to GroEL.[51] This solution was then diluted with D_2O to initiate H/DX. After various timepoints, a substrate analogue, ATP and GroES were added to release the substrate from GroEL. This release stimulated substrate folding, preventing any further H/DX of a subset of amide protons known to be buried in the hydrophobic core of folded DHFR.[55] The substrate was then purified, and the amide proton intensities were measured using ^{1}H–^{15}N HSQCs. In this way, the extent of H/DX for amide protons located in the hydrophobic core of DHFR prior to release from GroEL could be determined. Amide proton protection was quantified as the ratio between the exchange observed for a random coil and the exchange observed in the presence of GroEL. Analysis of the protection ratios revealed that the most protected amides are located in a central β-sheet found in natively folded DHFR, which indicated that the GroEL-bound structure is native-like.[51]

3.3.3 Complimentary NMR Experiments for Structural Elucidation of Large Proteins

In some cases, structural restraints derived from COSY, TOCSY and NOESY experiments are insufficient for complete structural elucidation; this is especially true for proteins and biomolecular complexes larger than 100 residues due to their rapid magnetization decay and consequent line broadening (Section 2.2 and discussed further below). In these cases, it is possible to employ one or more complementary techniques for structure refinement. This section will briefly describe additional NMR experiments that can be used to complement structural restraints derived from NOE experiments or from other biophysical data such as X-ray crystallography or X-ray scattering.

3.3.3.1 Residual Dipolar Couplings

Residual dipolar couplings (RDCs) are contributions to dipolar (through-space) couplings that arise when pairs of magnetically active nuclei undergoing dipole–dipole interactions are partially orientated in an external magnetic field.[56] Because resonance frequencies are influenced by the local magnetic field (*i.e.*, the nuclear spin states of neighboring nuclei), dipolar interactions cause reciprocal peak splitting. Signal splitting is one of the most important sources of information used for structural elucidation but is usually lost in isotropic solutions because Brownian motion results in rotational averaging of molecules such that dipolar interactions average to zero.

However, partial alignment of molecules relative to the external field causes incomplete averaging of dipolar couplings such that the rotationally-averaged dipolar interaction between pairs of magnetically-active nuclei is non-zero.[57] This alignment is usually achieved by suspending the molecule(s) of interest in liquid crystalline media such as polyethylene glycol. RDCs can provide rich information regarding the orientation of bonds relative to the external magnetic field and are thus highly complementary to distance constraints derived from NOEs.[58,59] The magnitude of RDCs is dependent on the internuclear distance and the bond orientation relative to the external magnetic field. The magnitude of the RDCs can then be used to estimate the bond angles relative to the external magnetic field (or between protein domains) and provide additional restraints to aid the structure determination.

3.3.3.2 Transverse Relaxation Optimized Spectroscopy Methods

Transverse relaxation optimized spectroscopy (TROSY) is an NMR experiment that was developed to increase the size limit of proteins that can be studied by solution NMR.[60] The linewidths of molecules in solution NMR increase in a manner dependent on the rate of molecular reorientation. This in turn is partly governed by the size of the molecule such that larger molecules tend to have shorter transverse nuclear spin (T_2) relaxation times, resulting in signal broadening and difficulty in interpreting the resulting spectra. The TROSY experiment selects and preserves the slowly relaxing components of the N–H spin-system multiplet resulting in sharper signals and thus better spectral resolution for large proteins and biomolecular complexes.[60] In contrast to typical HSQC spectra, decoupling is not applied for the TROSY HSQC, leading to the splitting of the ^{1}H–^{15}N amide signal into four peaks, one for each of the two possible spin states for each of the ^{1}H and ^{15}N nuclei. The transverse relaxation rate of these four signals is influenced by two main mechanisms: dipolar coupling (between ^{1}H and ^{15}N in the amide bond) and chemical shift anisotropy. The two mechanisms of relaxation are additive, but their signs are dependent on the nuclear spin state of each nucleus contributing to each signal. Consequently, one of the four amide component peaks has slower relaxation for both ^{1}H and ^{15}N, resulting in a sharp peak. This peak is selected for in the TROSY experiment, while the more rapidly decaying signals are removed.

RDCs and TROSY experiments were used in a combined approach to elucidate the structural ensemble of the nucleotide-binding domain (NBD) of Hsc70, the constitutively expressed, cytosolic form of HSP70.[61] HSP70 is a monomeric, ATP-dependent chaperone with a nucleotide-binding domain, a substrate binding domain and a C-terminal domain (CTD). Substrate binding is regulated in a nucleotide-dependent manner; in the ATP-bound state, the substrate affinity is low, while ATP hydrolysis leads to an increase in the substrate affinity due to a conformational change that positions the CTD over the substrate binding pocket.[62] It is thought that rapid cycles of substrate binding and release stimulate substrate folding.[63] Although biochemical and biophysical studies had been conducted previously, existing crystal structures could not explain the allostery occurring between Hsc70 domains during the cycles of substrate binding, nucleotide binding and nucleotide hydrolysis.[64] ^{1}H–^{15}N TROSY versions of the triple resonance experiments as discussed (Section 3.3.1) were used to assign the spectra of the NBD of Hsc70.[38,65] 3D NOESY–TROSY confirmed the overall fold and connectivity derived from the existing crystal structures. However, RDCs revealed substantial differences in the relative orientations of three of the domains in solution compared to their orientations in the crystal structures. The crystal structure whose back-calculated RDCs had the smallest rmsd from the experimental RDCs was then used in combination with the experimental RDCs to define a new solution model for the NBD of Hsc70. The deviations in the relative orientations of the various domains in the NMR structure compared to the crystal structure were proposed as a potential explanation for the domain rearrangements required for allosteric coupling.[61]

Selective isotope labelling (^{1}H and ^{13}C) of methyl groups of isoleucine, leucine and valine (ILV) while all other aliphatic protons are deuterated is often used in methyl TROSY experiments because it contributes to decreased linewidths and enhances signal discrimination resulting in a decrease in the complexity of the resulting spectra.[9,66] Rosenzweig and coworkers used heteronuclear multiple quantum coherence (HMQC) TROSY experiments in combination with selective ILV isotopic labelling to map the interactions between the *E. coli* HSP70 homolog DnaK and the *E. coli* chaperone ClpB.[67,68] Monomeric ILV-labelled ClpB was titrated with unlabelled DnaK (forming an approximately 100 kDa complex), and this identified part of the coiled-coil domain (CCD) of ClpB as the interaction surface. The reverse experiment was carried out with ILV-labelled DnaK and unlabelled hexameric ClpB (forming an ~650 kDa complex), demonstrating that ClpB binds to the NBD of DnaK. ^{1}H–^{15}N TROSY HSQCs confirmed that the NBD of DnaK binds to the CCD of ClpB. Paramagnetic relaxation enhancement (PRE) experiments (Section 3.2.3) using the NBD of DnaK and one of the two constructs of ClpB that both contained the CCD defined long-range distance restraints for the DnaK–ClpB complex. Structural restraints derived from chemical shift perturbations (CSPs) and PRE data were used to build structural models with HADDOCK,[69] delineating, at the residue level, the details of the interaction between the NBD of DnaK and the CCD of ClpB.

The application of TROSY to very large molecules (>100 kDa) is theoretically limited by the very rapid transverse relaxation that occurs when the INEPT sequence is used for magnetization transfer from the sensitive nucleus (^{1}H) to a less sensitive nucleus (*e.g.* ^{15}N).[26] Cross relaxation enhanced by polarization transfer (CRINEPT) is an alternative to the INEPT mechanism of magnetization transfer that can be used for very large molecules, further increasing the size limit for solution-state NMR spectroscopy up to approximately 1 MDa.[70] Horst and coworkers used a ^{1}H–^{15}N CRINEPT-HMQC experiment to directly observe GroEL-bound DHFR, revealing that the bound substrate populated an ensemble of conformers and that these conformers are at least partially disordered and unfolded while bound to GroEL.[71]

3.4 Exchange-based NMR Experiments for Elucidating the Structures and Interactions and Detecting NMR-invisible States

Time-dependent changes in the chemical environment of a nucleus give rise to changes in the shape and intensity of its peaks in an NMR spectrum. The rate of interconversion between these different chemical environments influences resonance frequencies, intensities, linewidths and the number of peaks that appear for a given spin system under investigation. In the slow exchange regime ($\geq s^{-1}$), two distinct peaks appear, representing two different chemical environments. In the fast exchange regime ($\leq \mu s^{-1}$), a single sharp peak is observed representing an average peak position between the two states. In the intermediate regime (μs^{-1} to s^{-1}), peak broadening is observed. This peak broadening is a rich source of information regarding the thermodynamics of the exchange process.[72] In recent decades, sophisticated NMR experiments have been developed for interrogating exchange processes occurring on timescales between μs^{-1} and s^{-1} even when the population of the major species is greater than 95% (Figure 3.2). Accessing chemical exchange processes occurring on these timescales is exceedingly useful for characterizing transient protein–protein interactions such as those that

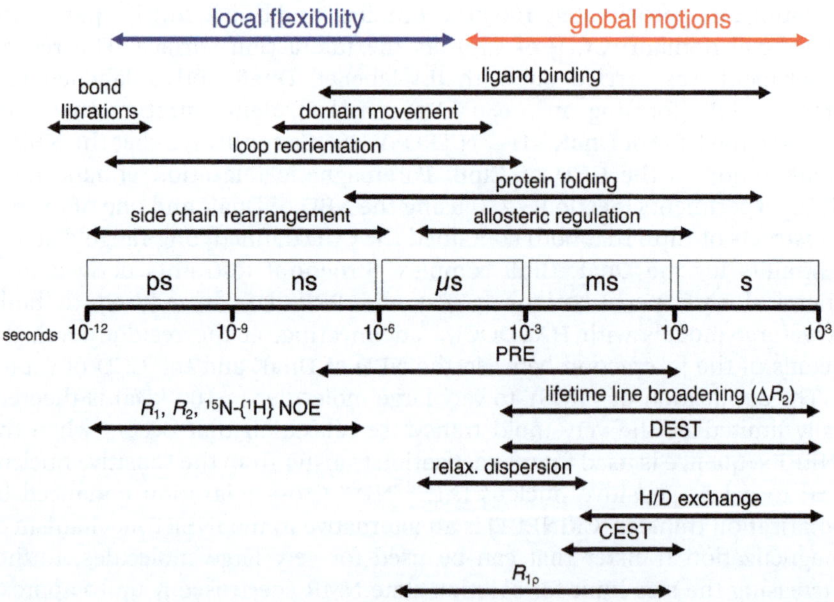

Figure 3.2 Exchange timescales in NMR spectroscopy. Important motions in proteins and biomolecules occur on the picosecond to second timescales. A variety of NMR experiments have been developed to access and characterize these motions.

typically occur between chaperones and their substrates.[73] This section will introduce several exchange-based NMR spectroscopy methods for detecting chemical exchange and provide several examples of how the data can be modelled to derive kinetic parameters governing the exchange. Though this section focuses on ^{15}N-based exchange experiments, the same methodologies have been applied to other nuclei such as ^{13}C.[74–78]

3.4.1 Nuclear Spin Relaxation and Heteronuclear NOEs

There are two mechanisms contributing to the decay of the net magnetization vector, spin–lattice (T_1) and spin–spin (T_2) mechanisms (Section 2.2).[79,80] Measurements of both types of relaxation are useful for interrogating protein dynamics on the picosecond to nanosecond timescale, while T_2 can also report on dynamics on the microsecond to millisecond timescale. Unless explicitly stated, the following discussion will focus on the relaxation of ^{15}N nuclei. T_1 is typically measured by enabling the nitrogen magnetization to return to the equilibrium state using inversion recovery approaches. In practice, it is more convenient to fit a signal that decays to zero; therefore, receiver phase cycling and a selective inversion pulse applied to the amide spins every successive scan result in a signal that decays to zero as T_1 approaches infinity. A typical T_2 experiment involves the rotation of the net magnetization vector from the $+z$ axis to the x–y plane. Individual nuclear spins comprising the net magnetization vector precess about the z-axis with frequencies depending on their local chemical environment. Because each nucleus experiences slightly different local chemical environments, the precession frequencies of each nucleus will also vary, resulting in time-dependent dephasing in the x–y plane. Relaxation experiments are carried out by acquiring multiple spectra with different delay times (the time allowed for dephasing in the x–y plane before detection). For both T_1 and T_2 experiments, signal intensities for each resonance are plotted as a function of the delay time and fit to a two-parameter (initial amplitude and T) exponential decay function to extract T_1 or T_2 relaxation times. This enables differences in relaxation rates to be identified at residue-level resolution. Generally, T_2 is correlated with molecular weight; as the size of the protein increases, T_2 becomes shorter meaning that the rate constant is increased ($T_2 = 1/R_2$). The relationship of T_1 with molecular weight is more complex.[3] Lifetime line-broadening is the measurement of the change in R_2 relaxation rates (ΔR_2) under conditions where chemical exchange occurs on the microsecond to millisecond timescale between two states with large differences in R_2 relaxation rates (such as might be expected if the two states have large differences in molecular weight) *versus* conditions where an exchange is suppressed.[81]

^{15}N–^{1}H heteronuclear NOE (hnNOE) experiments, which measure NOEs between backbone amide groups, are broadly complementary to T_1 and T_2 experiments in that they provide information regarding the dynamics of the N–H bond on the picosecond to nanosecond timescale.[82] Amide bonds that undergo very fast motion typically show decreased heteronuclear NOE intensities relative to the average NOE intensities observed across all residues due to inefficient magnetization transfer through space.

3.4.2 T_1 Relaxation in the Rotating Frame

Spin–lattice relaxation in the rotating frame, $T_{1\rho}$, may be thought of as T_1 relaxation in the presence of a radiofrequency field.[83] This experiment is similar to those used to measure T_2 relaxation in that magnetization is first rotated from the z-axis to the x–y plane with a hard radiofrequency pulse. However, in contrast to the T_2 experiment, a subsequent continuous wave radiofrequency pulse is applied along the y axis effectively "locking" the spins in the direction of the effective magnetic field. $T_{1\rho}$ experiments are useful for characterizing chemical or conformational exchange processes that occur on a wide range of time scales due to the ability to vary the strength of the effective field using on-resonance and off-resonance methods (Figure 3.2). Despite the sensitivity of $T_{1\rho}$ experiments to off-resonance effects, it is a useful approach for measuring R_2 since the off-resonance effects are easily corrected. Therefore, R_2 is readily calculated from $T_{1\rho}$ and T_1 measurements using the following equation:[84]

$$R_2 = \frac{\left[R_{1\rho} - R_1 \cos^2(\theta)\right]}{\sin^2(\theta)}$$

where θ is the angle between the effective spin-lock field and the external magnetic field.

3.4.3 Paramagnetic Relaxation Enhancement

Paramagnetic relaxation enhancement (PRE) is an NMR method generally used to aid in the structure determination or for identifying transient protein–protein interactions with exchange rates on the μs timescale[85] (Figure 3.2). As its name implies, it involves the enhancement of nuclear spin transverse relaxation rates (R_2) due to the proximity of nuclear spin systems to an unpaired electron of a paramagnetic center. In ideal cases, the PRE effect can be used to detect interactions between the paramagnetic agent and magnetically active nuclei at a distance of up to 35 Å away.[86] Paramagnetic labels commonly used for PRE experiments include nitroxide free radicals or some metal ions, for example, Mn^{2+}. The effect is due to dipolar interactions between the unpaired electron and the magnetically active nuclei.[87,88] For many PRE applications, the paramagnetic agent is typically covalently linked to the biomolecule of interest *via* disulfide chemistry.[89] However some proteins, such as metalloproteins, have intrinsic paramagnetic groups that can be used to enhance relaxation.[90]

In the solvent PRE experiment, the paramagnetic agent is included as a component of the solvent rather than intrinsic to the molecule of interest.[91,92] Consequently, the degree of relaxation enhancement induced by the paramagnetic agent correlates with the surface accessibility of exposed nuclei. Thus, solvent PREs can be useful for determining the residual structure (for example in disordered proteins) and long-range intramolecular or intermolecular interactions.[93] Another useful application of PRE experiments is investigating homotypic or heterotypic intermolecular interactions. These experiments usually involve mixing a protein that carries a paramagnetic label but which is not isotopically labelled with a binding partner that is isotopically labelled but which carries no paramagnetic label. Intermolecular interactions bring the paramagnetic label near the magnetically active nuclei, leading to an enhanced rate of nuclear spin transverse relaxation (R_2), enabling the intermolecular interactions to be mapped at residue-level resolution.

The measurement of the PRE effect relies on the measurement of effective relaxation rates (Γ_2) for both a paramagnetic sample and a diamagnetic control. Multiple approaches have been devised to accomplish this, and the pros and cons of the various methods are described in detail elsewhere.[86,94–96] Clore and coworkers proposed the use of a two-point measurement of ^1H_N transverse relaxation rates (R_2) of both paramagnetic and diamagnetic samples as a simple approach to determine Γ_2.[95] This involves recording HSQC (or TROSY) spectra with two different relaxation delay times on both a paramagnetic sample and a diamagnetic sample. The difference in the effective relaxation rate between the paramagnetic sample and the diamagnetic sample is proportional to $1/r^6$, where r refers to the internuclear distance between the nuclei of interest and the paramagnetic label.

Walti and co-workers used PRE experiments to define the GroEL binding site on a model substrate, the SH3 domain derived from Fyn tyrosine kinase (Fyn SH3).[97] EDTA–Mn^{2+} spin labels were placed at six different sites on GroEL, including on the exterior of the cavity. PRE effects were not observed when the PRE tags were placed on the exterior of the GroEL chamber, and only small effects were observed when the tags were placed at the top of the GroEL chamber, indicating that Fyn SH3 likely does not interact with the exterior of GroEL but may form a low affinity encounter complex at the top of the chamber. PRE tags placed within the GroEL chamber induced larger relaxation enhancements for Fyn SH3 at all three sites, consistent with the encapsulation of the substrate by GroEL. PRE enhancement was observed for residues of Fyn SH3 corresponding to three sites similar to those identified as interaction sites with three different chaperones, Spy, SurA and Skp.[98] Interestingly, a PRE tag placed on the flexible C-terminal tail of GroEL also induced PRE effects at all three interaction sites, suggesting that the C-terminal tail of GroEL can extend throughout the interior of the GroEL chamber and may make contacts with substrates that are only associated with the rim of GroEL.

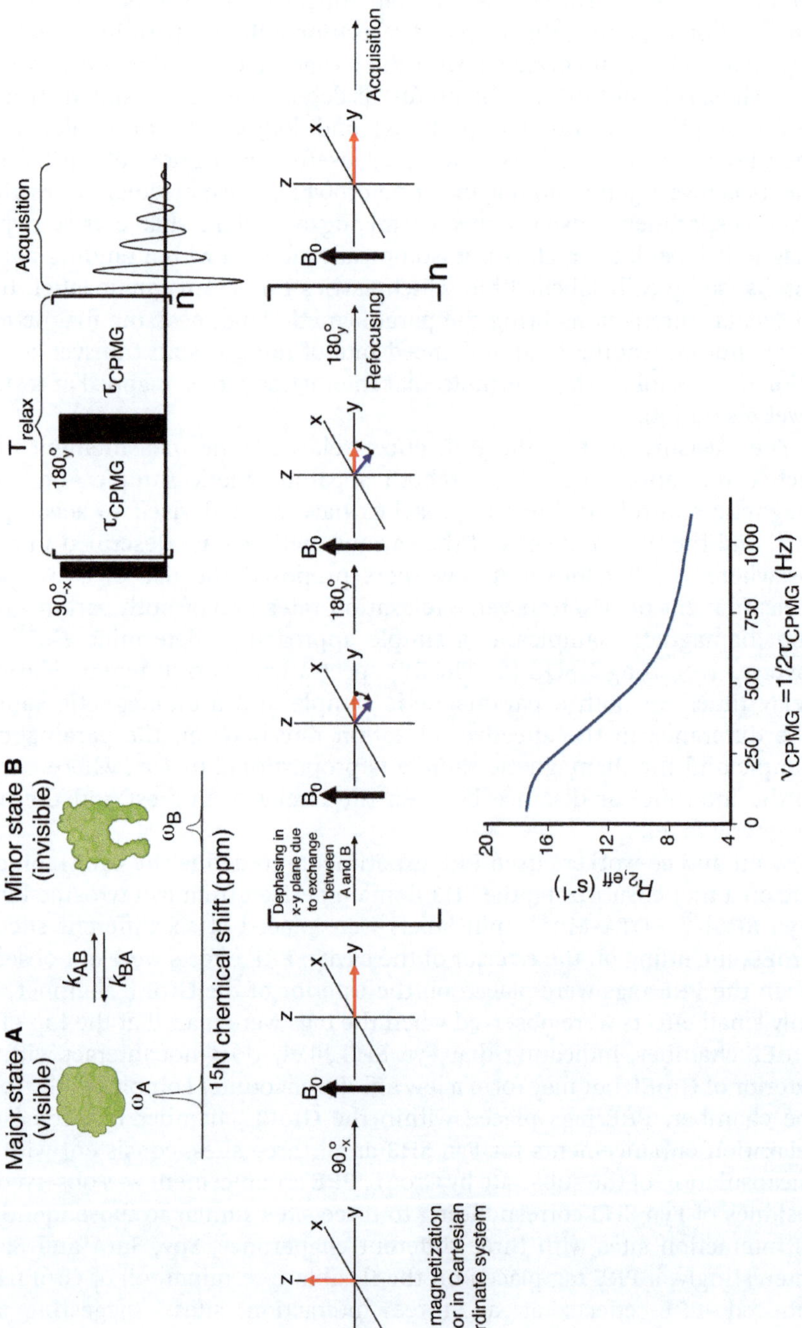
Major state A
(visible)
Minor state B
(invisible)
k_{AB}
k_{BA}
ω_A
ω_B
^{15}N chemical shift (ppm)
$90°_{-x}$
$180°_y$
τ_{CPMG}
τ_{CPMG}
T_{relax}
n
Acquisition
B_0
z
x
y
Net magnetization vector in Cartesian coordinate system
Dephasing in x-y plane due to exchange between A and B
Refocusing
$R_{2,eff}$ (s^{-1})
ν_{CPMG} =1/2τ_{CPMG} (Hz)
20
16
12
8
4
0
250
500
750
1000

3.4.4 CPMG Relaxation-dispersion Spectroscopy

CPMG-based relaxation dispersion (RD) experiments are a method for probing the dynamics of proteins on the millisecond timescale.[99] The key principle of CPMG RD is the spin-echo which, in its simplest form, comprises an initial 90°_{-x} pulse, a delay time, τ, followed by a 180°_{y} pulse and another identical delay time, τ (Figure 3.3).[100–102] During the first τ block, the nuclear spins acquire a phase with respect to *y*. The 180°_{y} pulse then inverts these phases about the *y* axis, which subsequently refocus during the second *t* block to yield a net phase with respect to the *y*-axis of zero at time 2τ. The signals are then detected, and in the case of perfect refocusing, strong signal intensities should be obtained. However, if the nuclei undergo chemical or conformational exchange resulting in a change in their magnetic environment before, during or after the 180°_{y} pulse, their precession frequencies will be altered, resulting in a non-zero phase with respect to *y* at time 2τ (Figure 3.3). This imperfect refocusing gives rise to signals of weaker intensity relative to signals obtained with perfect refocusing. In CPMG RD experiments, after the initial 90° pulse, multiple 180°_{y} pulses are applied, with the length of τ varying between each experiment. Varying τ alters the frequency of the spin echoes, which is reported as the CPMG frequency (ν_{CPMG}). If the spins under investigation do not undergo exchange on timescales comparable to ν_{CPMG}, refocusing is nearly perfect after each spin echo and the signal intensity is relatively unaffected by the changes in the delay times between 180°_{y} refocusing pulses. However, if the nuclei undergo chemical exchange on a timescale comparable to ν_{CPMG}, the efficiency of refocusing will be dependent on the delay times between 180°_{y} refocusing pulses, with shorter delay times leading to more efficient refocusing and higher signal intensities. R_2 relaxation profiles are collected for a range of different ν_{CPMG}, yielding relaxation dispersion profiles (R_2^{observed} as a function of ν_{CPMG}) (Figure 3.3). R_2^{observed} is related to the intrinsic transverse relaxation rate R_2 by[72]

$$R_2^{\text{observed}} = R_2 + R_{\text{ex}}$$

where R_{ex} is the contribution to R_2^{observed} governed by chemical exchange on the ms timescale. Flat relaxation dispersion profiles indicate that the

Figure 3.3 CPMG relaxation-dispersion NMR spectroscopy is useful for detecting chemical and conformational exchange occurring on the millisecond timescale between a major visible species and a minor invisible species that are separated by different chemical shifts. A 90°_{-x} pulse transfers magnetization to the *x*–*y* plane. Next, a sequence of a variable delay time, *t*, allows for dephasing in the *x*–*y* plane. This is followed by a 180°_{y} pulse and an identical *t* delay allowing for refocusing in the *x*–*y* plane. The t–180°_{y}–t sequence is repeated *n* times over a constant relaxation time, T_{relax}. This varies the ν_{CPMG}, which will affect the refocusing efficiency if chemical exchange occurs on a timescale comparable to ν_{CPMG}, and manifests as a reduction in R_2^{observed} as a function of ν_{CPMG}. Created with BioRender.com.

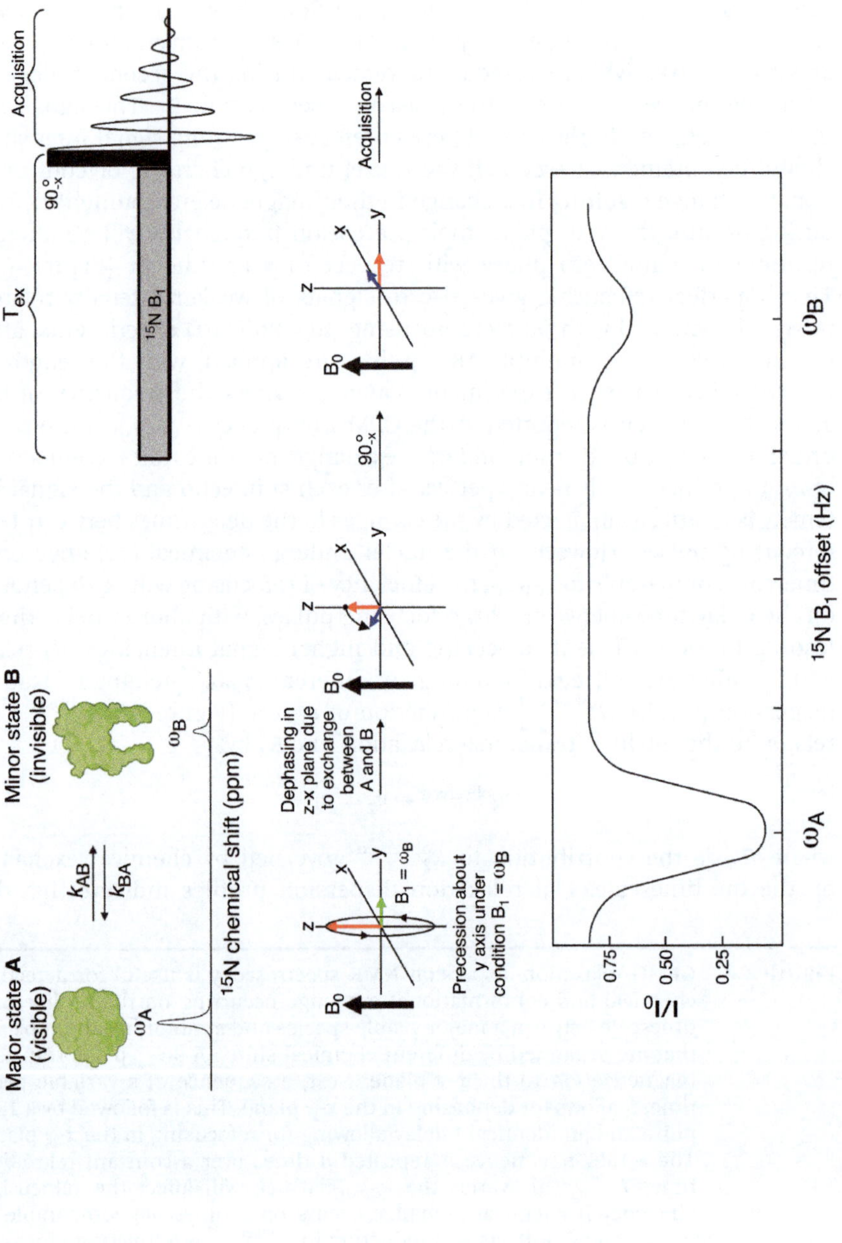
Major state A
(visible)
Minor state B
(invisible)
k_{AB}
k_{BA}
ω_A
ω_B
^{15}N chemical shift (ppm)
T_{ex}
^{15}N B_1
90°_{-x}
Acquisition
B_0
z
x
y
$B_1 = \omega_B$
Precession about
y axis under
condition $B_1 = \omega_B$
Dephasing in
z-x plane due
to exchange
between
A and B
90°_{-x}
Acquisition
I/I_0
0.75
0.50
0.25
ω_A
ω_B
^{15}N B_1 offset (Hz)

nucleus of interest does not undergo exchange on the ms timescale while a reduction in $R_2^{observed}$ with increasing ν_{CPMG} is generally indicative of chemical exchange occurring on this timescale. CPMG relaxation data, when collected with sufficient ν_{CPMG} and at two or more B_0 fields, can be fit to kinetic models to derive estimates of the difference in chemical shift between the exchanging states, the rate of exchange between the states and the relative populations of the states (Figure 3.3).

3.4.5 Saturation-transfer Spectroscopy: CEST and DEST

Two related techniques for investigating exchange processes on the ms to s timescale are chemical exchange saturation transfer (CEST) and dark-state exchange saturation transfer (DEST) spectroscopy.[103–105] CEST is useful for detecting exchange between a highly populated visible species and a lowly populated invisible species and requires that there is a frequency offset (chemical shift difference) between the two states. In the case of CEST, the NMR-visible and NMR-invisible species usually have similar transverse relaxation rates (Figure 3.4).

In contrast, DEST is useful for detecting exchange between an NMR-visible species and an NMR-invisible species with vastly different transverse relaxation rates (Figure 3.5). DEST is not particularly sensitive to the frequency difference between the visible and invisible states. Examples of systems that may be investigated by DEST include a protein or peptide generally less than 100 kDa interacting with a much larger species (>1 MDa) such as a membrane-bound receptor, large chaperones, the proteasome, ribosomes, large oligomeric species or protein aggregates, which have a reduced rate of molecular reorientation in solution relative to the NMR-visible species.

The distinguishing feature of the CEST or DEST experiment is the application of a weak radiofrequency pulse at a series of offsets from the ^{15}N resonance frequency of the NMR-visible species (Figures 3.4 and 3.5). These pulses are calibrated to not perturb the resonances of the NMR-visible species except for at the on-resonance condition. The magnetization of the NMR-invisible species will be selectively saturated, depending on the frequency offset. For the CEST experiment, 10–25 Hz saturation pulses are applied at sequential intervals across the spectral window. Attenuation of the invisible

Figure 3.4 CEST NMR spectroscopy is used to detect chemical or conformational exchange occurring on the millisecond–second timescale between a major visible species and a minor invisible species with different chemical shifts. A weak radiofrequency pulse of 10–25 Hz is applied across the spectral window prior to acquisition. Attenuation of the signal occurs when the pulse is on-resonance with either the major or the minor species and if the exchange between the two species is in the millisecond to the second regime. Created with BioRender.com.

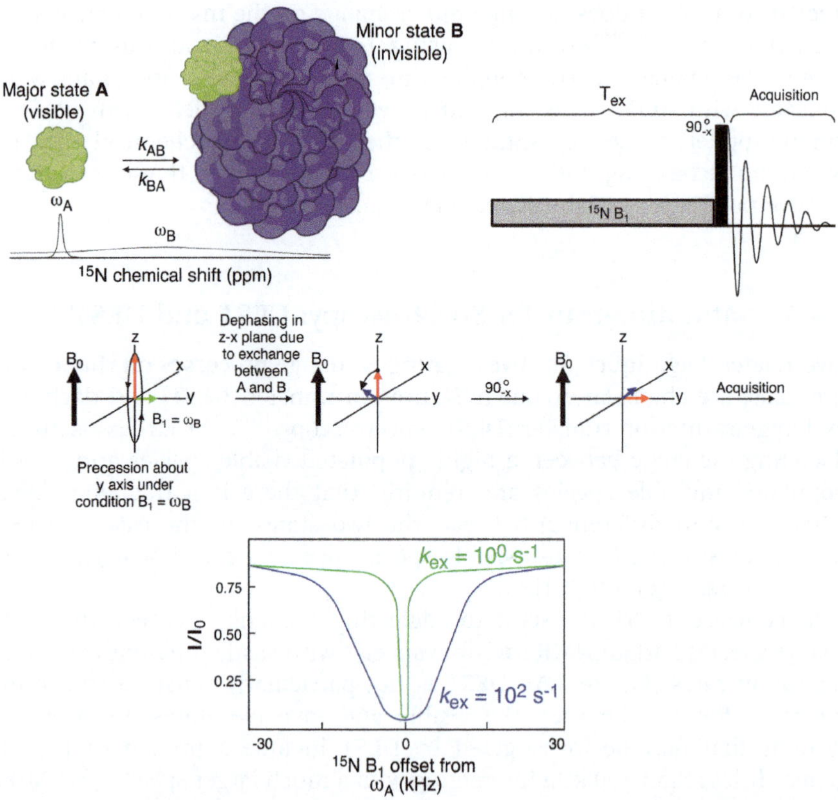

Figure 3.5 DEST NMR spectroscopy is used to detect chemical and conformational exchange occurring on the millisecond–second timescale between a major visible species and a minor invisible species with a difference in transverse relaxation rates, R_2. A weak radiofrequency pulse of 200–750 Hz is applied far off resonance, selectively saturating the resonances of the NMR-invisible species. Attenuation of the major species peak occurs if the exchange rate between the two species is on the millisecond–second timescale. The extent to which the peak intensity of the major species is attenuated is dependent on R_2 of the NMR-invisible species and on the exchange rate (k_{ex}) between the visible and invisible species. Created with BioRender.com.

minor species occurs if the saturation pulse is on-resonance with the minor species, and this will be detected as a reduction in the signal intensity at a resonance position off-set from the major (visible) species (Figure 3.4). In the DEST experiment, 200–750 Hz saturation pulses are applied far off-resonance (1–35 kHz) to selectively saturate the significantly broadened resonances of the larger NMR-invisible species (Figure 3.5). If chemical exchange occurs between the NMR-visible and NMR-invisible states on an appropriate timescale, saturation is transferred back to the NMR-visible state, attenuating the cross-peaks of the NMR-visible species. The extent to which each cross-peak is attenuated is dependent on the R_2 of the NMR-invisible species and the rate at which the exchange occurs (k_{ex}). The extent of signal attenuation as a function of the radiofrequency offset yields CEST or DEST profiles. The CEST and DEST profiles can be fit to kinetic models (discussed further below) to derive atomic-level details of the exchange between the NMR-visible species and NMR-invisible species. However, it is important to note that the utility of CEST and DEST, like all exchange-based NMR experiments, relies on chemical exchange between the two species occurring on a specific timescale. If the rate of exchange does not fall within the 10 ms to 1 second range, it must be perturbed by altering protein concentrations or by changing the temperature or solution conditions. While CEST and DEST were presented in the context of ^{15}N resonances, both techniques have been applied to ^{13}C and ^{13}C methyl nuclei.[76,77,103]

3.4.6 Cross-saturation Transfer

Cross-saturation transfer experiments were designed for identifying residues involved in a protein–protein interaction interface.[106] One of the binding partners is prepared without specific isotope labelling (*i.e.*, ^{14}N and ^{1}H isotopes) and the other binding partner is deuterated and ^{15}N labelled. The aliphatic protons of the unlabelled binding partner are saturated non-selectively using a radiofrequency pulse and spin-diffusion will transfer magnetization to saturate the aromatic and amide protons. The isotopically labelled binding partner is not saturated by the non-selective radiofrequency pulse due to the deuteration of the aliphatic resonances. However, at the complex interface, saturation can be transferred to back-exchanged amide protons but should not transfer across the whole molecule so long as the ^{1}H density is low (deuterated). Therefore, the attenuation of amide cross-peaks in the ^{1}H–^{15}N TROSY–HSQC spectra will reveal the residues involved in the interaction interface between the two molecules.

Park and co-workers investigated the interaction between p53 and HSP90 using cross-saturation transfer NMR experiments.[107] HSP90 exists as a homodimer and is thought to function downstream of the HSP40/HSP70 system in protein folding. The domain architecture of HSP90 is relatively well-conserved across the different isoforms and consists of an N-terminal ATP-binding domain, a charged linker region, the middle domain, which binds to the γ phosphate of ATP to help regulate ATPase activity and a C-terminal domain that mediates homodimerization.[108] An ATP-dependent conformational cycle regulates the HSP90 chaperone activity. In the "open conformation", the C-terminal domain of HSP90 forms inter-subunit contacts that mediate homodimerization.[108,109] In this state, a substrate can bind to the middle domain,[1] and ATP binds to the N-terminal domain.[108] ATP-binding promotes dimerization between the N-terminal domains, forming the "closed conformation".[108,109] ATP-hydrolysis promotes substrate folding.[1] Analysis of chemical shift perturbations in the ^{1}H–^{15}N TROSY–HSQC spectra of various truncated constructs of HSP90 in the presence of p53 was inconclusive, suggesting that the interaction between Hsp90 and p53 is weak/dynamic.[107] Cross-saturation transfer experiments were employed to better define the interaction site on the middle domain of Hsp90.[107] Although these experiments also indicated that the interaction between HSP90 and p53 is dynamic and of low affinity, there was a cluster of residues experiencing relatively large signal attenuation in the presence of p53 at the dimer interface of HSP90, close to the nucleotide-binding domain, indicating that this region may form the interaction site with p53 and be important for substrate folding.

3.4.7 Combining Multiple Experiments to Elucidate Structures, Interactions and Thermodynamics

The experiments discussed in Sections 3.4.1–3.4.6 present powerful approaches to interrogate the structural, dynamic and kinetic parameters governing chaperone–substrate interactions. Yet in some systems, when used individually, the experiments are insufficient for complete characterization of an exchanging system. In practice, it is often necessary to combine these experiments with a global analysis to elucidate more fully the details required to generate a structural model or mode of substrate interaction. This section will review some recent examples in which multiple NMR experiments were combined and analyzed globally to model three-dimensional structures and define the thermodynamic parameters governing chaperone–substrate interactions.

Karamanos and co-workers investigated the structure and dynamics of DNAJB6, a member of the DnaJ family of heat-shock proteins (also known as HSP40) using a range of solution NMR experiments.[110] Unlike other HSP40 family members, DNAJB6 exists as an equilibrium between monomers and polydisperse oligomers up to 1 MDa in size and is thus an extremely challenging target for NMR. To circumvent these challenges, they employed a stable monomeric form of DNAJB6 lacking the intrinsically disordered linker region. The ^{1}H–^{15}N TROSY–HSQC spectra of full-length DNAJB6 and the deletion mutant overlaid well, indicating that the deletion of the flexible linker did not result in large structural changes to the folded domains. The solution NMR spectra of the monomeric J-domain (JD) and the CTD deletion mutant were assigned using ^{1}H–^{15}N TROSY–HSQC versions of the triple resonance experiments (Section 3.3.1). Rotational correlation times, backbone amide RDCs, backbone chemical shifts, backbone amide PREs and methyl NOEs were used to solve the solution structure of monomeric DNAJB6 using CS-ROSETTA.[111] The solution structure revealed, in contrast to previously solved structures, that part of the flexible Gly/Phe linker forms a stable α-helix that docks onto and reorients with the JD in solution and therefore should be considered as part of the JD. Interestingly, the helix was found to interact with the same residues in the JD as those known to interact with DnaK (HSP70); therefore, this interaction was suggested to block the binding of DNAJB6 to DnaK.[45] Transient interactions between the JD and the CTD were predicted based on PREs existing between the fifth α-helix of the JD and the CTD. To define a model for the oligomerisation of DNAJB6, the concentration dependence of chemical shifts and ^{15}N-R_2 values were investigated. Larger changes in R_2 and concentration-dependent chemical shifts were observed for resonances in the CTD, implying that the CTD likely drives oligomerization. Further supporting this, residues closest to the C-terminus displayed concentration-dependent RD profiles, while the RD profiles of residues located in the JD were concentration independent. Concentration-dependent RD profiles are expected for residues participating in intermolecular interactions, while concentration-independent RD profiles are expected for residues involved only in intramolecular interactions. The CPMG RD data, ^{15}N-R_2 values and exchange-induced shifts were fit globally to a three-state model consisting of a sparsely populated monomeric closed conformation (in which the JD and the CTD interact), a highly populated monomeric open conformation (in which the JD and the CTD do not interact) and a sparsely populated oligomeric form in which the CTD makes intermolecular interactions. The global analysis yielded rate constants for the conversion between the various conformations, bound state R_2 values and molecular weight estimates of the oligomeric form. This work is therefore an excellent example of how solution NMR relaxation techniques can be applied to identify residual structural elements, conformational dynamics and sparsely populated conformations of a multidomain chaperone protein.

Libich and co-workers developed a novel approach for investigating the NMR-invisible, GroEL-bound state of the amyloid-beta 40 (Aβ40) peptide.[112] Previous investigations of substrates bound by GroEL were somewhat conflicting, likely due to the transient nature of the interactions. This work highlights the utility of several relaxation-based NMR techniques in identifying and characterising transient "dark-states". An interaction between GroEL and Aβ40 was detected by measuring the ^{15}N-R_2 values for Aβ40 in the presence and in the absence of GroEL. Comparison of the difference in the measurements (^{15}N-ΔR_2) indicated that the presence of GroEL in the samples induced line broadening of Aβ40. Line broadening can arise from chemical shift differences between free and GroEL-bound Aβ40 or from large increases in R_2 in the GroEL-bound state. Since small changes in chemical shift observed in the presence of GroEL were not correlated with the ^{15}N-ΔR_2 values induced by the addition of GroEL, the observed ^{15}N-ΔR_2 values were attributed to large differences in Aβ40 ^{15}N-R_2 in the free and GroEL-bound state. Further supporting this, the CPMG RD profiles indicated that the contribution of chemical exchange to the observed line broadening was extremely small, limited to less than 0.5 s^{-1}. Together, these observations suggested that the structure of Aβ40 is largely unchanged in the GroEL-bound state. ^{15}N DEST profiles, which provide information regarding residue-specific ^{15}N-R_2 values and dissociation rate constants, were obtained for Aβ40 in the presence of GroEL.

Global fitting of the ^{15}N-ΔR_2, CPMG RD, DEST and CSP data derived from each experiment to a two-state kinetic model using the McConnell equations enabled the extraction of the rate constants governing the exchange with a greater degree of accuracy than fits to any individual set of relaxation data. The McConnell equations describe all of the experimental observables yet the global fitting procedure was necessary because each relaxation experiment is influenced in different ways by the rate constants governing the binding of Aβ40 to GroEL.[113] This global analysis resulted in an estimated dissociation constant (K_D) of 70 μM and free and bound populations of ~98% and ~2%, respectively. In addition to deriving these rates and populations, the bound ^{15}N-R_2 values were found to vary in a residue-specific manner across the Aβ40 sequence. Two regions with the largest ^{15}N-R_2 values corresponded to the hydrophobic residues in Aβ40 and were predicted to be the GroEL-binding sites. Supporting the hypothesis that GroEL-bound Aβ40 remains mostly disordered, the magnitude of the ^{15}N-R_2 values was only 60–70% of that expected for a rigid peptide bond, indicating that the bound peptide retains significant flexibility or that the two regions of Aβ40 that bind to GroEL can do so independently of one another.

In a follow-up study, Walti and co-workers investigated the binding of Aβ42 to GroEL using a similar approach that was used for Aβ40.[114] As observed for the binding of Aβ40 to GroEL, there was a lack of correlation between lifetime line-broadening and exchange induced chemical shifts for Aβ42 in the presence of GroEL, likely indicating that the observed lifetime line broadening is caused by large increases in R_2 values in the bound state rather than large conformational changes induced by GroEL.[112] Similar to Aβ40, the population of GroEL-bound Aβ42 was calculated to be ~2% based on the global fitting of the four sets of relaxation experiments. The equilibrium dissociation constant was determined to be ~300 μM. Calculated ^{15}N-R_2 values for Aβ42 in the GroEL-bound state identified three regions of interaction with GroEL, the same two regions as identified for Aβ40 and the third being the hydrophobic residues 40–42 (41 and 42 are missing from Aβ40).[112] Despite this small difference in the binding mode, the ^{15}N-R_2 values for GroEL-bound Aβ42 indicated that it remains mostly disordered, consistent with the findings for Aβ40.

To further interrogate the substrate folding mechanism of GroEL, Libich and co-workers applied the global analysis of relaxation-based NMR experiment approach to investigate GroEL binding to a model substrate, Fyn SH3, that was engineered to exist in equilibrium between a folded state (F) and a folding intermediate (I).[115] Importantly, these two states interconvert on timescales accessible to CPMG RD (Section 3.4.4).[116] CPMG RD profiles revealed changes in both chemical shift and ^{15}N-R_2 values of Fyn SH3 in the presence of GroEL relative to those in the absence of GroEL, demonstrating that Fyn SH3 interacts with GroEL. The ^{15}N-DEST profiles revealed off-resonance saturation induced by GroEL. Global fitting of the relaxation data supported a model in which interconversion between the Fyn SH3 folded and intermediate states can occur both in solution and bound to GroEL and yielded the relevant rate and equilibrium dissociation constants, and the populations of the four states (folded, intermediate, GroEL-bound folded and GroEL-bound intermediate). Based on these results, GroEL-bound Fyn SH3 accounts for only ~1% of the total population of Fyn SH3, highlighting the sensitivity of relaxation-based NMR techniques for detecting otherwise invisible conformations of proteins in solution. The analysis also revealed that the exchange rate between the folded and intermediate states was accelerated when Fyn SH3 was bound to GroEL, implying that GroEL possesses some intrinsic unfoldase activity. The interactions between GroEL and Fyn SH3 were examined using a selective isotope labelling strategy in which Fyn SH3 was $^{15}N/^{1}H_{N}$, AILV-^{13}C methyl-protonated but deuterated at all ^{1}H aliphatic sites. ^{13}C-Methyl CPMG RD, lifetime line-broadening and DEST were recorded on the methyl-labelled samples under the same conditions used previously and in the presence and absence of GroEL. Global analysis of these data revealed a reduction in lifetime line-broadening and the rate of conversion between the GroEL-bound folded and intermediate states likely due to the deuterium isotope effect.

In a subsequent study, Libich and co-workers investigated GroEL binding to a C-terminally truncated Fyn SH3 that existed in equilibrium between the intermediate and unfolded states.[117] Chemical shifts for the sparsely populated, fully unfolded state were calculated from the CPMG RD profiles, yielding shifts consistent with those expected for a random coil. Lifetime line-broadening and broadening of the ^{15}N DEST profiles were observed when the Fyn SH3 truncation mutant was mixed with GroEL, and the results are consistent with the binding of the low molecular weight Fyn SH3 to the high molecular weight GroEL. CPMG RD profiles revealed differences between the apparently observed R_2 rates between the lowest and the highest CPMG field strengths in the presence and in the absence of GroEL. To fully account for these observations in the global fitting, a folded conformation of the Fyn SH3 mutant that can bind to but cannot unfold in the presence of GroEL was incorporated into the model. This suggested that the complete unfolding of the destabilized Fyn SH3 mutant does not occur within the GroEL cavity. When considered together, the studies by Libich *et al.* indicate that GroEL stabilizes the folding intermediate of Fyn SH3 relative to both the fully folded form and the fully unfolded form. They concluded that this is because the folding intermediate has an exposed hydrophobic surface that interacts with GroEL, while the radius of gyration of the fully unfolded conformation is likely too large for confinement within the GroEL cavity. Collectively, these studies reveal how GroEL participates in protein folding in an ATP-independent manner.

3.5 Future Directions and Perspectives

This chapter has outlined recent advances in biomolecular NMR that aid in the determination of the structure and chaperone–substrate interactions. Nevertheless, the use of NMR for structure determination has declined in recent years, in part because of advances in the cryo-EM technology that have enabled high-resolution structures of large molecular weight complexes to be resolved. Conversely, solution NMR typically suffers from size limitations due to increased linewidths that arise from larger molecular weight molecules and biomolecular complexes. However, new solution NMR methods continue to be developed for overcoming these size limitations. Furthermore, the traditional structure–function paradigm in protein biochemistry is increasingly challenged by the identification of functionally important conformations and interactions that cannot be interrogated by cryo-EM or crystallography. Weak and transient protein–protein interactions, intrinsically disordered regions and sparsely populated heterogenous conformations which constitute "dark-states" are extremely important for understanding how molecular chaperones function. For these reasons, biomolecular NMR will undoubtedly continue to be an important tool for further improving our understanding of chaperone structure and functions.

Future methodological advances in solution NMR should focus on improving our ability to observe how molecular chaperones operate in solution with substrates and cofactors that more closely recapitulate the *in vivo* scenario. For example, chaperones often function in larger complexes containing multiple different chaperone family members to assist with the folding or unfolding of specific client proteins. Conversely, some chaperones bind, stabilize and even disaggregate proteins from highly heterogenous and aggregated states, and these states remain extremely challenging to interrogate with all structural and biophysical approaches. To overcome these challenges, innovative strategies for interrogating individual molecules within much larger complexes are needed. Selective or segmental labelling strategies, including specific isotopic labelling of sidechains other than alanine, isoleucine, leucine and valine, could be extremely useful. It is also important for NMR spectroscopists to continue to improve the accessibility of advanced NMR methods to researchers lacking extensive training in NMR. With improved methodology and greater accessibility, NMR will be an essential tool for future studies aimed at uncovering the structural and dynamic underpinnings of molecular chaperone functions.

Abbreviations

Ab	Amyloid beta
CPMG	Carr–Purcell–Meiboom–Gill
CEST	Chemical exchange saturation transfer
CSP	Chemical shift perturbation
CCD	Coiled-coil domain
COSY	Correlated spectroscopy
CRINEPT	Cross-correlated relaxation-enhanced polarization transfer
Cryo-EM	Cryogenic electron microscopy
CTD	C-terminal domain
DEST	Dark-state exchange saturation transfer
DHFR	Dihydrofolate reductase

FT	Fourier transform
H/D exchange	Hydrogen–deuterium exchange
HSP	Heat shock protein
HMQC	Heteronuclear multiple quantum coherence
hnNOE	Heteronuclear NOE
HSQC	Heteronuclear single quantum coherence
INEPT	Insensitive nuclei enhanced by polarization transfer
IDP	Intrinsically disordered protein
IDR	Intrinsically disordered region
JD	J-domain
kDa	Kilodalton
MDa	Megadalton
NBD	N-terminal binding domain
NMR	Nuclear magnetic resonance
NOE	Nuclear Overhauser effect
NOSEY	Nuclear Overhauser effect spectroscopy
PRE	Paramagnetic relaxation enhancement
RD	Relaxation dispersion
RDC	Residual dipolar coupling
TOCSY	Total correlation spectroscopy
TROSY	Transverse relaxation optimized spectroscopy

Acknowledgements

DSL received the Shohet Family Fund for Ewing Sarcoma Research St. Baldrick's Scholar and acknowledges the support of the St. Baldrick's Foundation (634706). This study was funded in part by the Max and Minnie Tomerlin Voelcker Fund (Voelcker Foundation Young Investigator Grant to DSL) and the GCCRI Startup Funds (DSL). EES was supported by a CPRIT Research Training Award (RP170345).

References

1. F. U. Hartl, A. Bracher and M. Hayer-Hartl, *Nature*, 2011, **475**, 324–332.
2. B. M. Burmann and S. Hiller, *Prog. Nucl. Magn. Reson. Spectrosc.*, 2015, **86–87**, 41–64.
3. J. Cavanagh, W. Fairbrother, A. Palmer III, M. Rance and N. Skelton, *Protein NMR Spectroscopy*, Elsevier Academic Press, 2007.
4. A. Abragam, *The Principles of Nuclear Magnetism*, Oxford University Press, 1961.
5. E. H. Anderson, *Proc. Natl. Acad. Sci. U. S. A.*, 1946, **32**, 120–128.
6. J. Marley, M. Lu and C. Bracken, *J. Biomol. NMR*, 2001, **20**, 71–75.
7. N. K. Goto and L. E. Kay, *Curr. Opin. Struct. Biol.*, 2000, **10**, 585–592.
8. M. Kainosho, T. Torizawa, Y. Iwashita, T. Terauchi, A. Mei Ono and P. Guntert, *Nature*, 2006, **440**, 52–57.
9. V. Tugarinov and L. E. Kay, *J. Biomol. NMR*, 2004, **28**, 165–172.

10. A. P. French and E. F. Taylor, *An Introduction to Quantum Physics*, Routledge, 2018.
11. F. Bloch, *Phys. Rev.*, 1946, **70**, 460.
12. E. M. Purcell, H. C. Torrey and R. V. Pound, *Phys. Rev.*, 1946, **69**, 37.
13. I. I. Rabi, J. R. Zacharias, S. Millman and P. Kusch, *Phys. Rev.*, 1938, **53**, 318.
14. W. Heisenberg, in *Original Scientific Papers Wissenschaftliche Originalarbeiten*, Springer, 1985, pp. 478–504.
15. W. Aue, E. Bartholdi and R. R. Ernst, *J. Chem. Phys.*, 1976, **64**, 2229–2246.
16. C. Griesinger, O. Sørensen and R. Ernst, *J. Magn. Reson.*, 1987, **73**, 574–579.
17. H. Oschkinat, C. Griesinger, P. J. Kraulis, O. W. Sorensen, R. R. Ernst, A. M. Gronenborn and G. M. Clore, *Nature*, 1988, **332**, 374–376.
18. L. E. Kay, G. M. Clore, A. Bax and A. M. Gronenborn, *Science*, 1990, **249**, 411–414.
19. C. Griesinger, O. Sorensen and R. Ernst, *J. Am. Chem. Soc.*, 1987, **109**, 7227–7228.
20. E. Hahn and D. Maxwell, *Phys. Rev.*, 1952, **88**, 1070.
21. A. Bax and R. Freeman, *J. Magn. Reson.*, 1981, **44**, 542–561.
22. L. Braunschweiler and R. Ernst, *J. Magn. Reson.*, 1983, **53**, 521–528.
23. J. Jeener, B. Meier, P. Bachmann and R. Ernst, *J. Chem. Phys.*, 1979, **71**, 4546–4553.
24. S. Macura and R. Ernst, *Mol. Phys.*, 1980, **41**, 95–117.
25. J. Keeler, *Understanding NMR spectroscopy*, John Wiley & Sons, 2011.
26. G. A. Morris and R. Freeman, *J. Am. Chem. Soc.*, 1979, **101**, 760–762.
27. G. Batta, K. Kövér and C. Szántay Jr, *Methods for Structure Elucidation by High-Resolution NMR: Applications to Organic Molecules of Moderate Molecular Weight*, Elsevier, 1997.
28. R. Rosenzweig and L. E. Kay, *J. Am. Chem. Soc.*, 2016, **138**, 1466–1477.
29. E. R. Anil-Kumar, R. Ernst and K. Wüthrich, *Biochem. Biophys. Res. Commun.*, 1980, **64**, 2229–2246.
30. L. P. McIntosh and F. W. Dahlquist, *Q. Rev. Biophys.*, 1990, **23**, 1–38.
31. S. Grzesiek and A. Bax, *J. Am. Chem. Soc.*, 1992, **114**, 6291–6293.
32. S. Grzesiek and A. Bax, *J. Magn. Reson.*, 1992, **99**, 201–207.
33. L. E. Kay, M. Ikura, R. Tschudin and A. Bax, *J. Magn. Reson.*, 1990, **89**, 496–514.
34. S. Grzesiek and A. Bax, *J. Biomol. NMR*, 1993, **3**, 185–204.
35. S. Grzesiek, J. Anglister and A. Bax, *J. Magn. Reson.*, 1993, **101**, 114–119.
36. A. Bax and M. Ikura, *J. Biomol. NMR*, 1991, **1**, 99–104.
37. L. E. Kay, *J. Am. Chem. Soc.*, 1993, **115**, 2055–2057.
38. M. Salzmann, G. Wider, K. Pervushin, H. Senn and K. Wüthrich, *J. Am. Chem. Soc.*, 1999, **121**, 844–848.
39. N. B. Nillegoda, J. Kirstein, A. Szlachcic, M. Berynskyy, A. Stank, F. Stengel, K. Arnsburg, X. Gao, A. Scior, R. Aebersold, D. L. Guilbride, R. C. Wade, R. I. Morimoto, M. P. Mayer and B. Bukau, *Nature*, 2015, **524**, 247–251.

40. H. Rampelt, J. Kirstein-Miles, N. B. Nillegoda, K. Chi, S. R. Scholz, R. I. Morimoto and B. Bukau, *EMBO J.*, 2012, **31**, 4221–4235.
41. A. Scior, A. Buntru, K. Arnsburg, A. Ast, M. Iburg, K. Juenemann, M. L. Pigazzini, B. Mlody, D. Puchkov, J. Priller, E. E. Wanker, A. Prigione and J. Kirstein, *EMBO J.*, 2018, **37**, 282–299.
42. M. K. Greene, K. Maskos and S. J. Landry, *Proc. Natl. Acad. Sci. U. S. A.*, 1998, **95**, 6108–6113.
43. D. W. Summers, P. M. Douglas, C. H. Ramos and D. M. Cyr, *Trends Biochem. Sci.*, 2009, **34**, 230–233.
44. A. Szabo, T. Langer, H. Schroder, J. Flanagan, B. Bukau and F. U. Hartl, *Proc. Natl. Acad. Sci. U. S. A.*, 1994, **91**, 10345–10349.
45. T. Szyperski, M. Pellecchia, D. Wall, C. Georgopoulos and K. Wuthrich, *Proc. Natl. Acad. Sci. U. S. A.*, 1994, **91**, 11343–11347.
46. M. Pellecchia, T. Szyperski, D. Wall, C. Georgopoulos and K. Wuthrich, *J. Mol. Biol.*, 1996, **260**, 236–250.
47. S. W. Englander and L. Mayne, *Annu. Rev. Biophys. Biomol. Struct.*, 1992, **21**, 243–265.
48. A. Hvidt and K. Linderstrøm-Lang, *Biochim. Biophys. Acta*, 1954, **14**, 574–575.
49. G. Wagner and K. Wuthrich, *J. Mol. Biol.*, 1982, **160**, 343–361.
50. Y. Bai, J. S. Milne, L. Mayne and S. W. Englander, *Proteins*, 1993, **17**, 75–86.
51. M. S. Goldberg, J. Zhang, S. Sondek, C. R. Matthews, R. O. Fox and A. L. Horwich, *Proc. Natl. Acad. Sci. U. S. A.*, 1997, **94**, 1080–1085.
52. K. Braig, Z. Otwinowski, R. Hegde, D. C. Boisvert, A. Joachimiak, A. L. Horwich and P. B. Sigler, *Nature*, 1994, **371**, 578–586.
53. J. F. Hunt, A. J. Weaver, S. J. Landry, L. Gierasch and J. Deisenhofer, *Nature*, 1996, **379**, 37–45.
54. H. S. Rye, S. G. Burston, W. A. Fenton, J. M. Beechem, Z. Xu, P. B. Sigler and A. L. Horwich, *Nature*, 1997, **388**, 792–798.
55. B. J. Stockman, N. R. Nirmala, G. Wagner, T. J. Delcamp, M. T. DeYarman and J. H. Freisheim, *Biochemistry*, 1992, **31**, 218–229.
56. N. Tjandra, S. Grzesiek and A. Bax, *J. Am. Chem. Soc.*, 1996, **118**, 6264–6272.
57. N. Tjandra and A. Bax, *Science*, 1997, **278**, 1111–1114.
58. R. S. Lipsitz and N. Tjandra, *Annu. Rev. Biophys. Biomol. Struct.*, 2004, **33**, 387–413.
59. J. H. Prestegard, H. M. al-Hashimi and J. R. Tolman, *Q. Rev. Biophys.*, 2000, **33**, 371–424.
60. K. Pervushin, R. Riek, G. Wider and K. Wuthrich, *Proc. Natl. Acad. Sci. U. S. A.*, 1997, **94**, 12366–12371.
61. Y. Zhang and E. R. Zuiderweg, *Proc. Natl. Acad. Sci. U. S. A.*, 2004, **101**, 10272–10277.
62. W. Han and P. Christen, *Biochem. J.*, 2003, **369**, 627–634.
63. M. P. Mayer, *Trends Biochem. Sci.*, 2013, **38**, 507–514.
64. M. P. Mayer, D. Brehmer, C. S. Gassler and B. Bukau, *Adv. Protein Chem.*, 2001, **59**, 1–44.

65. M. Salzmann, K. Pervushin, G. Wider, H. Senn and K. Wuthrich, *Proc. Natl. Acad. Sci. U. S. A.*, 1998, **95**, 13585–13590.
66. V. Tugarinov and L. E. Kay, *J. Am. Chem. Soc.*, 2003, **125**, 13868–13878.
67. V. Tugarinov, P. M. Hwang, J. E. Ollerenshaw and L. E. Kay, *J. Am. Chem. Soc.*, 2003, **125**, 10420–10428.
68. R. Rosenzweig, S. Moradi, A. Zarrine-Afsar, J. R. Glover and L. E. Kay, *Science*, 2013, **339**, 1080–1083.
69. C. Dominguez, R. Boelens and A. M. Bonvin, *J. Am. Chem. Soc.*, 2003, **125**, 1731–1737.
70. R. Riek, G. Wider, K. Pervushin and K. Wuthrich, *Proc. Natl. Acad. Sci. U. S. A.*, 1999, **96**, 4918–4923.
71. R. Horst, E. B. Bertelsen, J. Fiaux, G. Wider, A. L. Horwich and K. Wuthrich, *Proc. Natl. Acad. Sci. U. S. A.*, 2005, **102**, 12748–12753.
72. P. J. Farber and A. Mittermaier, *Biophys. Rev.*, 2015, **7**, 191–200.
73. T. K. Karamanos and G. M. Clore, *Annu. Rev. Biophys.*, 2022, **51**, 223–246.
74. N. R. Skrynnikov, F. A. Mulder, B. Hon, F. W. Dahlquist and L. E. Kay, *J. Am. Chem. Soc.*, 2001, **123**, 4556–4566.
75. P. Lundstrom, P. Vallurupalli, T. L. Religa, F. W. Dahlquist and L. E. Kay, *J. Biomol. NMR*, 2007, **38**, 79–88.
76. G. Bouvignies and L. E. Kay, *J. Biomol. NMR*, 2012, **53**, 303–310.
77. N. L. Fawzi, D. S. Libich, J. Ying, V. Tugarinov and G. M. Clore, *Angew. Chem., Int. Ed.*, 2014, **53**, 10345–10349.
78. R. B. Hill, C. Bracken, W. F. DeGrado and A. G. Palmer, *J. Am. Chem. Soc.*, 2000, **122**, 11610–11619.
79. N. Bloembergen, E. Purcell and R. Pound, *Nature*, 1947, **160**, 475–476.
80. N. Bloembergen, E. M. Purcell and R. V. Pound, *Phys. Rev.*, 1948, **73**, 679.
81. N. L. Fawzi, J. Ying, D. A. Torchia and G. M. Clore, *J. Am. Chem. Soc.*, 2010, **132**, 9948–9951.
82. L. E. Kay, D. A. Torchia and A. Bax, *Biochemistry*, 1989, **28**, 8972–8979.
83. A. G. Redfield, *Phys. Rev.*, 1955, **98**, 1787.
84. D. G. Davis, M. E. Perlman and R. E. London, *J. Magn. Reson., Ser. B*, 1994, **104**, 266–275.
85. J. L. Battiste and G. Wagner, *Biochemistry*, 2000, **39**, 5355–5365.
86. G. M. Clore and J. Iwahara, *Chem. Rev.*, 2009, **109**, 4108–4139.
87. P. G. Schmidt and I. D. Kuntz, *Biochemistry*, 1984, **23**, 4261–4266.
88. I. Solomon and N. Bloembergen, *J. Chem. Phys.*, 1956, **25**, 261–266.
89. P. A. Kosen, *Methods Enzymol.*, 1989, **177**, 86–121.
90. I. Bertini, C. Luchinat and M. Piccioli, *Methods Enzymol.*, 2001, **339**, 314–340.
91. C. Hartlmuller, C. Gobl and T. Madl, *Angew. Chem., Int. Ed.*, 2016, **55**, 11970–11974.
92. C. Hartlmuller, E. Spreitzer, C. Gobl, F. Falsone and T. Madl, *J. Biomol. NMR*, 2019, **73**, 305–317.
93. J. R. Gillespie and D. Shortle, *J. Mol. Biol.*, 1997, **268**, 158–169.
94. J. Iwahara, C. D. Schwieters and G. M. Clore, *J. Am. Chem. Soc.*, 2004, **126**, 5879–5896.

95. J. Iwahara, C. Tang and G. M. Clore, *J. Magn. Reson.*, 2007, **184**, 185–195.
96. G. M. Clore, *Methods Enzymol.*, 2015, **564**, 485–497.
97. M. A. Walti, D. S. Libich and G. M. Clore, *J. Phys. Chem. Lett.*, 2018, **9**, 3368–3371.
98. L. He and S. Hiller, *Angew. Chem., Int. Ed.*, 2018, **57**, 5921–5924.
99. D. F. Hansen, P. Vallurupalli and L. E. Kay, *J. Phys. Chem. B*, 2008, **112**, 5898–5904.
100. H. Y. Carr and E. M. Purcell, *Phys. Rev.*, 1954, **94**, 630.
101. J. P. Loria, M. Rance and A. G. Palmer, *J. Am. Chem. Soc.*, 1999, **121**, 2331–2332.
102. S. Meiboom and D. Gill, *Rev. Sci. Instrum.*, 1958, **29**, 688–691.
103. P. Vallurupalli, G. Bouvignies and L. E. Kay, *J. Am. Chem. Soc.*, 2012, **134**, 8148–8161.
104. N. L. Fawzi, J. Ying, R. Ghirlando, D. A. Torchia and G. M. Clore, *Nature*, 2011, **480**, 268–272.
105. N. L. Fawzi, J. Ying, D. A. Torchia and G. M. Clore, *Nat. Protoc.*, 2012, 7, 1523–1533.
106. H. Takahashi, T. Nakanishi, K. Kami, Y. Arata and I. Shimada, *Nat. Struct. Biol.*, 2000, 7, 220–223.
107. S. J. Park, M. Kostic and H. J. Dyson, *J. Mol. Biol.*, 2011, **411**, 158–173.
108. M. M. Ali, S. M. Roe, C. K. Vaughan, P. Meyer, B. Panaretou, P. W. Piper, C. Prodromou and L. H. Pearl, *Nature*, 2006, **440**, 1013–1017.
109. A. K. Shiau, S. F. Harris, D. R. Southworth and D. A. Agard, *Cell*, 2006, **127**, 329–340.
110. T. K. Karamanos, V. Tugarinov and G. M. Clore, *Proc. Natl. Acad. Sci. U. S. A.*, 2019, **116**, 21529–21538.
111. Y. Shen, O. Lange, F. Delaglio, P. Rossi, J. M. Aramini, G. Liu, A. Eletsky, Y. Wu, K. K. Singarapu, A. Lemak, A. Ignatchenko, C. H. Arrowsmith, T. Szyperski, G. T. Montelione, D. Baker and A. Bax, *Proc. Natl. Acad. Sci. U. S. A.*, 2008, **105**, 4685–4690.
112. D. S. Libich, N. L. Fawzi, J. Ying and G. M. Clore, *Proc. Natl. Acad. Sci. U. S. A.*, 2013, **110**, 11361–11366.
113. H. M. McConnell, *J. Chem. Phys.*, 1958, **28**, 430–431.
114. M. A. Walti, J. Steiner, F. Meng, H. S. Chung, J. M. Louis, R. Ghirlando, V. Tugarinov, A. Nath and G. M. Clore, *Proc. Natl. Acad. Sci. U. S. A.*, 2018, **115**, E11924–E11932.
115. D. S. Libich, V. Tugarinov and G. M. Clore, *Proc. Natl. Acad. Sci. U. S. A.*, 2015, **112**, 8817–8823.
116. P. Neudecker, A. Zarrine-Afsar, W. Y. Choy, D. R. Muhandiram, A. R. Davidson and L. E. Kay, *J. Mol. Biol.*, 2006, **363**, 958–976.
117. D. S. Libich, V. Tugarinov, R. Ghirlando and G. M. Clore, *Biochemistry*, 2017, **56**, 903–906.

CHAPTER 4

Solution NMR Studies of Chaperone–Client Systems

C. WAN AND C. HUANG*

Ministry of Education Key Laboratory for Membrane-less Organelles and Cellular Dynamics, School of Life Sciences, Division of Life Science and Medicine, University of Science and Technology of China, Hefei, China
*Email: huangcd@ustc.edu.cn

4.1 Introduction

Protein homeostasis, also known as proteostasis, refers to the cellular balance of protein synthesis, folding and degradation and thus is crucial for cell survival.[1] This intricate machinery is achieved by extensive interplays among molecular chaperones, the ubiquitin–proteasome system (UPS) and the autophagy system.[2,3] As a key player in maintaining healthy proteostasis, molecular chaperones are a class of proteins that recognize various clients and subsequently assist in the proper folding of nascent polypeptides into their native structures, prevent misfolding or aggregation of unfolded proteins and even help in disassembling aggregates.[4–6] These vital roles played by molecular chaperones have implicated them in many human diseases including cancers, aging and especially neurodegenerative diseases such as Parkinson's disease and Alzheimer's disease.[7–10]

Central questions for fully understanding the action of molecular chaperones include how they selectively recognize and interact with a wide array of client proteins with no obvious consensus in sequence and how such seemingly promiscuous interactions are functionally coupled to their diverse cellular activities. The unique way in which chaperones function inherently demands a dynamic system, which requires conformational

New Developments in NMR No. 29
Biophysics of Molecular Chaperones: Function, Mechanisms and Client Protein Interactions
Edited by Sebastian Hiller, Maili Liu and Lichun He

Published by the Royal Society of Chemistry, www.rsc.org

plasticity for chaperones to accommodate various clients in their non-native states. Moreover, the chaperone–client interactions must be non-specific and often transient, and such dynamic capture-and-release binding manner and the resulting high conformational diversity are thought to be an efficient way to chaperone the clients.[11] An in-depth mechanistic understanding of the dynamic chaperone–client complexes demands atomic-resolution depictions of these dynamically associated protein assemblies. However, such dynamic protein complexes, which are characterized by low thermodynamic stability and low kinetic barrier, often demonstrate a significant amount of structural polymorphism or conformational heterogeneity. As a result, nowadays chaperone–client systems remain a daunting challenge for structural studies. Indeed, sample homogeneity is usually a key factor in obtaining high-resolution information for X-ray crystallography or cryogenic electron microscopy (cryo-EM), two major techniques employed in structural biology.

On the other hand, the highly dynamic interaction manner makes chaperone–client systems suitable targets for solution NMR spectroscopy, which has benefited greatly from recent advancements in both technical improvements and methodology developments. Indeed, recently chaperones have aroused great interest in NMR studies and we have witnessed a number of seminal NMR studies for chaperones or chaperone–client systems, which have provided unprecedented insights into the functional mechanism of chaperone systems. Here, we will first discuss the common challenges faced by solution NMR in studying chaperone–client systems and then summarize the newly developed strategies that have been made to overcome these difficulties. Finally, we will highlight some examples of chaperone–client systems studied using solution NMR.

4.2 Solution NMR – A Powerful Tool for Chaperone–Client Systems

As mentioned in Section 4.1, the intrinsic dynamic nature of chaperone–client systems has defined NMR, which is suited for studies in dynamic systems, the method of choice for mechanistic investigations of chaperones. The reasons are at least five-fold:

(i) Descriptions of chaperone–client systems at the atomic level are the first step to fully understand their function. Due to approximating the native environment, solution NMR is the most ideally suitable method for obtaining *in situ* atomic-level high-resolution information on chaperone–client systems.
(ii) The clients of chaperones are normally in their non-native or unfolded states, and solution NMR is a powerful tool for the analysis of structurally disordered biological molecules, which are typically recalcitrant for structural investigation by other methods.

(iii) Both chaperones and chaperone–client complexes often demonstrate highly dynamic properties in solutions; however, solution NMR can report detailed information on molecular dynamics at the atomic level, which is usually inaccessible by other means.
(iv) Most of the interactions between chaperones and clients are rather transient or weak. Solution NMR is a sensitive method for detecting this weak interaction even on a high-affinity-mixed background, in which the weaker interaction might be masked using other techniques.
(v) In the transient interactions between chaperones and various clients, the sparsely-populated excited states (NMR "invisible"), which are often closely related to functions, can be "visualized" by dynamic analysis toolkits of solution NMR techniques, as will be discussed below.

However, many chaperone studies involve proteins or protein–protein complexes with a molecular mass around or larger than 100 kDa, which are defined as large systems by the NMR standards. Indeed, compared to many other biophysical techniques such as fluorescence, NMR is by nature a rather insensitive method, which is largely due to the low Boltzmann distribution of nuclear spin states at thermal equilibrium. As the protein size increases, the inevitable reduction in molecular tumbling rate increases the transverse relaxation rate, mainly through ^{1}H–^{1}H dipolar interactions and chemical shift anisotropy (CSA), causing signal broadening and compromising sensitivity. This calls for new methodology developments to extend the molecular size barrier imposed on solution NMR spectroscopy, and numerous efforts from three parallel aspects have been made including (a) continuous improvement on the NMR spectrometer; (b) optimization of the protein sample isotope labelling strategy and (c) development of novel sophisticated pulse sequences.

Given the fact that the NMR signal-to-noise ratio increases nonlinearly (that is, proportional to $B^{3/2}$) with magnetic field B_0, achieving higher magnetic fields has long been harnessed to enhance NMR sensitivity. The introduction of superconducting magnets allowed the emergence of very high fields, nowadays reaching fields up to 1.2 GHz (28.2 T)[12] (see Figure 4.1). Generally speaking, with the power of modern NMR spectrometers, aided by a large number of multidimensional multinuclear NMR experiments designed for [^{13}C, ^{15}N]-labelled proteins, NMR spectroscopy has currently evolved into a routine tool to characterize proteins of less than 25 kDa. Replacement of a large number of protons within protein samples with deuterium, which is achieved by overexpressing target proteins in deuterated media, has been proved to be a powerful strategy to suppress the detrimental ^{1}H dipole–dipole interactions, therefore leading to a significant gain in spectral sensitivity. This, in turn, pushes the protein size limit studied by solution protein NMR to $\sim$50 kDa.

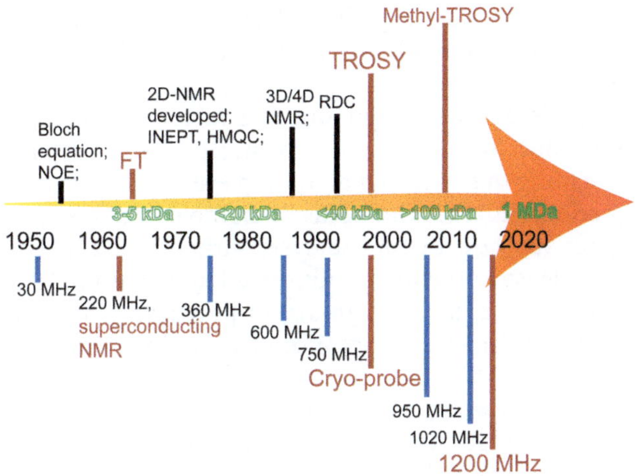

Figure 4.1 Timeline summarizing the solution NMR development.

To further extend the application of solution NMR to larger proteins, revolutionarily designed NMR experimental pulse sequences, known as transverse relaxation optimized spectroscopy (TROSY) sequences, were introduced by Wüthrich's group.[13] The TROSY approach exploits cross-correlation between dipolar and chemical shift anisotropy (CSA) relaxation mechanisms and strategically selects the most slowly relaxing component, therefore providing sharp and intense NMR signals for large biomolecules.[13] Since its introduction, the TROSY technique has soon been serving as a building block for many complex NMR experiments, including the triple-resonance experiments for assigning the protein backbone and side-chain nuclei, as well as for NOESY, and experiments for dynamic studies. The power of the TROSY-type experiment in large protein NMR studies has been elegantly demonstrated in its application to the backbone assignment studies of an octameric 110 kDa protein, which achieved a 20- to 50-fold higher sensitivity compared to the data obtained using conventional methods.[14]

Further efforts on optimization of the polarization transfer step led to the introduction of another notable NMR technique called CRINEPT, or cross relaxation-enhanced polarization transfer, which is based on combining cross-correlated relaxation-induced polarization transfer and polarization transfer *via* scalar couplings.[15] A remarkable example of CRINEPT application was represented in the study of a 900 kDa GroEL–GroES complex, in which the free and GroEL-bound forms of GroES were characterized by solution NMR.[16]

The pulse sequences of TROSY and CRINEPT were both originally developed for the amide group moieties on the protein backbone and soon the TROSY experiments for the side chain aromatic groups were also shown.[17] In practice, however, when the protein system under investigation is beyond the size of 100 kDa, the weak signals of N–H groups or aromatic side chains are normally no longer suitable for a detailed study. As a result, new NMR reporter nuclei are required.

Signal overlap is another major limiting factor for large protein NMR studies. The number of nuclei observed increases linearly along with the protein size, leading to severe signal overlap even with data recorded using a high magnetic field. This dramatically hampers resonance assignment and complicates subsequent data analysis. Although the spectral crowding problem can, at least in theory, be addressed by recording NMR data with high dimensionality *i.e.* larger than 3D, in practice, this is hardly achievable *via* conventional methods because of infeasibly long data collection time.

In Sections 4.3 and 4.4, we will discuss the current strategies to circumvent these difficulties in large protein studies, including sample labelling strategies, and some newly developed NMR pulses.

4.3 Isotope Labelling Strategies for Large Chaperone–Client Systems in Solution NMR

Sample preparation schemes are central for studying proteins by NMR spectroscopy and isotopic enrichment is one of the key milestones in developing modern NMR methods. Various isotopic labelling methods have been used for studying a number of proteins and protein–protein complexes. In this section, we will highlight various isotope labelling strategies routinely applied in protein NMR studies.

4.3.1 Uniform Isotopic Labelling

Uniform isotopic labelling is the conventional labelling strategy for studying biological molecules by NMR spectroscopy. In this, all nitrogen or carbon nuclei are replaced with ^{15}N or ^{13}C isotopes. The most cost-effective way to obtain uniformly isotopic proteins is to overexpress recombinant proteins using a T7-promoter-based *Escherichia coli* strain in a M9 minimal medium supplemented with ^{15}N ammonium salts and/or ^{13}C glucose as the sole nitrogen and carbon sources, respectively.[18,19] These isotopic nitrogen and carbon atoms are incorporated into the amino acids of target proteins with the help of the bacterial metabolic pathways. However, the prokaryotic *E. coli* expression system is not always feasible due to the limitations of post-translational modifications, inefficient disulfide bond formation and sometimes poor protein yield and/or folding.[20] For these reasons, other heterologous overexpression systems, such as yeast, insect, mammalian and cell-free systems, have been successfully applied for the production of isotope-labelled proteins, especially for the GPCR systems.[21–25] However, as these eukaryotic expression systems have not been widely used for chaperone–client system studies, we refer the interested readers to "Advanced isotopic labelling for the NMR investigation of challenging proteins and nucleic acids"[26] and references therein.

When the molecular weights of proteins increased up to 30–50 kDa, the slower overall tumbling of larger proteins in solution dramatically increases the relaxation rate because of the strong ^{1}H–^{1}H dipolar and ^{1}H-^{13}C or ^{1}H–^{15}N relaxation effects, resulting in a lower signal-to-noise ratio and sharply reduced spectra sensitivity. Using D_2O (^2H_2O) instead of ^1H_2O in the culture medium could substitute most non-exchangeable protons in proteins with deuterium, which has a significantly smaller gyromagnetic ratio than hydrogen, leading to suppression of spin diffusion and elimination of the main source of heteronuclear relaxation pathways.[13,27] Combining triple labelled proteins (U-^{2}H–^{13}C–^{15}N) and TROSY-type experiments, the size limits of the protein studied could be pushed to ~50 kDa and sometimes even up to ~100 kDa.

In some cases, partial deuteration (*e.g.* 50%) is more convenient than complete deuteration.[28,29] In fact, complete deuteration causes the loss of many valuable ^{1}H–^{1}H NOE signals, which are important for structural determination. Moreover, a number of pulse sequences based on proton resonance detection, which offers sensitivity enhancement, are unable to be applied to completely deuterated protein samples. As a result, uniformly deuterated protein samples are routinely prepared for early attempts, resonance assignment purposes or gathering information of amide protons; however, partial deuteration offers protein samples with more structural information at the price of sensitivity and spectral resolution.

4.3.2 Selective Isotopic Labelling

Selective isotopic labelling strategies could incorporate individual or groups of specific amino acid types into the target protein, dramatically alleviating the issue of signal overlapping and simplifying spectra for analysis. Compared with the above-mentioned uniform labelling scheme, the labelled sites could be tailored along the protein sequence. Currently, methyl groups are exploited as the primary reporters for the NMR study of proteins above 100 kDa including chaperone–client systems, and there are good reasons for it. First, methyl groups are ideal candidates for NMR studies of high molecular weight proteins, as they yield highly intense and narrow NMR signals by virtue of at least three factors: (a) the multiplicity of protons per methyl group, (b) the rapid rotation around the threefold methyl symmetry axis and (c) favorable relaxation *via* an extensive network of cross-correlated interactions that can be spectroscopically exploited. Indeed, due to the TROSY effect that involves the cancellation of intra-methyl dipolar relaxation interactions, for large proteins the resulting sensitivity gains from HMQC-type experiments can be as much as a factor of 3-fold more sensitive than their HSQC counterparts.[30] Moreover, the six occurring methyl-containing amino acids, that is, Ile, Val, Leu, Met,

Ala and Thr, representing up to 30–40% of the amino acid composition in proteins,[31] are enriched in the hydrophobic cores of proteins and accumulated at protein–protein interfaces.[32] Taken together, these auspicious properties entitle methyl groups to be effective and faithful NMR probes to investigate the protein structure and dynamics. Indeed, the methyl labelling approach has been successfully applied to many supramolecular systems and enabled the collection of NMR data with great quality, removing assignment ambiguities and leading to a blossoming era of NMR application in large protein complexes such as the 670 kDa 20S proteasome,[33] the 580 kDa ClpB disaggregase[34] and the 390 kDa RNA polymerase.[35]

Here, we provide a brief introduction about the commonly used selective isotopic labelling strategies which have been or have the potential to be applied to chaperone–client systems.

4.3.2.1 Methyl Labelling

After a long period of trial and error, the quest for an efficient and low-cost approach for isotopic labelling of methyl groups was finally settled, making use of α-keto acids as precursors for the production of a protein featuring ^{13}C-labelled methyl groups of Ile (δ1), Leu and Val residues with an otherwise highly deuterated background[36–38] (see Figure 4.2A). The essence of this approach relies on the biosynthetic pathway of the target methyl group, with the simplest cases involving amino acids that are the products of irreversible metabolic pathways. By following the same strategy, substantial advances have been achieved in the last two decades and the methyl labelling scheme has evolved into a rather exquisitely tuned strategy. For example, to further reduce the local proton concentration and thereby methyl relaxation rates, strategies have been introduced to label only one of the prochiral methyl groups in Leu/Val, in either a nonselective[39] or a stereospecific manner.[40–42] In addition, protocols for the labelling of leucine but not valine, and *vice versa*, have also been demonstrated.[43,44] Robust selective isotope labelling strategies for other methyl group-containing amino acids have also been developed, including Met,[45] Ala,[46] Thr,[47,48] and the γ2-methyl of Ile.[49] The methyl labelling strategies have been thoroughly reviewed elsewhere.[18,50] Notably, currently also commercially available precursors for isotopic labelling of methyl groups in the pattern of $^{13}CHD_2$ or $^{13}CH_2D$ have been shown to be very helpful for some certain NMR dynamics experiments, as we shall see next.

Meanwhile, some other labelling methods have also been introduced, aiming at increasing the signal sensitivity while reducing the spectral complexity for large protein NMR studies. One such approach is called SAIL, or stereo-array isotope labelling, featuring simultaneously isotopic labelling of methylene and methyl groups, as well as the six-membered aromatic rings in a stereo-selective manner.[51] Although the SAIL labelling strategy, compared to the methyl labelling method, can essentially provide more NMR

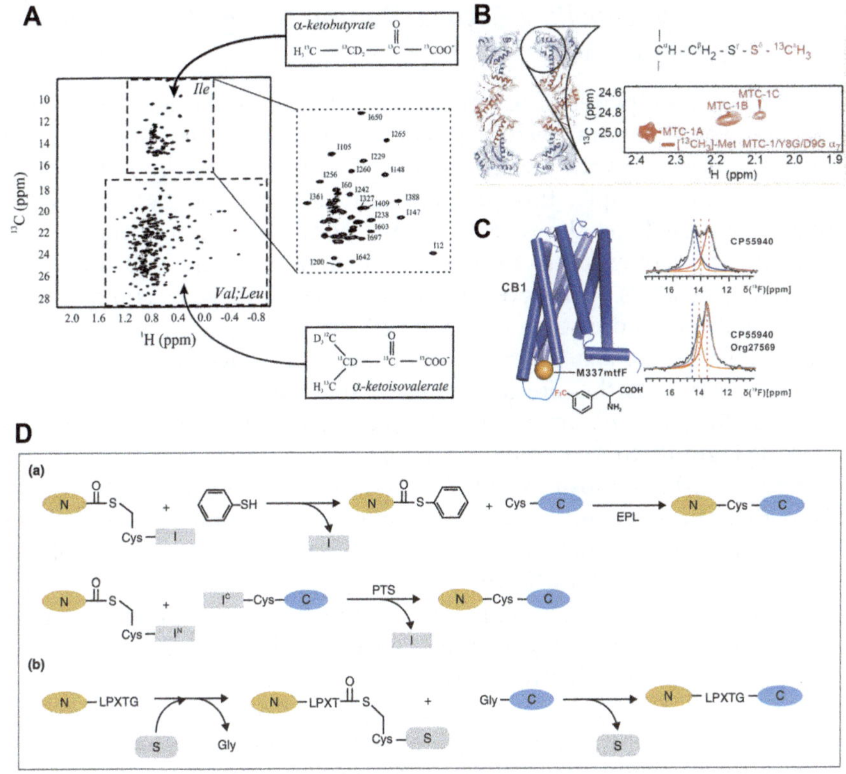

Figure 4.2 Labelling strategies for solution NMR. (A) A representative ^{1}H–^{13}C CT-HMQC spectrum of the Ile, Leu and Val methyls of protein (malate synthase G, MSG) labelled by α-keto acids.[39] Reproduced from ref. 39 with permission from American Chemical Society, Copyright 2003. The biosynthetic precursors and assignments for Ile residues are indicated. (B) ^{13}C-MMTS used as an NMR probe for studying the structure and dynamics of 20S proteasome.[53] Reproduced from ref. 53 with permission from American Chemical Society, Copyright 2011. Left: Cross-section side view of the $\alpha_7\beta_7\beta_7\alpha_7$ four ring structure of the full 20S proteasome (α_7-bule, β_7-red). Right: Chemical structure of MTC (up) and ^{1}H–^{13}C HMQC spectrum of U-^{1}H, Met-[$^{13}CH_3$] MTC-1/Y8G/D9G α_7. (C) ^{19}F NMR probe revealing the allosteric modulation mechanism of GPCR.[62] Reproduced from ref. 62 with permission from American Chemical Society, Copyright 2021. Left: Chemical structure of mtfF and ^{19}F labelling site in CB1. Right: 1D ^{19}F NMR spectra of CB1-NB with and without a ligand. (D) Schematic diagrams of (a) intein-based and (b) sortase-mediated protein ligation.[18] Reproduced from ref. 18 with permission from Elsevier, Copyright 2016.

probes and therefore additional information, its utility in practice has been largely limited due to the formidable cost associated with the cell-free expression protocol.

4.3.2.2 Methyl Labelling Beyond Natural Methyl Group Bearing Amino Acids

Methyl groups are powerful reporters of structure, motion and function in NMR studies of supramolecular protein assemblies. Their utility can be hindered, however, by the low abundance of methyl groups in protein areas of interest. In principle, this problem can be addressed by site-directed mutagenesis, such as the so-called "methionine scanning" method, in which the solvent-exposed residues were systematically substituted by methionine residues.[52] However, in certain cases mutagenesis may not be desired as it may cause some unwanted effects.

The thiol group of cysteine residue is unique due to its reactivity and thus frequently serves as the site for protein modification *via* various types of reactions. Religa *et al.* reported a simple method to covalently attach a ^{13}C-methyl group to the thiol group of a cysteine side-chain *via* using ^{13}C-methyl-methanethiosulfonate (^{13}C-MMTS)[53] (see Figure 4.2B). It was shown that the resultant side chain, which is structurally similar to methionine, gave rise to intense $^{13}CH_3$ NMR signals due to its favorable dynamic properties, making it therefore suitable for studies of very high molecular weight protein systems, such as the chaperone–client system.

The side chains of lysine residues have also been modified with the methyl group *via* a reductive methylation reaction. In dimethyl lysines, the intense $^{13}CH_3$ signal is generated by six protons from the methyl groups instead of two ε-NH_2 of lysine, which could be detected even at low protein concentrations.[54] Thus, the methyl-labelled lysine could be utilized as a highly sensitive probe to explore highly dynamic biological systems.

4.3.2.3 Aromatic Labelling

Although the aromatic side chain residues, including Phe, Trp, Tyr and His, are less abundant in proteins, their bulky side chains offer the largest number of distance constraints in typical globular proteins and often directly participate in the stabilization of the hydrophobic core, protein folding/unfolding and enzyme activity regulation. However, due to their heavily overlapped ^{1}H signals and the narrow dispersion of the carbon signals, as well as line broadening caused by large chemical shift anisotropy (CSA), the low spectral quality hampers their application in protein structural and dynamic analysis.[55] Recently, Wagner's lab used 2-^{13}C glycerol, 2-^{13}C pyruvate or 3-^{13}C pyruvate as the carbon source to successfully label their aromatic ring with alternate ^{13}C–^{12}C labelling patterns and obtained high-resolution aromatic ^{1}H–^{13}C TROSY HSQC spectra.[56] More importantly, the alternate aromatic ^{13}C labelling scheme is compatible with methyl labelling. Such labelling strategy has been successfully applied to a few large protein systems including the studies of chaperone SecB.[57]

4.3.2.4 Fluorine Labelling

A few lines of favorable properties have made ^{19}F an attractive probe in protein NMR studies:[58] (1) ^{19}F is a spin-1/2 nucleus with 100% natural abundance; (2) ^{19}F exhibits the second-highest gyromagnetic ratio (83% that of the proton), translating into high NMR detection sensitivity and strong dipolar couplings; (3) the range of ^{19}F chemical shift is over 400 ppm providing a sensitive means of studying subtle conformational changes in varied chemical environments and (4) ^{19}F is virtually absent in proteins that contribute few background signal overlaps. Incorporation of fluorine can be achieved *via* one of two ways: (a) cellular biogenesis, wherein the cell culture is supplemented with a fluorinated amino acid analog, or (b) chemical modification, where a fluorinated moiety is attached to a certain type of amino acid.[59,60] For the first approach, currently the majority of commercially available fluorinated amino acids are aromatics, such as the analogs of Tyr, Phe and Trp.[60] In the latter method, protein fluorination modification is achieved *via* a nucleophilic substitution mechanism, and the fluorine tag can be covalently linked to the labile side-chain of Cys and Lys, as well as Ser and Thr despite a lower labelling efficiency.[59] The attached trifluoromethyl group, which has a geometry identical to that of methyl groups, is an attractive probe in large protein NMR studies. Indeed, the utility of ^{19}F NMR has been represented in the studies of a 100 kDa protein system and GPCR[61–64] (see Figure 4.2C).

4.3.3 Segmental Isotopic Labelling and LEGO

In the NMR spectra of larger proteins, one of the major causes of spectral complexity is the more severe overlapped signals with increasing molecular weight. To reduce the spectral complexity, one important and useful tool is segmental isotope labelling, making arbitrary isotopic composition available. In this approach, the "NMR invisible" and "NMR visible" parts of a protein are produced separately and then ligate *via* an enzymatic or a chemical reaction. Three methods have been developed for the segmental isotope labelling, EPL (expressed protein ligation),[65,66] PTS (protein *trans*-splicing)[67,68] and sortase-mediated ligation[69] (see Figure 4.2D). EPL and PTS are based on the intein technology, while sortase-mediated ligation is based on the Sortase A enzyme. The intein technology relies on its splicing domains which excise themselves from the precursor protein without the requirement of any external cofactor or energy.[68,70–73] Sortase A is a transpeptidase that specifically recognizes an N-terminal LPXTG motif and then catalyzes the formation of a new peptide bond with the C-terminal part.[69] Another promising approach for alleviating the spectral complexity is the so-called LEGO,[74] in which several subunits could isotopically label with different schemes *via* genetically controlling distinct promotors in two plasmids. Indeed, the segmental labelling and LEGO approaches have been applied to large protein NMR studies, and we anticipate to see more applications to the chaperone–client systems in the near future.

4.3.4 Spin Labelling

Spin labelling is an efficient tool for studying the structure and dynamics of proteins. Traditionally, a spin probe with one or more unpaired electrons is attached to cysteine or lysine by *in vitro* chemical reactions.[75,76] The common spin probes including the methanethiosulfonate spin label (MTSL), 2,2,6,6-tetramethylpiperidine 1-oxyl (TEMPO), 1,4,7,10-tetraazacyclododecane-1,4,7-tris-acetic acid-10-maleimidoethylacetamide-Gd(III) (Gd(III)-DOTA) and 5- and 7-doxyl-stearic acids (5-DSA, 7-DSA) are commercially available. The incorporation of a spin label enables the measurements of long-range distance restraints and the detection of local transient conformation using PRE. Indeed, spin label has been applied to large protein investigations. More details can be seen in Section 4.4.2.

4.4 Advancement of NMR Techniques

NMR is a versatile technique. In addition to its well-known robust capability for atomic-resolution structure determination, NMR spectroscopy can report detailed information on molecular dynamics at the atomic level, which is usually inaccessible by other means. In the last two decades, a large number of NMR methods have been developed, which, in turn, provide an unprecedentedly deeper understanding of many fundamental cellular activities, including chaperone–client systems. Here we aim to briefly document some recent notable advances achieved for high-molecular-weight protein systems. We note here that this section is by no means a comprehensive review of NMR methodology, and some well-established methods, such as the residual dipolar coupling (RDC), will be intentionally omitted here, as they have not gained much popular application so far in chaperone–client studies. Moreover, as we will see next, choosing a method(s) for methyl side-chain dynamic measurements requires careful consideration of (a) the dynamic information required, (b) experimental sensitivity and (c) robustness in the evaluation of dynamics parameters and (d) as always, the experimental cost associated with the appropriate labelling scheme.

4.4.1 TROSY and CRINEPT

A NMR signal decays as a function of the transverse relaxation time (T_2) and the width of the resonance lines is inversely proportional to T_2. The larger the molecular mass, the shorter the T_2 and the broader the line becomes[77] (see Figure 4.3A). The accelerated transverse relaxation rate increasing with molecule size is the major obstacle for studies of large proteins such as chaperone–client systems. At a high magnetic field, T_2 is dominated by dipole–dipole coupling (DD), which is independent of the static magnetic field, and chemical shift anisotropy (CSA), which increases with a higher magnetic field.[78] Based on this principle, in 1997, Wüthrich's group introduced a new type of experiment, named transverse relaxation optimized spectroscopy (TROSY),[13] in which they used the constructive cancellation of transverse relaxation caused by ^{15}N chemical shift anisotropy and the NH dipole at a high magnetic field. Using TROSY-type experiments, the faster relaxing resonances are eliminated and the most slowly relaxing peak, which is the narrowest component in large proteins, is selectively observed. To date, the TROSY technique has been serving as building blocks for many complex experiments, and methyl-TROSY has become one of the most important TROSY-variants, which is routinely employed in studying chaperone–client systems.

Cross relaxation-enhanced polarization transfer (CRINEPT) is another notable NMR technique, which is based on combining cross-correlated relaxation-induced polarization transfer and polarization transfer *via* scalar coupling.[79] Because the rate of polarization transfer by cross-correlated relaxation is proportional to the rotational time, CRINEPT is suitable for studying very high-molecular-weight proteins. So far, several examples of CRINEPT application have been represented in the study of chaperones, including the 900 kDa GroEL–GroES complex[16] and the 200 kDa Hsp90–p53 system.[80]

4.4.2 PRE and PCS

Paramagnetic relaxation enhancement (PRE) is a well-established method providing long-range distance restraints in the range of 15–35 Å,[75] in good complementation to the classic NOE restraints, which describe the short contacts (<6 Å) in NMR structure determination. Thus, PRE calculations are always used to assist structural calculations, especially in large macromolecular systems in which only a limited number of unambiguous NOE crosspeaks could be resolved. In theory, PRE arises from magnetic dipolar interactions between the unpaired electrons of a paramagnetic center and the NMR active nuclei and results in an increase in transverse relaxation rates.[75] In general, PRE measurements require two sets of NMR signals recorded in the paramagnetic and diamagnetic states. The increased transverse relaxation rate manifests itself as line broadening of NMR signals, and the paramagnetic effect decreases with increasing distances from the paramagnetic center (see Figure 4.3B). In addition to its application on structure refinement, PRE has also been successfully applied to elucidate the transient interaction due to its sensitivity to the low population states in conformational ensembles.[81]

Pseudocontact shift (PCS) is another NMR observable that also provides long-range structural information. Different from the PRE effect that is proportional to r^{-6}, PCS displays an r^{-3} distance dependence resulting in a relatively longer detection range.[75,82] PCS is normally measured as the NMR

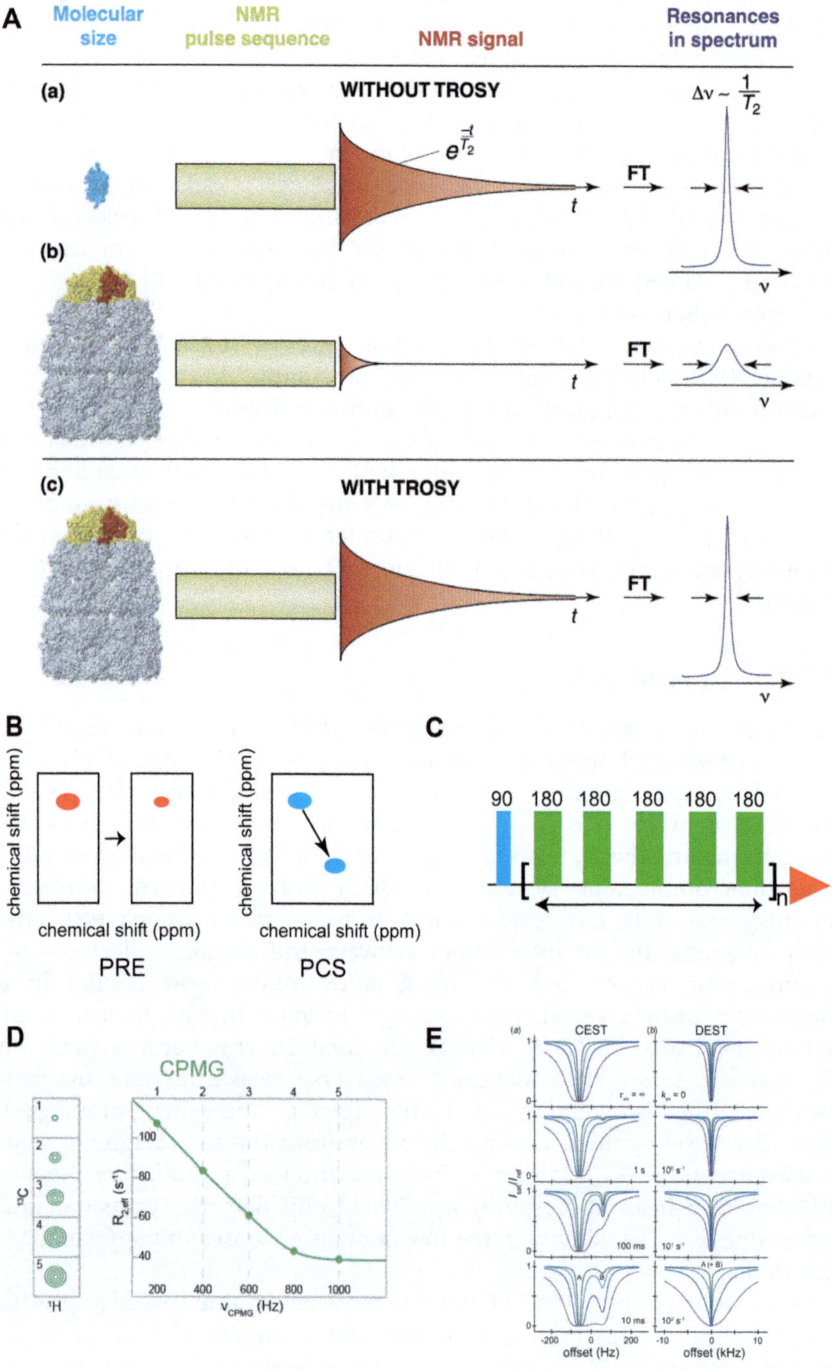
A
Molecular size
NMR pulse sequence
NMR signal
Resonances in spectrum
(a)
WITHOUT TROSY
$\Delta\nu \sim \frac{1}{T_2}$
$e^{\frac{-t}{T_2}}$
t
FT
ν
(b)
t
FT
ν
(c)
WITH TROSY
t
FT
ν
B
chemical shift (ppm)
chemical shift (ppm)
PRE
chemical shift (ppm)
chemical shift (ppm)
PCS
C
90 180 180 180 180 180
n
D
CPMG
1 2 3 4 5
^{13}C
^{1}H
$R_{2,eff}$ (s^{-1})
100
80
60
40
200 400 600 800 1000
ν_{CPMG} (Hz)
E
(a) CEST
(b) DEST
1 s
100 ms
10 ms
offset (Hz)
offset (kHz)

signal difference in chemical shift between a paramagnetic sample and a diamagnetic analog (see Figure 4.3B). Furthermore, as the PCS effect is correlated with the anisotropic magnetic susceptibility tensor $\Delta\chi$, PCS measurement can convert to the orientation information.[75,83] Taken together, PRE and PCS provide different types of restraints complementing the NOE information and the combination of these methods allows a more accurate description of the protein structure and dynamics.

4.4.3 CPMG

Relaxation dispersion NMR spectroscopy has been demonstrated to be an extremely effective tool to investigate the influence of slow (μs–ms) stochastic processes on the nuclear spin relaxation of biomolecules. Carr–Purcell–Meiboom–Gill (CPMG) relaxation dispersion experiments involve the application of a train of 180° pulses during a relaxation period at varying frequencies of pulse train to modulate the chemical/conformational exchange contribution to the observed transverse relaxation rate (see Figure 4.3C and D). For a molecular conformational exchange process, which is provided with exchange rates within the range between $\sim$100 s^{-1} and $\sim$6000 s^{-1} (ref. 84) and an at least 0.5% population of the excited state, CPMG relaxation dispersion experiments can yield sets of valuable information, including thermodynamic (the populations of (the exchanging states) and the kinetic exchange rate) measurements, as well as structural information based on chemical shifts,[85] bond vector orientations,[86] dihedral angles[87] and motion parameters[88] of the excited states. This allows structurally "visualizing" the conformations of the often transiently formed and "invisible" excited states.[89] For a detailed theoretical description of the CPMG relaxation dispersion method, we refer the readers to other excellent related reviews.[85]

As with many other NMR methods for dynamic measurements, the CPMG relaxation dispersion approach was originally developed for the measurement of amide moieties on protein backbones. Adapting this powerful method to methyl groups was first attempted *via* a ^{13}C-single quantum (SQ)-type pulse sequence.[90] Soon after, a multiple quantum (MQ)-based CPMG relaxation dispersion experiment was developed for measuring ^{13}C relaxation at methyl side-chain positions in high molecular weight proteins.[91]

Figure 4.3 Brief overview of some NMR methodology. (A) Schematic diagrams showing NMR spectra of small and large molecules in solution.[77] Reproduced from ref. 77 with permission from Elsevier, Copyright 2003. (B) Comparison of PRE and PCS experiments. (C) A representative CPMG pulse sequence. (D) Schematic representation of the CPMG experiment used to probe protein dynamics.[221] Reproduced from ref. 221 with permission from American Chemical Society, Copyright 2019. (E) Saturation profiles of CEST and DEST experiments.[96] Reproduced from ref. 96 with permission from Cambridge University Press, Copyright 2015.

This experiment benefits from a methyl-TROSY effect in which cancellation of intra-methyl dipole fields occurs, leading to methyl ^{13}C–^{1}H correlation spectra with much higher sensitivity (over fivefold improvement over its SQ-type counterpart), and therefore is ideally suited for applications to high-molecular-weight proteins. The powerful utility of this MQ methyl-TROSY relaxation dispersion approach has been demonstrated by its application to quite a few supramolecular protein systems, such as the 305 kDa ClpP protease[92] and a 350 kDa half-proteasome construct,[93] and provided fruitful insights into functional mechanisms.

4.4.4 Saturation Transfer (CEST and DEST)

Recently two closely related methods have been developed, CEST [94] and DEST,[95] for sensitive detection of the excited (minor) state that interconverts slowly with the observable (ground) state (see Figure 4.3E). These minor states are NMR invisible either because of their sparse population or due to the extreme line broadening effect caused by slow tumbling, the so-called "dark state".[96] Both CEST and DEST are, in principle, saturation-transfer-based experiments that were introduced in the early 1960s,[97] involving application of a weak off-field radiofrequency radiation scanning the spectrum while monitoring the effect on the observable state resonances. When the invisible state is irradiated by the radiofrequency pulse, the resultant magnetization will be transferred to its exchanging state *via* chemical exchange and therefore attenuate the NMR signals. Subsequently, the perturbation profile of visible peak intensities, which is a function of frequency offset and radiofrequency field strength, can be employed to quantitatively extract thermodynamic (population) and kinetic information (exchange rate) on the exchange process, as well as the chemical shift values for the invisible state (CEST experiment) or the transverse relaxation rates (DEST experiment). Despite many things in common, CEST and DEST experiments differ in prerequisites: the CEST experiment requires distinct chemical shifts of interconverting states, whereas DEST demands that dramatic difference in rotational correlation time;[98] these define their distinct application areas.

4.4.4.1 CEST

The approach of CEST has been recently established as a powerful utility in probing slow dynamics in biomolecules with the exchange rate at the slow end of the CPMG experiment (ranging from 50 s^{-1} to 400 s^{-1}), and with the population of excited state on the order of 1% or higher, thereby complementing well-established $R_{1\rho}$ and CPMG approaches.[99] The CEST experiment allows accurate determination of a set of thermodynamic and kinetic parameters, including the state population, exchange rate, and chemical shifts of excited states, with significant more accuracy than that obtained *via* the CPMG relaxation dispersion approach.[94]

The CEST methodology has been developed for dynamic studies of large proteins, making use of the methyl group probes with the labelling scheme of either $^{13}CH_3$[100,101] or $^{13}CHD_2$.[102] In particular, it has been shown that the labelling scheme of $^{13}CHD_2$ yielded much higher (up to 5–6 fold) sensitivity, thus much more advantageous over its $^{13}CH_3$ counterpart in application to proteins with high molecular weights.[102]

4.4.4.2 DEST

Unlike CEST wherein little requirement is imposed on the excited state relaxation properties, the application of the DEST experiment is restricted due to its requirement for a distinctly slower tumbling time (larger molecular weight) of the invisible state. This prerequisite, however, makes it a robust method in characterizing large biomolecular assemblies, especially in protein aggregation studies, as long as the exchange falls in the range of 500 μs^{-1} to 1 s^{-1}.[98] Indeed, as both backbone (^{15}N-) and methyl side-chain (^{13}C-) measurement-based methodologies have been developed, the DEST experiment has been employed to reveal a detailed aggregation mechanism of amyloid-β (Aβ) peptides into protofibrils with a molecular weight on the order of MDa,[95,103,104] as well as their interaction mechanism with a 780 kDa GroEL chaperonin.[105] One drawback of this method is that, unlike the CEST or CPMG relaxation dispersion experiment, which can provide direct information on the chemical shifts of the invisible state, it is often difficult for DEST to do so because of the broadness of the saturation transfer profiles.[106]

4.5 NMR Applications in Chaperone–Client Systems

With the help of ingenious isotope labelling schemes and sophisticated NMR techniques as well as advanced NMR instruments, structural and dynamic investigations of the challenging chaperone–client systems have been made possible. In this section, we will introduce some elegant NMR work on chaperone–client systems.

4.5.1 The Dynamic Skp–OMP System

In Gram-negative bacteria, there is a specialized chaperone network that enables protein quality control of outer membrane protein (OMP) biogenesis and transport.[107] In this machinery, two classical periplasmic chaperones, Skp (seventeen kilodalton protein) and SurA (survival factor A), work on two parallel pathways to regulate the transport of OMPs across the aqueous periplasm and eventually assist OMP insertion into the outer membrane by the β-barrel assembly machinery (BAM) complex.[107] Although the crystal structures of these chaperones have been resolved,[108–110] questions such as how they recognize their clients and how they protect clients in the unfolded or naturally folded form remain unanswered.

Since 2013, Hiller's lab has employed solution NMR as the principle technique to investigate the atomic structure and dynamics of the Skp–OMP system and made a complete description of the Skp–OmpX and Skp–tOmpA complexes.[111] Analysis of the ^{1}H–^{15}N TROSY HSQC spectra revealed that both the apo and the client-bound holo Skp featured a single, complete set of NMR resonance, suggesting each of the three Skp protomers samples an ensemble of fast-exchange averaging unstructured conformations with a short individual lifetime (<1 ms). Secondary-structure elemental analysis based on the backbone $^{13}C\alpha$ and $^{13}C\beta$ chemical shifts and spin relaxation revealed that the overall conformation of Skp is largely unperturbed upon binding to clients, except for a hinge formed by the two turns of helix α2.A and helix α3.A (see Figure 4.4A). The hinge shows a large amplitude of motions on a ps–ns timescale that facilitates a dynamic adaptation of Skp to its clients. Backbone structural and dynamic measurements, together with the temperature-dependent line-broadening analysis of the OMP polypeptides, demonstrated that the clients undergo a large dynamic rearrangement to be encapsulated in the cavity of Skp and adopt a disordered conformation with an ensemble of rapid, submillisecond interconverted conformers. Further time-dependent intramolecular PRE exchange measurements indicated that the global lifetime of the Skp–OMP complex is ∼2.6 h, in marked contrast to individual short-lived, nonspecific and transient contacts between Skp and substrates. In addition to PRE-derived long-range distance information, NMR spectroscopy allowed the detection of a few site-specific intermolecular short-range methyl-methyl NOEs for the Skp–tOmpA complex[112] (see Figure 4.4B). These data collectively allowed building of the complete structural and dynamic model of the Skp–OMP system.

Nowadays, more than 30 proteins have been identified as clients of Skp.[107] Therefore, one central question is how Skp adapts to recognize the clients with broadly-ranged sizes (from 150 residues to 700 residues). By combining solution NMR, molecular dynamic simulations and small-angle X-ray scattering (SAXS), Holdbrook *et al.* demonstrated that the hinges in the coiled-coil arms of Skp undergo conformational switch, which, in turn, extends the size of the center cavity to satisfy the size of the client through a "spring-loaded" mechanism.[113]

The functionality of the monomeric state of Skp is another controversy involved in the Skp chaperone system. Recently, Hiller's lab examined the monomer–trimer transition of Skp using solution NMR.[114] They demonstrated that the monomer–trimer transition of Skp corresponds to its disorder-to-order and inactive-to-active switching simultaneously. An atomic-resolution characterization of this transition revealed that the folding process occurred at the oligomerization interface *via* stapling of the N- and C-terminals of the alpha-helical coiled-coil arms together.[114]

Taken together, these fruitful investigations of the chaperone Skp and the Skp–OMP systems showed how the dynamic nature of chaperone and dynamic chaperone–client interactions play vital roles in function.

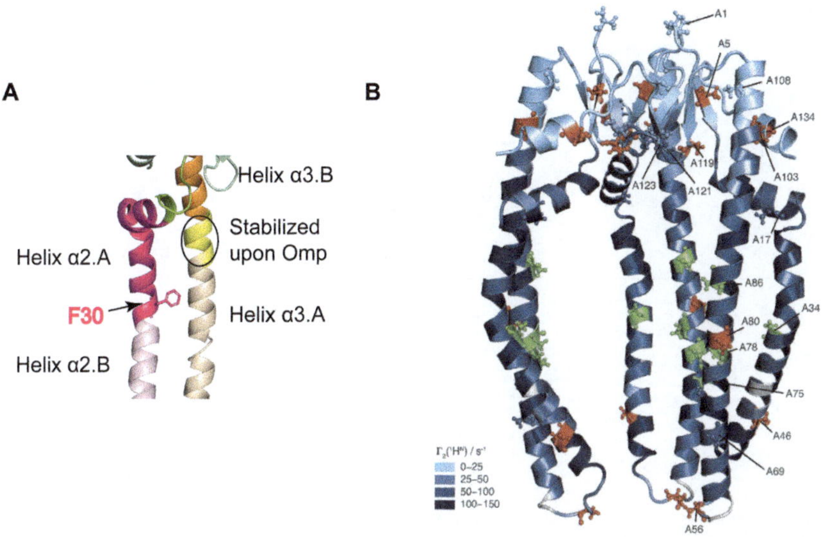

Figure 4.4 NMR characterization of the Skp–OMP system. (A) Overview of the structural elements in helices α2 and α3, based on the crystal structure of Skp (PDB 1SG2[109]). (B) Residues with significant intermolecular NOEs in the Skp–tOmpA complex (representative alanine residues colored by green ball and sticks) and intermolecular PREs in the Skp–OmpX complex (blue gradient for Γ_2-values) are indicated in the Skp crystal structure.[112] Reproduced from ref. 112 with permission from John Wiley & Sons, Copyright 2014.

4.5.2 The Dynamic Hsp70–Client Interaction and the Hsp70 Machinery

Hsp70 is a central molecular chaperone responsible for the maintenance of healthy cellular protein homeostasis.[115–117] It is a highly conserved chaperone family containing an N-terminal nucleotide-binding domain (NBD), a C-terminal client-binding domain (SBD) and a conserved linker in between[118] (see Figure 4.5A). Hsp70s cooperate with hsp40s and nucleotide exchange factors (NEF) in action and prefer to process unfolded clients in an ATP-driven chaperone cycle,[118,119] in which hsp70 undergoes enormous structural rearrangement. Despite that increasing numbers of high-resolution Hsp70 structures have been resolved, the mechanism governing the dynamic hsp70–client interaction remains poorly understood.

By a combination of a series of methyl TROSY-based NMR techniques, Kay's lab has provided seminal insights into the exquisite machinery of Hsp70.[120–124] In these studies, *E. coli* hsp70 (DnaK) and marginally stable clients were utilized to explore the dynamic hsp70–client interactions. The authors demonstrate that a model substrate, the human telomere repeat binding factor (1hTRF1) (see Figure 4.5B), bound to the ATP-, ADP- or nucleotide-free state of DnaK with similar structural properties, indicating that the client conformation was not directly associated with the hsp70

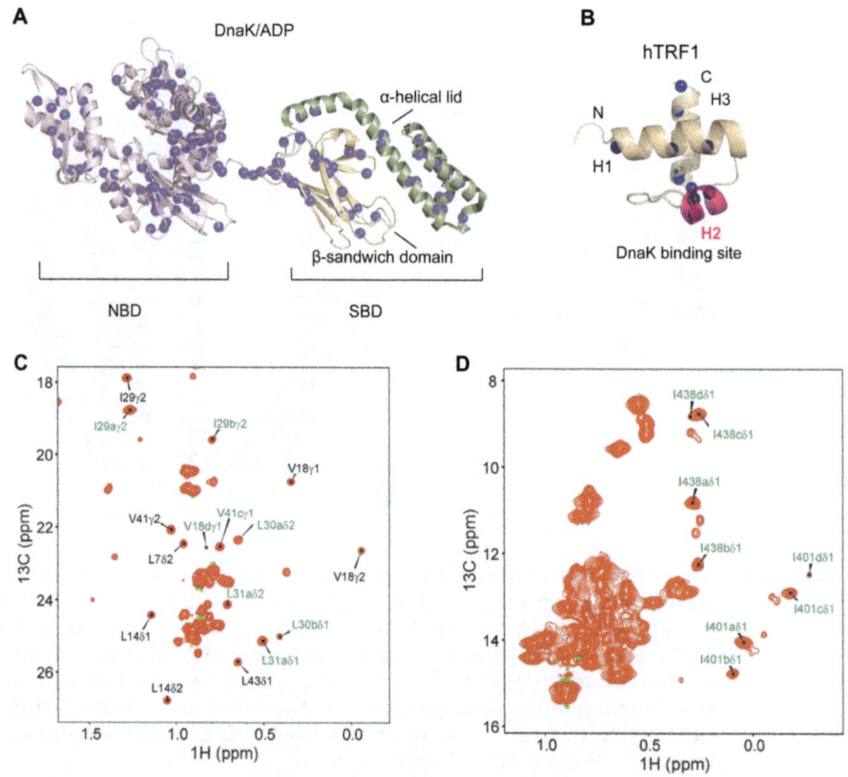

Figure 4.5 NMR characterization of the dynamic DnaK–client interaction. (A) Cartoon representations of DnaK/ADP (PDB: 2KHO[222]), highlighting the NBD, α-helical lid and β-sandwich domain of the SBD. The backbone nitrogen atoms of Ile, Leu, Val and Met residues are highlighted as blue spheres. (B) Cartoon representation of the hTRF1 substrate (PDB: 1BA5[223]). The DnaK binding site is colored magenta. The backbone nitrogen atoms of Ile, Leu, Val and Met residues are highlighted as blue spheres. (C) The Leu/Val regions of the ^{1}H–^{13}C HMQC spectrum of I_{γ}-LV-$^{13}CH_3$ hTRF1 bound to IM-$^{13}CH_3$ DnaK/ADP.[122] (D) The Ile region of the same HMQC spectrum shown in G.[122] Panels (C) and (D) reproduced from ref. 122 with permission from John Wiley & Sons, Copyright 2017.

nucleotide states.[120] Chemical shift-based analysis of the DnaK-bound hTRF1 demonstrated that the hsp70-bound client adopts a globally unfolded state with the local secondary structure in regions far from the binding site.[120] Further PRE and CEST-based experiments indicated that the transient long-range interactions between the unfolded client ensembles were greatly attenuated upon hsp70 binding.[123] These data revealed that, although hsp70 does not affect the overall conformation of the bound client, it disrupts the transient long-range contacts that generally exist in the intermediate conformations along the non-native protein folding pathways.

To gain more insights into the mechanisms of DnaK binding to hTRF1, hTRF1 was labelled as ILVM (Ile, Leu and Val) methyl groups and then titrated with U-^{2}H DnaK/ADP.[121] Surprisingly, upon the addition of DnaK/ADP, at least five peaks for the unique Ile residue (Ile29) and many 'redundant' peaks for Leu and Val methyl groups (assigned as Leu30, Leu31, Val18 and Val41) were observed in the ^{13}C–^{1}H HMQC spectra of hTRF1, whereas free hTRF1 showed only a single Ile methyl peak and normal amount of Leu and Val methyl peaks. Multiple conformations revealed that DnaK interacts with at least four distinct sites on hTRF1 in a slow exchange on the NMR observation timescale.

Next, the authors investigated the binding stoichiometry of the hTRF1/DnaK complex through a series of NMR titration experiments. Studies of *pro*R-LV-$^{13}CH_3$ hTRF1 titrated to IM-$^{13}CH_3$ DnaK/ADP revealed the presence of previously uncharacterized 1 : 2 and 3 : 1 hTRF1–DnaK complexes[122] (see Figure 4.5C and D). These data suggested that hsp70 acts as a promiscuous interactor to generate a highly heterogeneous hsp70-bound client ensemble for efficiently searching the optimal folding pathway.

A long-standing debate about the hsp70 binding mechanism, which conforms to passive 'conformation selection' or active 'induced fit', has been sustained for over a decade. Using magnetization exchange- and chemical exchange-based NMR methods, Kay and co-workers monitored the conformation exchanges of clients bound with DnaK to distinguish these two mechanisms.[124] In this study, they used a marginally stable client named R17*, the L90A mutant of the 17th domain of chicken-brain α-spectrin, to probe its interaction with DnaK. ^{13}C–^{1}H HMQC spectra showed five peaks in the Met methyl region, whereas only two Met residues M26 and M87 exist in R17*.[124] These peaks arising from the native and unfolded states were assigned by the recording temperature increase and mutational analysis. Upon the addition of I-$^{13}CH_3$ DnaK/ATP, a new Met signal appears and it was assigned to the bound state of M26. Because the peaks from the native, unfolded and bound states are well separated, M26 was chosen as a sensitive reporter for tracking the DnaK/R17* interaction process. ZZ-exchange magnetization transfer data showed only two cross-peaks connecting the bound and unfolded states of M26, with no cross-peak connecting the bound and native states of M26. Direct detection of R17* switching between the unfolded state and DnaK-bound state elegantly demonstrated that hsp70 selectively recognizes the unfolded state of its client instead of converting the native state to the unfolded state of the client, which are in support of the 'conformation selection' model.

Hsp70, with the help of hsp40s and NEFs, functions as an ATP-fueled allosteric machine to assist client folding. Using solution NMR, Gierasch's lab has contributed significantly to an understanding of the allosteric landscape of hsp70 triggered by ATP hydrolysis and/or client binding. By virtue of methyl TROSY NMR and the ATP hydrolysis defective mutant, DnaK T199A, which retains the ATP-induced conformational change, Gierasch's lab elucidated three distinct Hsp70 conformational ensembles (see Figure 4.6A) and revealed a delicate balance among these conformational states in detail.[125] The resulting allostery in Hsp70 was depicted as an energetic tug-of-war between intradomain interactions and interdomain interfaces, in which the allosterically active states in Hsp70 are functionally coupled, and provided an explanation of how substrates and co-chaperones modulate Hsp70 (see Figure 4.6B).

In a cell, the key to Hsp70's versatile functions is that Hsp70 requires precise control of not only the allosteric cycle, but also the selectively promiscuous binding to regulate its numerous substrate binding and release.[126,127] Previous peptide array data and mounting structural analysis for DnaK-peptide models have established that DnaK, the most studied protein in the Hsp70 family, favors binding to five to seven residue-long hydrophobic sequences containing some flanking positively charged

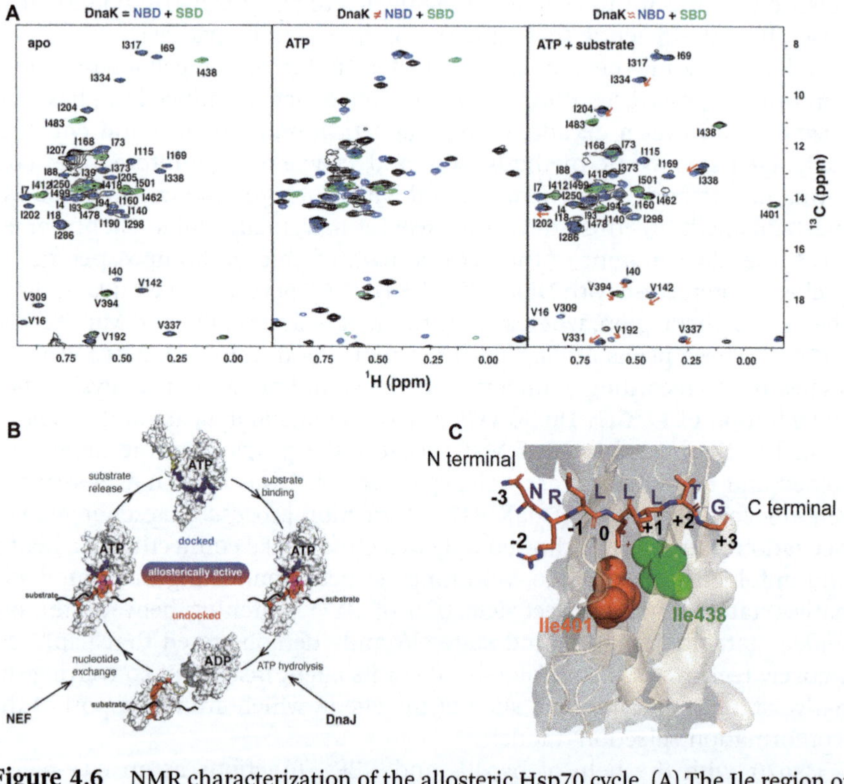

Figure 4.6 NMR characterization of the allosteric Hsp70 cycle. (A) The Ile region of the ^{1}H, ^{13}C-methyl-TROSY spectra of the two-domain DnaK constructs under different nucleotide conditions showing three distinct conformational ensembles.[125] (B) Illustration of the allosteric Hsp70 cycle.[125] Reproduced from ref. 125 with permission from Elsevier, Copyright 2012. (C) The side view of the crystal structure of the DnaK-model peptide complex (PDB: 1DKZ)[224] showing the five pockets in the binding cleft of the SBD.

residues[128–132] and owns a canonical binding cleft in the βSBD containing five pockets (labelled as −2, −1, 0, +1 and +2)[133,134] (see Figure 4.6C). However, little is known about their binding details and the origins of the selectively promiscuous binding. Recently the Gierasch lab investigated extensively the interaction between DnaK SBD and a 471-residue-long unfolded protein substrate, proPhoA. Combined with X-ray crystallography, methyl-TROSY, PRE and chemical cross-linking assays, they confirmed five DnaK binding sites in the full-length proPhoA and found different binding preferences for the pockets within the substrate binding cleft of the SBD.[135] Titration of unlabelled peptides encompassing the single binding site or the full-length proPhoA with a methyl-labelled DnaK SBD revealed the details of different binding modes and, importantly, demonstrated the coexistence of multiple binding modes for a particular client at the same time. These findings demonstrated the plasticity of the Hsp70 binding site and provided new insights into the promiscuous binding modes of chaperone–client systems.

Earlier investigations showed that fibrils of the Parkinson's disease-linked presynaptic α-synuclein can be dissolved by a synergistic action of Hsp70 and its cochaperones, Hsp40 and Hsp110,[136–139] but the underlying mechanism remains poorly understood. The Bukau group utilized the solution NMR technique to dissect the molecular events that lead to α-syn fibril disaggregation.[140] NMR titrations of ^{15}N-labelled α-syn with individual chaperones Hsp70, DNAJB1 (a class of Hsp40) and Hsp110 revealed that different chaperones bind to discrete sequences in α-synuclein, while chaperone combinations produced intensity profiles that essentially mirror the sum of the individual interactions. This information, together with extensive biochemical and biophysical data, allows the authors to propose a comprehensive model of the human HSP70-based amyloid disaggregation machinery.

4.5.3 The Complicated Hsp90–Client System

Heat-shock protein 90 (Hsp90) is a highly conserved protein, which is critical in the maturation, stabilization, activation and regulation of a number of transcription factors, kinases, hormone receptors and cancer-related proteins.[141–143] It is a homodimer with three domains in each protomer: an N-terminal nucleotide-binding domain, a middle domain and a C-terminal dimerization domain[144–146] (see Figure 4.7A). The N domain contains an ATP binding pocket which directly correlates to the weak ATPase activity of hsp90 and undergoes multiple conformational changes in the ATPase cycle.[147,148] A charged unstructured linker, for which the length varies dramatically among species, connects the N and M domains. The M domain has been proposed to recognize various clients and some co-chaperones and cooperate with the N domain to stimulate ATPase activity.[149] The C domain contains not only the primary site of dimerization but also the binding site of several TRP co-chaperones that recognize the MEEVD

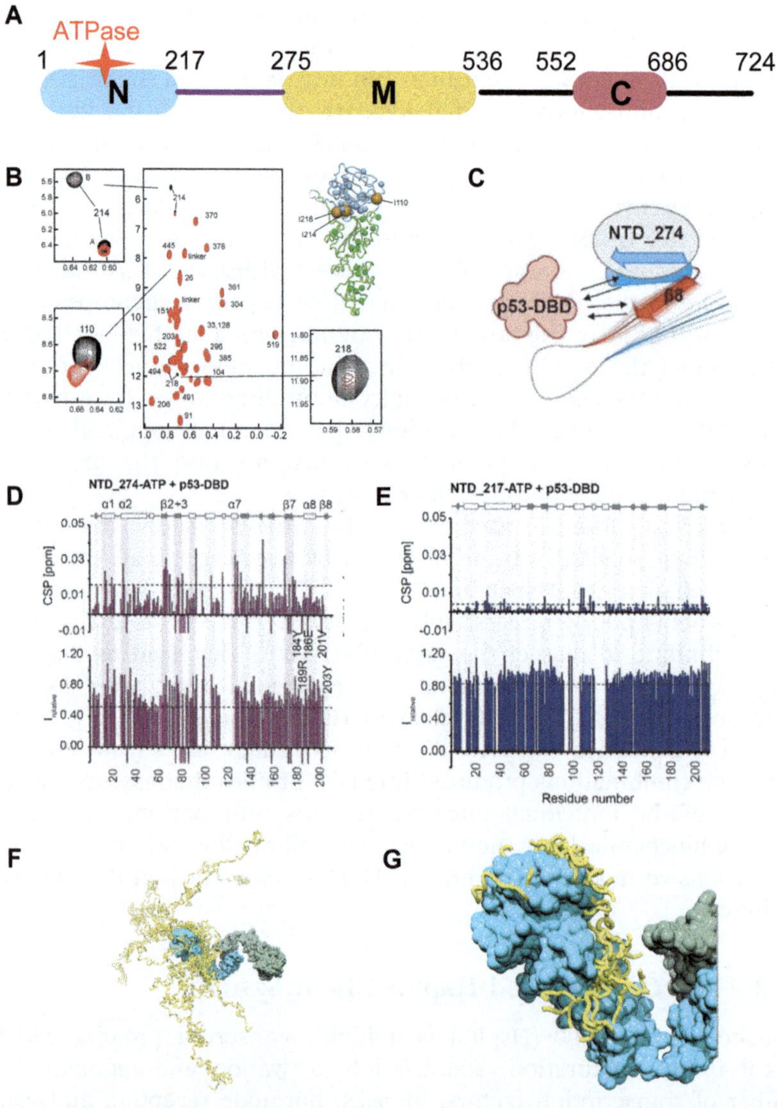

Figure 4.7 NMR characterization of a Hsp90–client system. (A) Domain structure of human hsp90β. (B) Superposition of the Ile region of the ^{1}H–^{13}C TROSY-HMQC spectra of Hsp90 NM in the absence (black) and in the presence of three-fold excess unlabelled p53 DBD (red).[155] The crystal structure of dimeric yeast Hsp82 (PDB: 2CGE)[144] with equivalent residues is shown in the inset. Reproduced from ref. 155 with permission from Elsevier, Copyright 2011. (C) Schematic diagram of the dynamic interaction of p53-DBD with Hsp90 NTD_274[157] (PDB: 1AM1).[225] (D and E) Chemical shift perturbation and intensity ratio plots of Hsp90 NTD_274 (I, purple) and Hsp90 NTD_217 (J, blue) constructs mixed with ATP upon binding of p53-DBD.[157] Panels (C–E) reproduced from ref. 157 with permission from John Wiley & Sons, Copyright 2020. (F and G) Models of the Hsp90–Tau complex obtained from NMR and SAXS data.[159] Reproduced from ref. 159 with permission from Elsevier, Copyright 2014.

motif.[150] Previous studies revealed that hsp90 adopts distinct conformations triggered by the ATP-coupled cycles, which are tightly fine-tuned by various co-chaperones and post-translational modifications.[150–153] However, the functional relationship between its ATPase activity and structural plastic features remains largely elusive. Moreover, more than 300 proteins have been identified as Hsp90 clients, differing dramatically in sizes, amino acid sequences, structures and functions.[153] The intrinsically dynamic nature of hsp90 and its miscellaneous clients, as well as the diversity of regulatory factors in the ATP-coupled cycles, make the hsp90–client system one of the most challenging chaperone–client systems.

Solution NMR has demonstrated its robustness in illuminating the intricate hsp90 machinery. Here we will make an overview of some representative studies of hsp90–client systems using solution NMR.

p53 protein, one of the best studied cancer-related proteins, is a natural substrate of hsp90. Back in 2002, Fersht's lab recorded the 2D ^{1}H, ^{15}N-CRINEPT–TROSY HSQC spectra of the p53 core domain in the presence of the perdeuterated hsp90[80] and proposed that the p53 core predominantly binds in an unfolded form to Hsp90. Subsequent titrations of the ^{15}N-labelled p53 core with an individual domain or two-domain combination constructs of hsp90, as well as the full-length hsp90 by Dyson's lab, confirmed the unfolded state of the p53 core upon binding to hsp90.[154] Further chemical shift perturbation analysis, hydrogen/deuterium exchange and fluorescence experiments demonstrated that the p53 core interacts with the N and M domains of hsp90 and exists in a dynamic molten globule-like ensemble.[154] Additionally, the ^{1}H,^{13}C-TROSY-HMQC spectra of various constructs of hsp90 with $^{13}CH_3$-labelled-Ile in the presence and in the absence of the p53 core showed slight chemical shift changes[155] (see Figure 4.7B), suggesting that the hsp90–p53 core interaction is highly dynamic. Intriguingly, another study by Kessler's group showed that p53 binds to the yeast hsp90 C-terminal domain in its native-like state using a combination of fluorescence polarization, MD and solution NMR.[156] Recently, Sattler's lab reported the regulatory role of the charged linker clamped by the N and M domains of hsp90 in the interaction with the p53 core.[157] The chemical shift perturbation analysis revealed that the dynamic arrangements of the linker alter the transient state of the strand β8 of the N domain, resulting in the strand β8 detachment from β2 and α8 and then the exposure of the p53 binding site (see Figure 4.7C–E). Thereby, the hsp90–p53 interaction seems not to follow a simple pattern and the dynamic details remain controversial.

The Hsp90–Tau system is another well-described Hsp90–substrate system. Tau is an intrinsically disordered protein implicated in a number of neurodegenerative diseases.[158] Rüdiger's group revealed an extended 106 Å-long binding interface on Hsp90 to prevent Tau aggregation using a combination of fluorescence spectroscopy, methyl-TROSY spectroscopy and SAXS[159] (see Figure 4.7F and G). This interface, which spans both the N and M domains of Hsp90, demonstrates multiple low-affinity contacts with Tau. Notably, the NMR signal of the apo and ATPγS-bound states of Hsp90 demonstrated resonance splitting upon Tau binding, suggesting that Tau binding breaks the symmetry of the hsp90 dimer. The authors further converged the NMR chemical shift perturbation analysis and SAXS data and finally built a highly dynamic structural model. Similar to the above-mentioned Skp–OMP system, this hsp90–tau complex also achieves the global high-affinity binding interaction *via* a large number of low-affinity contacts.

4.5.4 ClpB Works on the Protein Disaggregation and Reactivation

In collaboration with the Hsp70/DnaK chaperone system,[160,161] the (AAA+ ATPases associated with various cellular activities) ATPase ClpB acts as an ATP-dependent chaperone to mediate the resolubilization of protein aggregates.[162] Apparently, a clear picture of the ClpB–DnaK bichaperone complex structure is required to gain a full understanding of the mechanism that governs the protein aggregate reactivation process. However, structural characterization of the only transiently formed and dynamic supramolecular complex, such as the case of ClpB–DnaK, has been a huge challenge for any technique.

Rosenzweig *et al.* performed a series of methyl-TROSY-based NMR titrations to elucidate the binding interface that mediates the rather weak k_d (~30 μM) interaction between the 580 kDa ClpB and the 70 kDa DnaK.[163] Chemical shift perturbation mapping, together with biochemical experiments, revealed that the specific regions on the coil–coil domain (CCD) of ClpB and the N-terminal nucleotide-binding domain (NBD) of DnaK are responsible for the interaction.

In the next phase, PRE measurements were performed to obtain long-range distance constraints, which, together with information gained from chemical shift perturbation, allowed the authors to present a structural model for the 650 kDa dynamic ClpB–DnaK complex, and subsequently propose a mechanistic model for the protein aggregate reactivation by ClpB.

The regulatory role of the N-terminal domain (NTD) of ClpB was investigated in a follow-up study.[164] NMR titrations revealed that, in addition to a well-established tyrosine pore region on the first nucleotide-binding domain (NBD1), there is a second substrate binding groove located on the NTD of ClpB, which preferentially recognizes exposed hydrophobic stretches in unfolded or aggregated client proteins. Importantly, melting temperature measurement by NMR indicates that binding to the NTD stabilizes the unfolded conformation of a substrate, which discloses the role that the NTD plays in the protein disaggregation process.

To address the issue of signal overlap, an intein-based segmental labelling approach was implemented to prepare the 580 kDa ClpB NMR sample, whereby only the NTD portion of ClpB is isotopically enriched with

$^{13}CH_3$-methyl groups, whereas the rest of the molecule is fully deuterated (NMR invisible). The resulting methyl-TROSY spectra are of astonishing quality and enable a detailed study of the physiological role that the NTD plays within the context of full-length ClpB. Fast-dynamic measurements revealed that the NTD is loosely assembled, independent of the nucleotide state, to the rest of the ClpB molecule, as demonstrated by the rotational correlation time ($\sim$80 ns) that is significantly shorter than what is expected for a 580 kDa protein ($\sim$180 ns), and that the rotational correlation time and the methyl ^{1}H relaxation rates, R_{2S}, are unaltered regardless whether ADP or ATP is bound. Interestingly, the authors identified a few residues located on the linker connecting the NTD to NBD1 acting as faithful allosteric reporters for substrate engagement. This conveniently allows simultaneous characterization of the two distinct substrate-binding processes occurring in hexameric ClpB while only the NTD portion is isotopically labelled. The binding of the NTD to a substrate, unlike the tight and nucleotide-dependent binding of the tyrosine region on NBD1, demonstrates a similar affinity for either the ADP- or the ATP-bound form with a significantly weaker affinity.

4.5.5 The Trigger Factor (TF) Chaperone–PhoA Complex

The trigger factor (TF) chaperone is a highly abundant protein in bacteria, which prevents the aggregation or premature folding of nascent polypeptides and keeps proteins in the unfolded state during the earliest stages of protein folding.[165,166] Structurally, TF adopts an extended, "crouching dragon"-like shape with three major domains: a ribosome-binding domain (RBD), a peptidyl-prolyl-*cis*/*trans* isomerase domain (PPD) and a substrate-binding domain (SBD).[167,168] Previous reports found that TF chaperone is highly flexible on a number of different timescales and it undergoes a monomer–dimer exchange with an apparent k_d of 18 μM, while the dimeric TF adopts a head-to-tail orientation.[169]

Kalodimos's group has exploited recent advances in solution NMR and solved the structure of TF in complex with its client PhoA.[170] A unique feature of PhoA ($\sim$50 kDa) is that it remains unfolded under reducing conditions,[171] therefore offering a convenient probe in chaperone–substrate studies.[164,170] A series of NMR titrations revealed seven distinct TF-binding regions in PhoA and four substrate binding sites on the SBD and PPD of TF, both of which are composed primarily of hydrophobic and especially aromatic residues. The four TF sites are not a consecutive region but extend over more than 90 Å, thus combining up to a large nonpolar surface ($\sim$2300 Å^2) available for binding protein substrates. By virtue of simultaneously labelling six methyl-bearing (Ile, Leu, Val, Thr, Ala and Met) and aromatic (Phe, Tyr and Trp) residues, the structures of each TF molecule in complex with different regions of PhoA were determined and the highly resolved NMR spectra enabled following detailed studies. Chemical shift analysis and intermolecular NOE measurements demonstrated that some PhoA sites bind promiscuously to all TF sites, while other PhoA sites adopt a more selective manner. The different affinities obtained from diverse PhoA-peptides and distinct TF sites elucidated that the interaction of TF and PhoA is plastically dynamic and contains multiple weak transient states. ITC experiments and NMR titrations showed that the stoichiometric ratio of TF to PhoA is 3 : 1, indicating that the total complex molecular mass is up to ~200 kDa. ^{15}N-CPMG measurements of TF in complex with an individual PhoA-peptide revealed a highly dynamic TF–PhoA complex with both fast association ($k_{on} \sim 2 \times 10^7\ M^{-1}\ s^{-1}$) and dissociation ($k_{off} \sim 50\ s^{-1}$). Further mutagenesis assay, NMR titrations and light scattering monitored aggregation assays showed that TF acts as a powerful holdase when the low-populated, aggregation-prone unfolded state of a substrate is transiently exposed and recognized by TF chaperone.

This work presented the first high-resolution structures determined for a chaperone in complex with an unfolded protein and a general multiple binding mechanism for the action of molecular chaperones.

4.5.6 The SecB–PhoA Complex

SecB is a multitasking molecular chaperone and exhibits an unusually strong antifolding activity in bacteria.[172] It can prevent protein aggregation through interplays with other chaperones including TF, DnaK and GroEL.[173–175] In addition to its contribution to cellular proteostasis, SecB also functions in the Sec pathway, serving as a holdase to recognize the unfolded protein precursors and to maintain them in a translocation-competent state.[176–178] The crystal structure of SecB showed that it forms a tetramer (~70 kDa) organized as a dimer of dimers, with an overall rectangular, disk-like shape.[179,180] Over the last few decades numerous efforts have been made in the characterization of the interaction between chaperone SecB and its substrates. However, very limited success had been achieved at the atomic level.

Kalodimos's lab has applied advanced NMR technology to elucidate the structural basis for the antifolding activity of SecB.[57] Consistent with the dimer of dimer assembly model of SecB,[179,180] the NMR spectra revealed two sets of NMR resonances for many residues. Notably, the sensitivity of the NMR technology to the nucleus chemical environments permits discrimination between the two dimers of SecB based on their different structural packing pattern. A series of HSQC spectra were collected to monitor the refolding behaviors of two client proteins, maltose binding protein (MBP) and PhoA, in the presence of SecB, which confirmed the strong antifolding chaperone activity, as judged by narrow chemical shift dispersions.

Comprehensive NMR titrations combined with NOE analysis allow the authors to pinpoint the recognition sites on SecB and in its substrates, MBP and PhoA. The results collectively revealed that two long hydrophobic grooves serve as the binding interface of SecB. Similar to the case of TF, multiple separated binding sites are recognized by SecB in the unstructured client proteins, enriched in hydrophobic and bulky aromatic amino acids. Furthermore, high-resolution structure determination of the 120 kDa SecB–PhoA complex was achieved by iterative methyl-methyl NOE analysis of several SecB and PhoA complex samples bearing "orthogonal" labelling schemes.[57] The resulting complex structure was further validated by PRE measurement experiments, in which paramagnetic MTSL-spin labels were covalently attached to single-cysteine bearing PhoA variants and the PRE effect was monitored in the NMR spectra of SecB. The most remarkable feature of the complex is that, through simultaneous binding of multiple sites, PhoA wraps, *via* primarily hydrophobic contacts, around SecB in an overall arrangement that maximizes the interacting surface between the unfolded client protein and the chaperone. This provides a structural explanation for the mechanism that how chaperone SecB exerts its antifolding activity on its client proteins to keep them in a nonnative conformation. Furthermore, the structural plasticity of the SecB binding interface, as evidenced by an altered NOE pattern upon substrate binding, explains how chaperone recognizes and interacts with a wide array of substrate proteins with no obvious consensus in sequence.

4.5.7 Hsp40 Recognizes the Non-native Protein PhoA and the Diverse Hsp40 Family

Hsp40s, also called DnaJ heat shock proteins, are a class of important co-chaperones of hsp70 and play multi-faceted roles in neurodegenerative disease and cancer.[181] To understand how hsp40 recognizes and engages a non-native protein, Kalodimos's lab has characterized *Thermus thermophilus* class B Hsp40 (ttHsp40) and selected the unfolded PhoA as a model client.[182] High-quality NMR spectra of ^{13}C-labelled methyl-bearing ttHsp40 enabled near-complete backbone and methyl assignments in complex with PhoA. Similar to other non-native protein–chaperone systems, ttHsp40 interacts with PhoA in a dynamic fashion in a multivalent binding manner. NMR titration analysis identified two client-binding domains (CBDs) in ttHsp40 and as many as fifteen ttHsp40 recognition sites in PhoA, including seven strong sites and eight weak sites. Although ttHsp40 forms a dimer in solution, NMR and ITC analysis showed a stoichiometry of 1 : 1 ttHsp40/PhoA complex. NMR relaxation dispersion experiments indicated that dissociation and association events occurred between the individual PhoA sites and CBDs of ttHsp40 within an $\sim$100 ms timescale. A large number of intermolecular NOEs together with binding specificities allowed structural determination of the structures of seven individual PhoA–CBD complexes, and the dynamic exchange of the adjacent PhoA sites on the same CBD, as well as the PhoA sites on the same subunit that contains both CBD1 and CBD2, was extensively characterized. Finally, by combining with the step-by-step splicing of the various sub-complexes and the previously crystal structure of free ttHsp40, the entire complex structure of ttHsp40–PhoA was determined.

NMR data also revealed that in different Hsp40 species, the number of hsp40–client binding sites is varied. The variation in the number and composition of CBDs and the zinc finger might contribute to their distinct client selectivity and further specialized functions. As a specific example, the client-binding sites in class A Ydj1 CBD2 are blocked by an internal interaction with its C-tail.

Hsp40s work synergistically with hsp70. NMR analysis of the Hsp70/Hsp40 machinery showed that the C-tail of hsp70 interacts with the CBD of Hsp40, which overlaps the client protein binding site. Hsp40 can form complexes with varying stability of hsp70 and clients depending on the number of client-binding sites within the Hsp40. In addition, the activity of Hsp40, which aids the refolding of denatured proteins, can be altered by Hsp70, as verified in luciferase refolding and mutation experiments.

These structural data demonstrated how Hsp40s recognize non-native proteins and how hsp40s cooperate with hsp70 to assemble into the functional chaperone machinery. Compared with TF[170] and SecB,[57] the structural data showed a distinctive recognition mode of chaperone-non-native protein, in which the secondary structure of the client is altered upon Hsp40 binding.

Hsp40s constitute a large and diverse family with more than 40 different members in humans.[183,184] Intriguingly, despite similar structures, different Hsp40 species confer distinct activities in cells. The molecular mechanism behind this diverging behavior remains unknown. Recently, Faust *et al.* applied solution NMR to solve this puzzle.[185] Using a series of NMR titration experiments, the authors revealed that although in all types of Hsp40s, the interaction of the characteristic J-domain is responsible for the activation of Hsp70, in the type B Hsp40s, but not the other types, whose Hsp70-binding sites are intrinsically blocked by an adjacent glycine-phenylalanine rich region. A systematic combination of CSP analysis, RDC measurement and structural determination of DNAJB1$^{\text{JD-GF}}$ allows the authors to provide the molecular basis for this auto-inhibitory regulation: a helix within the disordered GF region, namely the helix V, docks onto the hsp70-binding sites in the J-domain and therefore prevents binding, and this inhibition can be released upon the interaction of a second binding site. Further disaggregation assays demonstrated that this autoinhibitory regulation machinery is essential for the targeting and clustering of Hsp70 onto the amyloid fibers. This study unraveled elegantly a class-specific autoinhibitory mechanism that how the class B Hsp40 regulates the targeting of hsp70 to clients. Indeed, in a following study from Rosenzweig's group, the authors used NMR spectroscopy, in combination with kinetic aggregation assays, to describe the different effects of two Hsp40 chaperones, DNAJA2 and DNAJB1, on tau amyloid-fiber formation and compare these to the small heat shock protein HSPB1.[186] Taken together, these results provide a molecular basis for how the diverse members of the Hsp40 exhibit distinct, yet complementary physiological roles.

4.5.8 Small Heat-shock Protein Hsp27 Modulates FUS Phase Separation

In response to cellular stresses to protect cells against damage, some proteins and nucleotides can reversibly assemble into membraneless organelles, which are highly dynamic and driven by protein phase separation.[187,188] Many canonical chaperones have been identified as essential components that anticipate this process.[187,189] However, the biological functions of chaperones within the membraneless organelles are not fully understood.

Stress granules (SGs), one of the best-studied membraneless organelles, are formed by the stalled translational complexes under stressful conditions.[187] Previous studies have shown that amyotrophic lateral sclerosis (ALS)-linked FUS mutants are associated with the assembly dynamics of SGs.[190–192] FUS possesses liquid, LLPS and solid states under different conditions, and the liquid-to-solid phase transition of FUS in SGs is related to neurodegenerative diseases.[193–196] How FUS is regulated and maintained in distinct phase states remains elusive.

Studies by Liu's group have linked the molecular chaperones to the phase transition of FUS in SGs[197,198] and used solution NMR to analyze the interaction of hsp27 and FUS in two distinct phase states.[197] The authors revealed that hsp27 could prevent the soluble FUS from liquid–liquid phase separation (LLPS) under normal cellular conditions, and the NMR titration results showed a global decrease of signal intensities and a little chemical shift perturbation of the serine residues in the FUS-LC domain, indicating that FUS-LC is a client of hsp27 with weak interaction. In the LLPS state, the authors applied the PRE experiment to characterize the intramolecular and intermolecular interactions between hsp27 and FUS-LC. Using different combinations of ^{15}N-labelled, MTSL-labelled and unlabelled FUS-LC protein samples, they confirmed extensive long-range intramolecular and intermolecular interactions within the LLPS state of FUS-LC, and hsp27 inhibits the LLPS state of FUS-LC *via* diminishing the inherent long-range, transient contacts. Moreover, under cellular stress, hsp27 is phosphorylated and the authors observed that the phosphorylated hsp27 efficiently holds FUS in the LLPS state against the solid-aggregation formation. Chemical shift perturbation analysis combined with site-directed mutagenesis allowed the identification of FUS LC binding sites on hsp27, which is located in the β5–β6–β7 interface and shapes as a shallow cavity. These data collectively revealed the mechanism how the small heat-shock protein hsp27 plays a switchable role in regulating different FUS states.

4.5.9 The Dynamic Chaperonin and Its Co-chaperonin

Different families of molecular chaperones exhibit distinctive activities and biological functions.[57] Some chaperones accelerate the folding of clients, while others delay it. The Hsp60 family (also called the chaperonin family) belongs to the latter. The *E. coli* Hsp60 homolog GroEL is one of the most well-studied chaperones in the Hsp60 family. It comprises two back-to-back heptameric rings and the large central cavity enclosed by each ring enables client binding.[199] Transiently capped by the co-chaperone GroES, GroEL encapsulates clients to trigger protein folding and undergoes ATP-dependent conformation rearrangements.[200–202] Moreover, hydrogen–deuterium exchange experiments[203,204] and NMR analysis[205,206] showed the bound client proteins are structurally flexible. However, fundamental questions such as how chaperonins help clients fold and how clients attain their correct states remain unanswered.

Clore's lab presented a series of quantitative studies to investigate the details of hsp60–client interaction using the solution NMR spectroscopy.[105,202,207–210] Aβ42 is one of the most common amyloid β peptides, which has been linked to the progression of Alzheimer's disease.[211,212] To understand the mechanism how Hsp60 safeguards neurons, Wälti and coworkers applied a variety of biophysical methods to characterize the dynamic interaction of GroEL with Aβ42.[207] NMR spectral analysis showed that the overall intensity of the backbone amide proton envelope of ^{15}N-labelled Aβ42 was reduced upon the addition of GroEL, indicating that the Aβ42-induced neurotoxic was notably inhibited. NMR titration experiments revealed that Aβ42 binds to GroEL as a monomer but not in the aggregate form. Assisted by serial EM and AFM observations, the decrease in NMR signal intensity at the time course of Aβ42 aggregation demonstrated that GroEL reduced the rate of protofibril and fibril formation. Moreover, the kinetic parameters of the interaction between GroEL and Aβ42 were measured by multinuclear relaxation-based NMR methods, including analysis of ^{15}N-lifetime line broadening (ΔR_2), dark state exchange saturation transfer (DEST), exchange-induced chemical shifts and Carr–Purcell–Meiboom–Gill (CPMG) relaxation dispersion. Global fitting of these combined relaxation data with a two-state exchange model, which comprised free and GroEL-bound Aβ42, indicated that the GroEL–Aβ42 complex was sparsely populated (2%) and short-lived ($\sim$1.1 ms), and the timescale of interconversion between bound and unbound states of GroEL–Aβ42 is shorter than the lifetime of the complex. In addition, the residue-specific transverse relaxation rate profile identified three client anchoring regions, which were situated in the hydrophobic GroEL-consensus binding sequences. This work provided deep insights into the dynamic interaction of GroEL with Aβ42 and established the mechanism how GroEL dynamically regulates Aβ42 aggregation.

Chaperonin functions as an ATP-dependent protein folding machine. Structural studies of chaperonin in different nucleotide states have captured a series of intuitive snapshots during the ATPase cycle, which show that chaperonin controls its ring chamber open and close in an ATP-dependent manner.[200,201] However, the conformational dynamics of the active ATP-fueled chaperonin, as well as the interactions with client proteins being refolded, remain elusive. Boisbouvier and coworkers applied solution NMR to investigate the dynamic action of a group II chaperonin from the hyperthermophilic *Pyrococcus horikoshii*, which has a molecular weight of ~1 MDa, and obtained high-quality methyl-TROSY spectra under physiologically relevant conditions (above 60 °C).[213] By combining selective valine and methionine methyl labelling with methyl–TROSY experiments, the authors monitored the structure and dynamics of the chaperonin at different nucleotide states and showed that ATP binding triggers closing of the chaperonin cavity. Two model proteins, hen egg white lysozyme (HEWL) and malate synthase G (MSG), were selected as model proteins to investigate the interaction of chaperonin with clients. NMR-detected thermal unfolding assay and PRE measurements showed that chaperonin interacts exclusively with the unfolded form of the client and its open state acts as a holdase of the unfolded client. It is noteworthy that to study the dynamic chaperonin–client system under steady working conditions, the authors implemented an ATP regeneration system directly inside the NMR sample. Moreover, interleaved NMR data collection mode allows to simultaneously probe the conformational states of the chaperonin and client, as well as the process of nucleotide binding and hydrolysis. Collectively, these data provide quantitative measures of the thermodynamics and kinetics of the transitions between the open and closed conformations and reveal how nucleotides and client protein binding regulate the transition between different conformations populated during the chaperonin functional cycle.

Recently Boisbouvier's group has performed a mechanistic study of a co-chaperone of chaperonin,[214] the type II co-chaperonin prefoldin (PFD), and probed how it inhibits fibrillation of the hormone islet amyloid polypeptide (IAPP), a 37 amino-acid long intrinsically disordered peptide that is associated with type II diabetes. NMR titrations, together with PRE measurements, allow mapping of multiple binding sites within IAPP and an extensive binding interface on PFD, indicative of an avidity-based dynamic binding that has been commonly observed in other chaperone–client systems. These data, in combination with other biochemical and biophysical measurements, provide a detailed molecular basis for the inhibition of IAPP fibril formation by the co-chaperonin prefoldin.

4.5.10 The Dynamic Mitochondrial TIM Chaperone Systems

Most mitochondrial proteins are synthesized as precursors in the cytosol and need to be imported into their intramitochondrial destination *via* specialized translocation systems.[215,216] As essential members, the small TIM (translocases of the inner membrane) proteins are responsible for the safe chaperoning of the hydrophobic membrane proteins through the aqueous mitochondrial intermembrane space (IMS).[215,217] The small TIM family in yeast is composed of five proteins: Tim8, Tim9, Tim10, Tim12 and Tim13, and these chaperones form hetero-hexameric complexes, composed of alternating subunits of Tim8 and Tim13 (TIM8·13), Tim9 and Tim10 (TIM9·10) or Tim9, Tim10 and Tim12 (TIM9·10·12).[217,218] Interestingly, the Tim chaperone complexes are assembled in a highly dynamic manner and the subunits are in continuous exchange between the hexameric state and monomeric forms. The dynamic assembly mechanism of alternate subunits in Tim chaperones and the structural basis of membrane protein chaperoning by the Tim complexes remain poorly understood.

Recently, two insightful NMR studies on the TIM chaperones have been carried out by the Schanda group.[219,220] In the first study, an integrated structural biology approach including NMR was employed to explore the molecular basis for membrane protein chaperoning by the TIM9·10 complex.[220] Weinhäupl and coworkers first used NMR to monitor the binding of ^{15}N and ALV-methyl-labelled TIM9·10 to different precursor proteins, Ggc1, Aac3, and a peptide fragment of the voltage-dependent anion channel (VDAC). A combination of resonance CSP analysis, PRE and NMR saturation-transfer measurements allows the authors to identify a universal binding site on TIM9·10, a clamp-like hydrophobic binding cleft formed between the N- and C-terminal helices, which could accommodate different clients. Of note, the similarity of apo and holo complex NMR spectra of TIM9·10 reveals that the hexameric chaperone remains structurally intact upon client binding. The binding interface was further validated by extensive mutagenesis. To gain in-depth insights into the structure and dynamics of TIM9·10 and the TIM9·10–client complex, a series of nuclear spin relaxation measurements were carried out, covering both the fast-exchange (ps–ns) and slow-exchange (μs–ms) dynamics regime. These data established that when staying bound on the surface of TIM9·10, the clients are in an unfolded, extended and highly dynamic state, adopting multiple inter-converting, short-lived conformations. These inter-conversion dynamics occur on a time scale of ~1 ms, as evidenced by the CPMG data analysis. Taken together, this study offers a complete description of the structure and dynamics of apo TIM9·10 and of the precursor protein–chaperone complex at atomic resolution.

The intricate balance between binding promiscuity *versus* specificity is the key to understanding the chaperone machinery. In the case of TIM complexes, a long-standing question is why two structurally similar chaperones, Tim9·10 and Tim8·13, demonstrate distinct client specificity. To address this, Sucec *et al.* studied chaperone complexes of TIM9·10 and TIM8·13 with two precursor proteins, the Ggc1 and the insertase component Tim23.[219] Competition experiments show that TIM8·13 and TIM9·10 interact differently with precursor proteins: the hydrophobic Ggc1 has a strong preference for TIM9·10, while the hydrophilic Tim23 fragment shows a slight preference for TIM8·13. Interestingly, NMR titration together with CSP analysis revealed that both TIM8·13 and TIM9·10 use a conserved hydrophobic cleft for membrane precursor protein binding, while the low affinity in Tim8·13 is consistent with the less hydrophobic nature of its binding region. Further NMR titrations of the Tim23 fragment with TIM8·13 and TIM9·10 show that hydrophilic fragments interact differently with these two chaperones, while TIM8·13 uses an additional hydrophilic face for Tim23 fragment binding. In the final step, the authors integrated the NMR data with other biophysical techniques and molecular dynamics (MD) simulations and provided a full structural and dynamic description of the chaperone–client complexes. This study exemplifies how the TIM chaperones fine-tune the balance of promiscuity (conserved hydrophobic cleft) *versus* specificity (additional hydrophilic or electrostatic interactions) in the chaperone–client interactions.

4.6 Conclusion and Outlook

Recent advances have pushed NMR spectroscopy into a new era, and this technique has been applied to many supraprotein systems with sizes not long ago thought to be impossible. There is no doubt that NMR can continuously contribute to the field of *de novo* biomolecular structure determinations, especially for proteins that are hard to crystallize, such as intrinsic disorder proteins. However, it has become cruelly clear that this technique is difficult to keep pace with X-ray crystallography or EM as a routine tool for the structural determination of a majority of proteins, particularly large molecules and complexes. Rather, the true power of NMR spectroscopy may lie in two aspects: (a) its ability to characterize the transiently formed biomolecular assemblies, which are of biological significance but hard to be studied by other biophysical techniques, such as the studies on chaperone–substrate complexes reviewed here and (b) its unique strength to study dynamics of biomolecules at a very broad time-scale range and even "visualize" the sparsely-populated ("invisible") excited states, which, sometimes rather than the ground-state, are often closely related to functions.

NMR is complementary to many other biophysical techniques. Therefore, although action alone is difficult to fulfill the role in routine structural determination, NMR spectroscopy can still yield many fruitful insights when in conjoined use with other tools, such as SAXS, Cryo-EM, EPR and FRET. Taken together, with recent and ongoing advances, we anticipate NMR spectroscopy will continue to contribute significantly to solving cutting-edge biological problems associated with chaperone–client systems.

Acknowledgements

This chapter was supported by the National Natural Science Foundation of China (31770807 and 31971144) (to Chengdong Huang) and the Anhui Provincial Natural Science Foundation (2208085MC47) (to Chanjuan Wan).

References

1. F. U. Hartl, A. Bracher and M. Hayer-Hartl, Molecular chaperones in protein folding and proteostasis, *Nature*, 2011, **475**(7356), 324–332.
2. K. Gadhave, N. Bolshette, A. Ahire, R. Pardeshi, K. Thakur and C. Trandafir, *et al.*, The ubiquitin proteasomal system: a potential target for the management of Alzheimer's disease, *J. Cell. Mol. Med.*, 2016, **20**(7), 1392–1407.

3. J. Macosek, G. Mas and S. Hiller, Redefining Molecular Chaperones as Chaotropes, *Front. Mol. Biosci.*, 2021, **8**, 683132.
4. H. Saibil, Chaperone machines for protein folding, unfolding and disaggregation, *Nat. Rev. Mol. Cell Biol.*, 2013, **14**(10), 630–642.
5. S. Hiller, Chaperone-Bound Clients: The Importance of Being Dynamic, *Trends Biochem. Sci.*, 2019, **44**(6), 517–527.
6. R. Freilich, T. Arhar, J. L. Abrams and J. E. Gestwicki, Protein-Protein Interactions in the Molecular Chaperone Network, *Acc. Chem. Res.*, 2018, **51**(4), 940–949.
7. D. Balchin, M. Hayer-Hartl and F. U. Hartl, In vivo aspects of protein folding and quality control, *Science*, 2016, **353**(6294), aac4354.
8. F. Chiti and C. M. Dobson, Protein Misfolding, Amyloid Formation, and Human Disease: A Summary of Progress Over the Last Decade, *Annu. Rev. Biochem.*, 2017, **86**, 27–68.
9. E. M. Sontag, R. S. Samant and J. Frydman, Mechanisms and Functions of Spatial Protein Quality Control, *Annu. Rev. Biochem.*, 2017, **86**, 97–122.
10. A. Wentink, C. Nussbaum-Krammer and B. Bukau, Modulation of Amyloid States by Molecular Chaperones, *Cold Spring Harbor Perspect. Biol.*, 2019, **11**(7), a033969.
11. S. Hiller and B. M. Burmann, Chaperone-client complexes: A dynamic liaison, *J. Magn. Reson.*, 2018, **289**, 142–155.
12. H. Schwalbe, Editorial: New 1.2 GHz NMR Spectrometers- New Horizons?, *Angew. Chem., Int. Ed.*, 2017, **56**(35), 10252–10253.
13. K. Pervushin, R. Riek, G. Wider and K. Wüthrich, Attenuated T_2 relaxation by mutual cancellation of dipole–dipole coupling and chemical shift anisotropy indicates an avenue to NMR structures of very large biological macromolecules in solution, *Proc. Natl. Acad. Sci. U. S. A.*, 1997, **94**(23), 12366–12371.
14. M. Salzmann, K. Pervushin, G. Wider, H. Senn and K. Wüthrich, NMR assignment and secondary structure determination of an octameric 110 kDa protein using TROSY in triple resonance experiments, *J. Am. Chem. Soc.*, 2000, **122**(31), 7543–7548.
15. R. Riek, G. Wider, K. Pervushin and K. Wüthrich, Polarization transfer by cross-correlated relaxation in solution NMR with very large molecules, *Proc. Natl. Acad. Sci. U. S. A.*, 1999, **96**(9), 4918–4923.
16. J. Fiaux, E. B. Bertelsen, A. L. Horwich and K. Wüthrich, NMR analysis of a 900K GroEL-GroES complex, *Nature*, 2002, **418**(6894), 207–211.
17. K. Pervushin, R. Riek, G. Wider and K. Wüthrich, Transverse relaxation-optimized spectroscopy (TROSY) for NMR studies of aromatic spin systems in C-13-labeled proteins, *J. Am. Chem. Soc.*, 1998, **120**(25), 6394–6400.
18. H. Zhang and H. van Ingen, Isotope-labeling strategies for solution NMR studies of macromolecular assemblies, *Curr. Opin. Struct. Biol.*, 2016, **38**, 75–82.

19. F. W. Studier and B. A. Moffatt, Use of Bacteriophage-T7 Rna-Polymerase to Direct Selective High-Level Expression of Cloned Genes, *J. Mol. Biol.*, 1986, **189**(1), 113–130.
20. A. Pandey, K. Shin, R. E. Patterson, X. Q. Liu and J. K. Rainey, Current strategies for protein production and purification enabling membrane protein structural biology, *Biochem. Cell Biol.*, 2016, **94**(6), 507–527.
21. D. Salom, P. Cao, Y. Yuan, M. Miyagi, Z. Feng and K. Palczewski, Isotopic labeling of mammalian G protein-coupled receptors heterologously expressed in Caenorhabditis elegans, *Anal. Biochem.*, 2015, **472**, 30–36.
22. L. Clark, I. Dikiy, D. M. Rosenbaum and K. H. Gardner, On the use of Pichia pastoris for isotopic labeling of human GPCRs for NMR studies, *J. Biomol. NMR*, 2018, **71**(4), 203–211.
23. B. Franke, C. Opitz, S. Isogai, A. Grahl, L. Delgado and A. D. Gossert, *et al.*, Production of isotope-labeled proteins in insect cells for NMR, *J. Biomol. NMR*, 2018, **71**(3), 173–184.
24. D. Milic and D. B. Veprintsev, Large-scale production and protein engineering of G protein-coupled receptors for structural studies, *Front. Pharmacol.*, 2015, **6**, 66.
25. R. Puthenveetil and O. Vinogradova, Solution NMR: A powerful tool for structural and functional studies of membrane proteins in reconstituted environments, *J. Biol. Chem.*, 2019, **294**(44), 15914–15931.
26. J. Boisbouvier and L. E. Kay, Advanced isotopic labeling for the NMR investigation of challenging proteins and nucleic acids, *J. Biomol. NMR*, 2018, **71**(3), 115–117.
27. M. Sattler and S. W. Fesik, Use of deuterium labeling in NMR: Overcoming a sizeable problem, *Structure*, 1996, **4**(11), 1245–1249.
28. K. Hiruma-Shimizu, H. Shimizu, G. S. Thompson, A. P. Kalverda and S. G. Patching, Deuterated detergents for structural and functional studies of membrane proteins: Properties, chemical synthesis and applications, *Mol. Membr. Biol.*, 2015, **32**(5–8), 139–155.
29. G. Bali, M. B. Foston, H. M. O'Neill, B. R. Evans, J. H. He and A. J. Ragauskas, The effect of deuteration on the structure of bacterial cellulose, *Carbohydr. Res.*, 2013, **374**, 82–88.
30. V. Tugarinov, P. M. Hwang, J. E. Ollerenshaw and L. E. Kay, Cross-correlated relaxation enhanced H-1-C-13 NMR spectroscopy of methyl groups in very high molecular weight proteins and protein complexes, *J. Am. Chem. Soc.*, 2003, **125**(34), 10420–10428.
31. S. Miller, J. Janin, A. M. Lesk and C. Chothia, Interior and Surface of Monomeric Proteins, *J. Mol. Biol.*, 1987, **196**(3), 641–656.
32. L. Lo Conte, C. Chothia and J. Janin, The atomic structure of protein-protein recognition sites, *J. Mol. Biol.*, 1999, **285**(5), 2177–2198.
33. R. Sprangers and L. E. Kay, Quantitative dynamics and binding studies of the 20S proteasome by NMR, *Nature*, 2007, **445**(7128), 618–622.

34. T. L. Religa, R. Sprangers and L. E. Kay, Dynamic Regulation of Archaeal Proteasome Gate Opening As Studied by TROSY NMR, *Science*, 2010, **328**(5974), 98–102.
35. J. Drogemuller, M. Strauss, K. Schweimer, M. Jurk, P. Rosch and S. H. Knauer, Determination of RNA polymerase binding surfaces of transcription factors by NMR spectroscopy, *Sci. Rep.*, 2015, **5**, 16428.
36. K. H. Gardner and L. E. Kay, Production and incorporation of N-15, C-13, H-2 (H-1-delta 1 methyl) isoleucine into proteins for multidimensional NMR studies, *J. Am. Chem. Soc.*, 1997, **119**(32), 7599–7600.
37. V. Tugarinov and L. E. Kay, Ile, Leu, and Val methyl assignments of the 723-residue malate synthase G using a new labeling strategy and novel NMR methods, *J. Am. Chem. Soc.*, 2003, **125**(45), 13868–13878.
38. V. Kurauskas, P. Schanda and R. Sounier, Methyl-specific isotope labeling strategies for NMR studies of membrane proteins, *Methods Mol. Biol.*, 2017, **1635**, 109–123.
39. V. Tugarinov and L. E. Kay, Ile, Leu, and Val methyl assignments of the 723-residue malate synthase G using a new labeling strategy and novel NMR methods, *J. Am. Chem. Soc.*, 2003, **125**(45), 13868–13878.
40. P. Gans, O. Hamelin, R. Sounier, I. Ayala, M. A. Dura and C. D. Amero, *et al.*, Stereospecific isotopic labeling of methyl groups for NMR spectroscopic studies of high-molecular-weight proteins, *Angew. Chem., Int. Ed.*, 2010, **49**(11), 1958–1962.
41. Y. R. Monneau, Y. Ishida, P. Rossi, T. Saio, S. R. Tzeng and M. Inouye, *et al.*, Exploiting E. coli auxotrophs for leucine, valine, and threonine specific methyl labeling of large proteins for NMR applications, *J. Biomol. NMR*, 2016, **65**(2), 99–108.
42. Y. Miyanoiri, Y. Ishida, M. Takeda, T. Terauchi, M. Inouye and M. Kainosho, Highly efficient residue-selective labeling with isotope-labeled Ile, Leu, and Val using a new auxotrophic E. coli strain, *J. Biomol. NMR*, 2016, **65**(2), 109–119.
43. Y. Miyanoiri, M. Takeda, K. Okuma, A. M. Ono, T. Terauchi and M. Kainosho, Differential isotope-labeling for Leu and Val residues in a protein by E. coli cellular expression using stereo-specifically methyl labeled amino acids, *J. Biomol. NMR*, 2013, **57**(3), 237–249.
44. R. J. Lichtenecker, N. Coudevylle, R. Konrat and W. Schmid, Selective isotope labelling of leucine residues by using alpha-ketoacid precursor compounds, *ChemBioChem*, 2013, **14**(7), 818–821.
45. M. Fischer, K. Kloiber, J. Hausler, K. Ledolter, R. Konrat and W. Schmid, Synthesis of a 13C-methyl-group-labeled methionine precursor as a useful tool for simplifying protein structural analysis by NMR spectroscopy, *ChemBioChem*, 2007, **8**(6), 610–612.
46. R. L. Isaacson, P. J. Simpson, M. Liu, E. Cota, X. Zhang and P. Freemont, *et al.*, A new labeling method for methyl transverse relaxation-optimized spectroscopy NMR spectra of alanine residues, *J. Am. Chem. Soc.*, 2007, **129**(50), 15428–15429.

47. K. Sinha, L. Jen-Jacobson and G. S. Rule, Specific labeling of threonine methyl groups for NMR studies of protein-nucleic acid complexes, *Biochemistry*, 2011, **50**(47), 10189–10191.
48. A. Velyvis, A. M. Ruschak and L. E. Kay, An economical method for production of 2H, 13CH3-threonine for solution NMR studies of large protein complexes: application to the 670 kDa proteasome, *PLoS One*, 2012, 7(9), e43725.
49. A. M. Ruschak, A. Velyvis and L. E. Kay, A simple strategy for 13C, 1H labeling at the Ile-gamma2 methyl position in highly deuterated proteins, *J. Biomol. NMR*, 2010, **48**(3), 129–135.
50. R. Kerfah, M. J. Plevin, R. Sounier, P. Gans and J. Boisbouvier, Methyl-specific isotopic labeling: a molecular tool box for solution NMR studies of large proteins, *Curr. Opin. Struct. Biol.*, 2015, **32**, 113–122.
51. M. Kainosho, T. Torizawa, Y. Iwashita, T. Terauchi, A. Mei Ono and P. Guntert, Optimal isotope labelling for NMR protein structure determinations, *Nature*, 2006, **440**(7080), 52–57.
52. M. C. Stoffregen, M. M. Schwer, F. A. Renschler and S. Wiesner, Methionine scanning as an NMR tool for detecting and analyzing biomolecular interaction surfaces, *Structure*, 2012, **20**(4), 573–581.
53. T. L. Religa, A. M. Ruschak, R. Rosenzweig and L. E. Kay, Site-directed methyl group labeling as an NMR probe of structure and dynamics in supramolecular protein systems: applications to the proteasome and to the ClpP protease, *J. Am. Chem. Soc.*, 2011, **133**(23), 9063–9068.
54. T. S. Chavan, S. Abraham and V. Gaponenko, Application of reductive 13C-methylation of lysines to enhance the sensitivity of conventional NMR methods, *Molecules*, 2013, **18**(6), 7103–7119.
55. U. Weininger, Optimal isotope labeling of aromatic amino acid side chains for NMR studies of protein dynamics, *Method Enzymol.*, 2019, **614**, 67–86.
56. A. G. Milbradt, H. Arthanari, K. Takeuchi, A. Boeszoermenyi, F. Hagn and G. Wagner, Increased resolution of aromatic cross peaks using alternate C-13 labeling and TROSY, *J. Biomol. NMR*, 2015, **62**(3), 291–301.
57. C. Huang, P. Rossi, T. Saio and C. G. Kalodimos, Structural basis for the antifolding activity of a molecular chaperone, *Nature*, 2016, **537**(7619), 202–206.
58. J. L. Kitevski-LeBlanc and R. S. Prosser, Current applications of F-19 NMR to studies of protein structure and dynamics, *Prog. Nucl. Magn. Reson. Spectrosc.*, 2012, **62**, 1–33.
59. J. L. Kitevski-LeBlanc and R. S. Prosser, Current applications of 19F NMR to studies of protein structure and dynamics, *Prog. Nucl. Magn. Reson. Spectrosc.*, 2012, **62**, 1–33.
60. E. N. Marsh and Y. Suzuki, Using 19F NMR to probe biological interactions of proteins and peptides, *ACS Chem. Biol.*, 2014, **9**(6), 1242–1250.

61. C. Li, G. F. Wang, Y. Wang, R. Creager-Allen, E. A. Lutz and H. Scronce, *et al.*, Protein 19F NMR in Escherichia coli, *J. Am. Chem. Soc.*, 2010, **132**(1), 321–327.
62. X. Y. Wang, D. S. Liu, L. Shen, F. H. Li, Y. Z. Li and L. Y. Yang, *et al.*, A Genetically Encoded F-19 NMR Probe Reveals the Allosteric Modulation Mechanism of Cannabinoid Receptor 1, *J. Am. Chem. Soc.*, 2021, **143**(40), 16320–16325.
63. J. J. Liu, R. Horst, V. Katritch, R. C. Stevens and K. Wüthrich, Biased signaling pathways in beta2-adrenergic receptor characterized by 19F-NMR, *Science*, 2012, **335**(6072), 1106–1110.
64. A. Manglik, T. H. Kim, M. Masureel, C. Altenbach, Z. Yang and D. Hilger, *et al.*, Structural Insights into the Dynamic Process of beta2-Adrenergic Receptor Signaling, *Cell*, 2015, **161**(5), 1101–1111.
65. T. W. Muir, Semisynthesis of proteins by expressed protein ligation, *Annu. Rev. Biochem.*, 2003, **72**, 249–289.
66. L. Berrade and J. A. Camarero, Expressed protein ligation: a resourceful tool to study protein structure and function, *Cell. Mol. Life Sci.*, 2009, **66**(24), 3909–3922.
67. M. Muona, A. S. Aranko, V. Raulinaitis and H. Iwai, Segmental isotopic labeling of multi-domain and fusion proteins by protein trans-splicing in vivo and in vitro, *Nat. Protoc.*, 2010, **5**(3), 574–587.
68. T. Yamazaki, T. Otomo, N. Oda, Y. Kyogoku, K. Uegaki and N. Ito, *et al.*, Segmental isotope labeling for protein NMR using peptide splicing, *J. Am. Chem. Soc.*, 1998, **120**(22), 5591–5592.
69. H. Y. Mao, S. A. Hart, A. Schink and B. A. Pollok, Sortase-mediated protein ligation: A new method for protein engineering, *J. Am. Chem. Soc.*, 2004, **126**(9), 2670–2671.
70. T. W. Muir, D. Sondhi and P. A. Cole, Expressed protein ligation: A general method for protein engineering, *Proc. Natl. Acad. Sci. U. S. A.*, 1998, **95**(12), 6705–6710.
71. K. Severinov and T. W. Muir, Expressed protein ligation, a novel method for studying protein-protein interactions in transcription, *J. Biol. Chem.*, 1998, **273**(26), 16205–16209.
72. T. C. Evans, J. Benner and M. Q. Xu, Semisynthesis of cytotoxic proteins using a modified protein splicing element, *Protein Sci.*, 1998, **7**(11), 2256–2264.
73. A. Nanda, S. S. Nasker, A. Mehra, S. Panda and S. Nayak, Inteins in Science: Evolution to Application, *Microorganisms*, 2020, **8**(12), 2004.
74. M. Mund, J. H. Overbeck, J. Ullmann and R. Sprangers, LEGO-NMR Spectroscopy: A Method to Visualize Individual Subunits in Large Heteromeric Complexes, *Angew. Chem., Int. Ed.*, 2013, **52**(43), 11401–11405.
75. G. M. Clore and J. Iwahara, Theory, Practice, and Applications of Paramagnetic Relaxation Enhancement for the Characterization of Transient Low-Population States of Biological Macromolecules and Their Complexes, *Chem. Rev.*, 2009, **109**(9), 4108–4139.

76. M. Melanson, A. Sood, F. Torok and M. Torok, Introduction to spin label electron paramagnetic resonance spectroscopy of proteins, *Biochem. Mol. Biol. Educ.*, 2013, **41**(3), 156–162.
77. C. Fernandez and G. Wider, TROSY in NMR studies of the structure and function of large biological macromolecules, *Curr. Opin. Struct. Biol.*, 2003, **13**(5), 570–580.
78. C. M. Quinn, M. Wang and T. Polenova, NMR of Macromolecular Assemblies and Machines at 1 GHz and Beyond: New Transformative Opportunities for Molecular Structural Biology, *Methods Mol. Biol.*, 2018, **1688**, 1–35.
79. R. Riek, K. Pervushin and K. Wüthrich, TROSY and CRINEPT: NMR with large molecular and supramolecular structures in solution, *Trends Biochem. Sci.*, 2000, **25**(10), 462–468.
80. S. Rudiger, S. M. Freund, D. B. Veprintsev and A. R. Fersht, CRINEPT-TROSY NMR reveals p53 core domain bound in an unfolded form to the chaperone Hsp90, *Proc. Natl. Acad. Sci. U. S. A.*, 2002, **99**(17), 11085–11090.
81. G. M. Clore, C. Tang and J. Iwahara, Elucidating transient macromolecular interactions using paramagnetic relaxation enhancement, *Curr. Opin. Struct. Biol.*, 2007, **17**(5), 603–616.
82. D. A. Gell, A. H. Kwan, J. P. Mackay, NMR spectroscopy in the analysis of protein-protein interactions, *Modern Magnetic Resonance*, Springer, 2017, pp. 1339–1346.
83. M. A. Hass and M. Ubbink, Structure determination of protein-protein complexes with long-range anisotropic paramagnetic NMR restraints, *Curr. Opin. Struct. Biol.*, 2014, **24**, 45–53.
84. P. Vallurupalli, G. Bouvignies and L. E. Kay, Increasing the exchange time-scale that can be probed by CPMG relaxation dispersion NMR, *J. Phys. Chem. B*, 2011, **115**(49), 14891–14900.
85. A. G. Palmer 3rd, C. D. Kroenke and J. P. Loria, Nuclear magnetic resonance methods for quantifying microsecond-to-millisecond motions in biological macromolecules, *Methods Enzymol.*, 2001, **339**, 204–238.
86. P. Vallurupalli, D. F. Hansen, E. Stollar, E. Meirovitch and L. E. Kay, Measurement of bond vector orientations in invisible excited states of proteins, *Proc. Natl. Acad. Sci. U. S. A.*, 2007, **104**(47), 18473–18477.
87. D. F. Hansen, P. Neudecker, P. Vallurupalli, F. A. Mulder and L. E. Kay, Determination of Leu side-chain conformations in excited protein states by NMR relaxation dispersion, *J. Am. Chem. Soc.*, 2010, **132**(1), 42–43.
88. D. F. Hansen, P. Vallurupalli and L. E. Kay, Measurement of methyl group motional parameters of invisible, excited protein states by NMR spectroscopy, *J. Am. Chem. Soc.*, 2009, **131**(35), 12745–12754.
89. G. Bouvignies, P. Vallurupalli, D. F. Hansen, B. E. Correia, O. Lange and A. Bah, *et al.*, Solution structure of a minor and transiently formed state of a T4 lysozyme mutant, *Nature*, 2011, **477**(7362), 111–114.

90. N. R. Skrynnikov, F. A. Mulder, B. Hon, F. W. Dahlquist and L. E. Kay, Probing slow time scale dynamics at methyl-containing side chains in proteins by relaxation dispersion NMR measurements: application to methionine residues in a cavity mutant of T4 lysozyme, *J. Am. Chem. Soc.*, 2001, **123**(19), 4556–4566.
91. D. M. Korzhnev, K. Kloiber, V. Kanelis, V. Tugarinov and L. E. Kay, Probing slow dynamics in high molecular weight proteins by methyl-TROSY NMR spectroscopy: application to a 723-residue enzyme, *J. Am. Chem. Soc.*, 2004, **126**(12), 3964–3973.
92. R. Sprangers, A. Gribun, P. M. Hwang, W. A. Houry and L. E. Kay, Quantitative NMR spectroscopy of supramolecular complexes: dynamic side pores in ClpP are important for product release, *Proc. Natl. Acad. Sci. U. S. A.*, 2005, **102**(46), 16678–16683.
93. R. Sprangers and L. E. Kay, Quantitative dynamics and binding studies of the 20S proteasome by NMR, *Nature*, 2007, **445**(7128), 618–622.
94. P. Vallurupalli, G. Bouvignies and L. E. Kay, Studying "invisible" excited protein states in slow exchange with a major state conformation, *J. Am. Chem. Soc.*, 2012, **134**(19), 8148–8161.
95. N. L. Fawzi, J. Ying, R. Ghirlando, D. A. Torchia and G. M. Clore, Atomic-resolution dynamics on the surface of amyloid-beta protofibrils probed by solution NMR, *Nature*, 2011, **480**(7376), 268–272.
96. N. J. Anthis and G. M. Clore, Visualizing transient dark states by NMR spectroscopy, *Q. Rev. Biophys.*, 2015, **48**(1), 35–116.
97. S. Forsén and R. A. Hoffman, Study of moderately rapid chemical exchange reactions by means of nuclear magnetic double resonance, *J. Chem. Phys.*, 1963, **39**(11), 10.
98. N. J. Anthis and G. M. Clore, Visualizing transient dark states by NMR spectroscopy, *Q. Rev. Biophys.*, 2015, **48**(1), 35–116.
99. L. E. Kay, New Views of Functionally Dynamic Proteins by Solution NMR Spectroscopy, *J. Mol. Biol.*, 2016, **428**(2 Pt A), 323–331.
100. G. Bouvignies and L. E. Kay, A 2D 13C-CEST experiment for studying slowly exchanging protein systems using methyl probes: an application to protein folding, *J. Biomol. NMR*, 2012, **53**(4), 303–310.
101. G. Bouvignies and L. E. Kay, Measurement of proton chemical shifts in invisible states of slowly exchanging protein systems by chemical exchange saturation transfer, *J. Phys. Chem. B*, 2012, **116**(49), 14311–14317.
102. E. Rennella, R. Huang, A. Velyvis and L. E. Kay, 13CHD2-CEST NMR spectroscopy provides an avenue for studies of conformational exchange in high molecular weight proteins, *J. Biomol. NMR*, 2015, **63**(2), 187–199.
103. N. L. Fawzi, J. Ying, D. A. Torchia and G. M. Clore, Kinetics of amyloid beta monomer-to-oligomer exchange by NMR relaxation, *J. Am. Chem. Soc.*, 2010, **132**(29), 9948–9951.
104. N. L. Fawzi, D. S. Libich, J. Ying, V. Tugarinov and G. M. Clore, Characterizing methyl-bearing side chain contacts and dynamics mediating

amyloid beta protofibril interactions using 13C(methyl)-DEST and lifetime line broadening, *Angew. Chem., Int. Ed.*, 2014, **53**(39), 10345–10349.
105. D. S. Libich, N. L. Fawzi, J. Ying and G. M. Clore, Probing the transient dark state of substrate binding to GroEL by relaxation-based solution NMR, *Proc. Natl. Acad. Sci. U. S. A.*, 2013, **110**(28), 11361–11366.
106. N. L. Fawzi, J. Ying, D. A. Torchia and G. M. Clore, Probing exchange kinetics and atomic resolution dynamics in high-molecular-weight complexes using dark-state exchange saturation transfer NMR spectroscopy, *Nat. Protoc.*, 2012, **7**(8), 1523–1533.
107. G. Mas, J. Thoma and S. Hiller, The Periplasmic Chaperones Skp and SurA, *Subcell. Biochem.*, 2019, **92**, 169–186.
108. E. Bitto and D. B. McKay, Crystallographic structure of SurA, a molecular chaperone that facilitates folding of outer membrane porins, *Structure*, 2002, **10**(11), 1489–1498.
109. I. P. Korndorfer, M. K. Dommel and A. Skerra, Structure of the periplasmic chaperone Skp suggests functional similarity with cytosolic chaperones despite differing architecture, *Nat. Struct. Mol. Biol.*, 2004, **11**(10), 1015–1020.
110. T. A. Walton and M. C. Sousa, Crystal structure of Skp, a prefoldin-like chaperone that protects soluble and membrane proteins from aggregation, *Mol. Cell*, 2004, **15**(3), 367–374.
111. B. M. Burmann, C. W. Wang and S. Hiller, Conformation and dynamics of the periplasmic membrane-protein-chaperone complexes OmpX-Skp and tOmpA-Skp, *Nat. Struct. Mol. Biol.*, 2013, **20**(11), 1265–1272.
112. M. Callon, B. M. Burmann and S. Hiller, Structural Mapping of a Chaperone-Substrate Interaction Surface, *Angew. Chem., Int. Ed.*, 2014, **53**(20), 5069–5072.
113. D. A. Holdbrook, B. M. Burmann, R. G. Huber, M. V. Petoukhov, D. I. Svergun and S. Hiller, *et al.*, A Spring-Loaded Mechanism Governs the Clamp-like Dynamics of the Skp Chaperone, *Structure*, 2017, **25**(7), 1079–1088.e3.
114. G. Mas, B. M. Burmann, T. Sharpe, B. Claudi, D. Bumann and S. Hiller, Regulation of chaperone function by coupled folding and oligomerization, *Sci. Adv.*, 2020, **6**(43), eabc5822.
115. W. Meng, E. M. Clerico, N. McArthur and L. M. Gierasch, Allosteric landscapes of eukaryotic cytoplasmic Hsp70s are shaped by evolutionary tuning of key interfaces, *Proc. Natl. Acad. Sci. U. S. A.*, 2018, **115**(47), 11970–11975.
116. A. Finka, S. K. Sharma and P. Goloubinoff, Multi-layered molecular mechanisms of polypeptide holding, unfolding and disaggregation by HSP70/HSP110 chaperones, *Front. Mol. Biosci.*, 2015, **2**, 29.
117. A. Sekhar and L. E. Kay, An NMR View of Protein Dynamics in Health and Disease, *Annu. Rev. Biophys.*, 2019, **48**, 297–319.
118. M. P. Mayer and B. Bukau, Hsp70 chaperones: cellular functions and molecular mechanism, *Cell. Mol. Life Sci.*, 2005, **62**(6), 670–684.

119. E. R. Zuiderweg, E. B. Bertelsen, A. Rousaki, M. P. Mayer, J. E. Gestwicki and A. Ahmad, Allostery in the Hsp70 chaperone proteins, *Top. Curr. Chem.*, 2013, **328**, 99–153.
120. A. Sekhar, R. Rosenzweig, G. Bouvignies and L. E. Kay, Mapping the conformation of a client protein through the Hsp70 functional cycle, *Proc. Natl. Acad. Sci. U. S. A.*, 2015, **112**(33), 10395–10400.
121. R. Rosenzweig, A. Sekhar, J. Nagesh and L. E. Kay, Promiscuous binding by Hsp70 results in conformational heterogeneity and fuzzy chaperone-substrate ensembles, *eLife*, 2017, **6**, e28030.
122. A. Sekhar, J. Nagesh, R. Rosenzweig and L. E. Kay, Conformational heterogeneity in the Hsp70 chaperone-substrate ensemble identified from analysis of NMR-detected titration data, *Protein Sci.*, 2017, **26**(11), 2207–2220.
123. A. Sekhar, R. Rosenzweig, G. Bouvignies and L. E. Kay, Hsp70 biases the folding pathways of client proteins, *Proc. Natl. Acad. Sci. U. S. A.*, 2016, **113**(20), E2794–E2801.
124. A. Sekhar, A. Velyvis, G. Zoltsman, R. Rosenzweig, G. Bouvignies and L. E. Kay, Conserved conformational selection mechanism of Hsp70 chaperone-substrate interactions, *eLife*, 2018, **7**, e32764.
125. A. Zhuravleva, E. M. Clerico and L. M. Gierasch, An interdomain energetic tug-of-war creates the allosterically active state in Hsp70 molecular chaperones, *Cell*, 2012, **151**(6), 1296–1307.
126. M. P. Mayer and L. M. Gierasch, Recent advances in the structural and mechanistic aspects of Hsp70 molecular chaperones, *J. Biol. Chem.*, 2019, **294**(6), 2085–2097.
127. E. M. Clerico, W. L. Meng, A. Pozhidaeva, K. Bhasne, C. Petridis and L. M. Gierasch, Hsp70 molecular chaperones: multifunctional allosteric holding and unfolding machines, *Biochem. J.*, 2019, **476**, 1653–1677.
128. J. Van Durme, S. Maurer-Stroh, R. Gallardo, H. Wilkinson, F. Rousseau and J. Schymkowitz, Accurate Prediction of DnaK-Peptide Binding *via* Homology Modelling and Experimental Data, *PLoS Comput. Biol.*, 2009, **5**(8), e1000475.
129. A. M. Fourie, J. F. Sambrook and M. J. H. Gething, Common and Divergent Peptide Binding Specificities of Hsp70 Molecular Chaperones, *J. Biol. Chem.*, 1994, **269**(48), 30470–30478.
130. S. Rudiger, L. Germeroth, J. SchneiderMergener and B. Bukau, Substrate specificity of the DnaK chaperone determined by screening cellulose-bound peptide libraries, *EMBO J.*, 1997, **16**(7), 1501–1507.
131. G. C. Flynn, T. G. Chappell and J. E. Rothman, Peptide Binding and Release by Proteins Implicated as Catalysts of Protein Assembly, *Science*, 1989, **245**(4916), 385–390.
132. S. Blondelguindi, A. M. Fourie, J. F. Sambrook and M. J. H. Gething, Peptide-Dependent Stimulation of the Atpase Activity of the Molecular Chaperone Bip Is the Result of Conversion of Oligomers to Active Monomers, *J. Biol. Chem.*, 1993, **268**(17), 12730–12735.

133. M. Zahn, N. Berthold, B. Kieslich, D. Knappe, R. Hoffmann and N. Strater, Structural Studies on the Forward and Reverse Binding Modes of Peptides to the Chaperone DnaK, *J. Mol. Biol.*, 2013, **425**(14), 2463–2479.
134. X. T. Zhu, X. Zhao, W. F. Burkholder, A. Gragerov, C. M. Ogata and M. E. Gottesman, *et al.*, Structural analysis of substrate binding by the molecular chaperone DnaK, *Science*, 1996, **272**(5268), 1606–1614.
135. E. M. Clerico, A. K. Pozhidaeva, R. M. Jansen, C. Ozden, J. M. Tilitsky and L. M. Gierasch, Selective promiscuity in the binding of E. coli Hsp70 to an unfolded protein, *Proc. Natl. Acad. Sci. U. S. A.*, 2021, **118**(41), e2016962118.
136. J. Winkler, J. Tyedmers, B. Bukau and A. Mogk, Hsp70 targets Hsp100 chaperones to substrates for protein disaggregation and prion fragmentation, *J. Cell Biol.*, 2012, **198**(3), 387–404.
137. J. Shorter and S. Lindquist, Hsp104 catalyzes formation and elimination of self-replicating Sup35 prion conformers, *Science*, 2004, **304**(5678), 1793–1797.
138. J. Shorter, The mammalian disaggregase machinery: Hsp110 synergizes with Hsp70 and Hsp40 to catalyze protein disaggregation and reactivation in a cell-free system, *PLoS One*, 2011, **6**(10), e26319.
139. X. Gao, M. Carroni, C. Nussbaum-Krammer, A. Mogk, N. B. Nillegoda and A. Szlachcic, *et al.*, Human Hsp70 Disaggregase Reverses Parkinson's-Linked alpha-Synuclein Amyloid Fibrils, *Mol. Cell*, 2015, **59**(5), 781–793.
140. A. S. Wentink, N. B. Nillegoda, J. Feufel, G. Ubartaite, C. P. Schneider, P. De Los Rios, J. Hennig, A. Barducci and B. Bukau, Molecular dissection of amyloid disaggregation by human HSP70, *Nature*, 2020, **587**(7834), 483–488.
141. M. Taipale, I. Krykbaeva, M. Koeva, C. Kayatekin, K. D. Westover and G. I. Karras, *et al.*, Quantitative analysis of HSP90-client interactions reveals principles of substrate recognition, *Cell*, 2012, **150**(5), 987–1001.
142. C. Huang and C. G. Kalodimos, Structures of Large Protein Complexes Determined by Nuclear Magnetic Resonance Spectroscopy, *Annu. Rev. Biophys.*, 2017, **46**, 317–336.
143. C. Prodromou, Mechanisms of Hsp90 regulation, *Biochem. J.*, 2016, **473**(16), 2439–2452.
144. M. M. Ali, S. M. Roe, C. K. Vaughan, P. Meyer, B. Panaretou and P. W. Piper, *et al.*, Crystal structure of an Hsp90-nucleotide-p23/Sba1 closed chaperone complex, *Nature*, 2006, **440**(7087), 1013–1017.
145. D. E. Dollins, J. J. Warren, R. M. Immormino and D. T. Gewirth, Structures of GRP94-Nucleotide complexes reveal mechanistic differences between the hsp90 chaperones, *Mol. Cell*, 2007, **28**(1), 41–56.
146. L. H. Pearl and C. Prodromou, Structure and mechanism of the Hsp90 molecular chaperone machinery, *Annu. Rev. Biochem.*, 2006, **75**, 271–294.
147. K. A. Krukenberg, F. Forster, L. M. Rice, A. Sali and D. A. Agard, Multiple conformations of E-coli Hsp90 in solution: Insights into the conformational dynamics of Hsp90, *Structure*, 2008, **16**(5), 755–765.

148. L. H. Pearl, The HSP90 Molecular Chaperone-An Enigmatic ATPase, *Biopolymers*, 2016, **105**(8), 594–607.
149. T. O. Street, L. A. Lavery and D. A. Agard, Substrate Binding Drives Large-Scale Conformational Changes in the Hsp90 Molecular Chaperone, *Mol. Cell*, 2011, **42**(1), 96–105.
150. F. H. Schopf, M. M. Biebl and J. Buchner, The HSP90 chaperone machinery, *Nat Rev Mol Cell Bio.*, 2017, **18**(6), 345–360.
151. M. Radli and S. G. D. Rudiger, Dancing with the Diva: Hsp90-Client Interactions, *J. Mol. Biol.*, 2018, **430**(18 Pt B), 3029–3040.
152. M. Mollapour and L. Neckers, Post-translational modifications of Hsp90 and their contributions to chaperone regulation, *Biochim. Biophys. Acta*, 2012, **1823**(3), 648–655.
153. S. J. Backe, R. A. Sager, M. R. Woodford, A. M. Makedon and M. Mollapour, Post-translational modifications of Hsp90 and translating the chaperone code, *J. Biol. Chem.*, 2020, **295**(32), 11099–11117.
154. S. J. Park, B. N. Borin, M. A. Martinez-Yamout and H. J. Dyson, The client protein p53 adopts a molten globule-like state in the presence of Hsp90, *Nat. Struct. Mol. Biol.*, 2011, **18**(5), 537–541.
155. S. J. Park, M. Kostic and H. J. Dyson, Dynamic Interaction of Hsp90 with Its Client Protein p53, *J. Mol. Biol.*, 2011, **411**(1), 158–173.
156. F. Hagn, S. Lagleder, M. Retzlaff, J. Rohrberg, O. Demmer, K. Richter, J. Buchner and H. Kessler, Structural analysis of the interaction between Hsp90 and the tumor suppressor protein p53, *Nat. Struct. Mol. Biol.*, 2011, **18**(10), 1086–1093.
157. A. Lopez, A. R. Elimelech, K. Klimm and M. Sattler, The Charged Linker Modulates the Conformations and Molecular Interactions of Hsp90, *ChemBioChem*, 2021, **22**(6), 1084–1092.
158. L. M. Ittner, Y. D. Ke, F. Delerue, M. A. Bi, A. Gladbach and J. van Eersel, *et al.*, Dendritic Function of Tau Mediates Amyloid-beta Toxicity in Alzheimer's Disease Mouse Models, *Cell*, 2010, **142**(3), 387–397.
159. G. E. Karagoz, A. M. S. Duarte, E. Akoury, H. Ippel, J. Biernat, T. M. Luengo, M. Radli, T. Didenko, B. A. Nordhues, D. B. Veprintsev, C. A. Dickey, E. Mandelkow, M. Zweckstetter, R. Boelens, T. Madl and S. G. Rüdiger, Hsp90-Tau Complex Reveals Molecular Basis for Specificity in Chaperone Action, *Cell*, 2014, **156**(5), 963–974.
160. P. Goloubinoff, A. Mogk, A. P. Zvi, T. Tomoyasu and B. Bukau, Sequential mechanism of solubilization and refolding of stable protein aggregates by a bichaperone network, *Proc. Natl. Acad. Sci. U. S. A.*, 1999, **96**(24), 13732–13737.
161. J. R. Glover and S. Lindquist, Hsp104, Hsp70, and Hsp40: a novel chaperone system that rescues previously aggregated proteins, *Cell*, 1998, **94**(1), 73–82.
162. D. A. Parsell, A. S. Kowal, M. A. Singer and S. Lindquist, Protein disaggregation mediated by heat-shock protein Hsp104, *Nature*, 1994, **372**(6505), 475–478.

163. R. Rosenzweig, S. Moradi, A. Zarrine-Afsar, J. R. Glover and L. E. Kay, Unraveling the mechanism of protein disaggregation through a ClpB-DnaK interaction, *Science*, 2013, **339**(6123), 1080–1083.
164. R. Rosenzweig, P. Farber, A. Velyvis, E. Rennella, M. P. Latham and L. E. Kay, ClpB N-terminal domain plays a regulatory role in protein disaggregation, *Proc. Natl. Acad. Sci. U. S. A.*, 2015, **112**(50), E6872–E6881.
165. L. Ferbitz, T. Maier, H. Patzelt, B. Bukau, E. Deuerling and N. Ban, Trigger factor in complex with the ribosome forms a molecular cradle for nascent proteins, *Nature*, 2004, **431**(7008), 590–596.
166. A. Hoffmann, B. Bukau and G. Kramer, Structure and function of the molecular chaperone Trigger Factor, *Bba-Mol Cell Res.*, 2010, **1803**(6), 650–661.
167. T. Arhar, A. Shkedi, C. M. Nadel and J. E. Gestwicki, The interactions of molecular chaperones with client proteins: why are they so weak?, *J. Biol. Chem.*, 2021, **297**(5), 101282.
168. T. Maier, L. Ferbitz, E. Deuerling and N. Ban, A cradle for new proteins: trigger factor at the ribosome, *Curr. Opin. Struct. Biol.*, 2005, **15**(2), 204–212.
169. H. Patzelt, G. Kramer, T. Rauch, H. J. Schonfeld, B. Bukau and E. Deuerling, Three-state equilibrium of Escherichia coli trigger factor, *Biol. Chem.*, 2002, **383**(10), 1611–1619.
170. T. Saio, X. Guan, P. Rossi, A. Economou and C. G. Kalodimos, Structural basis for protein antiaggregation activity of the trigger factor chaperone, *Science*, 2014, **344**(6184), 1250494.
171. J. C. Bardwell, K. McGovern and J. Beckwith, Identification of a protein required for disulfide bond formation in vivo, *Cell*, 1991, **67**(3), 581–589.
172. A. Sala, P. Bordes and P. Genevaux, Multitasking SecB chaperones in bacteria, *Front. Microbiol.*, 2014, **5**, 666.
173. E. Deuerling, A. Schulze-Specking, T. Tomoyasu, A. Mogk and B. Bukau, Trigger factor and DnaK cooperate in folding of newly synthesized proteins, *Nature*, 1999, **400**(6745), 693–696.
174. V. R. Agashe, S. Guha, H. C. Chang, P. Genevaux, M. Hayer-Hartl and M. Stemp, *et al.*, Function of trigger factor and DnaK in multidomain protein folding: increase in yield at the expense of folding speed, *Cell*, 2004, **117**(2), 199–209.
175. M. J. Kerner, D. J. Naylor, Y. Ishihama, T. Maier, H. C. Chang and A. P. Stines, *et al.*, Proteome-wide analysis of chaperonin-dependent protein folding in Escherichia coli, *Cell*, 2005, **122**(2), 209–220.
176. F. U. Hartl, S. Lecker, E. Schiebel, J. P. Hendrick and W. Wickner, The binding cascade of SecB to SecA to SecY/E mediates preprotein targeting to the E. coli plasma membrane, *Cell*, 1990, **63**(2), 269–279.
177. J. Zhou and Z. Xu, Structural determinants of SecB recognition by SecA in bacterial protein translocation, *Nat. Struct. Biol.*, 2003, **10**(11), 942–947.

178. P. Bechtluft, R. G. van Leeuwen, M. Tyreman, D. Tomkiewicz, N. Nouwen and H. L. Tepper, *et al.*, Direct observation of chaperone-induced changes in a protein folding pathway, *Science*, 2007, **318**(5855), 1458–1461.
179. Z. Xu, J. D. Knafels and K. Yoshino, Crystal structure of the bacterial protein export chaperone secB, *Nat. Struct. Biol.*, 2000, 7(12), 1172–1177.
180. C. Dekker, B. de Kruijff and P. Gros, Crystal structure of SecB from Escherichia coli, *J. Struct. Biol.*, 2003, **144**(3), 313–319.
181. A. Mitra, L. A. Shevde and R. S. Samant, Multi-faceted role of HSP40 in cancer, *Clin. Exp. Metastasis*, 2009, **26**(6), 559–567.
182. Y. Jiang, P. Rossi and C. G. Kalodimos, Structural basis for client recognition and activity of Hsp40 chaperones, *Science*, 2019, **365**(6459), 1313–1319.
183. H. H. Kampinga and E. A. Craig, The HSP70 chaperone machinery: J proteins as drivers of functional specificity, *Nat. Rev. Mol. Cell Biol.*, 2010, **11**(8), 579–592.
184. R. Kityk, J. Kopp and M. P. Mayer, Molecular Mechanism of J-Domain-Triggered ATP Hydrolysis by Hsp70 Chaperones, *Mol. Cell*, 2018, **69**(2), 227–237.e4.
185. O. Faust, M. Abayev-Avraham, A. S. Wentink, M. Maurer, N. B. Nillegoda and N. London, *et al.*, HSP40 proteins use class-specific regulation to drive HSP70 functional diversity, *Nature*, 2020, **587**(7834), 489–494.
186. R. Irwin, O. Faust, I. Petrovic, S. G. Wolf, H. Hofmann and R. Rosenzweig, Hsp40s play complementary roles in the prevention of tau amyloid formation, *eLife*, 2021, **10**, e69601.
187. S. Jain, J. R. Wheeler, R. W. Walters, A. Agrawal, A. Barsic and R. Parker, ATPase-Modulated Stress Granules Contain a Diverse Proteome and Substructure, *Cell*, 2016, **164**(3), 487–498.
188. J. R. Buchan and R. Parker, Eukaryotic stress granules: the ins and outs of translation, *Mol. Cell*, 2009, **36**(6), 932–941.
189. D. Mateju, T. M. Franzmann, A. Patel, A. Kopach, E. E. Boczek and S. Maharana, *et al.*, An aberrant phase transition of stress granules triggered by misfolded protein and prevented by chaperone function, *EMBO J.*, 2017, **36**(12), 1669–1687.
190. R. R. Sama, C. L. Ward, L. J. Kaushansky, N. Lemay, S. Ishigaki and F. Urano, *et al.*, FUS/TLS assembles into stress granules and is a prosurvival factor during hyperosmolar stress, *J. Cell. Physiol.*, 2013, **228**(11), 2222–2231.
191. X. Zhang, F. Wang, Y. Hu, R. Chen, D. Meng and L. Guo, *et al.*, In vivo stress granule misprocessing evidenced in a FUS knock-in ALS mouse model, *Brain*, 2020, **143**(5), 1350–1367.
192. D. M. Baron, L. J. Kaushansky, C. L. Ward, R. R. Sama, R. J. Chian and K. J. Boggio, *et al.*, Amyotrophic lateral sclerosis-linked FUS/TLS alters stress granule assembly and dynamics, *Mol. Neurodegener.*, 2013, **8**, 30.

193. A. Patel, H. O. Lee, L. Jawerth, S. Maharana, M. Jahnel and M. Y. Hein, *et al.*, A Liquid-to-Solid Phase Transition of the ALS Protein FUS Accelerated by Disease Mutation, *Cell*, 2015, **162**(5), 1066–1077.
194. M. Ramaswami, J. P. Taylor and R. Parker, Altered ribostasis: RNA-protein granules in degenerative disorders, *Cell*, 2013, **154**(4), 727–736.
195. T. Murakami, S. Qamar, J. Q. Lin, G. S. Schierle, E. Rees and A. Miyashita, *et al.*, ALS/FTD Mutation-Induced Phase Transition of FUS Liquid Droplets and Reversible Hydrogels into Irreversible Hydrogels Impairs RNP Granule Function, *Neuron*, 2015, **88**(4), 678–690.
196. A. Aguzzi and M. Altmeyer, Phase Separation: Linking Cellular Compartmentalization to Disease, *Trends Cell Biol.*, 2016, **26**(7), 547–558.
197. Z. Liu, S. Zhang, J. Gu, Y. Tong, Y. Li and X. Gui, *et al.*, Hsp27 chaperones FUS phase separation under the modulation of stress-induced phosphorylation, *Nat. Struct. Mol. Biol.*, 2020, **27**(4), 363–372.
198. J. Gu, Z. Liu, S. Zhang, Y. Li, W. Xia and C. Wang, *et al.*, Hsp40 proteins phase separate to chaperone the assembly and maintenance of membraneless organelles, *Proc. Natl. Acad. Sci. U. S. A.*, 2020, **117**(49), 31123–31133.
199. D. Thirumalai and G. H. Lorimer, Chaperonin-mediated protein folding, *Annu. Rev. Biophys. Biomol. Struct.*, 2001, **30**, 245–269.
200. H. R. Saibil, Chaperone machines in action, *Curr. Opin. Struct. Biol.*, 2008, **18**(1), 35–42.
201. D. H. Chen, D. Madan, J. Weaver, Z. Lin, G. F. Schroder and W. Chiu, *et al.*, Visualizing GroEL/ES in the act of encapsulating a folding protein, *Cell*, 2013, **153**(6), 1354–1365.
202. D. S. Libich, V. Tugarinov and G. M. Clore, Intrinsic unfoldase/foldase activity of the chaperonin GroEL directly demonstrated using multinuclear relaxation-based NMR, *Proc. Natl. Acad. Sci. U. S. A.*, 2015, **112**(29), 8817–8823.
203. R. Zahn, C. Spitzfaden, M. Ottiger, K. Wüthrich and A. Pluckthun, Destabilization of the complete protein secondary structure on binding to the chaperone GroEL, *Nature*, 1994, **368**(6468), 261–265.
204. C. V. Robinson, M. Gross, S. J. Eyles, J. J. Ewbank, M. Mayhew and F. U. Hartl, *et al.*, Conformation of GroEL-bound alpha-lactalbumin probed by mass spectrometry, *Nature*, 1994, **372**(6507), 646–651.
205. R. Horst, E. B. Bertelsen, J. Fiaux, G. Wider, A. L. Horwich and K. Wüthrich, Direct NMR observation of a substrate protein bound to the chaperonin GroEL, *Proc. Natl. Acad. Sci. U. S. A.*, 2005, **102**(36), 12748–12753.
206. E. Koculi, R. Horst, A. L. Horwich and K. Wüthrich, Nuclear magnetic resonance spectroscopy with the stringent substrate rhodanese bound to the single-ring variant SR1 of the E. coli chaperonin GroEL, *Protein Sci.*, 2011, **20**(8), 1380–1386.

207. M. A. Walti, J. Steiner, F. Meng, H. S. Chung, J. M. Louis and R. Ghirlando, *et al.*, Probing the mechanism of inhibition of amyloid-beta(1-42)-induced neurotoxicity by the chaperonin GroEL, *Proc. Natl. Acad. Sci. U. S. A.*, 2018, **115**(51), E11924–E11932.
208. M. A. Walti, S. A. Kotler and G. M. Clore, Probing the Interaction of Huntingtin Exon-1 Polypeptides with the Chaperonin Nanomachine GroEL, *ChemBioChem*, 2021, **22**(11), 1985–1991.
209. D. S. Libich, V. Tugarinov, R. Ghirlando and G. M. Clore, Confinement and Stabilization of Fyn SH3 Folding Intermediate Mimetics within the Cavity of the Chaperonin GroEL Demonstrated by Relaxation-Based NMR, *Biochemistry*, 2017, **56**(7), 903–906.
210. M. A. Walti, T. Schmidt, D. T. Murray, H. Wang, J. E. Hinshaw and G. M. Clore, Chaperonin GroEL accelerates protofibril formation and decorates fibrils of the Het-s prion protein, *Proc. Natl. Acad. Sci. U. S. A.*, 2017, **114**(34), 9104–9109.
211. K. F. Young, S. H. Pasternak and R. J. Rylett, Oligomeric aggregates of amyloid beta peptide 1-42 activate ERK/MAPK in SH-SY5Y cells *via* the alpha7 nicotinic receptor, *Neurochem. Int.*, 2009, 55(8), 796–801.
212. M. Costanzo and C. Zurzolo, The cell biology of prion-like spread of protein aggregates: mechanisms and implication in neurodegeneration, *Biochem. J.*, 2013, **452**(1), 1–17.
213. G. Mas, J. Y. Guan, E. Crublet, E. C. Debled, C. Moriscot, P. Gans, G. Schoehn, P. Macek, P. Schanda and J. Boisbouvier, Structural investigation of a chaperonin in action reveals how nucleotide binding regulates the functional cycle, *Sci. Adv.*, 2018, **4**(9), eaau4196.
214. R. Torner, T. Kupreichyk, L. Gremer, E. C. Debled, D. Fenel, S. Schemmert, P. Gans, D. Willbold, G. Schoehn, W. Hoyer and J. Boisbouvier, Structural basis for the inhibition of IAPP fibril formation by the co-chaperonin prefoldin, *Nat. Commun.*, 2022, **13**(1), 2363.
215. E. Ceh-Pavia, M. P. Spiller and H. Lu, Folding and biogenesis of mitochondrial small Tim proteins, *Int. J. Mol. Sci.*, 2013, **14**(8), 16685–16705.
216. S. A. Paschen and W. Neupert, Protein import into mitochondria, *IUBMB Life*, 2001, **52**(3–5), 101–112.
217. C. M. Koehler, The small Tim proteins and the twin Cx3C motif, *Trends Biochem. Sci.*, 2004, **29**(1), 1–4.
218. K. Weinhaupl, Y. Wang, A. Hessel, M. Brennich, K. Lindorff-Larsen and P. Schanda, Architecture and assembly dynamics of the essential mitochondrial chaperone complex TIM9.10.12, *Structure*, 2021, **29**(9), 1065–1073.e4.
219. I. Sucec, Y. Wang, O. Dakhlaoui, K. Weinhaupl, T. Jores, D. Costa, A. Hessel, M. Brennich, D. Rapaport, K. Lindorff-Larsen, B. Bersch and P. Schanda, Structural basis of client specificity in mitochondrial membrane-protein chaperones, *Sci. Adv.*, 2020, **6**(51), eabd0263.
220. K. Weinhaupl, C. Lindau, A. Hessel, Y. Wang, C. Schutze, T. Jores, L. Melchionda, B. Schönfisch, H. Kalbacher, B. Bersch, D. Rapaport,

M. Brennich, K. Lindorff-Larsen, N. Wiedemann and P. Schanda, Structural Basis of Membrane Protein Chaperoning through the Mitochondrial Intermembrane Space, *Cell*, 2018, **175**(5), 1365–1379.e25.
221. Z. K. Boswell and M. P. Latham, Methyl-Based NMR Spectroscopy Methods for Uncovering Structural Dynamics in Large Proteins and Protein Complexes, *Biochemistry*, 2019, **58**(3), 144–155.
222. E. B. Bertelsen, L. Chang, J. E. Gestwicki and E. R. Zuiderweg, Solution conformation of wild-type E. coli Hsp70 (DnaK) chaperone complexed with ADP and substrate, *Proc. Natl. Acad. Sci. U. S. A.*, 2009, **106**(21), 8471–8476.
223. T. Nishikawa, A. Nagadoi, S. Yoshimura, S. Aimoto and Y. Nishimura, Solution structure of the DNA-binding domain of human telomeric protein, hTRF1, *Structure*, 1998, **6**(8), 1057–1065.
224. X. Zhu, X. Zhao, W. F. Burkholder, A. Gragerov, C. M. Ogata and M. E. Gottesman, *et al.*, Structural analysis of substrate binding by the molecular chaperone DnaK, *Science*, 1996, **272**(5268), 1606–1614.
225. C. Prodromou, S. M. Roe, R. O'Brien, J. E. Ladbury, P. W. Piper and L. H. Pearl, Identification and structural characterization of the ATP/ADP-binding site in the Hsp90 molecular chaperone, *Cell*, 1997, **90**(1), 65–75.

CHAPTER 5

Preparing Chaperone–Client Protein Complexes for Biophysical and Structural Studies

I. SUČEC[a] AND P. SCHANDA*[b]

[a] Univ. Grenoble Alpes, CEA, CNRS, Institut de Biologie Structurale (IBS), 71, avenue des martyrs, F-38044 Grenoble, France; [b] Institute of Science and Technology Austria, Am Campus 1, A-3400 Klosterneuburg, Austria
*Email: paul.schanda@ist.ac.at

5.1 Introduction

Many proteins are prone to misfolding or unfolding and aggregation in the aqueous environment of the cell, either due to their complex ternary structure, cellular localization, molecular crowding, cellular stress, or their intrinsic hydrophobic nature.[1,2] Molecular chaperones are involved in the biogenesis of such proteins and in ensuring their intactness throughout their life cycle.[3–5] While there are many different types of chaperones with various cellular roles, essentially all of them are characterized by a basic "holdase" function, *i.e.* the ability to bind to proteins that are generally in some non-native conformation – often these are called "client" proteins. Chaperones may have more "active" roles in addition to the basic function: "foldases" assist their client protein in folding to its native state (reviewed in Balchin *et al.*, 2020)[6] and "disaggregases" help in dissociating protein aggregates (reviewed in Mattoo and Goloubinoff, 2014).[7] Understanding how

New Developments in NMR No. 29
Biophysics of Molecular Chaperones: Function, Mechanisms and Client Protein Interactions
Edited by Sebastian Hiller, Maili Liu and Lichun He

Published by the Royal Society of Chemistry, www.rsc.org

these complexes form, what (residual) structure the client proteins have when bound, and how the client protein may eventually be released again from the chaperone is essential for resolving questions regarding the biogenesis of cells and organelles.

A detailed study of chaperone–client complexes at the molecular level is challenging for several reasons. The interactions involved in the complex formation are often highly dynamic, *i.e.* binding often relies on the avidity of multiple interactions (see reviews Burmann and Hiller, 2015; L. He and Hiller, 2019; Hiller, 2019; Macošek *et al.*, 2021; and Sučec *et al.*, 2021).[8–11] Furthermore, chaperones themselves are often large, up to the megadalton range, and can adopt multiple states.[12,13] All these factors make their experimental characterization difficult. Moreover – and this is the focus of the present chapter – chaperone–client complex formation for any experimental study turns out to be challenging. The proteins to be bound to the chaperone are in many cases aggregation prone. This is particularly the case for membrane proteins in aqueous solutions,[14,15] but it applies also to many soluble proteins in the state in which they interact with chaperones, *i.e.* in partially unfolded states.[16–18] An experimental way to keep aggregation-prone proteins in solution is to add a denaturant, but under such conditions, many chaperones are also destabilized, which impairs complex formation. Moreover, shielding of the client by the denaturant also hampers complex formation. Any experimental work that aims to characterize chaperone–client complexes *in vitro* needs to find a set of conditions that enable the formation of the complex. In this chapter, we review the possibilities for such complex formation, starting from some general considerations and highlighting the different possibilities with recent examples from structural and biophysical studies. The cases we present are to be considered as example cases, with the aim to show the range of methods that have been proposed, rather than an exhaustive review of all chaperone–client complexes that have been reported.

5.2 Client–Chaperone Complex Formation

5.2.1 General Considerations

In vivo, client–chaperone complexes form either at specific locations where a client protein emerges – for example at the exit of the ribosome tunnel, or at the exit of translocation pores, or they form in a situation where the chaperone is rather abundant compared to the client. Examples of the former are the trigger factor chaperone, associated with the ribosome.[19] Trigger factor's association with the ribosome provides a shielded environment for the newly synthesized peptides where a co-translational folding of client domains could occur, with possibly several chaperone molecules bound to the client polypeptide chain.[20] Another example is the TIM chaperones that localize at the membrane translocation pore TOM in mitochondria, presumably in a position ready to bind membrane-protein precursor polypeptides, as they emerge in the intermembrane space.[21] Some chaperones are abundant in the cell; for example, the family of 70 kDa heat-shock proteins (Hsp70) on its own is estimated to correspond to up to 3% of the total protein mass in eukaryotic cells under non-stress conditions.[22,23] Under stress conditions, the cellular chaperone concentration or activity can be upregulated, such as in the case of many heat-shock proteins[24,25] and the Spy chaperone.[26] In cases other than binding nascent polypeptide chains, stress response chaperones are upregulated as a response to misfolding or aggregation of client proteins that become structurally unstable or denatured due to stress conditions such as heat, pH changes, osmotic and oxidative stress, and nutrient starvation,[27] leading to increased chaperone cellular levels or their activation.

These cellular conditions and environments facilitate the encounter of chaperones and client proteins and the formation of complexes with affinities that range all the way from low-affinity transient interactions to tight binding with nanomolar affinity.

For an *in vitro* preparation procedure, the complex formation starts from a very different point, namely, in general, from purified proteins (chaperone, client protein). The question is, thus, how one can find suitable conditions for the encounter of the two proteins and complex formation under native conditions.

Formation of a complex between a chaperone (C) and a client protein (P) requires that, under the chosen conditions, the P–C complex is thermodynamically the most stable state or that it is a kinetically trapped state. The desired outcome, namely the formation of the P–C complex, is challenged by other reactions that the protein P may undergo with itself (intramolecular processes such as folding and unfolding), with other copies of itself (such as oligomerization or, quite commonly, aggregation), or with other components in the system (*e.g.* other protein binding partners, a lipid bilayer, or small molecules). In order to understand how the chaperone–client protein complex may form, it is instructive to understand these possible reactions. A simplified scheme of this set of reactions is shown in Figure 5.1. The rate of these reactions differs from one chaperone–client pair to another. Therefore, understanding this set of reactions provides a way of seeing the possibilities that the experimentalist has to prepare P–C complexes.

Any protein P exists in an aqueous solution in an equilibrium between folded conformations (P_{folded}) and possibly a multitude of different partially unfolded intermediate states (P_{int}) and an ensemble of unfolded states ($P_{unfolded}$; Figure 5.1). The distinction we make here among folded, partially unfolded, and fully unfolded is somewhat arbitrary, as this is rather a continuum of states, and even the "folded state" is to be seen as a dynamic ensemble of conformations. For the sake of discussion, we keep this simplified scheme here. The kinetics and thermodynamics of this equilibrium on the continuum from folded to unfolded are protein dependent. Many proteins are able to fold only in a proper environment or in the presence of proper binding partners – this is particularly true for the vast majority of membrane proteins, which are unable to fold in an aqueous solution.

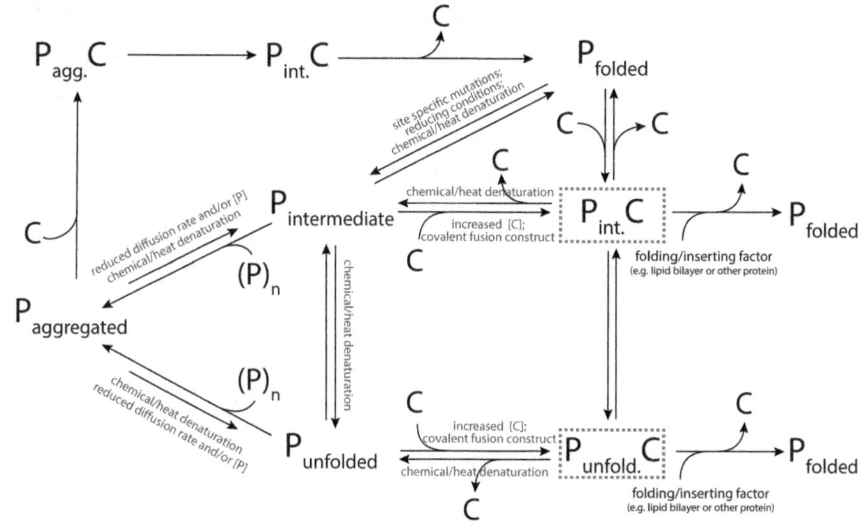

Figure 5.1 Schematic overview of the reactions that may occur in a system composed of a chaperone (C) and a client protein (P). The preparation of complexes of chaperones and client proteins may proceed by direct binding of a chaperone to a folded state or to (partially or fully) unfolded states. In the formed complexes, the protein P differs generally from the folded state, and the complexes are depicted by green dashed boxes. The distinction between an "unfolded" client and a partially folded one ("intermediate" or "int.") is arbitrary and shown here to denote that there is a range of possible residual structures in chaperone-bound clients. Several other reactions involving P can occur and compete with the P–C complex formation. These are the folding of P to a state that is not binding competent, the aggregation of P or, if available, the interaction of P with binding partners or a membrane (right in this scheme). Note that also the apoprotein in its folded or partially unfolded state may bind to such an additional binding partner (not shown). The various reaction equilibria and their kinetics can be modulated by changing the experimental conditions or by modifying the client protein. The effect of some of these conditions on the reaction equilibria is indicated near the arrow. The condition shifts the equilibrium in the direction of the arrow. For example, the rate of aggregate formation can be reduced by immobilizing P (Section 5.3.5) or reducing the concentration of P; the folding/unfolding equilibrium can be shifted towards unfolded states by *e.g.* denaturing conditions, reducing conditions, or mutations of P (Section 5.3.3). From one P/C system to another, the relative importance of these reactions, *i.e.* the relative rate constants and thermodynamic constants, can vary largely.

It is primarily the partially or fully unfolded states that are prone to aggregation: by the encounter of several copies of the protein ($(P)_n$) an aggregated state, $P_{aggregated}$, may form. The (partially or fully) unfolded states are generally more prone to aggregation than the folded states. The reason for this finding is that in the folded states of globular proteins, hydrophobic parts are mostly buried within the core, and they may become exposed upon unfolding.

Likewise, the polypeptides that finally get inserted into a lipid bilayer as membrane proteins are most often highly hydrophobic. The aggregated state that brings together the hydrophobic parts is often highly stable.

Molecular chaperones enter this picture of protein folding, unfolding, aggregation, and possibly insertion (into a membrane or a complex) because they can interact with one or several of these states and protect them from aggregation. They may also play an active role by lowering energy barriers in folding processes.[3,28] From the practical standpoint of this chapter, namely for the experimental preparation of chaperone–client complexes, it is important to understand where the chaperones intervene, what the relevant kinetic rate constants are, compared to the competing processes, and how the chaperone–client complexes can be thermodynamically favored. Figure 5.1 schematizes the possible reactions generally, and the following sections then discuss the situation for several chaperone–client complexes, highlighting which practical experimental approach was successful for obtaining complexes.

Most chaperones bind to parts of proteins that are (partly) unfolded and expose hydrophobic residues;[29] this realization is linked to reports of chaperones that bind to locally "frustrated" sites in proteins,[30,31] which are prone to local unfolding. P–C complex formation, thus, tends to be favored when the P_{folded} state is disfavored compared to the (partially) unfolded ones. For the preparation of P–C complexes that start with purified soluble protein P, it is, thus, often helpful to increase the population of (partially) unfolded states. *In vitro*, destabilization of P_{folded} can be achieved by different means. One way is by acting on the chemical properties of the protein sequence itself, namely through destabilizing mutations or by truncation of the sequence. Such changes disrupt interactions that form in the folded state. Similarly, the disruption of intramolecular disulfide bonds can equally favor the unfolded states. Other factors that can be utilized to act on the $P_{folded}/P_{int}/P_{unfolded}$ equilibrium are the addition of a chemical denaturant or the application of a high or low temperature and pH or high hydrostatic pressure.

These different approaches – either the mutation/truncation of P or the change in buffer conditions, temperature, or pressure – differ in an important aspect: all the changes in solvent, temperature, or pressure can also affect the chaperone (*e.g.* unfold it) and its ability to bind to client proteins. Thus, while such changes under solvent conditions may achieve the desired increase in $[P_{unfolded}]$ or $[P_{int}]$, the formation of the complex (P_{int}–C, $P_{unfolded}$–C) may be disfavored, *i.e.* either the reaction $P_{unfolded} + C \rightarrow P_{unfolded}\text{–}C$ or $P_{int} + C \rightarrow P_{int}\text{–}C$ has a very low reaction rate or its equilibrium is on the left-hand side of these equations. Forming the P–C complex, thus, requires the removal of the conditions that denature/destabilize the chaperone (such as the denaturant or high temperature).

A practically very important reaction is the one that converts soluble proteins into insoluble aggregates. For the formation of a soluble P–C complex, this reaction is a major threat. The essential question for *in vitro* complex formation is whether the formation of aggregates is thermodynamically and/or kinetically favored over the formation of the P–C complex. The coupled equilibrium among aggregated, free, and chaperone-bound protein thus has a few important parameters: the kinetics and thermodynamics of the aggregation of soluble P, and the kinetics and thermodynamics of P–C complex formation. The thermodynamic stabilization of the soluble form over the aggregated form can be achieved by the addition of a denaturant or high hydrostatic pressure. However, as already discussed, if such conditions are incompatible with keeping the chaperone intact, this strategy does not lead to the productive formation of the P–C complex. Therefore, in most cases, one can rather act upon the kinetics of aggregate formation *vs.* the kinetics of complex formation, as discussed in the following.

As aggregation is an intermolecular process involving many copies of P, the kinetics of formation of aggregates is concentration dependent. It can, therefore, be slowed down when the concentration, [P], is reduced, or the rate of diffusion of P is reduced, such that the encounter of copies of P is slowed down. As the kinetics of P–C formation is also concentration dependent (namely on both [P] and [C]), a rather high [C] and low [P] help in favoring the reaction $P + C \rightarrow P\text{–}C$ over the aggregation reaction $P + P_n \rightarrow P_{n+1}$. The experimentalist's options in this sense include drip-diluting the protein from a denatured state (*e.g.* from a solution containing urea or other chaotropes) to a buffer that contains the chaperone and sustains the chaperone's integrity. Another possibility is to have the protein synthesized in the presence of a chaperone, such that a nascent protein would encounter chaperones before having the chance to encounter other copies of P and aggregate. This strategy is exploited in *in vitro* protein synthesis systems containing chaperones.[32–34]

The other parameter that the experimentalist can influence in order to slow down the aggregation process is limiting P's diffusion coefficient and thus limiting the probability of encountering copies of P. A useful approach that exploits this principle is based on immobilizing the protein to *e.g.* a resin *via* an affinity tag. Doing so, the encounter of proteins is strongly reduced (the diffusion coefficient is essentially infinitely small) and added chaperone C can bind with P, without having the competing process of aggregation of P.

The P–C complexes have often a substantial degree of flexibility, as revealed by numerous atomic-level studies reviewed in ref. 11 and 35. The client protein can exist in a range of conformational states when bound to C, from fully unfolded to largely folded (see the equilibrium of P_{int}–C and $P_{unfolded}$–C in Figure 5.1).

The release of P from the P–C complex is an important parameter that determines if biochemical, biophysical, and structural studies of P–C can be performed. The different states of P have different affinities to C. Although not strictly necessary, it can be assumed that less native-like states of P have a higher propensity to bind to C. Accordingly, the off-rate, $P\text{–}C \rightarrow P_{int} + C$ or $P\text{–}C \rightarrow P_{unfolded} + C$, depends on the conformation of the client protein. Rapid off-rate and slow on rate are obviously counter-productive for obtaining large quantities of a stable P–C complex. As discussed above, the off-rate is enhanced and the on-rate is reduced under conditions that destabilize the chaperone (*e.g.* with denaturants).

One way of increasing the rate of P–C complex formation (P + C → P–C) is by covalently binding/fusing the client protein to the chaperone; this is best done by introducing a flexible linker between the sequences of P and C, such that the interactions between a chaperone and a client protein are minimally perturbed. In such fusion constructs, the off-rate is not necessarily altered as compared to a non-covalent protein complex. However, as the distance between C and P is short, the probability of an encounter is enhanced. As an additional effect, the covalent fusion can hamper the aggregation of P, even in its detached state, due to steric hindrance.

For completeness, Figure 5.1 (right) also sketches an additional reaction that can occur, particularly in the cell. The protein may become inserted into a complex with a cognate binding partner, or inserted into a membrane, and this reaction may strongly stabilize its folded state and thus essentially remove it from the equilibrium. For example, a membrane-protein precursor that gets inserted into the membrane is generally not bound by soluble chaperones any more. While this may be an important reaction in the cell, experimentally it is, of course, simply avoided by not having the respective binding partner or lipid bilayer present.

5.2.2 Overview of the Possibilities to Prepare Chaperone–Client Complexes

Following these basic considerations, we can enumerate several strategies which, in principle, can be successful in generating P–C complexes. In the simplest case, mixing of a purified chaperone sample with a purified client protein may lead to the spontaneous formation of a complex; this approach may need destabilizing mutations in the client protein and/or covalently linking the client to its chaperone. Reducing conditions may also be considered as a similar strategy that impacts the client protein by disrupting disulfide bonds, but not the chaperone (unless it contains disulfide bonds). Alternatively, heating the mixed sample or applying pressure may allow the formation of a complex, provided that the chaperone remains intact at the elevated temperature/pressure. Another possibility is to choose the solvent conditions to either destabilize the (soluble) protein or to extract it from the aggregate state in which it may have been produced. The denatured protein can then be mixed with the chaperone, and simultaneous or subsequent removal of the denaturing conditions may allow the complex formation. This removal can be achieved in different ways (*e.g.* sudden dilution and dialysis; discussed below). Finally, complex formation may be achieved directly in the host organism in which the protein is produced, and the intact chaperone–client complex may be extracted. Somewhat similarly, the client protein may be captured upon production, using a cell-free (*in vitro*) protein synthesis system.

Rationally designing the most appropriate complex-formation strategy may, in principle, work, based on the knowledge of the kinetics and thermodynamics of the reactions outlined in Figure 5.1. However, it is not often the case that these details are known, and trying a few different schemes is common. The following sections discuss these different strategies with reported examples from the literature.

5.3 Different Complex-formation Approaches in Practice

5.3.1 Forming Complexes in Solution by Mixing Chaperones and Their Soluble Client Proteins

5.3.1.1 Hsp70 (DnaK) Complexes

In some cases, the preparation of P–C complexes can be achieved in a straightforward manner by mixing a chaperone and a client. An example of such a case is the complex formed by DnaK (bacterial Hsp70) and the human telomere repeat binding factor 1 (hTRF1).[36] In solution, hTRF1 exists in an equilibrium of folded and unfolded states, but the folded state largely dominates; at 35 °C, *ca.* 4% are unfolded[37] and at 25 °C only *ca.* 0.3% of the population is unfolded.[36] There are two equilibria present in solution under the conditions chosen by the authors (25 °C, concentrations of *ca.* 0.5–0.6 mM): (i) folding/unfolding is heavily skewed towards the folded state with a population of 99.7% and rate constants of *ca.* 1 s^{-1} for the unfolding process and 288 s^{-1} for the folding process and (ii) binding/release to DnaK occurs with a dissociation constant in the low micromolar range ($K_d \sim 18$ μM, $k_{on} = (1.1 \pm 0.2) \times 10^6$ M^{-1} s^{-1}, $k_{off} = 20.4 \pm 0.2$ s^{-1}). The binding is proposed to proceed from the native unbound state (N) *via* the unfolded free state (U), to the bound state.[36] In the complex hTRF1 has some residual secondary structure, and it seems plausible that DnaK can bind to hTRF1 when the latter has some residual helicity. In this particular case, the sample preparation involved mixing of hTRF1 with an excess of DnaK (for the experiments in which hTRF1 was isotope-labeled for NMR) or DnaK with an excess of hTRF1 (for NMR studies of DnaK).

Sekhar, Kay, and co-workers used similar sample-preparation strategies – namely mixing of a chaperone and a client – to resolve an important mechanistic question: does complex formation proceed *via* the selection of the unfolded state that pre-exists in solution or does the chaperone rather bind to the folded state and then induce unfolding of the client in the bound state? The two schemes are sketched in Figure 5.2, following the general scheme of Figure 5.1. (Hereby, we considered that the aggregation propensity can be ignored, *i.e.* the client is largely soluble.) Sekhar *et al.* used two different proteins, one with a native α-helical fold and one with a native β-sheet fold, to resolve this question for both classes of proteins. They used marginally stable and slowly folding proteins, which allows dissecting the

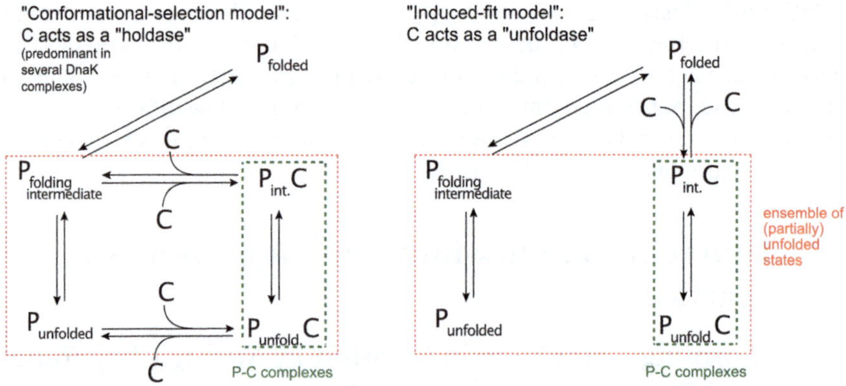

Figure 5.2 Schematic overview of the relevant reactions that can lead to the formation of a complex of a chaperone (C) with a soluble client protein (P) that has negligible aggregation propensity. P and C are mixed and the inherent property of P to expose partially unfolded parts and the ability of C to recognize and bind these parts lead to complex formation. For the case of DnaK, discussed in Section 5.3.1, the left mechanism, namely the selection of (partially or fully) unfolded states by the chaperone, is dominant.[38]

folding/unfolding equilibrium and the binding/release process. In an elegant combination of NMR dynamics experiments, such as zz-exchange and CEST, they demonstrated that conformational selection (Figure 5.2, left) is dominant.[38] Thus, the chaperone preferentially binds to the pre-existing unfolded state(s).

Similarly to working with full-length clients, another option is to use fragments thereof, which has been done by the same group:[39] using soluble peptides of the hTRF1 client and several NMR labeling schemes, multiple chaperone binding sites on the client were detected. A comparison of DnaK-bound state populations of these different binding sites (peptides) indicated that multiple DnaK chaperones bind to the hTRF1 client. Conceptually, what happens when working with fragments of clients is that the folding/unfolding equilibrium is heavily skewed towards the unfolded state – the fragment is often unable to fold – and hence the equilibrium $N \leftrightarrow U$ is eliminated from the reaction scheme.

5.3.1.2 Complexes of the Chaperone Spy with Im7

The client–chaperone recognition and interaction of the stress-induced Spy foldase[26] has been studied with the small soluble protein Im7. Im7 functions by binding to its cognate partner colicin E7, but in isolation Im7 tends to (partially) unfold, due to a number of locally frustrated sites. For this reason, Im7 has been used as a model protein in a number of protein-folding studies. Conditions and mutants of Im7 have been reported in which Im7 remains soluble, either in partially folded or unfolded states.[40–42] Complexes with Spy can be formed by mixing the protein with the chaperone.[43–45] Wild-type Im7 is in a partially unfolded state when it interacts with Spy, in which the residues experiencing local frustrations drive the sequence-unspecific binding to Spy.[31,43] In addition to the wild-type Im7 client, site-specific mutations were introduced that were shown to prevent Im7 folding,[46] while preserving its solubility during the production and purification. Mixing of this soluble but folding-incompetent Im7 variant with the chaperone was successful to prepare samples suitable for crystallography[44] and for NMR studies.[43] While crystallography is challenging to interpret in this case of a highly disordered client,[47] NMR showed that a combination of hydrophobic and charged residues appears to be required for binding to the unfolded client. The ITC and stopped-flow experiments showed that Spy binds to both folded and partially folded clients with similar affinities (3–10 μM).[45] In these examples, the concentration of the complex was high enough for NMR (several hundred μM) and ITC studies.

In kinetic measurements, chemically denatured wild-type Im7 was used and its chaperone-bound folding was followed upon dilution into different concentrations of the Spy chaperone.[45] In other assays, chemically denatured client malate-dehydrogenase (MDH) and DTT-reduced α-lactalbumin (α-LA) were used as fast-aggregating clients to probe the anti-aggregating activity of wild-type Spy[48] and its variants.[49] The aggregation was followed by light scattering upon diluting denatured clients in the presence or in the absence of wild-type Spy or Spy variants.

5.3.1.3 Hsp90 and Hsp110 Complexes

The 90 kDa heat-shock protein (Hsp90) is another well-studied system in which complexes could be formed by mixing with clients. Street *et al.* probed the binding and complex formation of the protein Δ131Δ (131-residue fragment of staphylococcal nuclease; globally unfolded protein fragment) to Hsp90. Fluorescently labeled Δ131Δ was mixed with Hsp90, and the binding was followed by fluorescence polarization anisotropy titration experiments, with increasing concentrations of the binding partners (up to 50 μM). Such straightforward titration experiments are possible due to the solubility of both the client and the chaperone. The dissociation constant was found to be in the low μM range (comparable to Spy–Im7 above), with a 1 : 1 stoichiometry; binding involves a partially structured region within the unfolded client.[50] Rüdiger and co-workers studied the complex formed by Hsp90 and the intrinsically disordered protein tau[51] using NMR methods and electron paramagnetic resonance (EPR).[52] Hsp90 and tau were mixed at concentrations of tens to hundreds of micromolar, which is possible due to the relatively high solubility and comparably low aggregation-propensity of tau (as compared to other client proteins described below).

A recent example investigated how the Hsp100 chaperone ClpB, a disaggregase, interacts with a client protein. A 20-fold excess of α-casein, an intrinsically disordered client protein, was mixed and incubated with an

ATPase-deficient variant of ClpB (0.7 mg mL^{-1}) and structurally investigated by cryo-EM.[53] Deville *et al.* prepared different mutants of ClpB at concentrations in the range of 1.6 mg mL^{-1} to 8 mg mL^{-1} and incubated them with casein (in excess). By comparing the structural features of the mutant complexes with the client protein, they proposed a mechanism of ATPase initiation and client binding by the disaggregase.[54] Recent work by Yin *et al.* investigated the interactions of ClpB, DnaK, and the client casein, *i.e.* a bi-chaperoning system, using cryo-EM.[55] For preparing the complexes, a simple incubation of the components proved successful.

Taken together, for many soluble client proteins, generating complexes with chaperones is achieved by mixing solutions of the two binding partners. The possibility of obtaining complexes spontaneously provides opportunities for measuring the binding affinity. The complex formation is facilitated by an inherent propensity of the client protein to unfold without aggregate formation (in the client's wild-type form or by mutations). At least in some documented cases (*e.g.* DnaK discussed above), the unfolded state is selected by the chaperone. If the binding affinity for the unfolded client is sufficiently high, then even a small proportion of unfolded client suffices to generate large quantities of P–C complex in the coupled set of reactions of unfolding and binding.

5.3.2 Making a Client Protein Bind by Adjusting the Sample Conditions

5.3.2.1 Hsp60 (Thermosome) Complexes Formed at High Temperatures

If the population of unfolded clients is very low under native conditions, then it may be feasible to change the sample conditions to promote unfolding. For example, if the chaperone has a high tolerance to heat, then it may be straightforward to identify client proteins that become unfolded – and prone to chaperone-binding – at temperatures where the chaperone is still functional. This strategy has been chosen for making complexes of the Hsp60 chaperone from the hyperthermophilic archaeon *Pyrococcus horikoshii* and two mesophilic client proteins (lysozyme and malate synthase G). *P. horikoshii*, an organism that is found in deep-ocean hydrothermal vents, has a growth optimum beyond 90 °C. Its Hsp60, also called thermosome, forms a *ca.* 1 MDa large assembly from 16 copies of the 60 kDa large subunits, which assemble into two large barrel-like chambers. Mas *et al.* studied the binding of lysozyme to thermosome.[56] In the absence of a chaperone, heat unfolds lysozyme with a midpoint unfolding temperature of *ca.* 70 °C. Folding and unfolding occur on a time scale of milliseconds in this temperature range. The unfolding is reversible: cooling the sample restored the folded state, without significant accumulation of aggregates. Mas *et al.* then mixed thermosome with lysozyme in an NMR tube at concentrations in the tens-of-micromolar range. At ambient temperature the interaction is negligible whereas at high temperatures, lysozyme binds to the chaperone, evidenced by *e.g.* translational diffusion measurements that probe the size of the client protein. This finding shows that the chaperone binds preferentially (or exclusively) to the unfolded state. Interestingly, NMR showed that the midpoint temperature of unfolding decreases by *ca.* 5 °C in the presence of a chaperone. This suggests that the presence of the chaperone shifts the equilibrium ($P_{folded}/P_{unfolded}$) towards the unfolded state. The Hsp60–lysozyme$_{unfolded}$ interaction at high temperatures is dynamic: NMR data show that binding and release are fast on the NMR time scale, *i.e.* in the sub-millisecond range. Cooling down the sample shifts the equilibrium of lysozyme to its folded state, which does not interact with the chaperone any more. Repeating the experiment at different ratios of lysozyme to thermosome (*i.e.* titration) allows measuring the proportion of lysozyme bound to thermosome, through NMR chemical shifts. This allowed the authors to determine the dissociation constant ($K_d \sim 1.6$ μM); NMR measurements of isotope-labeled thermosome allowed mapping of the interaction site. The short lifetime of this dynamic complex strongly contrasts with other cases discussed in Sections 5.3.4 and 5.3.5, whose lifetimes are more than six orders of magnitude longer.

When applied to a different client protein, the 83 kDa large malate synthase G (MSG), the same strategy with the same chaperone resulted in a somewhat different behavior.[56] First, when MSG alone is heated, it aggregates at elevated temperatures. When a mixture of MSG and *P. horikoshii* Hsp60 is heated, the chaperone protects MSG from aggregation (at least in part). The complex, once formed at a high temperature, can be brought to a low temperature without dissociation, *i.e.* the complex formation is irreversible, or, more precisely, the off-rate of the client is very low. This might at least in part be explained by the size of the client (83 kDa), which essentially fills the cavity of the chaperone cage. A further difference between these two cases is that MSG is much more aggregation prone than lysozyme.

5.3.2.2 *Trigger Factor Complexes Obtained* via *Temperature or Disulfide-bond Breakage*

Another example of applying heat for client-selective denaturing is the study of the holdase function of the trigger factor (TF) where a temperature-sensitive variant of maltose binding protein (MBP) showed reduced aggregation at high temperatures (50 °C) in the presence of TF.[20] The TF chaperone does not bind to the folded client (MBP variant at lower temperatures); however, by increasing the temperature partially unfolded states of the client protein become more populated, leading to complex formation. In the same study, with a different client protein, reducing conditions were used to keep the client unfolded and chaperone-binding prone: the client protein, wild-type PhoA protein, contains four cysteine residues and, *in vivo,* folds in the oxidizing periplasmic environment. NMR was used to study the state of PhoA under reducing conditions. It was shown that several regions had only low (20–60%) secondary structure propensities.[20] Isotopically (^{15}N) labeled full-length PhoA and the PhoA fragments were titrated under reducing conditions with the unlabeled TF. This approach allowed identification of the binding site(s) of the chaperone.[20] These studies were successful because the TF chaperone is insensitive to the reducing agent because it lacks cysteine residues. This approach has a drawback, however, as the resulting sample may not be homogeneously reduced (see discussion in Section 5.3.3). In addition to NMR experiments, isothermal titration calorimetry (ITC) with the fragments of the client protein and multiangle light scattering (MALS) experiments of the complex were performed to determine the stoichiometry of the complex and the binding affinity. The chaperone appeared monomeric upon interaction with the client and the dissociation constant was between 25 and 200 μM.[20] The same two client proteins (PhoA and MBP) were used for the characterization of client-protein binding by the chaperone SecB under similar conditions (reducing buffer or introducing mutations to make the client soluble and unfoldable). The resulting complexes were studied by NMR, surface plasmon resonance (SPR), and bio-layer interferometry (BLI).[57]

Taken together, in cases where the client does not spontaneously bind (unlike those presented in Section 5.3.1), harsher conditions for the client protein (high temperature and reducing conditions) are used. By choosing chaperones from hyperthermophilic organisms or chaperones that do not rely on disulfide bonds, such conditions may be found.

5.3.3 Mutating Client Proteins to Make Them Chaperone-binding Prone (or Reduce Their Aggregation Propensity)

As already mentioned above in some of the examples (Im7 in Section 5.3.1 and MBP in Section 5.3.2), site-specific mutations are used to stabilize unfolded conformations of the client in order to increase the population of chaperone-binding competent states. The phosphatase A (PhoA) protein, encountered in Section 5.3.2.2, contains native disulfides which are important for folding. The binding-competent unfolded state can be enhanced under reducing conditions[20,57] or by mutating the cysteines to *e.g.* serines. The cysteine-less variant of full-length PhoA (proPhoAS4) shows no indication of residual structure based on its NMR spectrum, in contrast to the reduced form of a wild-type protein, for which some residual secondary structure propensity has been observed.[58] The reduction of the cysteines in a wild-type PhoA may not be complete, and it may change over time as the reducing agent gets consumed (oxidized). Consequently, a mixture of unfolded and partially folded species of the client protein is present in solution, and the unfolded portion of the client has a tendency to decrease over time.[58] The cysteine-less variant has been used for studies of binding to the substrate-binding domain (SBD) of DnaK.[58] The practical drawbacks of a cysteine-less variant, however, can be that (i) its production requires solubilization and purification from inclusion bodies and (ii) it may allow only for short time storing at low concentrations, because aggregation from the unfolded state is more rapid than if the protein was folded during storage.[58]

Mutating cysteines may also be of use in a different context: the formation of intermolecular disulfide bonds may enhance the aggregation of client proteins. For forming complexes of membrane proteins, *i.e.* highly aggregation-prone polypeptides, Weinhäupl *et al.* developed a pull-down approach with the chaperone TIM9·10 described in Section 5.3.5. The chaperone contains essential disulfide bonds while the two client proteins chosen by the authors (ADP/ATP carrier; GDP/GTP carrier) contain four cysteines and one cysteine, respectively.[34] The authors found that the presence of native cysteines enhances the propensity of the client protein to aggregate. Employing reducing conditions is not an option because the chaperone unfolds when its disulfides are broken. Hence, in this case, mutating the native cysteines in the client proteins to Ala or Ser was a successful solution for complex formation without accumulating aggregates (see Section 5.3.5).

5.3.4 Complex Formation upon Removal of a Denaturant

Some client proteins may not be kept in aqueous solution at all, *i.e.* the aggregation reaction may be predominant. For such proteins one needs to find conditions in which they can encounter (in solution) the chaperone proteins; of course, these conditions shall not be denaturing, so as to maintain the chaperone in a folded and active state. Membrane-protein (MP) precursors on the way to their membranes (*e.g.* bacterial outer-membrane proteins or MPs of organelles) are generally such highly-insoluble polypeptides. Bacterial outer membrane proteins (OMPs) with a native β-barrel fold require purification under chemically denaturing conditions from inclusion bodies. Chemical denaturation at high concentrations of chaotropic agents such as 8 M urea or 6 M guanidine-hydrochloride implies denaturing of all proteins present in solution and shielding of all (hydrophobic) intramolecular and intermolecular interactions. Consequently, the denaturing agent needs to be removed for the client–chaperone complex formation. It has been shown that drop-wise dilution (drip dilution) of denatured Omp into a solution containing a native chaperone of OMPs, Skp, leads to successful complex formation.[59] Drip dilution is a very fast dilution of the denaturant by dropwise addition of a denatured sample into a typically 100 times bigger volume of the refolding buffer over a period of 1 h, keeping the unfolded protein at minimal concentrations during this procedure.[60] In the denaturing buffer, the OMP is unfolded, but as soon as the OMP is rapidly dissolved in the chaperone-containing buffer at a very low denaturant concentration, the OMP may bind to the chaperone, but it may also aggregate. (The folding reaction is negligible, as membrane proteins generally do not fold to a defined structure outside a membrane.) The relative rate constants of aggregation *vs.* binding, thus, determine whether a complex forms. The aggregation rate constant is temperature dependent, and aggregation is slowed at low temperatures.[61] The aggregation furthermore depends on the concentration of the client protein; diluting into a sufficiently large volume and good mixing during dilution are, therefore, helpful to suppress aggregation. The rate constant of P–C complex formation is dependent on the concentrations of both the client protein and the chaperone. The drip dilution method implies high dilution of the client protein concentration too, causing an additional decrease in the aggregation rate due to reduced diffusion-collision probability of the unfolded client molecules. However, to ensure that most of the chaperone is in the client bound state, excess of denatured client needs to be added in a drop-wise manner until precipitation due to a lack of free chaperone.[59] In the preparation of Skp–OMP complexes, the Hiller group has added an excess of OMP (solubilized in 6 M guanidine) in a drop-wise manner to Skp in native buffer (such as 25 mM MES, 150 mM NaCl, pH 6.5), under stirring, until precipitation appeared.[59,62] The solution was then centrifuged before analysis by structural and biophysical methods. In some special cases, the assembly buffer in which the client is drip-diluted can contain a reducing agent such as DTT (see Section 5.3.2 for retaining the client binding-prone under chaperone compatible conditions) if working with a disulfide bond-free chaperone. Once the chaperone–client complex is formed, the assembly buffer can be exchanged for the analysis compatible buffer without a chemical reductant.[63]

Essentially the same method was applied for forming complexes of trigger factor (TF) with OmpA. Here, the purification of the client protein was performed under denaturing conditions, and complexes were formed by diluting the client sample 20-fold in the presence of a chaperone under reducing conditions to keep the OmpA unfolded and TF-binding prone.[20]

It is noteworthy that the kinetics and thermodynamics of complex formation between chaperones and highly aggregation-prone polypeptides are very different from those of the soluble clients (described in Sections 5.3.1–5.3.3). For Skp complexes with several β-barrel client proteins of the OMP family, the dissociation constant was consistently in the one- to two-digit nanomolar range.[64] The lifetime of the Skp–OmpA complex is *ca.* 2.6 hours,[59] which is more than six orders of magnitude longer than those of complexes formed by *e.g.* thermosome and lysozyme, and more than three orders of magnitude longer than those of the DnaK complexes (Section 5.3.1).

In a similar way, a GroEL/ES-client complex was obtained for cryo-EM studies capturing chaperone 'in action' with an encapsulated client. The clients used in this study were denatured in acidic urea buffer and mixed with an open-state GroEL/ES chaperone solution under reducing conditions, by drip-diluting the denaturant 50-fold, after which ATP was added to ensure the encapsulation of the client.[65]

Complex formation by the drip dilution method requires that the binding is faster than aggregation and that the complexes are stable. The method works for forming complexes of bacterial outer-membrane proteins with different chaperones (Skp, SurA, and trigger factor). We found that in another case of membrane-protein precursors as client proteins, the drip dilution method fails: the case of mitochondrial-carrier membrane proteins and their native chaperones, the hetero-hexameric TIM9·10 chaperone. Drip dilution of different guanidine-solubilized mitochondrial carriers into a solution of TIM9·10 (testing a range of optimization parameters) has not produced a complex; rather, the client protein precipitated (not published). A reason for this failure may be that mitochondrial carrier proteins (α-helical in their native state) are more hydrophobic than OMPs. This higher hydrophobicity may lead to faster aggregation compared to OMPs. Additionally, TIM chaperones disassemble above approx. 100 mM guanidine-hydrochloride (unpublished data). The guanidine sensitivity of the chaperone might result in the destabilization of the chaperone at the locally high denaturant concentration as the drop gets diluted into a chaperone-containing buffer.

In a Förster resonance energy transfer (FRET) study, DnaK, introduced in Section 5.3.1, has been studied in complex with rhodanese.[66] Of interest is that the rhodanese–DnaK complex is strictly dependent on the presence of the J-domain protein DnaJ (which is essential for ATP hydrolysis of Hsp70 chaperones) and ATP. Kellner *et al.* prepared samples from rhodanese that were denatured in 4 M guanidinium buffer and then diluted the protein into a buffer without a denaturant, but with DnaK, DnaJ, GrpE (a nucleotide exchange factor), and ATP. The particularity of single-molecule FRET is its ability to work at very low concentrations (25 to 75 pM in this study).

5.3.4.1 Complex Formation by Dialysis

Another method for the formation of complexes from a denaturant solution is dialysis. In such a procedure, one would mix the client protein and the chaperone in the presence of a denaturant and dialyze out the denaturant. At the initially high denaturant concentration, not only the client protein is solubilized but, most often, the chaperone is also perturbed; moreover, the shielding by the chaotropic agent likely hampers the interaction between the client and the chaperone. Dialysis may, thus, work if during the time-dependent reduction of denaturant concentration the chaperone's integrity and binding capacity are restored before the aggregation of the client protein becomes dominant. This approach has been reported to be successful *e.g.* for the DegQ chaperone/protease.[67,68] In the case of the above-described TIM9·10 chaperone with mitochondrial carriers, dialysis approaches proved unsuccessful: after dialysis, we could only obtain the aggregated client protein and free chaperone (unpublished). It is likely that this is due to the fact that the hetero-hexameric chaperone is disassembled at a concentration of *ca.* 100 mM guanidine, likely much less than what is needed to keep mitochondrial carriers in solution.

5.3.5 Hampering Aggregation by Client Immobilization: Complex Formation with a Pull-down Approach

As outlined above, drip dilution and dialysis were unsuccessful in generating complexes of TIM9·10 and mitochondrial carriers (such as the ADP/ATP carrier Aac1 and the GDP/GTP carrier Ggc1). Weinhäupl *et al.* have developed an approach for complex formation that strongly reduces aggregation and allows obtaining homogeneous and long-lived complexes, stable over days to weeks[34,69] (Figure 5.3). The method comprises first the binding of an unfolded client protein in a denaturant (such as 6 M guanidine) on an affinity column, *via* an affinity tag on the client-protein construct (His_6-tag). As the individual molecules are attached to the beads at a fixed location, they are unable to diffuse and encounter each other and thus form multimeric aggregates. Then, the buffer is exchanged for a more native buffer without a denaturant ("binding buffer"). The concentration of the denatured client

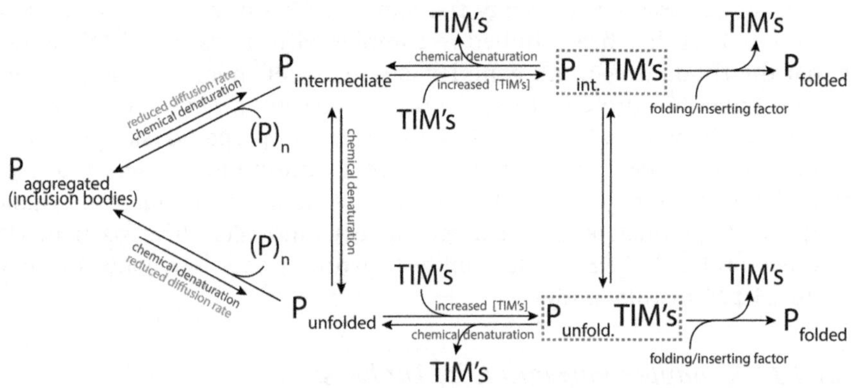

Figure 5.3 Scheme of different approaches for the mitochondrial intermembrane space chaperone complex formation. The method developed for complex formation with mitochondrial membrane proteins[34] comprises three steps. In the first step (red) hydrophobic client protein is unfolded from the inclusion bodies using a strong chemical denaturant that prevents the formation of protein aggregates and keeps the client in an unfolded state. However, a strong denaturant prevents any possible interaction with the client due to the chaperone's instability under such conditions. In the second step (green), the client is bound to the affinity column, which reduces the local concentration of the client and its diffusion rate, and the rate of client aggregation is reduced once the denaturant is removed. In the third step, the denaturant is removed and saturating concentration of the chaperone is added (blue) to the bound client increasing the rate of complex formation. In a second method, the client is produced by a cell-free (*in vitro*) expression system which contains the chaperone, in addition to the usual components required for *in vitro* translation. In this approach, there are no aggregation-prone client species (*i.e.* unfolded and free client) as long as there are free chaperones in the cell-free reaction solution.

loaded on the column is kept low: on a 5 ml affinity resin approximately 2.5 mg of unfolded client is loaded. This helps to reduce the aggregation rate. Immediately after the removal of the denaturant, an excess of chaperone, typically 2 times higher compared to the client protein, is passed through the column. The non-bound chaperone is washed off the column, and the complex is then eluted (*e.g.* with imidazole). In a slightly modified version of this protocol, the denaturant was not removed before the addition of the chaperone. Instead, a 2-fold excess of chaperone (compared to the loaded client protein) was added in one-fifth of the column volume and the denaturant concentration gradually decreased by diluting the flow-through solution 1 : 1 with binding buffer before passing it onto the column again. This process was repeated until the denaturant concentration was below 0.05 M (roughly in a total volume of 10 times the column volume). In this latter approach, the chaperone is presumably disassembled in the beginning and when the denaturant concentration is below a critical level (for TIM9·10 this is of the order of 100 mM), the chaperone assembles and is able to bind to the client protein.

For TIM chaperones and mitochondrial carriers, both variants work, and the former (namely removing the denaturant from the column before adding the chaperone) is experimentally somewhat simpler. After desalting and concentrating the eluted complex, the final concentration of the chaperone complex was of the order of 150 μM, well suited for structural studies. The sample was stable for days of measurement at up to 35 °C. The lifetime of the complex was *ca.* 4 hours, *i.e.* when apo-chaperone was added to the P–C complex, it took 4 hours to transfer the client from one chaperone to another. No significant aggregation of the client was observed under these conditions. Interestingly, however, gel filtration experiments of the complex were unsuccessful: the TIM9·10 complex with mitochondrial carriers dissociates during passage on a size exclusion column (unpublished data). In contrast, with another client protein, Tim23, the P–C complexes of both TIM9·10 and a homologous chaperone, TIM8·13, could be analyzed by gel filtration and coupled SEC-MALS (see supplemental figures S12 and S13 in Sučec *et al.*, 2020).[69] Characterization of these complexes has also been performed using other biophysical/biochemical methods, such as analytical ultra-centrifugation, small-angle X-ray scattering, NMR-detected translational diffusion experiments, or SDS-PAGE.[34,69] Note that with the drip-dilution method, one can determine dissociation constants K_d, by following *e.g.* fluorescence spectra upon titrating the client protein into a chaperone solution.[64] In the pulldown method described here, determining affinities or stoichiometries is not possible.

It is worth noting that this approach to complex formation was not successful for the TIM chaperones with a precursor of the outer membrane β-barrels. None of the above methods (drip dilution, dialysis, and pull-down) resulted in complex formation. The β-barrel client protein was found to aggregate under all these conditions (reported in Weinhäupl *et al.*, 2018).[34] The authors were able to generate complexes of weak affinity (K_d in the low mM range) only with a β-turn fragment of the β-barrel client, but only if the β-turn was stabilized by cyclization of the peptide. A likely explanation for this observation is that in a cyclic β-turn element the side chains that are hydrophobic and hydrophilic point to the two opposite faces of the β-turn element. In other words, such an element creates one hydrophilic and one hydrophobic face – and the latter presumably binds. In a totally disordered polypeptide, the hydrophilic/hydrophobic side chains point randomly in different directions, which likely impedes the chaperone binding.

5.3.6 Purifying P–C Complexes from the Cell

Direct purification of the P–C complex from the host organism would be biologically most relevant for biophysical studies and characterization. In order for this approach to be successful, the stability of the P–C complex must be sufficiently high such that the complex survives all steps from cell disruption to purification. This approach was shown to be successful in the case of the Hsp60 chaperone GroEL/ES[70] and the bacterial periplasmic chaperone/protease DegP with outer-membrane proteins.[71] The majority of DegP purified from *E. coli* is co-purified as a 12-mer and 24-mer with bound OMPs (OmpA, OmpC, OmpF, and LamB). These complexes could be separated from the apo-chaperone by ion exchange chromatography. The yield of the purified complex was sufficient for electron microscopy structural studies to determine the active chaperone state.[72] However, this approach is often discarded for client–chaperone sample preparation due to the heterogeneity of the sample, where chaperones with different client proteins can be present. In a study by Fei *et al.*, a subsequent refolding of GroEL was applied after chaperone purification to reduce unwanted client-bound particles.[70] However, co-purification of chaperone–client particles may in fact be highly desirable, in particular if coupled to cryo-EM, where the separation into free and client-loaded classes can be done post acquisition.

Co-expression of the client and substrate protein from the same expression plasmid in the bacterial cells was applied to obtain the trigger factor in a complex with the S7 (ribosomal subunit protein) and OmpA client proteins for crystallography studies.[73] This was done after identifying the TF-substrate proteome in *E. coli* by co-purification with TF and liquid chromatography–tandem mass spectrometry (LC–MS/MS) analysis. The complexes of the trigger factor formed during co-expression could be co-purified on the affinity column with only one binding partner carrying the affinity tag, and its stability is sufficient for characterization by size-exclusion chromatography and analytical ultracentrifugation with a yield high enough for crystallographic studies.[73]

Baculovirus co-expression of human Hsp90 complexes in the insect cells was used for the production of chaperone–cochaperone–client complex for (cryo-) electro-microscopy studies. With the addition of a mimetic of the ATP γ-phosphate (molybdate) during the complex co-purification, the closed state of an Hsp90–Cdc37–Cdk4 ternary complex was captured with the client in the semi-folded state.[74,75] Interestingly, Verba *et al.* also reported that simple mixing of proteins individually produced and purified did not result in successful complex formation. This may be due to modifications that occur after complex formation or additional factors required for the complex formation.[75]

5.3.7 Capturing Emerging Client Proteins in a Cell-free System

Another method of chaperone–client complex formation, with a client being highly aggregation prone, is during *in vitro* protein synthesis (cell-free expression) of a client,[33] in the presence of a purified chaperone.[32,34] *In vitro* protein synthesis using a bacterial cell extract in the presence of recombinantly produced and purified TIM9·10 chaperone significantly improved the solubility of mitochondrial carrier precursors (Mpc3, Mpc2, and Ggc1) in a chaperone concentration-dependent manner. The majority of the hydrophobic carrier precursor aggregate during the cell-free expression in the absence of detergent or chaperone. From these experiments, it was concluded that the mitochondrial pyruvate carriers (Mpc) depend on the TIM9·10 chaperone during their import. In addition, binding to the previously established hydrophobic binding site on the chaperone could be confirmed.[32] A high local concentration of chaperone in the cell-free reaction mixture enables clients to capture on its exit of the ribosome, reducing the probability of client aggregation.

The protein yield in cell-free expression systems is often of the order of 0.5 mg ml^{-1}. Constant improvements in the methodology enable the production of 'difficult-to-obtain' proteins (some eukaryotic, post-translationally modified, membrane or 'toxic-for-the-cell' proteins)[76] as well as specific labeling schemes for NMR studies, including deuteration.[77] In parallel with the reduction of costs, this method is becoming used more often to obtain samples for structural biology studies.[78] However, it often requires extensive optimization of the reaction conditions that is protein dependent, time consuming, and often expensive.

5.3.8 Fusing Chaperone and Client into a Single Polypeptide Chain

An additional approach to generate complexes is to fuse a client protein (or a fragment) to a chaperone, *i.e.* to generate a single polypeptide containing both proteins, separated by a suitable linker. Conceptually, this approach shifts the equilibrium (P/P–C) towards the bound form because of the proximity; furthermore, the presence of the chaperone may hamper the aggregation reaction and may also hamper the folding of the client protein. This approach was used, for example, for stabilizing a predominantly unfolded variant of the maltose binding protein (MBP) to trigger factor separated by a 25-residue linker for a binding site(s) verification.[20]

5.3.9 Chaperones Bound to Protein Aggregates

A particularly interesting process is how chaperones bind to pre-existing protein aggregates. Chaperones have been shown to play key roles in suppressing protein aggregation by interacting with aggregation-prone proteins at different stages, including in oligomeric and fibrillized forms. In one recent example, the interaction of J-domain proteins (Hsp40) with amyloid fibrils formed by the protein tau has been monitored. From a preparation standpoint such samples are, arguably simpler than the above examples that aim to generate a soluble sample: the pre-formed amyloid fibers were incubated with the chaperone and monitored by negative-stain EM.[79]

5.4 Concluding Remarks

We have outlined here a number of strategies that can be successful for preparing chaperone complexes with client proteins, with the aim to characterize these complexes structurally and functionally. These strategies are summarized in the scheme of Figure 5.4. The properties of the various client proteins and chaperones vastly differ, and so do the preparation schemes, affinities, and complex lifetimes. In our view, it is not possible to make precise predictions about which of the protocols will be successful, but general trends nonetheless emerge. If the client protein is largely soluble, then not aggregation of P, but rather the stability of the folded state of P may

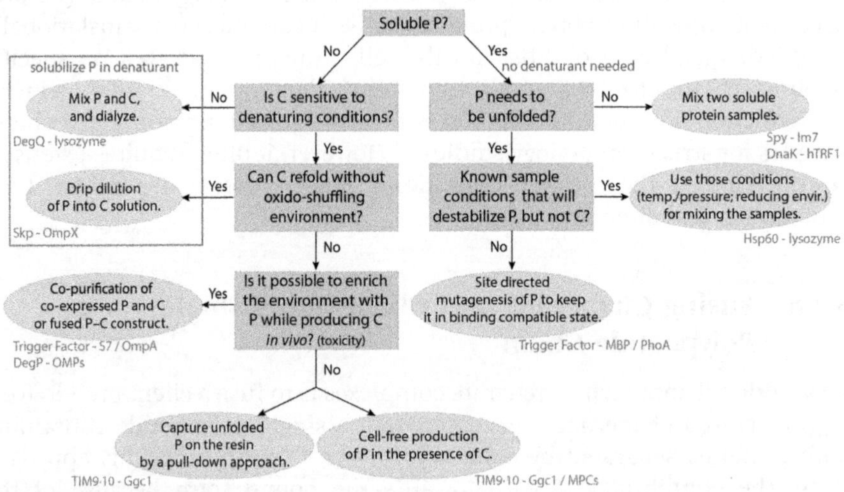

Figure 5.4 Overview of the main methods used for preparing chaperone–client complexes. Examples of these different cases are given by the name of the involved proteins. See the main text for details.

hamper successful P–C complex formation. In such cases, destabilizing P by mutations or appropriate sample conditions (denaturing, heating, and possibly also by applying pressure) is a promising route, provided the chaperone resists. The studied cases suggest that the binding and release reactions in such complexes are fast, often on millisecond time scales.

If the client protein is highly aggregation prone, then protocols that start with a denatured state are the way to go. Although the number of reported cases is small, we propose that the most aggregation-prone client proteins may not be amenable to the drip-dilution approach, and they may rather need *e.g.* a pull-down approach[34] or a cell-free production or isolation from the native environment.

The steady improvement of methods for the structural and dynamical study of chaperone complexes, including cryo-EM, solution- and solid-state NMR, and optical spectroscopies such as single-molecule FRET comes jointly with a better understanding of the sample preparation protocols. Biochemistry, biophysics, and structural-biology tools as well as *in vivo* experiments have allowed understanding of the fascinating mechanisms of chaperone function at the atomic level, which is not only of fundamental interest but also of biomedical importance.

Acknowledgements

We thank Dr Beate Bersch (IBS Grenoble) and Undina Guillerm (IST Austria) for critical reading of the manuscript.

References

1. J. F. Díaz-Villanueva, R. Díaz-Molina and V. García-González, *Int. J. Mol. Sci.*, 2015, **16**, 17193–17230.
2. R. J. Ellis and A. P. Minton, *Biol. Chem.*, 2006, **387**, 485–497.
3. F. U. Hartl, A. Bracher and M. Hayer-Hartl, *Nature*, 2011, **475**, 324–332.
4. Y. E. Kim, M. S. Hipp, A. Bracher, M. Hayer-Hartl and F. U. Hartl, *Annu. Rev. Biochem.*, 2013, **82**, 323–355.
5. T. Sinnige, A. Yu and R. I. Morimoto, *Adv. Exp. Med. Biol.*, 2020, **1243**, 53–68.
6. D. Balchin, M. Hayer-Hartl and F. U. Hartl, *FEBS Lett.*, 2020, **594**, 2770–2781.
7. R. U. Mattoo and P. Goloubinoff, *Cell. Mol. Life Sci.*, 2014, **71**, 3311–3325.
8. B. M. Burmann and S. Hiller, *Prog. Nucl. Magn. Reson. Spectrosc.*, 2015, **86–87**, 41–64.
9. L. He and S. Hiller, *ChemBioChem*, 2019, **20**, 2803–2806.
10. J. Macošek, G. Mas and S. Hiller, *Front. Mol. Biosci.*, 2021, **8**, 1–11.
11. I. Sučec, B. Bersch and P. Schanda, *Front. Mol. Biosci.*, 2021, **8**, 1–24.
12. A. Bracher and F. U. Hartl, in *Encyclopedia of Biological Chemistry*, ed. W. J. Lennarz and M. D. Lane, Academic Press, Waltham, 2nd edn, 2013, pp. 456–460.

13. L. Skjærven, J. Cuellar, A. Martinez and J. M. Valpuesta, *FEBS Lett.*, 2015, **589**, 2522–2532.
14. M. Heyden, J. A. Freites, M. B. Ulmschneider, S. H. White and D. J. Tobias, *Soft Matter*, 2012, **8**, 7742–7752.
15. B. Schiffrin, D. J. Brockwell and S. E. Radford, *BMC Biol.*, 2017, **15**, 1–16.
16. A. P. Capaldi, C. Kleanthous and S. E. Radford, *Nat. Struct. Biol.*, 2002, **9**, 209–216.
17. A. Espargaró, V. Castillo, N. S. de Groot and S. Ventura, *J. Mol. Biol.*, 2008, **378**, 1116–1131.
18. A. Tsirigotaki, K. E. Chatzi, M. Koukaki, J. De Geyter, A. G. Portaliou, G. Orfanoudaki, M. F. Sardis, M. B. Trelle, T. J. D. Jørgensen, S. Karamanou and A. Economou, *Structure*, 2018, **26**, 695–707.e695.
19. L. Ferbitz, T. Maier, H. Patzelt, B. Bukau, E. Deuerling and N. Ban, *Nature*, 2004, **431**, 590–596.
20. T. Saio, X. Guan, P. Rossi, A. Economou and C. G. Kalodimos, *Science*, 2014, **344**, 1250494.
21. T. Shiota, K. Imai, J. Qiu, V. L. Hewitt, K. Tan, H.-H. Shen, N. Sakiyama, Y. Fukasawa, S. Hayat, M. Kamiya, A. Elofsson, K. Tomii, P. Horton, N. Wiedemann, N. Pfanner, T. Lithgow and T. Endo, *Science*, 2015, **349**, 1544–1548.
22. A. Finka, R. U. Mattoo and P. Goloubinoff, *Cell Stress Chaperones*, 2011, **16**, 15–31.
23. A. Finka and P. Goloubinoff, *Cell Stress Chaperones*, 2013, **18**, 591–605.
24. J. C. Young, V. R. Agashe, K. Siegers and F. U. Hartl, *Nat. Rev. Mol. Cell Biol.*, 2004, **5**, 781–791.
25. R. Bakthisaran, R. Tangirala and C. M. Rao, *Biochim. Biophys. Acta, Proteins Proteomics*, 2015, **1854**, 291–319.
26. S. Quan, P. Koldewey, T. Tapley, N. Kirsch, K. M. Ruane, J. Pfizenmaier, R. Shi, S. Hofmann, L. Foit, G. Ren, U. Jakob, Z. Xu, M. Cygler and J. C. Bardwell, *Nat. Struct. Mol. Biol.*, 2011, **18**, 262–269.
27. A. Alam, J. E. Bröms, R. Kumar and A. Sjöstedt, *Front. Mol. Biosci.*, 2021, **8**, 668910.
28. K. Chakraborty, M. Chatila, J. Sinha, Q. Shi, B. C. Poschner, M. Sikor, G. Jiang, D. C. Lamb, F. U. Hartl and M. Hayer-Hartl, *Cell*, 2010, **142**, 112–122.
29. S. Rüdiger, A. Buchberger and B. Bukau, *Nat. Struct. Biol.*, 1997, **4**, 342–349.
30. L. He, B. Bardiaux, M. Ahmed, J. Spehr, R. König, H. Lünsdorf, U. Rand, T. Lührs and C. Ritter, *Proc. Natl. Acad. Sci. U. S. A.*, 2016, **113**, E272–E281.
31. L. He and S. Hiller, *Angew. Chem., Int. Ed.*, 2018, **57**, 5921–5924.
32. H. Rampelt, I. Sučec, B. Bersch, P. Horten, I. Perschil, J. C. Martinou, M. van der Laan, N. Wiedemann, P. Schanda and N. Pfanner, *BMC Biol.*, 2020, **18**, 2.

33. B. Schneider, F. Junge, V. A. Shirokov, F. Durst, D. Schwarz, V. Dotsch and F. Bernhard, *Methods Mol. Biol.*, 2010, **601**, 165–186.
34. K. Weinhäupl, C. Lindau, A. Hessel, Y. Wang, C. Schütze, T. Jores, L. Melchionda, B. Schönfisch, H. Kalbacher, B. Bersch, D. Rapaport, M. Brennich, K. Lindorff-Larsen, N. Wiedemann and P. Schanda, *Cell*, 2018, **175**, 1365–1379.e1325.
35. S. Hiller and B. M. Burmann, *J. Magn. Reson.*, 2018, **289**, 142–155.
36. A. Sekhar, R. Rosenzweig, G. Bouvignies and L. E. Kay, *Proc. Natl. Acad. Sci. U. S. A.*, 2015, **112**, 10395–10400.
37. S. Gianni, N. R. Guydosh, F. Khan, T. D. Caldas, U. Mayor, G. W. N. White, M. L. DeMarco, V. Daggett and A. R. Fersht, *Proc. Natl. Acad. Sci. U. S. A.*, 2003, **100**, 13286–13291.
38. A. Sekhar, A. Velyvis, G. Zoltsman, R. Rosenzweig, G. Bouvignies and L. E. Kay, *eLife*, 2018, 7, e32764.
39. R. Rosenzweig, A. Sekhar, J. Nagesh and L. E. Kay, *eLife*, 2017, **6**, e28030.
40. A. P. Capaldi, M. C. R. Shastry, C. Kleanthous, H. Roder and S. E. Radford, *Nat. Struct. Biol.*, 2001, **8**, 68–72.
41. C. T. Friel, D. A. Smith, M. Vendruscolo, J. Gsponer and S. E. Radford, *Nat. Struct. Mol. Biol.*, 2009, **16**, 318–324.
42. S. B. M. Whittaker, G. R. Spence, J. Günter Grossmann, S. E. Radford and G. R. Moore, *J. Mol. Biol.*, 2007, **366**, 1001–1015.
43. L. He, T. Sharpe, A. Mazur and S. Hiller, *Sci. Adv.*, 2016, **2**, 1–12.
44. S. Horowitz, L. Salmon, P. Koldewey, L. S. Ahlstrom, R. Martin, S. Quan, P. V. Afonine, H. van den Bedem, L. Wang, Q. Xu, R. C. Trievel, C. L. Brooks 3rd and J. C. Bardwell, *Nat. Struct. Mol. Biol.*, 2016, **23**, 691–697.
45. F. Stull, P. Koldewey, J. R. Humes, S. E. Radford and J. C. A. Bardwell, *Nat. Struct. Mol. Biol.*, 2016, **23**, 53–58.
46. C. L. Pashley, G. J. Morgan, A. P. Kalverda, G. S. Thompson, C. Kleanthous and S. E. Radford, *J. Mol. Biol.*, 2012, **416**, 300–318.
47. J. Wang, *Nat. Struct. Mol. Biol.*, 2018, **25**, 989–990.
48. F. Stull, P. Koldewey, J. R. Humes, S. E. Radford and J. C. Bardwell, *Nat. Struct. Mol. Biol.*, 2016, **23**, 53–58.
49. W. He, J. Zhang, V. Sachsenhauser, L. Wang, J. C. A. Bardwell and S. Quan, *J. Biol. Chem.*, 2020, **295**, 14488–14500.
50. T. O. Street, L. A. Lavery and D. A. Agard, *Mol. Cell*, 2011, **42**, 96–105.
51. G. E. Karagoz, A. M. Duarte, E. Akoury, H. Ippel, J. Biernat, T. Moran Luengo, M. Radli, T. Didenko, B. A. Nordhues, D. B. Veprintsev, C. A. Dickey, E. Mandelkow, M. Zweckstetter, R. Boelens, T. Madl and S. G. Rudiger, *Cell*, 2014, **156**, 963–974.
52. S. Weickert, M. Wawrzyniuk, L. H. John, S. G. D. Rüdiger and M. Drescher, *Sci. Adv.*, 2020, **6**, eaax6999.
53. C. Deville, M. Carroni, K. B. Franke, M. Topf, B. Bukau, A. Mogk and H. R. Saibil, *Sci. Adv.*, 2017, **3**, e1701726.
54. C. Deville, K. Franke, A. Mogk, B. Bukau and H. R. Saibil, *Cell Rep.*, 2019, **27**, 3433–3446.e3434.

55. Y. Yin, X. Feng, H. Yu, A. Fay, A. Kovach, M. S. Glickman and H. Li, *Cell Rep.*, 2021, **35**, 109166.
56. G. Mas, J. Y. Guan, E. Crublet, E. C. Debled, C. Moriscot, P. Gans, G. Schoehn, P. Macek, P. Schanda and J. Boisbouvier, *Sci. Adv.*, 2018, **4**, 1–10.
57. C. Huang, P. Rossi, T. Saio and C. G. Kalodimos, *Nature*, 2016, **537**, 202–206.
58. E. M. Clerico, A. K. Pozhidaeva, R. M. Jansen, C. Ozden, J. M. Tilitsky and L. M. Gierasch, *Proc. Natl. Acad. Sci. U. S. A.*, 2021, **118**(41), e2016962118.
59. B. M. Burmann, C. Wang and S. Hiller, *Nat. Struct. Mol. Biol.*, 2013, **20**, 1265–1272.
60. R. R. Burgess, *Methods Enzymol.*, 2009, **463**, 259–282.
61. Y. Xie and D. B. Wetlaufer, *Prot. Sci.*, 1996, **5**, 517–523.
62. M. Callon, B. M. Burmann and S. Hiller, *Angew. Chem., Int. Ed.*, 2014, **53**, 5069–5072.
63. J. Thoma, B. M. Burmann, S. Hiller and D. J. Muller, *Nat. Struct. Mol. Biol.*, 2015, **22**, 795–802.
64. J. Qu, C. Mayer, S. Behrens, O. Holst and J. H. Kleinschmidt, *J. Mol. Biol.*, 2007, **374**, 91–105.
65. D.-H. Chen, D. Madan, J. Weaver, Z. Lin, G. F. Schröder, W. Chiu and H. S. Rye, *Cell*, 2013, **153**, 1354–1365.
66. R. Kellner, H. Hofmann, A. Barducci, B. Wunderlich, D. Nettels and B. Schuler, *Proc. Natl. Acad. Sci. U. S. A.*, 2014, **111**, 13355–13360.
67. H. Malet, F. Canellas, J. Sawa, J. Yan, K. Thalassinos, M. Ehrmann, T. Clausen and H. R. Saibil, *Nat. Struct. Mol. Biol.*, 2012, **19**, 152–157.
68. R. V. Mauldin and R. T. Sauer, *Nat. Chem. Biol.*, 2013, **9**, 90–96.
69. I. Sučec, Y. Wang, O. Dakhlaoui, K. Weinhäupl, T. Jores, D. Costa, A. Hessel, M. Brennich, D. Rapaport, K. Lindorff-Larsen, B. Bersch and P. Schanda, *Sci. Adv.*, 2020, **6**, eabd0263.
70. X. Fei, X. Ye, N. A. LaRonde and G. H. Lorimer, *Proc. Natl. Acad. Sci. U. S. A.*, 2014, **111**, 12775–12780.
71. T. Krojer, K. Pangerl, J. Kurt, J. Sawa, C. Stingl, K. Mechtler, R. Huber, M. Ehrmann and T. Clausen, *Proc. Natl. Acad. Sci. U. S. A.*, 2008, **105**, 7702–7707.
72. T. Krojer, J. Sawa, E. Schäfer, H. R. Saibil, M. Ehrmann and T. Clausen, *Nature*, 2008, **453**(7197), 885–890.
73. E. Martinez-Hackert and W. A. Hendrickson, *Cell*, 2009, **138**, 923–934.
74. C. K. Vaughan, U. Gohlke, F. Sobott, V. M. Good, M. M. Ali, C. Prodromou, C. V. Robinson, H. R. Saibil and L. H. Pearl, *Mol. Cell*, 2006, **23**, 697–707.
75. K. A. Verba, R. Y. R. Wang, A. Arakawa, Y. Liu, M. Shirouzu, S. Yokoyama and D. A. Agard, *Science*, 2016, **352**, 1542–1547.
76. X. Jin and S. H. Hong, *Biochem. Eng. J.*, 2018, **138**, 156–164.
77. L. Imbert, R. Lenoir-Capello, E. Crublet, A. Vallet, R. Awad, I. Ayala, C. Juillan-Binard, H. Mayerhofer, R. Kerfah, P. Gans, E. Miclet and

J. Boisbouvier, in *Structural Genomics*, ed. Y. W. Chen and C. B. Yiu, Humana, Methods in Molecular Biology, New York, NY, 2nd edn, 2021, pp. 127–149.
78. T. Terada, S. Kusano, T. Matsuda, M. Shirouzu and S. Yokoyama, in *Advanced Methods in Structural Biology*, 2016, ch. 5, pp. 83–102.
79. R. Irwin, O. Faust, I. Petrovic, S. G. Wolf, H. Hofmann and R. Rosenzweig, *eLife*, 2021, **10**, e69601.

CHAPTER 6

NMR Study of the Structure and Dynamics of Chaperone–Client Complexes

T. SAIO[a,b]

[a] Division of Molecular Life Science, Institute of Advanced Medical Sciences, Tokushima University, Tokushima 770-8503, Tokushima, Japan; [b] Fujii Memorial Institute of Medical Sciences, Institute of Advanced Medical Sciences, Tokushima University, Tokushima 770-8503, Tokushima, Japan
Email: saio@tokushima-u.ac.jp

6.1 Introduction

Structural information on chaperone–client complexes at atomic resolution provides deep insights into the mechanisms underlying the effects of molecular chaperones on client proteins. Due to recent technical developments in the field of structural biology, increasing numbers of high-resolution molecular chaperone structures are being elucidated. Indeed, a number of crystal and cryo-electron microscopy (EM) structures have been reported for molecular chaperones.[1,2] Unveiling the distinct shapes and architectures of molecular chaperones has extended our understanding of molecular chaperone functions. For instance, recent studies on the molecular chaperones GroEL[3,4] and tailless complex polypeptide 1 ring complex (TRiC), also known as chaperonin containing tailless complex polypeptide 1 (CCT),[5] have shown that these chaperones have chamber structures undergoing structural changes coupled with ATP-binding and hydrolysis and thus implied incorporation and isolation of client proteins. Other examples include

New Developments in NMR No. 29
Biophysics of Molecular Chaperones: Function, Mechanisms and Client Protein Interactions
Edited by Sebastian Hiller, Maili Liu and Lichun He

Published by the Royal Society of Chemistry, www.rsc.org

prefoldin[6] and Skp[7] whose crystal structures have been shown to be basket-like, implying the incorporation of unfolded client proteins within the basket-like structure. Moreover, the cradle shape of the ribosome-associated molecular chaperone trigger factor (TF) implies that TF covers the ribosome exit tunnel and waits for the nascent protein to be captured by the cradle cavity.[8,9]

In addition to structural analyses of molecular chaperones, recent studies have investigated the structure of chaperone–client complexes. However, direct observation of client proteins bound to chaperones is technically challenging. CryoEM is a powerful tool for studying chaperone–client complexes and mapping the location of client proteins in relation to the corresponding chaperones. Early CryoEM studies were reported for GroEL in 1994[10] and prefoldin in 2002.[11] Although these studies provided information on the position of client proteins with respect to their corresponding chaperones, the limited resolution of cryoEM at that time and the dynamic properties of client protein binding to chaperones precluded a detailed understanding of the mechanisms underlying chaperone–client recognition. Cross-linking mass spectroscopy (MS) also allows the mapping of client proteins on the molecular chaperones and can be used to determine the regions in the client proteins that are recognized by molecular chaperones, as demonstrated for TF,[12] SurA,[13,14] HdeA/B,[15] and peroxiredoxin.[16] Although cross-linking MS provides peptide-level information on protein interactions, higher resolutions are required to unveil the detailed mechanisms underlying the recognition of client proteins by molecular chaperones. In general, X-ray crystallography and cryoEM are the most powerful techniques for determining protein structures at atomic resolution. However, obtaining structural information on chaperone–client complexes remains technically challenging, predominantly due to the dynamic properties of chaperone–client complexes. As molecular chaperones typically target client proteins in unfolded or partially folded states, client proteins in complex with molecular chaperones are often highly mobile and flexible. Moreover, the interaction between molecular chaperones and client proteins is often relatively weak and undergoes fast on–off exchange.[17] These dynamic features of chaperone–client complexes make the use of X-ray crystallography and cryoEM for structure determination challenging.

To investigate the structures of dynamic protein complexes, nuclear magnetic resonance (NMR) represents one of the most effective tools.[18–21] As NMR provides information on nuclei of proteins dissolved in solution, NMR provides information, in principle, at atomic resolution. Moreover, NMR can be used to study protein interactions and dynamics in addition to structure determination. Furthermore, weak interactions with K_D values up to mM order can be investigated by NMR.[22] These versatile properties have led to the increasing use of NMR in structural studies of chaperone–client complexes. This chapter reviews recent important NMR structural studies on chaperone–client complexes.

6.2 Technical Developments in the Structural Study of Chaperone–Client Complexes

6.2.1 NMR as a Tool for Structure Determination of Chaperone–Client Complexes

Although NMR has many properties that allow its use for the structural analysis of chaperone–client complexes, a major drawback of this technique is the limited size of target proteins that can be analyzed. In standard NMR analysis, the size of target proteins that can be analyzed is typically limited to smaller than ~30 kDa. However, recent technical developments, including those in isotope labeling schemes, pulse programs, hardware, and data analysis software, have allowed detailed structural studies of proteins larger than 30 kDa, with studies of several hundred kDa proteins now feasible.[18,19,23,24]

Isotope labeling represents one of the most important recent technical developments in NMR.[23,24] In contrast to the conventional approach where proteins are uniformly labeled with ^{1}H, ^{13}C, and ^{15}N, selective labeling of specific atoms, particularly ^{1}H- and ^{13}C-labeling of the methyl groups on a deuterated background, improves the sensitivity and resolution of NMR spectra of large proteins.[24] Using this labeling scheme, proteins up to approximately 100 kDa[25–27] and even 1 MDa[25–27] can be investigated using NMR. Atom-selective labeling is not limited to methyl-bearing residues and can be applied to all 20 amino acid types using the stereo-array isotope labeling (SAIL) technique.[23] The SAIL technique allows more detailed analyses of the protein structure and dynamics.[31,32] A more detailed description of isotope labeling is provided in Chapter 4.

Technical developments in data analysis tools have also increased the utility of NMR. The assignment of NMR resonances is required to obtain structural information, but resonance assignment is not always straightforward. In particular, resonance assignment is generally performed by site-directed mutation and/or structure-based assignment using the nuclear Overhauser effect (NOE) when large proteins are labeled using methyl-selective labeling. Although NOE analysis and resonance assignment are time-consuming, several programs have been developed to facilitate automated resonance assignment.[31,33–35] Although these automated programs require further development before fully replacing manual assignments, automated assignment already has utility in assisting the analysis.

6.2.2 Preparation of Unfolded Proteins for Structural Studies

As described above, recent developments in NMR techniques have increased the size of target proteins that can be analyzed, with the size of many molecular chaperones now well within the range of advanced NMR analysis. However, a further obstacle to the structural study of chaperone–client proteins is the preparation and handling of unfolded client proteins.

For structure determination by NMR, protein samples need to be stable for weeks at high concentrations (>0.1 mM). Molecular chaperones generally recognize fully or partially unfolded proteins that are prone to aggregation and degradation. Proteins can be maintained in unfolded states in the presence of high concentrations (6–8 M) of chaotropic agents including urea and guanidine hydrochloride. However, chaotropic agents that can unfold chaperones need to be removed during the preparation of chaperone–client complexes. Many intrinsically disordered proteins (IDPs) are soluble in water without chaotropic agents; however, IDPs generally have less population of hydrophobic residues in their amino acid sequences and are therefore often poor representatives of folding intermediates. On the other hand, several folded proteins including luciferase,[16,36,37] DapA,[38] and malate dehydrogenase (MDH)[10,39] are often denatured to be used as unfolded client proteins. Although these proteins have utility in refolding assays and single molecular observations, their use in high-resolution NMR structural studies was precluded presumably due to low solubility and stability.

Although most of the folded proteins suffer from low solubility and stability in the denatured unfolded state, several proteins have been recently reported to be highly soluble and stable in the unfolded state. One example is alkaline phosphatase (PhoA) which has frequently been used as a client protein in high-resolution structural analyses of chaperone–client complexes.[27,40–42] PhoA is an approximately 50 kDa (471 amino acid residues) periplasmic enzyme containing two intramolecular disulfide bonds in its native state. PhoA requires oxidizing conditions for folding, similar to several other periplasmic proteins.[43] Thus, one of the advantages of the use of PhoA in the structural study of chaperone complexes is that PhoA stays unfolded under reducing conditions without chaotropic reagents.[40] The high solubility and stability of unfolded PhoA underlie its utility in NMR studies. Other commonly used client proteins in NMR analyses include the maltose-binding protein (MBP),[27,41–42,44,45] immunity protein 7 (Im7),[46–48] Drk SH3,[49,50] outer-membrane proteins (OMPs),[51] and hTRF1.[52,53] As several mutants representing partially folded and fully unfolded states have been described for Im7, these mutants have been used to compare interactions between chaperones and client proteins at varying folding stages. Drk SH3 undergoes fold–unfold exchange in slow exchange regimes in the NMR timescale, allowing the observation of two sets of NMR signals corresponding to the unfolded and folded states on NMR spectra.[50,54] The use of these unfolded proteins as client proteins increases the utility of NMR for structural analyses of chaperone–client complexes.

6.3 Structural Features of Chaperone–Client Complexes

As described in Section 6.2, recent technical developments in solution NMR and the identification of soluble and stable unfolded client proteins have accelerated the structural study of molecular chaperones in complex with client proteins. The structural features of chaperone–client complexes and the mechanisms underlying the effects of molecular chaperones on client proteins unveiled by the structural studies are described below.

6.3.1 Hydrophobic Stretches of Unfolded Proteins are Recognized by Molecular Chaperones

Details of the mechanisms underlying chaperone–client interactions have been elucidated by studies on TF,[40] SecB,[27] and Hsp40[41] (see Figure 6.1). In all three cases, high-resolution structures of chaperones in complex with unfolded client proteins were determined by solution NMR exploiting methyl-selective labeling and methyl-TROSY. Of these, the structure of TF in complex with unfolded PhoA was reported first.[40]

The structure of the TF–PhoA complex reveals that the TF chaperone recognizes multiple hydrophobic stretches of the client protein PhoA, consisting of 5–10 amino acid residues (see Figure 6.1). One TF molecule recognizes 100–150 amino acid residues of PhoA. The recognition of the hydrophobic regions of the client protein by TF suggests that the chaperone protects the hydrophobic regions and prevents nonspecific contact with other molecules, thereby preventing aggregation. The interaction between TF and PhoA is mediated by hydrophobic interactions, with contributions from hydrogen bonds and charge interactions. In particular, bulky hydrophobic aliphatic residues including Ile, Leu, and Val and aromatic residues are recognized by TF. TF uses the four isolated client-binding sites to capture multiple sites of PhoA and keeps the protein in an extended and elongated conformation. The client-binding sites on TF identified by this structural study correspond to those suggested by previous cross-linking experiments for TF and the nascent protein emerging from the ribosome.[12] Moreover, these regions recognized by TF are highly hydrophobic and are supposed to be folded into the core of PhoA in the native state (see Figure 6.1), suggesting that recognition of these hydrophobic segments by TF prevents PhoA from folding. Interestingly, similar recognition patterns have been observed for other chaperones including SecB[27] and Hsp40[41] (see Figure 6.1). The shared binding regions on the client protein, which consist of hydrophobic amino acids and are eventually folded into the hydrophobic core of the protein in its native state, imply that SecB and Hsp40 also keep the client protein in an unfolded and extended state. Indeed, biochemical assays have demonstrated that SecB and Hsp40 function as "holdases" in preventing the folding of the client protein.[27,41] Moreover, the shared binding sites on client proteins among multiple chaperones also suggest that recognition of client proteins by chaperones can be promiscuous. Note that shared binding sites for multiple chaperones were also observed for the recognition of the client protein, Im7, by molecular chaperones Spy, SurA, and Skp.[55]

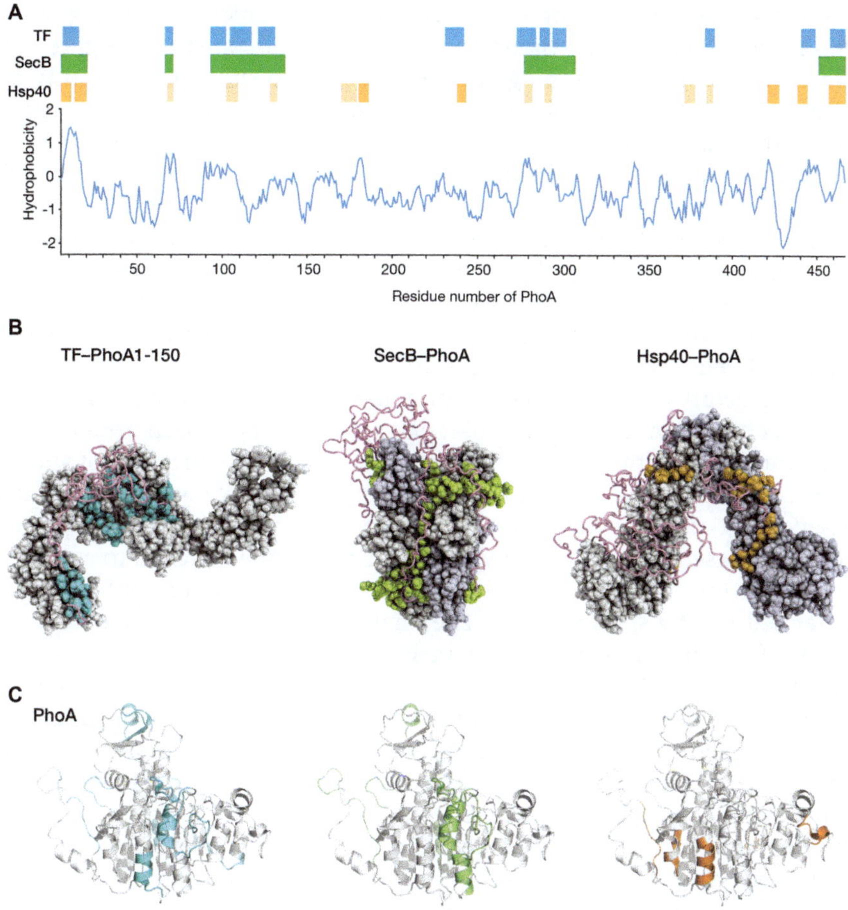

Figure 6.1 Recognition of unfolded PhoA by molecular chaperones TF, SecB, and Hsp40. (A) Hydrophobicity plot of PhoA with the regions recognized by TF, SecB, and Hsp40 indicated by cyan, green, and orange boxes, respectively. In the case of Hsp40, weak interaction sites are shown in pale orange. A hydrophobicity score (Roseman algorithm, window = 9) higher than zero denotes increased hydrophobicity. (B) Structures of chaperone–client complexes. Solution structures are shown for TF–PhoA1-150 (PDB ID: 2MLY),[40] SecB-PhoA (PDB ID: 5JTL),[27] and Hsp40-PhoA (PDB ID: 6PSI),[41] in which chaperone molecules are shown as sphere models and PhoA is shown as a ribbon model. TF is shown in gray. Subunits of the SecB tetramer and the Hsp40 dimer are shown in gray and pale blue. The amino acid residues of PhoA recognized by the chaperones are shown in a sphere model. (C) Mapping of the chaperone-binding sites of PhoA on the structure of PhoA (PDB ID: 1Y6V). Only one subunit of the dimeric PhoA is shown. The amino acid residues that are recognized by TF, SecB, and Hsp40 are shown in cyan, green, and orange, respectively.

Because the structures of TF in a complex with three different regions of PhoA were determined,[40] the structures can be compared (see Figure 6.2). Among the four client-binding sites on TF, the two sites, sites A and B, are highlighted in Figure 6.2 in which the binding sites holding three distinct PhoA stretches are represented. Comparison of the structures for site A and site B in a complex with distinct PhoA stretches demonstrates that the recognition is not strict but there appears to be weak selectivity for particular amino acids (see Figure 6.2). For example, certain pockets in the client-binding site prefer bulky aliphatic amino acids such as Leu and aromatic amino acids. This "semi-selective" recognition of amino acid type by molecular chaperones may be important for molecular chaperones to recognize many different protein sequences in the cell.

The structures of chaperone–client complexes have also been elucidated for other molecular chaperones. Several approaches have been used to study Spy in complex with the client protein Im7.[48,55,56] Visualization of the dynamic Spy–Im7 complex has been attempted using an X-ray crystallography-based approach in which the position of Im7 with respect to Spy was evaluated based on anomalous scattering from iodine attached to a specific residue of Im7 that was used to select the minimal set of conformations generated by MD simulations.[56] The analysis identified multiple conformations of Im7 in complex with Spy, including unfolded, partially folded, and native-like states,[56] which is consistent with the results of a previous kinetic study from the same group demonstrating client protein folding proceeds while the client protein binds on the Spy chaperone.[48] In the same year, an NMR study of the Spy–Im7 complex was reported.[56] This NMR study used paramagnetic relaxation enhancement that provides long-range (<20 Å) distance information and demonstrated that Im7 has more compact conformations when in complex with Spy. This spatial compaction of the client protein in complex with the chaperone was suggested to allow Spy to facilitate the folding of the client protein.

Several other NMR structural studies of chaperone–client complexes have been reported for Hsp90,[57,58] TIM8·13, TIM9·10,[59] and Skp,[51] highlighting the utility of NMR in structural studies of molecular chaperones.

6.3.2 Dynamic Features of Chaperone–Client Complexes

Although high-resolution structures of chaperone–client complexes for TF, SecB, and Hsp40 have demonstrated that the chaperone captures the client protein and maintains it in an extended and elongated conformation, NMR data have also revealed that chaperone–client complexes are dynamic and undergo fast on–off (bind–release) exchange. NMR relaxation dispersion experiments have been performed to investigate the binding kinetics of TF with the unfolded client protein PhoA.[40] This study has demonstrated that the complex undergoes on–off exchange at a rate of 1299–1499 per second.[40] Fast on–off exchange during client recognition has also been shown for SecB and Hsp40 using surface plasmon resonance and biolayer interferometry (BLI).[27,41]

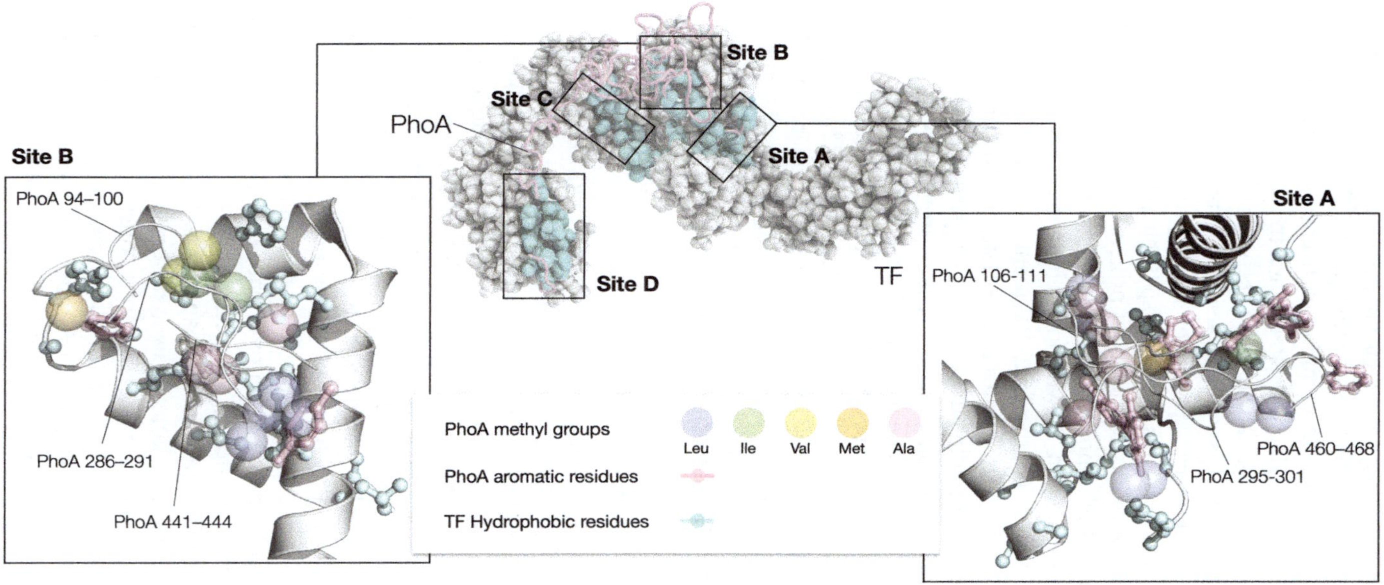

Figure 6.2 Magnified views of the structure of TF in complex with a client protein, PhoA. Recognition of multiple PhoA segments by TF *via* the client-binding sites A and B. For clarity, only one coordinate is shown for TF. The hydrophobic residues of TF consisting of the client-binding sites of TF are shown in cyan as a ball-stick model. The methyl groups of PhoA are shown as spheres. The aromatic residues of PhoA are shown in pink as a ball-stick model. This figure was prepared from structures of TF in complex with PhoA1-150 (PDBID: 2 mly), PhoA220-310 (PDBID: 2 mlx), and PhoA365-471 (PDBID: 2 mlz).[40]

Moreover, dynamic interactions between chaperone and client proteins have been reported for several other molecular chaperones including Skp,[51] Spy,[55] and Hsp89.[57] Interestingly, it has also been reported that client protein conformations are dynamic when in complex with chaperones. Such dynamic conformations have been reported for Hsp40 when in complex with PhoA, with hydrophobic stretches of PhoA shown to be recognized by multiple sites on Hsp40 and the complex comprising an ensemble of multiple conformational states.[41] Thus, hydrophobic stretches of the client protein appear to interact with multiple client-binding sites on the Hsp40 chaperone. The binding of multiple hydrophobic segments of the client protein was also observed in the TF–PhoA complex, in which shorter PhoA fragments were found to bind to multiple client-binding sites on TF with low selectivity; however, binding became more selective as the length of the fragment increased with the overall conformation becoming more stabilized and defined.[40] Other examples of dynamic binding of client proteins to molecular chaperones have been shown for the Spy chaperone in complex with the client protein Im7[55] and the Skp chaperone in complex with the client protein OmpX.[51] In these cases, hopping of the client protein on the chaperone was very fast. As the conformational exchange was too fast to build up NOE signals, PRE was used instead of NOE to obtain structural information. The absence of intermolecular NOEs for chaperone–client complexes in the case of Skp and Spy highlights the difference in the dynamic properties of Spy and Skp compared to Hsp40,[41] TF,[40] and SecB[27] showing intermolecular NOEs with the client proteins.

6.3.3 Activity–Kinetics Relationship

As described in Section 6.3.2, chaperone–client interactions are mediated by dynamic interactions and fast on–off exchanges. Furthermore, the kinetic properties of chaperone–client interaction are diverse. Recent studies indicate that the kinetic properties of chaperone–client interactions play important roles in the distinct functional properties of chaperones.[2,25,27] One example of diversity in the functional and kinetic properties of molecular chaperones can be seen in the comparison between TF and SecB. During refolding of MBP, TF weakly promotes the folding of MBP whereas SecB strongly prevents refolding.[27] However, structural studies of SecB and TF have demonstrated that both chaperones bind to similar regions on the client protein, PhoA, with comparable affinity in the low-micromolar range (see Figure 6.1),[25,27] thereby providing no clear explanation for the observed functional differences between the two chaperones. On the other hand, significant differences in binding kinetics were observed,[27] with kinetic studies demonstrating that SecB has an association rate constant (k_{on}) in the order of 10^6, whereas TF has a k_{on} in the order of 10^5. It was proposed that the variations of k_{on} can explain the variation of functional properties. Both chaperones recognize hydrophobic stretches and maintain the client protein in an unfolded and extended state on the chaperone. However, given the dynamic properties of the chaperone–client complexes, the chaperones occasionally release the client protein thereby allowing the client protein to undergo spontaneous folding (see Figure 6.3). If the chaperone has a moderate k_{on}, the client protein may have sufficient time to undergo (even partial) folding, whereas a higher k_{on} decreases the time for client protein folding. These data suggest there is an "activity–kinetics relationship" in which chaperones with a higher k_{on} are expected to have stronger holdase activity.[2] Furthermore, this relationship highlights the importance of chaperone concentration as the association rate is determined by the product of k_{on} and concentration. This phenomenon was demonstrated by the refolding of MBP in the presence of varying chaperone concentrations, with higher chaperone concentrations resulting in a stronger holdase activity.[27] This finding implies that the intracellular concentration of chaperones should also be considered when evaluating the functional properties of chaperones. The intracellular concentrations of chaperones are known to be diverse. For instance, the intracellular concentrations of TF, DnaK, and SecB (as a tetramer) in *E. coli* are reportedly ~49 μM,[60,61] ~15 μM,[62] and ~2 μM,[63] respectively. Given that chaperone localization determines the local concentration, studies of the localization and local concentrations of chaperones will be anticipated in addition to detailed *in vitro* assays and structural studies to fully understand the chaperone system.

Furthermore, the spontaneous folding rate of client proteins has been shown to contribute to the activity–kinetics relationship, with the refolding of MBP mutants having varying spontaneous folding rates.[27] In the refolding of wild-type MBP having a relatively fast folding rate, TF showed foldase activity. On the other hand, in the refolding of MBP mutants having a slower folding rate, TF showed holdase activity. Thus, the effect of the chaperone is determined by the balance between the spontaneous folding rate of the client protein and chaperone–client-binding kinetics. Interestingly, it was reported that attachment of the signal sequence decreases the protein folding rate,[64] so that the chaperone can exert stronger holdase activity to keep the client protein in an unfolded state for efficient protein translocation through the channel. More recently, it has been shown that even mature

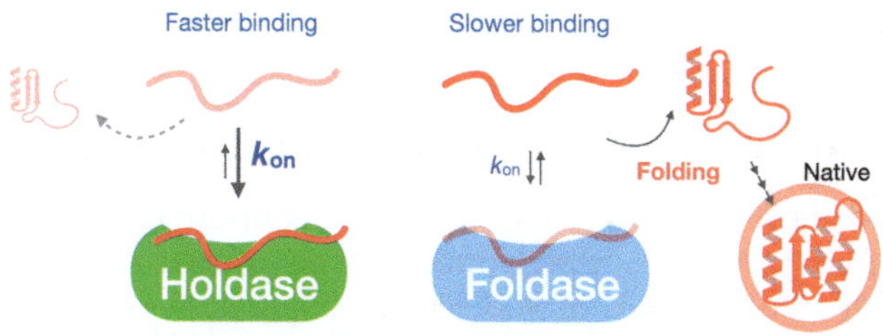

Figure 6.3 Schematic representation of the activity–kinetics relationship. Chaperones with faster association rates with client proteins have stronger holdase activity, whereas chaperones with moderate association rates function as foldases.[25,27] The intrinsic folding rate of the client protein affects the balance between holdase and foldase functions; a chaperone exerts stronger holdase activity on client proteins with slower folding rates.

regions of secretary proteins have a slower folding rate compared to their cytosolic structural homologs, implying that the amino acid sequences of mature regions are designed to have a folding rate optimized for recognition by holdase.[65] Thus, secretary proteins appear to function in optimizing the folding rate to increase binding and prevent folding by chaperones, thereby maintaining an unfolded state that allows secretion through membrane channels. The contribution of folding kinetics to the binding selectivity of SecB, a chaperone responsible for post-translational protein translocation, was proposed as "kinetic partitioning" 30 years ago;[66] however, the general importance of kinetics has recently been revisited.

An activity–kinetics relationship has also been reported for the TF monomer and dimer.[25] TF forms a homodimer in solution with a dissociation constant of approximately 2 μM in an anti-parallel orientation.[25,67,68] The stopped-flow experiment showed that the TF dimer has a faster association rate compared to the monomer.[25] As expected from the kinetic properties, the TF dimer showed stronger holdase activity, demonstrating that even the same chaperone can modulate its activity through oligomerization. Although the detailed mechanism for kinetic modulation coupled with oligomerization remains unclear, one key feature is that the client-binding sites of TF are assembled by dimerization to form a large continuous surface that is thought to promote the binding of the client protein. Interestingly, SecB has a faster association rate to the client protein than TF and forms a stable tetramer in solution.[27] In addition to SecB, many other chaperones including Hsp40 (dimer),[41] Spy (dimer),[46] Skp (trimer),[69] GroEL (14-mer), and Clp chaperones (hexamer) are known to form oligomers. Furthermore, zinc-dependent activation of *Thermus thermophilus* TF[70] has recently been reported to involve oligomerization.[71] Thus, oligomerization may represent a general strategy employed by chaperones to optimize binding kinetics required for specific roles and functions in the chaperone network. Although the recognition of client proteins by chaperones is known to be promiscuous, binding kinetics can be modulated by oligomerization leading to divergent functional properties of chaperones. This idea is consistent with a recent study demonstrating the functional importance of oligomerization of the Skp chaperone.[72]

6.3.4 Catalytic Domains of Molecular Chaperones

As described above, binding kinetics play an important role in the functional diversity of chaperones; however, mechanisms other than simple binding contribute to the function of molecular chaperones. Several chaperones possess catalytic domains. TF,[8,73] PrsA,[74] and SurA[75,76] possess PPIase domains and several ER chaperones including PDI and P5 possess disulfide isomerase domains[77–79] (see Figure 6.4). Furthermore, Hsp90 has several accessory chaperones that include FKBP-type PPIases containing the tetratricopeptide repeat (TPR) domains.[80–82] For instance, FKPB51 and FKBP52 are known to bind to Hsp90 through a TPR domain.[83] These catalytic

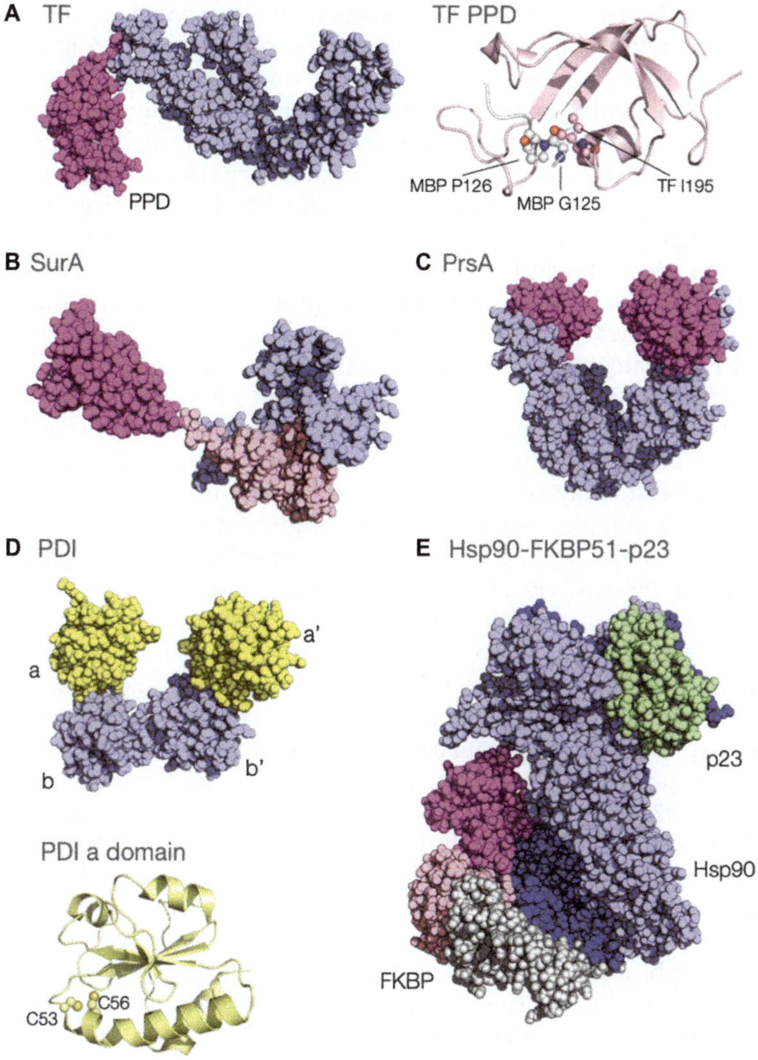

Figure 6.4 Catalytic domains of molecular chaperones. Sphere representations of the structure of TF (PDBID: 1w26)[8] and TF PPD in complex with the MBP peptide (PDBID: 5zr0)[45] (A), SurA with two PPIase domains (PDBID: 1 m5y)[76] (B), PrsA dimer (PDBID: 4wo7)[74] (C), PDI with two catalytic domains (PDBID: 4ekz)[77] (D), and a Hsp90–FKBP51–p23 complex in which the Hsp90 dimer binds to one FKBP51 and one p23 protein (PDBID: 7l7i)[82] (E). PPIase domains are shown in magenta. PPIase-like domains with a similar architecture to PPIase domains without activity are shown in pale pink. The catalytic domains of PDI are shown in yellow. In panels (A and D), the ribbon representations of the TF PPIase domain (pink) in complex with the client protein, MBP (gray), and PDI domain (yellow) are shown in which the residues involved in enzymatic activity are shown as a ball-stick model. In panel (E), Hsp90 subunits are shown in blue and pale blue, p23 is shown in green, and FKBP is shown in magenta (FK1), pink (FK2), and white (TPR).

domains and catalytic accessory proteins may also contribute to the regulation of protein folding. PPIase catalyzes the *cis/trans* isomerization of the peptide bond between Xaa and Pro.[84] The transition between the *cis* and *trans* forms of peptidyl–prolyl bonds, in which the omega dihedral angle is approximately 0° or 179°, respectively, occurs slowly in the absence of PPIases due to the high energy barrier imposed by the partial double bond character of the peptide bond. On the other hand, the native fold of the protein often requires either a *cis* or a *trans* conformation of the peptidyl–prolyl bond. Accordingly, the isomerization process is often the rate-limiting step of protein folding. The mechanism of proline *cis/trans* isomerization by PPIase chaperones has been examined extensively using the TF PPIase domain (PPD) in biochemical and structural studies. Biochemical studies of TF PPD and its homologous protein human FKBP12 have indicated that conserved phenylalanine residues in TF PPD (F197) and FKBP12 (F98) are essential for PPIase activity.[85,86] Several other conserved residues in TF PPD have since been analyzed.[87] However, the detailed mechanisms underlying PPIase activity remain unclear. Structural studies of PPIase in complex with client peptides have been performed including studies reporting the crystal structure of *Plasmodium vivax* FK506 binding protein 35 (PvFKBP35) in complex with a tetrapeptide substrate[88] and SlyD in complex with tetrapeptide[90] and 15 residue-long peptides.[90] These crystal structures provide detailed views that contribute to our understanding of the recognition of substrate peptides by PPIase. Recent structural and dynamics studies of TF PPD have provided information on substrate recognition by determining the solution structure of TF PPD in complex with a substrate protein (MBP fragment 237–265) (see Figure 6.4(A)) and visualizing the intermediate transition state of the *cis/trans* isomerization by MD simulations.[45] MD simulations showed that an intermolecular hydrogen bond between the amide group of TF PPD I195 and the carbonyl oxygen of the prolyl peptide bond is formed in the intermediate *syn* state. This hydrogen bond is expected to stabilize the *syn* state and thus lower the energy barrier between the *cis* and *trans* states. This *cis/trans* exchange of the client protein on TF PPD was measured in a Carr–Purcell–Meiboom–Gill (CPMG) relaxation dispersion experiment that measured the isomerization rate of PPIase as approximately 740 s^{-1} at 35 °C,[45] indicating that the exchange is much faster than with uncatalyzed isomerization (approximately 0.01 s^{-1}).[91]

Given the conserved architecture of FKBP-type PPIases, the observed mechanisms underlying the function of TF PPD may apply to other FKBP-type PPIases including human FKBPs. In contrast to TF PPD which comprises additional domains including a substrate-binding domain that allows tethering of substrate proteins,[40] human FKBPs often accompany the Hsp89 chaperone through recognition using the TPR domain[57,82] (see Figure 6.4). The accompanied chaperone and TPR domains are expected to work cooperatively with the PPIase domains to regulate the folding and/or activity of client proteins. For instance, Hsp90 is known to bind to several PPIases including FKBP51, FKBP52, FKBP38, and Cyp40.[93–95] These PPIases bind to the C-terminal tail of Hsp90 through their TPR domain with varying affinities.[95,96] Hsp90 is posited to exchange its PPIase partner under certain conditions, although the functional importance of this switching has yet to be fully elucidated.

6.4 Perspective/Conclusion

Molecular chaperones recognize immature, unfolded proteins through relatively weak and dynamic recognition and alter the folding properties of client proteins. As described above, information regarding chaperone–client complex structures, dynamics, and interactions from solution NMR has increased the understanding of the mechanisms responsible for the functions of molecular chaperones. Owing to recent technical developments, NMR can now be applied to large protein complexes up to several hundreds of kDa. Given the advantages of NMR in the analysis of dynamic and weak complexes at atomic resolution, NMR is one of the best tools for investigating the mechanisms underlying molecular chaperone functions. Moreover, further technical developments in NMR are anticipated. One major limitation in the current NMR structural study is a high dependence on NOEs for providing short-range ($\sim$5 Å) distance restraints. Short-range distance information is important for understanding local protein–protein interactions but provides limited global information regarding the overall structure of proteins. Obtaining global and long-range structural information is important, particularly for the structural study of multi-domain proteins and protein complexes. Paramagnetic probes have recently been developed that provide long-range structural information.[96–98] Paramagnetic probes exploit nitroxide spin probes, lanthanide ions, and Mn^{2+} to obtain long-range distance information (<20 Å) from paramagnetic relaxation enhancement (PRE),[99] distance and angle information (<40 Å) from pseudocontact shift (PCS),[3,10,95,96] and angle information from residual dipolar coupling.[101] Long-range paramagnetic information can be a powerful addition to the structural analysis of chaperone–client and multi-chaperone complexes.[51,57] Future technical developments are expected to further expand the utility and advantages of NMR in the structural study of chaperones.

Acknowledgements

This work was supported by funding from the JSPS KAKENHI (JP18H05229, JP19H0495, JP20H03199, JP20KK0156, JP21H05093, JP 22K18361, and JP22H02560), the MEXT Grant-in-Aid for Transformative Research Areas (B) (JP21H05094 and JP21H05093), the AMED (JP22ek0109437h and JP22wm0425004h), and the JST FOREST Program (JPMJFR204W). This work was also partially supported by the Takeda Science Foundation Grant, The Kato Memorial Trust for Nambyo Research, Mochida Memorial Foundation for Medical and Pharmaceutical Research, The Naito Foundation, and The Yukihiko Miyata Memorial Trust for ALS Research.

References

1. N. A. D. Bascos and S. J. Landry, *Int. J. Mol. Sci.*, 2019, **20**, 6195.
2. S. Kawagoe, K. Ishimori and T. Saio, *Int. J. Mol. Sci.*, 2022, **23**, 2485.
3. P. B. Sigler, Z. Xu, H. S. Rye, S. G. Burston, W. A. Fenton and A. L. Horwich, *Annu. Rev. Biochem.*, 1998, **67**, 581–608.
4. K. Braig, Z. Otwinowskitt, H. Hegdett, D. C. Boisvert, A. Joachimiakt, A. L. Horwich and P. B. Sigler Tt, *Nature*, 1994, **371**, 578–586.
5. C. Dekker, S. M. Roe, E. A. McCormack, F. Beuron, L. H. Pearl and K. R. Willison, *EMBO J.*, 2011, **30**, 3078–3090.
6. R. Siegert, M. R. Leroux, C. Scheufler, F. U. Hartl and I. Moarefi, *Cell*, 2000, **103**, 621–632.
7. T. A. Walton and M. C. Sousa, *Mol. Cell*, 2004, **15**, 367–374.
8. L. Ferbitz, T. Maier, H. Patzelt, B. Bukau, E. Deuerling and N. Ban, *Nature*, 2004, **431**, 590–596.
9. A. Hoffmann, B. Bukau and G. Kramer, *Biochim. Biophys. Acta, Mol. Cell Res.*, 2010, **1803**, 650–661.
10. S. Chen, A. M. Roseman, A. S. Hunter, S. P. Wood, S. G. Burston, N. A. Ranson, A. R. Clarke and H. R. Saibil, *Nature*, 1994, **371**, 261–264.
11. J. Martín-Benito, J. Boskovic, P. Gómez-Puertas, J. L. Carrascosa, C. T. Simons, S. A. Lewis, F. Bartolini, N. J. Cowan and J. M. Valpuesta, *EMBO J.*, 2002, **21**, 6377–6386.
12. F. Merz, D. Boehringer, C. Schaffitzel, S. Preissler, A. Hoffmann, T. Maier, A. Rutkowska, J. Lozza, N. Ban, B. Bukau and E. Deuerling, *EMBO J.*, 2008, **27**, 1622–1632.
13. A. N. Calabrese, B. Schiffrin, M. Watson, T. K. Karamanos, M. Walko, J. R. Humes, J. E. Horne, P. White, A. J. Wilson, A. C. Kalli, R. Tuma, A. E. Ashcroft, D. J. Brockwell and S. E. Radford, *Nat. Commun.*, 2020, **11**, 1–16.
14. D. C. Marx, A. M. Plummer, A. M. Faustino, T. Devlin, M. A. Roskopf, M. J. Leblanc, H. J. Lessen, B. T. Amann, P. J. Fleming, S. Krueger, S. D. Fried and K. G. Fleming, *Proc. Natl. Acad. Sci. U. S. A.*, 2020, **117**, 28026–28035.
15. S. Zhang, D. He, Z. Lin, Y. Yang, H. Song and P. R. Chen, *Acc. Chem. Res.*, 2017, **50**, 1184–1192.
16. F. Teixeira, E. Tse, H. Castro, K. A. T. Makepeace, B. A. Meinen, C. H. Borchers, L. B. Poole, J. C. Bardwell, A. M. Tomás, D. R. Southworth and U. Jakob, *Nat. Commun.*, 2019, **10**, 1–14.
17. T. Arhar, A. Shkedi, C. M. Nadel and J. E. Gestwicki, *J. Biol. Chem.*, 2021, **297**, 101282.
18. C. Huang and C. G. Kalodimos, *Biochem., Mol. Biol. Biophys.*, 2017, **46**, 317–336.
19. Y. Jiang and C. G. Kalodimos, *J. Mol. Biol.*, 2017, **429**, 2667–2676.
20. S. Hiller, *Trends Biochem. Sci.*, 2019, **44**, 517–527.
21. K. Takeuchi, K. Baskaran and H. Arthanari, *J. Magn. Reson.*, 2019, **306**, 195–201.

22. M. P. Williamson, *Prog. Nucl. Magn. Reson. Spectrosc.*, 2013, **73**, 1–16.
23. M. Kainosho, T. Torizawa, Y. Iwashita, T. Terauchi, A. M. Ono and P. Güntert, *Nature*, 2006, **440**, 52–57.
24. V. Tugarinov, V. Kanelis and L. E. Kay, *Nat. Protoc.*, 2006, **1**, 749–754.
25. T. Saio, S. Kawagoe, K. Ishimori and C. G. Kalodimos, *eLife*, 2018, **7**, e35731.
26. H. Nanaura, H. Kawamukai, A. Fujiwara, T. Uehara, Y. Aiba, M. Nakanishi, T. Shiota, M. Hibino, P. Wiriyasermkul, S. Kikuchi, R. Nagata, M. Matsubayashi, Y. Shinkai, T. Niwa, T. Mannen and N. Morikawa, *Nat. Commun.*, 2021, 1–12.
27. C. Huang, P. Rossi, T. Saio and C. G. Kalodimos, *Nature*, 2016, **537**, 7619.
28. R. Sprangers and L. E. Kay, *Nature*, 2007, **445**, 618–622.
29. G. Mas, J. Y. Guan, E. Crublet, E. C. Debled, C. Moriscot, P. Gans, G. Schoehn, P. Macek, P. Schanda and J. Boisbouvier, *Sci. Adv.*, 2018, **4**, 1–10.
30. D. F. Gauto, L. F. Estrozi, C. D. Schwieters, G. Effantin, P. Macek, R. Sounier, A. C. Sivertsen, E. Schmidt, R. Kerfah, G. Mas, J.-P. Colletier, P. Güntert, A. Favier, G. Schoehn, P. Schanda and J. Boisbouvier, *Nat. Commun.*, 2019, **10**, 2697.
31. M. Takeda, T. Ikeya, P. Güntert and M. Kainosho, *Nat. Protoc.*, 2007, **2**, 2896–2902.
32. Y. Miyanoiri, M. Takeda, T. Terauchi and M. Kainosho, *Biochim. Biophys. Acta, Gen. Subj.*, 2020, **1864**, 129439.
33. I. Pritišanac, J. M. Würz, T. R. Alderson and P. Güntert, *Nat. Commun.*, 2019, **10**, 1–12.
34. I. Pritišanac, T. R. Alderson and P. Güntert, *Prog. Nucl. Magn. Reson. Spectrosc.*, 2020, **118–119**, 54–73.
35. Y. R. Monneau, P. Rossi, A. Bhaumik, C. Huang, Y. Jiang, T. Saleh, T. Xie, Q. Xing and C. G. Kalodimos, *J. Biomol. NMR*, 2017, **50**, 1–13.
36. H. Wegele, S. K. Wandinger, A. B. Schmid, J. Reinstein and J. Buchner, *J. Mol. Biol.*, 2006, **356**, 802–811.
37. Z. N. Scholl, W. Yang and P. E. Marszalek, *J. Biol. Chem.*, 2014, **289**, 28607–28618.
38. F. Georgescauld, K. Popova, A. J. Gupta, A. Bracher, J. R. Engen, M. Hayer-Hartl and F. U. Hartl, *Cell*, 2014, **157**, 922–934.
39. E. Basha, G. J. Lee, B. Demeler and E. Vierling, *Eur. J. Biochem.*, 2004, **271**, 1426–1436.
40. T. Saio, X. Guan, P. Rossi, A. Economou and C. G. Kalodimos, *Science*, 2014, **344**, 6184.
41. Y. Jiang, P. Rossi and C. G. Kalodimos, *Science*, 2019, **365**, 1313–1319.
42. R. Rosenzweig, P. Farber, A. Velyvis, E. Rennella, M. P. Latham, L. E. Kay and A. L. Horwich, *Proc. Natl. Acad. Sci. U. S. A.*, 2015, **112**, E6872–E6881.
43. J. C. Bardwell, K. McGovern and J. Beckwith, *Cell*, 1991, **67**, 581–589.
44. A. Mashaghi, G. Kramer, P. Bechtluft, B. Zachmann-Brand, A. J. M. Driessen, B. Bukau and S. J. Tans, *Nature*, 2013, **500**, 98–101.

45. S. Kawagoe, H. Nakagawa, H. Kumeta, K. Ishimori and T. Saio, *J. Biol. Chem.*, 2018, **293**, 15095–15106.
46. L. He, T. Sharpe, A. Mazur and S. Hiller, *Sci. Adv.*, 2016, **2**, 1–11.
47. S. Horowitz, L. Salmon, P. Koldewey, L. S. Ahlstrom, R. Martin, S. Quan, P. V. Afonine, H. van den Bedem, L. Wang, Q. Xu, R. C. Trievel, C. L. Brooks and J. C. A. Bardwell, *Nat. Struct. Mol. Biol.*, 2016, **23**, 691–697.
48. F. Stull, P. Koldewey, J. R. Humes, S. E. Radford and J. C. A. Bardwell, *Nat. Struct. Mol. Biol.*, 2016, **23**, 53–58.
49. O. Zhang, L. E. Kay, J. P. Olivier and J. D. Forman-Kay, *J. Biomol. NMR*, 1994, **4**, 845–858.
50. J. H. Lee, D. Zhang, C. Hughes, Y. Okuno, A. Sekhar and S. Cavagnero, *Proc. Natl. Acad. Sci. U. S. A.*, 2015, **112**, E4206–E4215.
51. B. M. Burmann, C. Wang and S. Hiller, *Nat. Struct. Mol. Biol.*, 2013, **20**, 1265–1272.
52. A. Sekhar, R. Rosenzweig, G. Bouvignies and L. E. Kay, *Proc. Natl. Acad. Sci. U. S. A.*, 2015, **112**, 10395–10400.
53. R. Rosenzweig, A. Sekhar, J. Nagesh and L. E. Kay, *eLife*, 2017, **6**, e28030.
54. M. Tollinger, N. R. Skrynnikov, F. A. A. Mulder, J. D. Forman-Kay and L. E. Kay, *J. Am. Chem. Soc.*, 2001, **123**, 11341–11352.
55. L. He, T. Sharpe, A. Mazur and S. Hiller, *Sci. Adv.*, 2016, **2**, e1601625.
56. S. Horowitz, L. Salmon, P. Koldewey, L. S. Ahlstrom, R. Martin, S. Quan, P. V. Afonine, H. Van Den Bedem, L. Wang, Q. Xu, R. C. Trievel, C. L. Brooks and J. C. A. Bardwell, *Nat. Struct. Mol. Biol.*, 2016, **23**, 691–697.
57. J. Oroz, B. J. Chang, P. Wysoczanski, C.-T. Lee, Á. Pérez-Lara, P. Chakraborty, R. V Hofele, J. D. Baker, L. J. Blair, J. Biernat, H. Urlaub, E. Mandelkow, C. A. Dickey and M. Zweckstetter, *Nat. Commun.*, 2018, **9**, 4532.
58. G. E. Karagöz, A. M. S. Duarte, E. Akoury, H. Ippel, J. Biernat, T. Morán Luengo, M. Radli, T. Didenko, B. A. Nordhues, D. B. Veprintsev, C. A. Dickey, E. Mandelkow, M. Zweckstetter, R. Boelens, T. Madl and S. G. D. Rüdiger, *Cell*, 2014, **156**, 963–974.
59. I. Sučec, Y. Wang, O. Dakhlaoui, K. Weinhäupl, T. Jores, D. Costa, A. Hessel, M. Brennich, D. Rapaport, K. Lindorff-Larsen, B. Bersch and P. Schanda, *Sci. Adv.*, 2020, **6**, 1–16.
60. S. A. Teter, W. A. Houry, D. Ang, T. Tradler, D. Rockabrand, G. Fischer, P. Blum, C. Georgopoulos and F. U. Hartl, *Cell*, 1999, **97**, 755–765.
61. H. Patzelt, G. Kramer, T. Rauch, H.-J. Schönfeld, B. Bukau and E. Deuerling, *Biol. Chem.*, 2002, **383**, 1611–1619.
62. H. El-Samad, H. Kurata, J. C. Doyle, C. A. Gross and M. Khammash, *Proc. Natl. Acad. Sci. U. S. A.*, 2005, **102**, 2736–2741.
63. B. T. Findik and L. L. Randall, *PLoS One*, 2017, **12**, 1–11.
64. S. Park, G. Liu, T. B. Topping, W. H. Cover and L. L. Randall, *Science*, 1988, **239**, 1033–1035.

65. A. Tsirigotaki, K. E. Chatzi, M. Koukaki, J. De Geyter, A. G. Portaliou, G. Orfanoudaki, M. F. Sardis, M. B. Trelle, T. J. D. Jørgensen, S. Karamanou and A. Economou, *Struct. Des.*, 2018, **26**, 695–707.e5.
66. S. J. S. Hardy and L. L. Randall, *Science*, 1991, **251**, 439–443.
67. L. Morgado, B. M. Burmann, T. Sharpe, A. Mazur and S. Hiller, *Nat. Commun.*, 2017, **8**, 1992.
68. C.-T. Huang, Y.-C. Lai, S.-Y. Chen, M.-R. Ho, Y.-W. Chiang and S.-T. D. Hsu, *Magn. Reson.*, 2021, **2**, 375–386.
69. B. M. Burmann, C. Wang and S. Hiller, *Nat. Struct. Mol. Biol.*, 2013, **20**, 1265–1272.
70. R. Suno, H. Taguchi, R. Masui, M. Odaka and M. Yoshida, *J. Biol. Chem.*, 2004, **279**, 6380–6384.
71. H. Zhu, M. Matsusaki, T. Sugawara, K. Ishimori and T. Saio, *Biology*, 2021, **10**, 1106.
72. G. Mas, B. M. Burmann, T. Sharpe, B. Claudi, D. Bumann and S. Hiller, *Sci. Adv.*, 2020, **6**, eabc5822.
73. A. Hoffmann, B. Bukau and G. Kramer, *Biochim. Biophys. Acta, Mol. Cell Res.*, 2010, **1803**, 650–661.
74. R. P. Jakob, J. R. Koch, B. M. Burmann, P. A. M. Schmidpeter, M. Hunkeler, S. Hiller, F. X. Schmid and T. Maier, *J. Biol. Chem.*, 2014, **290**, 3278–3292.
75. X. Xu, S. Wang, Y. X. Hu and D. B. McKay, *J. Mol. Biol.*, 2007, **373**, 367–381.
76. E. Bitto and D. B. McKay, *Structure*, 2002, **10**, 1489–1498.
77. C. Wang, W. Li, J. Ren, J. Fang, H. Ke, W. Gong, W. Feng and C. Wang, *Antioxid. Redox Signaling*, 2013, **19**, 36–45.
78. M. Matsusaki, S. Kanemura, M. Kinoshita, Y. H. Lee, K. Inaba and M. Okumura, *Biochim. Biophys. Acta, Gen. Subj.*, 2020, **1864**, 129338.
79. M. Okumura, S. Kanemura, M. Matsusaki, M. Kinoshita, T. Saio, D. Ito, C. Hirayama, H. Kumeta, M. Watabe, Y. Amagai, Y.-H. Lee, S. Akiyama and K. Inaba, *Structure*, 2021, **29**, 1–14.
80. S. D. Hanes, *Biochim. Biophys. Acta, Gen. Subj.*, 2015, **1850**, 2017–2034.
81. A. Brinker, C. Scheufler, F. Von Der Mülbe, B. Fleckenstein, C. Herrmann, G. Jung, I. Moarefi and F. Ulrich Hartl, *J. Biol. Chem.*, 2002, **277**, 19265–19275.
82. K. Lee, A. C. Thwin, C. M. Nadel, E. Tse, S. N. Gates, J. E. Gestwicki and D. R. Southworth, *Mol. Cell*, 2021, **81**, 3496–3508.
83. C. L. Storer, C. A. Dickey, M. D. Galigniana, T. Rein and M. B. Cox, *Trends Endocrinol. Metab.*, 2011, **22**, 481–490.
84. P. A. M. Schmidpeter and F. X. Schmid, *J. Mol. Biol.*, 2015, **427**, 1609–1631.
85. G. Kramer, G. Kramer, H. Patzelt, T. Rauch, T. A. Kurz, S. Vorderwülbecke, B. Bukau and E. Deuerling, *J. Biol. Chem.*, 2004, **279**, 14165–14170.
86. M. T. DeCenzo, S. T. Park, B. P. Jarrett, R. A. Aldape, O. Futer, M. A. Murcko and D. J. Livingston, *Protein Eng.*, 1996, **9**, 173–180.

87. C.-P. Liu, Q.-M. Zhou, D.-J. Fan and J.-M. Zhou, *Int. J. Biochem. Cell Biol.*, 2010, **42**, 890–901.
88. R. Alag, A. M. Balakrishna, S. Rajan, I. A. Qureshi, J. Shin, J. Lescar, G. Grüber and H. S. Yoon, *Eukaryotic Cell*, 2013, **12**, 627–634.
89. C. Löw, P. Neumann, H. Tidow, U. Weininger, C. Haupt, B. Friedrich-Epler, C. Scholz, M. T. Stubbs and J. Balbach, *J. Mol. Biol.*, 2010, **398**, 375–390.
90. E. M. Quistgaard, U. Weininger, Y. Ural-Blimke, K. Modig, P. Nordlund, M. Akke and C. Löw, *BMC Biol.*, 2016, **14**, 82.
91. J. F. Brandts, H. R. Halvorson and M. Brennan, *Biochemistry*, 1975, **14**, 4953–4963.
92. S. D. Hanes, *Biochim. Biophys. Acta, Gen. Subj.*, 2015, **1850**, 2017–2034.
93. K. L. I. M. Blundell, M. Pal, S. M. Roe, L. H. Pearl and C. Prodromou, *PLoS One*, 2017, **12**, 1–15.
94. F. Pirkl and J. Buchner, *J. Mol. Biol.*, 2001, **308**, 795–806.
95. J. C. Young, W. M. J. Obermann and F. U. Hartl, *J. Biol. Chem.*, 1998, **273**, 18007–18010.
96. T. Saio and K. Ishimori, *Biochim. Biophys. Acta, Gen. Subj.*, 2020, **1864**, 129332.
97. C. Nitsche and G. Otting, *Prog. Nucl. Magn. Reson. Spectrosc.*, 2017, **98–99**, 20–49.
98. Q. Miao, C. Nitsche, H. Orton, M. Overhand, G. Otting and M. Ubbink, *Chem. Rev.*, 2021, **122**, 9571–9642.
99. G. M. Clore and J. Iwahara, *Chem. Rev.*, 2009, **109**, 4108–4139.
100. M. Allegrozzi, I. Bertini, M. B. L. Janik, Y.-M. Lee, G. Liu and C. Luchinat, Lanthanide-Induced Pseudocontact Shifts for Solution Structure Refinements of Macromolecules in Shells up to 40 Å from the Metal Ion, *J. Am. Chem. Soc.*, 2000, **122**(17), 4154–4161.
101. X. Wang, J. P. Kirkpatrick, H. M. M. Launay, A. de Simone, D. Häussinger, C. M. Dobson, M. Vendruscolo, L. D. Cabrita, C. A. Waudby and J. Christodoulou, *Sci. Rep.*, 2019, **9**, 1–9.

CHAPTER 7

Single Molecule Fluorescence Methods for Molecular Chaperones and Their Client Interactions

T. MONDOL, B. SOHMEN AND T. HUGEL*

Institute of Physical Chemistry and Signalling Research Centers BIOSS and CIBSS, University of Freiburg, Germany
*Email: thorsten.hugel@pc.uni-freiburg.de

7.1 Introduction

7.1.1 Single Molecule Fluorescence for Biomolecular Machines

Single molecule fluorescence is an efficient and versatile technique that is extensively used in cutting-edge biophysical and biological research. It has paved the way for understanding the complexity of dynamic biological systems and their interactions in real time. Single molecule fluorescence techniques have revealed the mechanisms of many fundamental biological processes including DNA replication,[1–3] transcription,[4–8] protein biosynthesis,[9–11] protein–nucleic acid interaction,[12,13] protein dynamics and folding,[14–16] molecular motors,[17–20] intrinsically disordered proteins[21,22] and biomolecular machines such as kinesin,[23] F_0F_1-ATP synthase[24] and helicase.[25] A unique advantage of single molecule methods lies in studying the individual biomolecules without relying on ensemble averaging. This

New Developments in NMR No. 29
Biophysics of Molecular Chaperones: Function, Mechanisms and Client Protein Interactions
Edited by Sebastian Hiller, Maili Liu and Lichun He

Published by the Royal Society of Chemistry, www.rsc.org

way, they provide key information about the heterogeneous population of biomolecules.

Single molecule FRET (smFRET) has emerged as a powerful tool for the determination of distances, structures and dynamics of biomolecules with Angstrom precision.[26–28] FRET (Förster resonance energy transfer)[29] is a non-radiative energy transfer between a donor and an acceptor dye, given that they are only a few nanometers apart (see Figure 7.1(A)). In particular, smFRET identifies distinct conformational states and transient interactions between biomolecules. In conventional two color smFRET, two fluorescent dyes are attached at distinct positions of the protein of interest, and the FRET efficiency is monitored in real time. However, in the case of more complex biological systems containing multi-protein interaction partners, two-color smFRET methods may not be sufficient to obtain a detailed understanding of such systems. Multicolor smFRET is ideal to investigate multi-component protein systems by monitoring multiple distances between dyes simultaneously.[30–32] For this purpose, three or four components can be labelled with different fluorescent dyes and the resulting network of FRET pairs can be used to observe the conformational dynamics of a complex system.[33]

In the case of biological molecular machines such as kinesin, myosin, helicases or the F_0F_1-ATPase, two-color smFRET is sufficient, as the rotation angle or the distance moved by the protein can be measured to observe the sequential succession of many states. For example, Figure 7.1 shows the translocation and DNA unwinding activity by the DnaB-family helicase with a single base pair resolution monitored using smFRET.[25] The DNA substrate is a 40 base pair duplex DNA which is attached to a polymer passivated surface using biotin–neutravidin coupling. The authors have monitored single DNA molecule unwinding by G40P helicase in real time by measuring FRET between a donor and an acceptor dye located on the DNA. DNA unwinding by G40P helicase reduces the FRET efficiency resulting in an increased donor emission and decreased acceptor emission as shown in Figure 7.1(C). Many studies of this kind have shown that smFRET is an extremely powerful tool for understanding biological systems with high temporal and spatial resolution. Nevertheless, smFRET has only sparsely been used to explore chaperone systems as detailed in the following sections.

7.1.2 Single Molecule Fluorescence to Study Chaperone Machineries

All organisms contain specialized protein networks that maintain protein homeostasis and protect the proteome from the severe effects of heat and other stresses. Chaperones serve as the guardian of the proteome and function as the hub of eukaryotic protein homeostasis by mediating protein folding and protein degradation and by dissolving protein aggregates. In bacteria, the GroEL–GroES chaperonin system is essentially responsible for

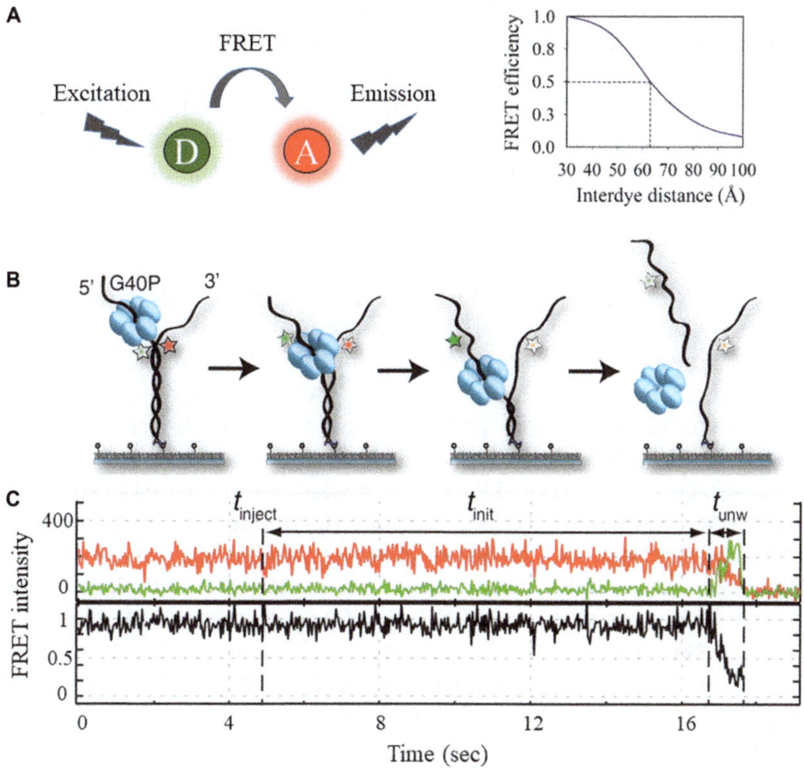

Figure 7.1 (A) Left: Schematic illustration of FRET from a donor (green) to an acceptor (red) dye. Right: FRET serves as a molecular ruler as the FRET efficiency strongly depends on the interdye distance. The highest sensitivity of FRET is obtained at distances around the Förster radius, which is indicated as a vertical dashed line (typically ∼60 Å). (B) Schematic of the DNA unwinding mechanism of the DnaB-family helicase G40P. The donor (Cy3)–acceptor (Cy5) FRET pair is indicated with green and red stars, respectively. After binding G40P helicase unwinds the DNA and gradually separates the FRET pair which finally results in strand separation and helicase unloading. (C) Example of a TIRF trace that shows the donor (green), acceptor (red) and FRET efficiency signal (black) of a labelled DNA unwinded by the helicase G40P. At t_{inject} G40P is injected, and t_{init} indicates the unwinding initiation time, *i.e.* the time from protein injection until unwinding starts. t_{unw} is the unwinding time, *i.e.* the time from the moment the FRET efficiency decreases until strand separation. (B) and (C) are reproduced from ref. 25, https://doi.org/10.7554/eLife.42001, under the terms of the CC BY 4.0 license, https://creativecommons.org/licenses/by/4.0/.

protein folding. The GroEL–GroES system binds and encapsulates nonnative proteins for folding. Several single molecule fluorescence techniques have significantly contributed to understanding the protein folding mechanism by the GroEL–GroES system.[34–36] Hsp70 systems have also been investigated with single molecule fluorescence techniques on the millisecond to minute timescale. These experiments are described in Section 7.2.1. in detail. Here we elaborate on the molecular chaperone heat shock protein 90 (Hsp90) as a model system because it is currently the most investigated chaperone and therefore an ideal example of what can be learned by single molecule fluorescence. Hsp90 assists in folding, stabilization and maturation of more than 200 structurally and functionally diverse clients that function as signal transducers, kinases and transcription factors.[37] Cochaperones are key regulators which trigger Hsp90 functions including recruitment of specific clients to Hsp90 and stimulation or inhibition of the ATPase activity of Hsp90.[38] They assist Hsp90 in regulating diverse cellular processes including signal transduction, cell cycle control, DNA repair, gene regulation and neurodegeneration through their interaction with clients. Structurally, Hsp90 is a homodimer in which each monomer consists of an N-terminal (N) domain containing the ATP binding pocket, a middle (M) domain where most of the clients bind and a C-terminal (C) domain which is responsible for Hsp90 dimerization.[39–42] In the absence of ATP, Hsp90 mainly adopts a V-shaped open conformation with dissociated N-terminal and middle domains.[42,43] In the presence of ATP, Hsp90 shifts towards the closed state, where the N-domains dimerize and ATP is hydrolysed.[44,45] After ATP hydrolysis, the N domains are assumed to dissociate, ADP and inorganic phosphate are released and Hsp90 adopts a more open conformation. Nevertheless, the open–close dynamics of the Hsp90 dimer is only weakly coupled to ATP hydrolysis.[45]

Hsp90 and a large group of cochaperones collaborate to recruit and assemble diverse client proteins to form dynamic multi-protein complexes. Decades of research led to characterization of the interactions of clients with Hsp90 and its cochaperones. X-ray crystallography and cryo-electron microscopy provided information on the static structure of Hsp90 and several Hsp90–client–cochaperone complexes with atomic resolution. These static structures are like snapshots of the highly dynamic rearrangements that occur during the course of a chaperone's action. However, these snapshots miss important information on the high flexibility of these conformations. In addition, for a chaperone there is usually no clear continuously changing reaction coordinate (*e.g.* an increasing displacement along a track), while moving through the chaperone cycle. This is in contrast to motor proteins, in which, for example, the distance moved by the protein can be picked as an easy-to-monitor coordinate. Therefore the analysis of chaperones in non-equilibrium is more complicated. To overcome this challenge, in the past decade, smFRET-based kinetic assays were developed, to follow the highly dynamic Hsp90 conformational cycle along various reaction coordinates.[46,47] In Section 7.2, we demonstrate the potential of smFRET and related techniques to investigate the dynamics of Hsp90 and to study its interaction with client and cochaperone proteins. We like to point out that NMR spectroscopy is a powerful technique for quantifying protein dynamics and has been used to investigate the Hsp90 structure, dynamics and client interactions.[48–51] NMR is ideal to determine local dynamics,[52] while we focus here on complementary, larger (molecule-spanning) dynamics.

7.2 Investigation of Chaperone Dynamics Across Timescales

Visualizing the structure of chaperones inspired scientists to think about how their interplay mediates regulation, protein stabilization and communication. However, for a full understanding structural information needs to be complemented by dynamics information. Elucidation of chaperone function requires techniques that cover protein dynamics on a wide range of length- and timescales. Making oneself aware of the many orders of timescales involved (several minutes down to nanoseconds or even picoseconds) provides a first sense of how chaperones achieve this wide variety of complex tasks in our cells. Here we address how to access the different time regimes for chaperones going from slow to very fast dynamics using single molecule fluorescence methods.

7.2.1 Dynamics on the Millisecond to Minutes Timescale

Dynamics on the millisecond to minute timescale have mainly been investigated for Hsp70 and Hsp90 chaperone proteins.

For Hsp70, Mapa *et al.* have investigated the interaction between the N-terminal nucleotide binding domain (NBD) and the substrate binding domain (SBD) by smFRET for both yeast mitochondrial Hsp70 (Ssc1) and bacterial Hsp70 (DnaK).[53] These experiments were performed on a two-color confocal microscope using pulsed interleaved excitation (see Figure 7.2). A schematic of the confocal detection volume in smFRET measurements is shown in Figure 7.2(A). Confocal detection allows monitoring of protein dynamics at fast timescales which is limited by the diffusion time of a single molecule through the confocal volume (typically around 1 ms). In this study, protein constructs with dyes at positions 448 and 590 located in the base and lid of the SBD (Ssc1 448, 590) were used to monitor SBD opening and closing dynamics (see Figure 7.2(B) and (C)).[53] The authors demonstrated that in the ATP state of Ssc1, the SBD is in an open state wherein the α helical lid (SBDα, yellow colored) is detached from the β-sheet containing base (SBDβ, orange colored) causing a low FRET efficiency at 0.2 (as shown in the FRET histogram in Figure 7.2(B)). In the ADP bound state of Ssc1, the FRET efficiency histogram (see Figure 7.2(C)) is broad and shifted to a higher FRET efficiency indicating a more closed conformational state of the SBD. In this state, the NBD–SBD domains are mostly separated (indicated by the structure of Hsp70 homolog DnaK in Figure 7.2(C)). Using the cochaperone J protein Mdj1, the authors could show that Mdj1 promotes the conversion of the ATP form of Ssc1 into the ADP form.[53] Sikor *et al.* have studied the conformational dynamics of Ssc1 using the same label positions over minutes by total internal reflection fluorescence (TIRF) microscopy.[54] In this study, Ssc1 was encapsulated in vesicles and immobilized on the surface to monitor smFRET. The authors found that Ssc1 undergoes dynamic switching between a closed (high FRET) and an open (intermediate FRET) conformational state in the presence of ADP over tens of seconds, as indicated in Figure 7.2(D). Additionally, a recent study used multicolor smFRET to compare the conformational cycle of different Hsp70 family chaperones.[55]

For Hsp90, many smFRET studies have been performed revealing a highly dynamic picture that involves several different N-terminal open and closed states.[45] One open[47] and one closed[39] structure of yeast Hsp90 are shown in Figure 7.2(F) and (G). Schmid *et al.*[46,56] have characterized the Hsp90 conformational dynamics by TIRF microscopy (Figure 7.2(E)–(I)). In this study, one Hsp90 monomer was labelled with the donor fluorophore at position 61 and the other Hsp90 monomer with the acceptor fluorophore at position 385. The Hsp90 dimer was stabilized by an artificial C-terminal zipper motif. *In vivo* biotinylation of the C-terminal end of Hsp90 allowed its surface immobilization on a polyethylene glycol (PEG) passivated microscope slide using biotin–neutravidin coupling (Figure 7.2(E)). Figure 7.2(H) shows TIRF

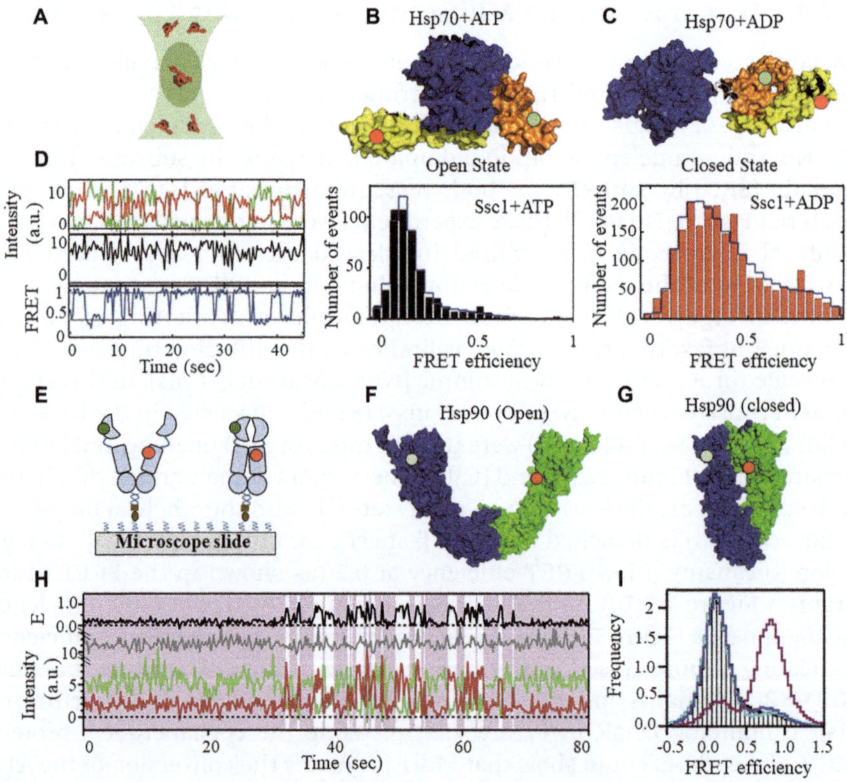

smFRET traces of Hsp90. Hsp90 in the absence of nucleotide is mainly in the open state (low FRET efficiency), but also closes from time to time (high FRET efficiency) on a wide range of timescales.[56,57] Figure 7.2(I) shows the conformational state distribution of Hsp90 for different nucleotide conditions. Interestingly, compared to the apo condition, no significant change of the Hsp90 FRET efficiency distribution is observed upon addition of ADP or ATP, while the non-hydrolysable nucleotide analog AMP-PNP (purple signal) shifts Hsp90 towards a closed state. Additionally, further studies have shown that Hsp90 adopts several 'hidden' states, *i.e.* states which are not directly distinguishable by the FRET efficiency, but from different transition kinetics.[46,56] For the study of these hidden states, we refer the reader to the literature.

Figure 7.2 (A–D) SmFRET measurements on Hsp70, Ssc1. (A) Schematics of a confocal smFRET measurement in solution. (B) Crystal structure of Hsp70 in the ATP bound state (Hsp70 homolog DnaK, PDB: 4B9Q) showing the open state of the SBD, where the lid (SBDα) is detached from SBDβ resulting in a low FRET efficiency. In the lower panel, the FRET efficiency histogram of Ssc1 is shown (black histogram). The NBD domain of Hsp70 is highlighted in blue. The SBD subdomain of Hsp70 includes SBDα (yellow colored) and SBDβ (orange colored). The green and red circles indicate the approximate position of the donor (Atto532) and acceptor (Atto647N) fluorophores at positions 448 and 590 in the SBD of Ssc1. (C) Structure of Hsp70 in the ADP bound state (PDB: 2KHO) showing both the subdomains SBDα and SBDβ in close proximity resulting in a high FRET efficiency. In the lower panel, the FRET efficiency histogram is shown. The heterogeneous, broad FRET population indicates a more closed SBD (red histogram). FRET histograms of (B) and (C) are reproduced from ref. 53 with permission from Elsevier, Copyright 2010. (D) smFRET TIRF data of Ssc1 in the presence of ADP. Green trace: donor signal, red trace: acceptor signal, black trace: total intensity and blue trace: FRET efficiency. This figure is reproduced from ref. 54 with permission from John Wiley & Sons, Copyright 2013. (E–I) SmFRET measurements on yeast Hsp90. (E) Schematic of a labelled Hsp90 dimer attached to a PEG passivated microscope slide for TIRF measurements. (F) Open structure of yeast Hsp90.[47] (G) Closed structure of yeast Hsp90[39] (PDB: 2CG9). The donor (Atto550, green sphere) and acceptor (Atto647N, red sphere) dyes are at positions 61 and 385, respectively. (H) smFRET trace of yeast Hsp90 recorded in real time in the absence of a nucleotide. Green trace: donor signal, red trace: acceptor signal, gray trace: total intensity, and black trace: FRET efficiency. Colored or white overlay indicates the hidden Markov model (HMM) derived low FRET (open) or high FRET (closed) states respectively. This figure is reproduced from ref. 56 with permission from AIP Publishing, Copyright 2018. (I) Nucleotide-dependent population of Hsp90's open and closed conformational states. FRET efficiency histogram of Hsp90 in the absence or in the presence of ADP, ATP or AMP-PNP. Hsp90 apo: black; Hsp90 in the presence of ADP: light blue; Hsp90 in the presence of ATP: dark blue; and Hsp90 in the presence of AMP-PNP: purple. This figure is reproduced from ref. 46 with permission from Elsevier, Copyright 2016.

7.2.2 Dynamics on the Microsecond to Millisecond Timescale

Structural dynamics on the microsecond to millisecond timescale can also be probed by measuring FRET between two fluorophores attached to the protein of interest. More precisely, the dynamics is revealed by the widths of the FRET efficiency distributions. Recently, the dynamic structure of Hsp90 has been resolved by confocal-based smFRET experiments using more than 100 FRET pairs across the entire Hsp90 dimer in solution. Besides the mean distances between these FRET pairs, the spatial fluctuations on the microsecond and millisecond timescales could be determined.[47] In combination with fluorescence lifetime experiments and fluorescence correlation spectroscopy (FCS), these dynamics could be attributed to conformational changes following ATP hydrolysis.[58]

Photoinduced electron transfer (PET) enriches the potential of single molecule methods because it occurs between common dyes and amino acids frequently found in the protein of interest. Schulze *et al.* implemented a PET-based reporter system to monitor site-specific conformational changes in yeast Hsp90 (see Figure 7.3).[59] They also used a confocal setup (Figure 7.3(A)) and added labels into the N-terminal domain of Hsp90 (Figure 7.3(B)). This allowed them to unravel the N-terminal lid dynamics of Hsp90 with sub-millisecond time resolution. As the PET process is distance-dependent, conformational protein dynamics affect the quenching kinetics and are reflected in the FCS signal. Hence, the PET reporters translate conformational dynamics into a signal measurable by PET-FCS (see Figure 7.3(C) and (D)). Schulze and co-workers showed that the N terminal β-strand and the lid are highly dynamic and not rigid structures, with rate constants of $1000 \pm 150\ s^{-1}$ for β-strand unfolding (see Figure 7.3(C)) and $1500 \pm 70\ s^{-1}$ for lid unfolding (see Figure 7.3(D)). Binding of ATP accelerated lid release (Figure 7.3(D), inset).

7.2.3 Dynamics on the Nanosecond Timescale

On the one hand, many studies have revealed insights into local picosecond fluctuations of chaperones mainly by NMR spectroscopy (see other chapters in this book). On the other hand, global micro- to millisecond conformational changes in several chaperones have been characterized (see above). However, the nanosecond timescale is largely missing. This is a fundamental gap because it remains unclear, if nanosecond dynamics is the underlying mechanism of large-scale rearrangements in chaperone systems, or if these fast dynamics are only local. Experimentally, extracting nanosecond chaperone dynamics constitutes a particular challenge because its timescale is technically difficult to access and often entangled with global rotational diffusion. Seminal work to investigate nanosecond protein dynamics by nsFCS was contributed by the Schuler lab based on intrinsically disordered proteins. For example, in 2007, it became possible to detect chain reconfiguration times of

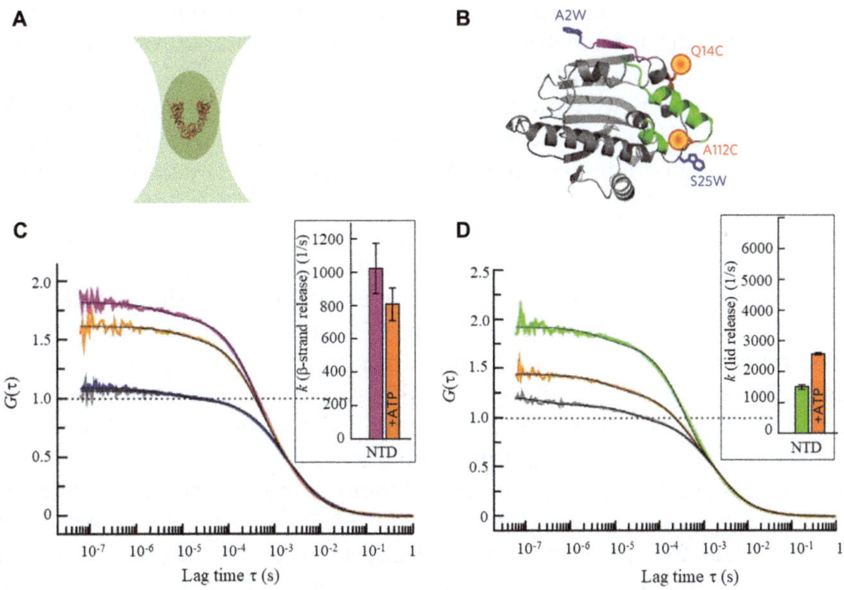

Figure 7.3 (A) Schematic of confocal PET-FCS measurement. (B) Structure of Hsp90's N-terminus. The N terminal β-strand is indicated in magenta and the lid is highlighted in green. The Q14C-A2W reporter is designed to probe the β-strand. The A112C-S25W reporter is designed to probe the lid. The dyes AttoOxa11 are indicated as red spheres, and tryptophans (Trp) are indicated as blue sticks. (C) Autocorrelation functions ($G(\tau)$) of Q14C-A2W on the β-strand are shown. Magenta: no nucleotide; orange: ATP; gray: no Trp; and blue: Trp zipper motif. The inset shows the rate constants of the N-terminal domain β-strand release (magenta) and the effect of ATP (orange). (D) As in (C), now for the A112C-S25W on the lid. Green: no nucleotide; orange: ATP; and gray: no Trp. The inset shows the rate constants of the N-terminal domain lid release (green) and the effect of ATP (orange) on the lid release. The figures (B–D) are reproduced from ref. 59 with permission from Springer Nature, Copyright 2016.

the cold shock protein on the 40–50 ns timescale by combining start–stop experiments with conventional FCS experiments.[60] Technical improvements and a theoretical framework to analyse correlation functions calculated from fluorescence intensity fluctuations by Gopich *et al.*[61] allowed the investigation of photoinduced electron transfer and FRET in polyproline peptides.[62] By now nsFCS emerged as an established technique that provides unique insights into peptides and proteins.[63–65]

For Hsp90, these fast timescales could recently be resolved by combining FCS with time-resolved anisotropy experiments. This way, molecule-spanning motions on the 100 ns timescale could be disentangled from the rotation of the overall Hsp90 dimer.[66] Interestingly, the nature of these internal chaperone dynamics is neither describable by pure Brownian diffusion nor by harmonic oscillations. In combination with MD simulations and neutron scattering experiments, they rather turn out to represent a complex pattern of delocalized diffusive modes distributed over the whole chaperone. The mode distribution over the entire Hsp90 dimer provides a good intra-connectivity within Hsp90. This could be the basis for its allosteric actions and its variable and fast response to external triggering factors.

These fast chaperone dynamics can be accessed based on a confocal single molecule fluorescence setup. With polarization- and color-sensitive detection at picosecond time-resolution, important single molecule parameters such as the FRET efficiency, labelling stoichiometry, lifetime and anisotropy can be calculated which is referred to as multiparameter fluorescence detection (MFD). In addition, FCS analyses the similarity of the fluorescence signal over time to reveal fluctuations at characteristic timescales. Its strength is the detection of kinetics over a broad range of timescales. With sufficient measurement time (several hours) and fluorescent labels of high molecular brightness, kinetics on the nanosecond timescale become detectable. To characterize and identify these fluctuations, FCS can be complemented by information from FRET[67,68] or time-resolved anisotropy.[69] Anisotropy experiments selectively probe rotational nanosecond dynamics by measuring the temporal decorrelation of the sample fluorescence polarization. A combination of these methods enables to disentangle and quantify local and global (rotational) dynamics of the protein of interest.

7.2.4 The Dynamic Picture of Hsp90 Across Timescales

An overview of Hsp90's dynamics measured by different single molecule fluorescence techniques is given in Figure 7.4. Hsp90's large N-terminal

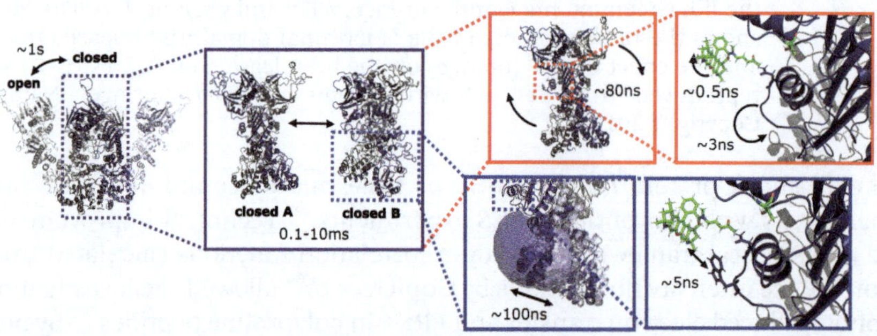

Figure 7.4 Overview of accessible timescales to study chaperone dynamics by single molecule fluorescence techniques (from left to right): the N-terminal open–closed dynamics of Hsp90 (~s), transitions between conformational substates – here two different closed states of Hsp90 (~μs–ms), subdomain fluctuations (~100 ns), global rotation (~80 ns), local loop dynamics (~5 ns) and dye self-rotation (<ns). Contributions from translational and rotational dynamics can be determined separately by complementary fluorescence techniques which are a pre-requisite to extract internal chaperone dynamics.

open–closed dynamics occur on the second timescale as monitored by *e.g.* single molecule TIRF experiments (Section 7.2.1). Interconversion between conformational substates – here two closed states – involves dynamics up to milliseconds as determined by the FRET efficiency and fluorescence lifetime analysis (Section 7.2.2). On the 100 ns timescale, global rotation and subdomain fluctuations were disentangled by FCS and anisotropy experiments (Section 7.2.3). Loop fluctuations were detected on the low nanosecond timescale and are well separated from dye-self rotation occurring on the sub-nanosecond timescale.[66]

Measuring the protein dynamics on this wide range of timescales and assigning them to certain rotational and translational modes opens new possibilities to study allosteric communication pathways in chaperone systems. For example, recently an allosteric signal transfer starting from the nucleotide-binding site could be determined for Hsp90. Fluorescence-based measurements as described above together with all-atom molecular dynamics (MD) simulations showed how ATP hydrolysis leads to large conformational rearrangements on hierarchical timescales from sub-nanoseconds to milliseconds.[58] Recent results show that studying chaperone nanosecond dynamics by single molecule methods has left the fledgling stage. Ultra-fast dynamics have also been reported for important cochaperones of Hsp90, such as p23.[70] This shows that measuring nanosecond dynamics is essential to support results from MD simulations and moreover to obtain a deeper understanding of chaperone function and regulation.

7.3 Chaperone–Cochaperone–Client Interactions

As far as we know, chaperone–cochaperone–client interactions have not been investigated with single molecule fluorescence, apart from some first studies on the Hsp90 system (see below). Such studies will be very rewarding, as they enable transition from a structure–function description to a dynamics–structure–function understanding. In the following, we will detail this transition and describe opportunities on the example of the Hsp90 system.

Hsp90 is associated with a wide range of various cochaperones and clients. Previous studies have explained Hsp90's versatile client recruitment by focusing on how cochaperones affect the conformational equilibrium in Hsp90's ATPase cycle and interact with clients.[38,71] Despite several advances in the field, the client recognition process of Hsp90 is not yet fully understood. In this regard, also the role of clients is still unclear because a commonly shared Hsp90 binding motif could not yet be identified. Currently, we are investigating conformational as well as interaction dynamics on all timescales, because the Hsp90 machinery is not a static complex, but very dynamic in two ways. First, it continuously assembles and disassembles. Second, the different components undergo (coupled) conformational changes themselves. Therefore, a thorough understanding of the Hsp90 protein machinery requires a dynamic, time-resolved view of protein organization at high spatial and temporal resolution. We are convinced that this will also be the case for many other chaperone machineries.

Here, we will use the example of the Hsp90–Cdc37–kinase interactions to demonstrate how multicolor smFRET can reveal basic principles of the dynamics of assembly, disassembly and conformational changes. Cdc37 is the cell division cycle protein 37, which links a wide spectrum of client kinases to Hsp90 and is required for stable kinase folding or activation.[72,73] Cdc37 has been identified to act as a sensor to distinguish between client and non-client kinases and subsequently recruits the client-kinases directly to Hsp90.[74] The structure of the Hsp90–Cdc37–kinase complex,[40,75] containing the cyclin-dependent kinase 4 (CDK4), has been determined. Cdk4 kinase regulates transition through the G_1 phase of the cell cycle and deregulation of Cdk4 is associated with cancers.[76] Together with standard bulk experiments, the following directed cycle is conceivable: first, the kinase binds to Cdc37, and then the Cdc37–kinase complex binds to Hsp90 in its open form and initiates its closing. In the final complex, the kinase is processed and released as an activated kinase.[71] An alternative dynamic model to this strictly directional cycle is depicted in Figure 7.5. It shows an equilibrium

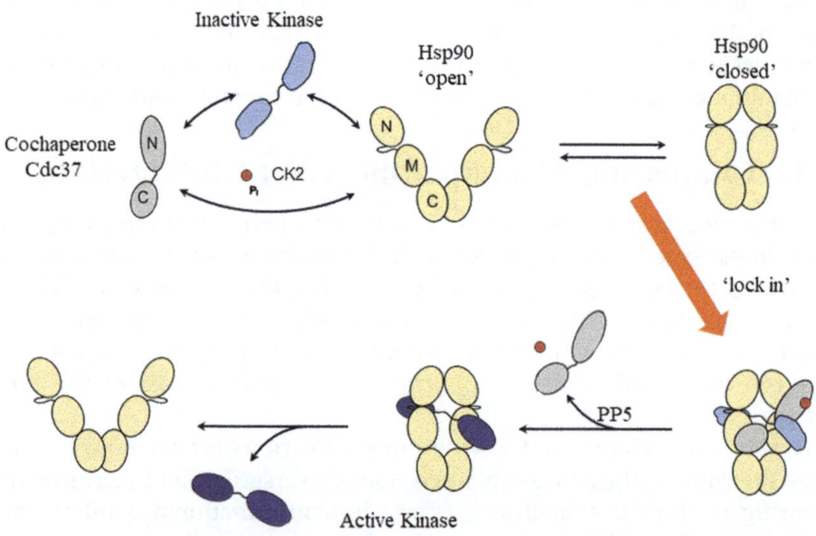

Figure 7.5 Dynamic model for the interaction of Hsp90 with Cdc37 and kinase: Cdc37, the inactive kinase and Hsp90 (open or closed state) might interact dynamically. Once all components meet in the correct conformation, they might form ('lock in') a stable complex (bottom right). At some point, phosphatase PP5 binding and ATP hydrolysis could then lead to the displacement of Cdc37 and the release of activated kinase. This is an alternative, largely unexplored, model compared to the standard strictly directional models (see main text for details). Note: Cdc37 phosphorylation by casein kinase 2 (CK2) and dephosphorylation by PP5 are required for Cdc37 function. P_i (red circle) indicates inorganic phosphate.

between several states, namely Hsp90 in its N-terminal open or closed state with Cdc37 and/or kinase bound or unbound (top row). When at some point all three interaction partners meet in the correct conformation they will lock into this state (orange arrow) and kinase processing (activation) follows.

Currently, there is no evidence for one or the other model, although they are fundamentally different. Such evidence is difficult to obtain, because with standard bulk experiments, the (equilibrium) kinetics are determined by bringing the system out of equilibrium and observing its relaxation back to equilibrium or the other way around. From such experiments, the steady-state kinetics can only be obtained in a perfect two-state system. However, this is usually not the case, especially not in multi-component protein systems. In addition, such synchronization for bulk experiments produces strong interference with the system. These problems can be overcome with single molecule methods. In particular, multicolor smFRET enables us to follow the association and dissociation kinetics and its coordination with conformational changes in real time. That this is indeed possible has already been demonstrated and discussed for the Hsp90 multi-component system, comprising the cochaperone Sba1 and a nucleotide.[33]

This study has shown that the Hsp90 system is very dynamic and not directional, making a clear sequence of conformations and assembly for the Hsp90–Cdc37–kinase system also unlikely. Such diverse, not strictly directional dynamics have also been reported for the Hsp70 system.[77] Is it possible to distinguish such a model from a clearly successive model? Indeed it is possible with single molecule experiments, because the rates (k_{i+}, k_{i-}) for all i transitions in both directions (+/−) can be determined at steady state, which is crucial to determine the free energy per cycle:[78,79]

$$\oint G = k_{\mathrm{B}} T \ln \frac{\Pi k_{i-}}{\Pi k_{i+}}$$

Here, T is the temperature, and k_{B} is the Boltzmann constant. For molecular motors usually strong directionality is observed,[80,81] whereas for Hsp90 on its own no directionality has been seen,[46] and for Hsp90 in complex with Sba1 only very weak directionality was observed.[33] Experiments of this kind will allow us to determine the coordination or independence of interactions as well as their coupling to ATP hydrolysis.

7.4 Summary and Outlook

The experiments above have shown that insights into chaperone dynamics and its regulation can be obtained with single molecule fluorescence-based methods. This will likely result in many important new findings ensuing the great success of the application of single molecule methods to motor proteins. Such experiments are not only important for a fundamental understanding of chaperone systems, but also should lead to new approaches for the development of specific drugs. An understanding of the assembly and disassembly dynamics of chaperones, cochaperones and clients, as well as their interplay with conformational changes – as *e.g.* anticipated in Figure 7.5 – might show new ways to just inhibit the interaction of the chaperone with a single defined client (*e.g.* by targeting defined conformational states or dynamics).

Finally, the examples above show that timescales at least from nanoseconds to minutes are involved in chaperone dynamics and likely in the regulation of chaperone activity. Regulation of nanosecond local fluctuations within a domain could *e.g.* affect larger millisecond fluctuations. This could eventually lead to the selection of states on timescales of seconds. At first sight, this complicates the description of chaperones, but in the long term, this might enable us to control chaperones in a much more defined and precise way. For example, interfering with certain fast modes (nanoseconds) in the chaperone might selectively affect the interaction with one client, while another client might be more susceptible to changes on slower (millisecond or second) timescales. Finally, for other clients, only the total time spent in one or the other conformational state might be relevant. Future studies will clarify, if and to which extent the effects on different timescales are coupled, but likely all of them are regulated in a chaperone system.

Several challenges still lie ahead in the application of single molecule fluorescence to chaperone systems. Such single molecule experiments usually only work at low concentrations of labelled proteins. In this regard, protein fusion[82] controlled microfluidic mixing[83,84] and small compartments (zero mode waveguides[85] or vesicles[86]) will be helpful. The transition towards high-throughput and *in vivo* studies will be even more challenging, but first steps in these directions are also taken.[87,88] We are convinced that the unprecedented data obtainable by single molecule fluorescence-based methods and further developments will increase the impact of these powerful techniques.

References

1. S. A. Stratmann and A. M. van Oijen, *Chem. Soc. Rev.*, 2014, **43**, 1201–1220.
2. S. Ticau, L. J. Friedman, N. A. Ivica, J. Gelles and S. P. Bell, *Cell*, 2015, **161**, 513–525.
3. J. E. Graham, K. J. Marians and S. C. Kowalczykowski, *Cell*, 2017, **169**, 1201–1213.e17.
4. A. N. Kapanidis, E. Margeat, S. O. Ho, E. Kortkhonjia, S. Weiss and R. H. Ebright, *Science*, 2006, **314**, 1144–1147.
5. F. Wang, S. Redding, I. J. Finkelstein, J. Gorman, D. R. Reichman and E. C. Greene, *Nat. Struct. Mol. Biol.*, 2013, **20**, 174–181.
6. T. T. Harden, K. S. Herlambang, M. Chamberlain, J.-B. Lalanne, C. D. Wells, G.-W. Li, R. Landick, A. Hochschild, J. Kondev and J. Gelles, *Nat. Commun.*, 2020, **11**, 448.

7. Y. Alhadid, S. Chung, E. Lerner, D. J. Taatjes, S. Borukhov and S. Weiss, *Protein Sci.*, 2017, **26**, 1278–1290.
8. J. Andrecka, B. Treutlein, M. A. I. Arcusa, A. Muschielok, R. Lewis, A. C. M. Cheung, P. Cramer and J. Michaelis, *Nucleic Acids Res.*, 2009, **37**, 5803–5809.
9. C. E. Perez and R. L. Gonzalez, *Curr. Opin. Chem. Biol.*, 2011, **15**, 853–863.
10. A. Petrov, G. Kornberg, S. O'Leary, A. Tsai, S. Uemura and J. D. Puglisi, *Curr. Opin. Struct. Biol.*, 2011, **21**, 137–145.
11. A. Ferguson, L. Wang, R. B. Altman, D. S. Terry, M. F. Juette, B. J. Burnett, J. L. Alejo, R. A. Dass, M. M. Parks, C. T. Vincent and S. C. Blanchard, *Mol. Cell*, 2015, **60**, 475–486.
12. M. J. Comstock, K. D. Whitley, H. Jia, J. Sokoloski, T. M. Lohman, T. Ha and Y. R. Chemla, *Science*, 2015, **348**, 352–354.
13. D. L. Jones, P. Leroy, C. Unoson, D. Fange, V. Ćurić, M. J. Lawson and J. Elf, *Science*, 2017, **357**, 1420–1424.
14. B. Schuler and H. Hofmann, *Curr. Opin. Struct. Biol.*, 2013, **23**, 36–47.
15. M. Margittai, J. Widengren, E. Schweinberger, G. F. Schröder, S. Felekyan, E. Haustein, M. König, D. Fasshauer, H. Grubmüller, R. Jahn and C. A. M. Seidel, *Proc. Natl. Acad. Sci. U. S. A.*, 2003, **100**, 15516–15521.
16. M. Pirchi, G. Ziv, I. Riven, S. S. Cohen, N. Zohar, Y. Barak and G. Haran, *Nat. Commun.*, 2011, **2**, 493.
17. H. Park, E. Toprak and P. R. Selvin, *Q. Rev. Biophys.*, 2007, **40**, 87–111.
18. E. J. G. Peterman, H. Sosa and W. E. Moerner, *Annu. Rev. Phys. Chem.*, 2004, **55**, 79–96.
19. S. Niekamp, N. Stuurman, N. Zhang and R. D. Vale, *Proc. Natl. Acad. Sci. U. S. A.*, 2021, **118**(31), e2101391118.
20. C. J. Bustamante, Y. R. Chemla, S. Liu and M. D. Wang, *Nat. Rev. Methods Primers*, 2021, **1**, 25.
21. B. Schuler, A. Soranno, H. Hofmann and D. Nettels, *Annu. Rev. Biophys.*, 2016, **45**, 207–231.
22. A. C. M. Ferreon, C. R. Moran, Y. Gambin and A. A. Deniz, in *Single Molecule Tools: Fluorescence Based Approaches, Part A*, ed. N. G. Walter, Academic Press, 2010, vol. 472, pp. 179–204.
23. A. Yildiz, M. Tomishige, R. D. Vale and P. R. Selvin, *Science*, 2004, **303**, 676–678.
24. M. Diez, B. Zimmermann, M. Börsch, M. König, E. Schweinberger, S. Steigmiller, R. Reuter, S. Felekyan, V. Kudryavtsev, C. A. M. Seidel and P. Gräber, *Nat. Struct. Mol. Biol.*, 2004, **11**, 135–141.
25. M. Schlierf, G. Wang, X. S. Chen and T. Ha, *eLife*, 2019, **8**, e42001.
26. E. Lerner, T. Cordes, A. Ingargiola, Y. Alhadid, S. Chung, X. Michalet and S. Weiss, *Science*, 2018, **359**, eaan1133.
27. B. Hellenkamp, S. Schmid, O. Doroshenko, O. Opanasyuk, R. Kühnemuth, S. Rezaei Adariani, B. Ambrose, M. Aznauryan, A. Barth, V. Birkedal, M. E. Bowen, H. Chen, T. Cordes, T. Eilert, C. Fijen, C. Gebhardt, M. Götz, G. Gouridis, E. Gratton, T. Ha, P. Hao,

C. A. Hanke, A. Hartmann, J. Hendrix, L. L. Hildebrandt, V. Hirschfeld, J. Hohlbein, B. Hua, C. G. Hübner, E. Kallis, A. N. Kapanidis, J.-Y. Kim, G. Krainer, D. C. Lamb, N. K. Lee, E. A. Lemke, B. Levesque, M. Levitus, J. J. McCann, N. Naredi-Rainer, D. Nettels, T. Ngo, R. Qiu, N. C. Robb, C. Röcker, H. Sanabria, M. Schlierf, T. Schröder, B. Schuler, H. Seidel, L. Streit, J. Thurn, P. Tinnefeld, S. Tyagi, N. Vandenberk, A. M. Vera, K. R. Weninger, B. Wünsch, I. S. Yanez-Orozco, J. Michaelis, C. A. M. Seidel, T. D. Craggs and T. Hugel, *Nat. Methods*, 2018, **15**, 669–676.

28. W. R. Algar, N. Hildebrandt, S. S. Vogel and I. L. Medintz, *Nat. Methods*, 2019, **16**, 815–829.
29. T. Förster, *Ann. Phys.*, 1948, **437**, 55–75.
30. M. Götz, P. Wortmann, S. Schmid and T. Hugel, in *Single-Molecule Enzymology: Fluorescence-Based and High-Throughput Methods*, ed. M. Spies and Y. R. Chemla, Academic Press, 2016, vol. 581, pp. 487–516.
31. N. K. Lee, A. N. Kapanidis, H. R. Koh, Y. Korlann, S. O. Ho, Y. Kim, N. Gassman, S. K. Kim and S. Weiss, *Biophys. J.*, 2007, **92**, 303–312.
32. S. Lee, J. Lee and S. Hohng, *PLoS One*, 2010, **5**, 1–9.
33. C. Ratzke, B. Hellenkamp and T. Hugel, *Nat. Commun.*, 2014, **5**, 4192.
34. H. Hofmann, F. Hillger, S. H. Pfeil, A. Hoffmann, D. Streich, D. Haenni, D. Nettels, E. A. Lipman and B. Schuler, *Proc. Natl. Acad. Sci. U. S. A.*, 2010, **107**, 11793–11798.
35. Y. Takei, R. Iizuka, T. Ueno and T. Funatsu, *J. Biol. Chem.*, 2012, **287**, 41118–41125.
36. A. J. Gupta, S. Haldar, G. Miličić, F. U. Hartl and M. Hayer-Hartl, *J. Mol. Biol.*, 2014, **426**, 2739–2754.
37. L. H. Pearl, C. Prodromou and P. Workman, *Biochem. J.*, 2008, **410**, 439–453.
38. M. M. Biebl and J. Buchner, *Cold Spring Harbor Perspect. Biol.*, 2019, **11**(9), a034017.
39. M. M. U. Ali, S. M. Roe, C. K. Vaughan, P. Meyer, B. Panaretou, P. W. Piper, C. Prodromou and L. H. Pearl, *Nature*, 2006, **440**, 1013–1017.
40. K. A. Verba, R. Y.-R. Wang, A. Arakawa, Y. Liu, M. Shirouzu, S. Yokoyama and D. A. Agard, *Science*, 2016, **352**, 1542–1547.
41. C. Prodromou, S. M. Roe, R. O'Brien, J. E. Ladbury, P. W. Piper and L. H. Pearl, *Cell*, 1997, **90**, 65–75.
42. A. K. Shiau, S. F. Harris, D. R. Southworth and D. A. Agard, *Cell*, 2006, **127**, 329–340.
43. C. Prodromou, B. Panaretou, S. Chohan, G. Siligardi, R. O'Brien, J. E. Ladbury, S. M. Roe, P. W. Piper and L. H. Pearl, *EMBO J.*, 2000, **19**, 4383–4392.
44. M. Hessling, K. Richter and J. Buchner, *Nat. Struct. Mol. Biol.*, 2009, **16**, 287–293.
45. M. Mickler, M. Hessling, C. Ratzke, J. Buchner and T. Hugel, *Nat. Struct. Mol. Biol.*, 2009, **16**, 281–286.
46. S. Schmid, M. Götz and T. Hugel, *Biophys. J.*, 2016, **111**, 1375–1384.

47. B. Hellenkamp, P. Wortmann, F. Kandzia, M. Zacharias and T. Hugel, *Nat. Methods*, 2017, **14**, 174–180.
48. T. Didenko, A. M. S. Duarte, G. E. Karagöz and S. G. D. Rüdiger, *Biochim. Biophys. Acta, Mol. Cell Res.*, 2012, **1823**, 636–647.
49. A. Lopez, V. Dahiya, F. Delhommel, L. Freiburger, R. Stehle, S. Asami, D. Rutz, L. Blair, J. Buchner and M. Sattler, *Sci. Adv.*, 2021, 7, eabl7295.
50. G. E. Karagöz, A. M. S. Duarte, E. Akoury, H. Ippel, J. Biernat, T. M. Luengo, M. Radli, T. Didenko, B. A. Nordhues, D. B. Veprintsev, C. A. Dickey, E. Mandelkow, M. Zweckstetter, R. Boelens, T. Madl and S. G. D. Rüdiger, *Cell*, 2014, **156**, 963–974.
51. B. M. Burmann and S. Hiller, *Prog. Nucl. Magn. Reson. Spectrosc.*, 2015, **86–87**, 41–64.
52. S. Hiller, *Trends Biochem. Sci.*, 2019, **44**, 517–527.
53. K. Mapa, M. Sikor, V. Kudryavtsev, K. Waegemann, S. Kalinin, C. A. M. Seidel, W. Neupert, D. C. Lamb and D. Mokranjac, *Mol. Cell*, 2010, **38**, 89–100.
54. M. Sikor, K. Mapa, L. V. von Voithenberg, D. Mokranjac and D. C. Lamb, *EMBO J.*, 2013, **32**, 1639–1649.
55. L. Voith von Voithenberg, A. Barth, V. Trauschke, B. Demarco, S. Tyagi, C. Koehler, E. A. Lemke and D. C. Lamb, *Proc. Natl. Acad. Sci. U. S. A.*, 2021, **118**(33), e2025578118.
56. S. Schmid and T. Hugel, *J. Chem. Phys.*, 2018, **148**, 123312.
57. W. Ye, M. Götz, S. Celiksoy, L. Tüting, C. Ratzke, J. Prasad, J. Ricken, S. V. Wegner, R. Ahijado-Guzmán, T. Hugel and C. Sönnichsen, *Nano Lett.*, 2018, **18**, 6633–6637.
58. S. Wolf, B. Sohmen, B. Hellenkamp, J. Thurn, G. Stock and T. Hugel, *Chem. Sci.*, 2021, **12**, 3350–3359.
59. A. Schulze, G. Beliu, D. A. Helmerich, J. Schubert, L. H. Pearl, C. Prodromou and H. Neuweiler, *Nat. Chem. Biol.*, 2016, **12**, 628–635.
60. D. Nettels, I. V. Gopich, A. Hoffmann and B. Schuler, *Proc. Natl. Acad. Sci. U. S. A.*, 2007, **104**, 2655–2660.
61. I. V. Gopich, D. Nettels, B. Schuler and A. Szabo, *J. Chem. Phys.*, 2009, **131**, 095102.
62. D. Haenni, F. Zosel, L. Reymond, D. Nettels and B. Schuler, *J. Phys. Chem. B*, 2013, **117**, 13015–13028.
63. M. F. Nüesch, M. T. Ivanović, J.-B. Claude, D. Nettels, R. B. Best, J. Wenger and B. Schuler, *J. Am. Chem. Soc.*, 2022, **144**, 52–56.
64. G. Krainer, P. Gracia, E. Frotscher, A. Hartmann, P. Gröger, S. Keller and M. Schlierf, *Biophys. J.*, 2017, **113**, 1280–1289.
65. A. Borgia, M. B. Borgia, K. Bugge, V. M. Kissling, P. O. Heidarsson, C. B. Fernandes, A. Sottini, A. Soranno, K. J. Buholzer, D. Nettels, B. B. Kragelund, R. B. Best and B. Schuler, *Nature*, 2018, **555**, 61–66.
66. B. Sohmen, C. Beck, T. Seydel, I. Hoffmann, B. Hermann, M. Nuesch, M. Grimaldo, F. Schreiber, S. Wolf, F. Roosen-Runge and T. Hugel, *arXiv*, preprint, arXiv:2110.10483, 2022.
67. T. Torres and M. Levitus, *J. Phys. Chem. B*, 2007, **111**, 7392–7400.

68. S. Felekyan, S. Kalinin, H. Sanabria, A. Valeri and C. A. M. Seidel, *ChemPhysChem*, 2012, **13**, 1036–1053.
69. F. Hillger, D. Hänni, D. Nettels, S. Geister, M. Grandin, M. Textor and B. Schuler, *Angew. Chem., Int. Ed.*, 2008, **47**, 6184–6188.
70. M. M. Biebl, A. Lopez, A. Rehn, L. Freiburger, J. Lawatscheck, B. Blank, M. Sattler and J. Buchner, *Nat. Commun.*, 2021, **12**, 828.
71. F. H. Schopf, M. M. Biebl and J. Buchner, *Nat. Rev. Mol. Cell Biol.*, 2017, **18**, 345–360.
72. L. H. Pearl, *Curr. Opin. Genet. Dev.*, 2005, **15**, 55–61.
73. M. Taipale, I. Krykbaeva, M. Koeva, C. Kayatekin, K. D. Westover, G. I. Karras and S. Lindquist, *Cell*, 2012, **150**, 987–1001.
74. D. Keramisanou, A. Aboalroub, Z. Zhang, W. Liu, D. Marshall, A. Diviney, R. W. Larsen, R. Landgraf and I. Gelis, *Mol. Cell*, 2016, **62**, 260–271.
75. C. K. Vaughan, U. Gohlke, F. Sobott, V. M. Good, M. M. U. Ali, C. Prodromou, C. V. Robinson, H. R. Saibil and L. H. Pearl, *Mol. Cell*, 2006, **23**, 697–707.
76. P. J. Day, A. Cleasby, I. J. Tickle, M. O'Reilly, J. E. Coyle, F. P. Holding, R. L. McMenamin, J. Yon, R. Chopra, C. Lengauer and H. Jhoti, *Proc. Natl. Acad. Sci. U. S. A.*, 2009, **106**, 4166–4170.
77. A. Mashaghi, S. Bezrukavnikov, D. P. Minde, A. S. Wentink, R. Kityk, B. Zachmann-Brand, M. P. Mayer, G. Kramer, B. Bukau and S. J. Tans, *Nature*, 2016, **539**, 448–451.
78. J. Howard, *Mechanics of Motor Proteins and the Cytoskeleton*, Sinauer Associates, Sunderland, MA, 2001.
79. T. L. Hill, *Free Energy Transduction and Biochemical Cycle Kinetics*, Springer, New York, NY, 1989.
80. M. Rief, R. S. Rock, A. D. Mehta, M. S. Mooseker, R. E. Cheney and J. A. Spudich, *Proc. Natl. Acad. Sci. U. S. A.*, 2000, **97**, 9482.
81. T. Hugel and C. Lumme, *Curr. Opin. Biotechnol*, 2010, **21**, 683–689.
82. X. Chen, J. L. Zaro and W.-C. Shen, *Adv. Drug Delivery Rev.*, 2013, **65**, 1357–1369.
83. R. Kellner, H. Hofmann, A. Barducci, B. Wunderlich, D. Nettels and B. Schuler, *Proc. Natl. Acad. Sci. U. S. A.*, 2014, **111**, 13355–13360.
84. B. Hellenkamp, J. Thurn, M. Stadlmeier and T. Hugel, *J. Phys. Chem. B*, 2018, **122**, 11554–11560.
85. S. Patra, J.-B. Claude, J.-V. Naubron and J. Wenger, *Nucleic Acids Res.*, 2021, **49**(21), 12348–12357.
86. I. Cisse, B. Okumus, C. Joo and T. Ha, *Proc. Natl. Acad. Sci. U. S. A.*, 2007, **104**, 12646–12650.
87. M. Segal, A. Ingargiola, E. Lerner, S. Chung, J. A. White, A. Streets, S. Weiss and X. Michalet, *Methods*, 2019, **169**, 21–45.
88. S. Tyagi, V. VanDelinder, N. Banterle, G. Fuertes, S. Milles, M. Agez and E. A. Lemke, *Nat. Methods*, 2014, **11**, 297–300.

CHAPTER 8

Visualization of Chaperone Mediated Protein Folding Using X-ray Crystallography

K. WU,[a] S. HOROWITZ*[b] AND J. C. A. BARDWELL*[a]

[a] Department of Molecular, Cellular, and Developmental Biology and Howard Hughes Medical Institute, University of Michigan, Ann Arbor, MI, USA; [b] Department of Chemistry & Biochemistry, Knoebel Institute for Healthy Aging, University of Denver, Denver, CO, USA
*Emails: jbardwel@umich.edu; Scott.Horowitz@du.edu

8.1 Introduction

Enzymes cause clearly defined chemical changes in well-defined substrates. Chaperones, on the other hand, generally cause poorly defined alterations in poorly defined substrates. This is in part because of the heterogeneous and partially disordered nature of protein folding intermediates with which chaperones generally interact, making many high-resolution biophysical approaches such as crystallography difficult. X-ray crystallography has been an invaluable tool for the determination of protein structures in general. In the field of chaperones, while most known chaperones have had their initial structures determined by X-ray crystallography, there is a relative paucity of X-ray structural data on one of the central questions in chaperone biology: how do chaperones recognize and help proteins to fold? We will here discuss the cases of the chaperonins, GroEL and CCT/TRiC, and the chaperones, Hsp70, trigger factor and Spy. Conventional crystallography techniques have provided good insight into Hsp70 client binding and mechanism, but so far they have provided less information on chaperonins and the trigger factor,

New Developments in NMR No. 29
Biophysics of Molecular Chaperones: Function, Mechanisms and Client Protein Interactions
Edited by Sebastian Hiller, Maili Liu and Lichun He

Published by the Royal Society of Chemistry, www.rsc.org

due to the dynamic nature of their client binding. For Spy, a new X-ray crystallography method of solving ensemble structures was developed that takes this dynamic nature into account, a method that might be also applicable to other chaperone–client complexes. Below we discuss a subset of crystallography studies of chaperone:client interactions. This is not an exhaustive list but represents a grouping that illustrates both the advantages and disadvantages of crystallography as an approach for these cases as illustrated by different chaperones with different client-binding properties.

8.2 Hsp70

Heat shock protein 70 (Hsp70) is universally conserved in evolution and plays important roles in protein homeostasis, cellular trafficking and solubilization of aggregates and amyloid fibrils.[1–3] Despite having very versatile functions, Hsp70 proteins act using similar mechanistic principles, driven in part by the structural features they share.

Hsp70 is composed of two domains, a substrate-binding domain and a nucleotide-binding domain, which are joined by a flexible linker. The substrate binding domain consists of a β-sandwich subdomain, an α-helical lid subdomain and an unstructured C-terminal tail.[4] The substrate binding specificity of Hsp70 has been examined through peptide-based screening approaches.[5,6] From these screening studies, a heptapeptide NRLLLTG emerged as a high-affinity Hsp70 binding ligand and this peptide was used to obtain a co-crystal structure of a Hsp70–client peptide complex.[4] In this complex, the β-sandwich subdomain of Hsp70 harbors a polypeptide binding cleft that is covered by a helical lid subdomain. The closed conformation of this substrate-bound Hsp70 structural model suggests that the helical lid subdomain would need to lift off the β-sandwich subdomain in order to allow substrate release[7] (see Figure 8.1). This crystal structure revealed a client binding cleft of ∼15 Å in length, long enough to accommodate polypeptides of 5–7 amino acids in length that contain centrally located hydrophobic residues. Hydrophobic residues are often buried within the core of folded proteins but become exposed often when proteins are unfolded under stress conditions and are exposed during protein synthesis. The promiscuous binding of Hsp70 to these hydrophobic-rich regions helps explain how Hsp70 can selectively bind to and protect its client proteins from aggregation. Additional co-crystal structures of Hsp70 in complex with various peptides demonstrate that the same peptide can bind in this cleft in both forward and reverse orientation, highlighting the promiscuous nature of Hsp70's peptide binding properties.[8] NMR studies further showed that such heterogeneous Hsp70–client ensembles with a short 53-residue client exist at equilibrium in solution and enable them to overcome kinetic traps during protein folding processes.[9]

Though Hsp70 on its own can associate with its client proteins, its substrate binding affinity in the cell is tuned by its cooperation with at least two co-chaperones, Hsp40 and a nucleotide-exchange factor. In the ATP-bound state,

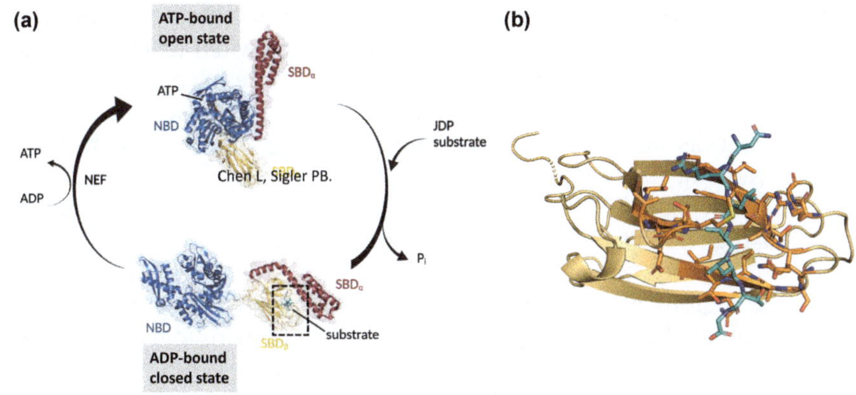

Figure 8.1 Hsp70 chaperone cycle. (a) J-domain protein (JDP) recognizes and delivers unfolded client substrates to the ATP-bound open state of Hsp70, which has a low substrate binding affinity. Binding of a substrate and JDP synergistically stimulates ATP hydrolysis and induces a large conformational switch to an ADP-bound closed state with a strong substrate binding affinity. Nucleotide exchange factors (NEFs) then trigger ADP dissociation and ATP association, thereby converting Hsp70s back to an ATP-bound open state and completing its chaperone cycle. Structures of Hsp70 in the open state (PDB: 4B9Q) and closed state (PDB: 2KHO) are represented as cartoons and individual domains are colored differently. Substrate-binding domains (SBDs) with two subdomains: an α-helical lid subdomain (SBD_{α}), light orange, and a β-sandwich subdomain (SBD_{β}), raspberry; nucleotide-binding domain (NBD, blue); the substrate is represented as cyan sticks. (b) Zoom-in image of the substrate-binding site of the substrate binding domain in complex with the NRLLLTG peptide (PDB: 1DKX). To illustrate the peptide binding site on the substrate binding domain, the α-helical lid subdomain is removed. The side chains of residues involved in substrate binding are shown as dark orange sticks and the NRLLLTG peptide is shown as cyan sticks.

Hsp70 adopts an open conformation, where its β-sandwich subdomain and the helical lid subdomain dissociate from each other and dock onto Hsp70's nucleotide-binding domain.[10] This ATP-induced conformational change opens up Hsp70's substrate binding cleft, resulting in an increase in both client dissociation and association rate constants. Since the increase in the dissociation rate is greater than the association rate, the result is a net decrease in binding affinity toward its client proteins.[11,12] To efficiently trap client proteins on Hsp70s, Hsp40s (also called J-domain proteins) are required to stimulate the hydrolysis of ATP,[13] thereby inducing a large conformational change in Hsp70 which transitions from an open conformation to a closed conformation and strengthens its binding affinity to client proteins.[4,14] The molecular details of how Hsp40s associate with Hsp70s have been recently revealed by a crystal structure of Hsp70 solved in complex with a J domain.[15] This structural model helps explain how Hsp40 stimulates Hsp70's ATPase activity and locks the client into the closed conformation of Hsp70.

To complete its chaperone cycle, Hsp70 needs to release clients. It does this by exchanging the bound ADP with ATP which acts to switch its conformation into the ATP-bound open conformation. However, since the basal nucleotide-exchange rate of Hsp70 is very low, the assistance of the nucleotide exchange factors to accelerate this process is required to make this happen on physiologically relevant timescales.

The current model of the Hsp70 chaperone cycle is mainly derived from studies using short peptides as Hsp70 clients. While these studies illustrate how Hsp70 can associate with short peptides and likely with the unstructured regions of a highly unfolded client, the prevailing model might not fully demonstrate how Hsp70s bind to the region with folded structures. More and more evidence shows that Hsp70s not only bind to extended stretches of client proteins but also associate with misfolded proteins, native proteins or even amyloids.[16] It is hard to imagine how Hsp70s accommodate all these various types of client proteins within the client binding pocket revealed by the earlier crystal structures of Hsp70–peptide complexes. Two independent studies show that the Hsp70's lid subdomain does not fully close upon binding to folded client proteins.[17,18] Considering the variety of functions Hsp70s play in cells, structural flexibility of the client binding site should be essential for Hsp70s to accommodate a broad spectrum of their client proteins.

8.3 Chaperonins

Chaperonins are large double doughnut-shaped protein assemblies that contain an inner chamber important for the protein folding reaction. In chaperonins, proteins can enter these cavities in a partially folded or misfolded state where they have a chance to refold before they are released back into the surrounding environment. This Hsp60 class of chaperones consists of multiple varieties of chaperonins present within the different kingdoms of life; the two best studied are GroEL and CCT/TRiC. GroEL became an early model system for chaperone function, as it was the first heat shock protein demonstrated to have a protein folding function.[19] CCT/TRiC is of special interest to the medical community, as it is the primary chaperonin in humans, and therefore a potential player in a number of protein misfolding diseases. Although different in shape and size, GroEL and CCT/TRiC have many mechanistic aspects in common, including ATP-driven conformational changes that alter the inner surface of the chaperonin and permit client entry and exit.[7]

Given their important roles, it is unsurprising that obtaining structural data of how chaperonins interact with clients and aid in their folding has been of high priority. However, the structural biology challenge with a chaperonin, especially for crystallography, is considerably greater than for Hsp70. While Hsp70 primarily binds to a short linear peptide that when crystallized in a single state presumably is informative of the structure of a natural chaperone–client complex, substrate proteins found inside the chaperonin barrel can be in the process of folding. Making the problem more difficult, the folding protein inside the chaperonin is thought to largely not be in direct contact with the chaperonin inner surface, making it less likely to be in a single definable state. In crystallography terms, if the folding protein is in a different conformational state in each unit cell in the crystal, the result will be an averaging out of the client protein signal that makes traditional model building impossible. Still, two studies of note looked at clients bound inside chaperonins, albeit one accidentally.

In 1999, Chen and Sigler published a structure of GroEL bound to a peptide mimic of a ligand.[20] To determine which peptide to use, the authors first performed a phage display experiment to find tight-binding peptides to use in crystallization trials. Co-crystallization of GroEL and the tightest binding peptide yielded crystals that diffracted to 2.0 Å (see Figure 8.2(a) and (b)). The density of the peptide could clearly be seen and modeled. The peptide was bound to the surface of the apical domain on the interior of the protein in a small β-hairpin structure. The binding interactions included a hydrophobic patch as well as polar contacts.

Although a milestone in chaperonin studies, this crystal structure comes with a sizable problem: GroEL's function is not to bind to a peptide in a

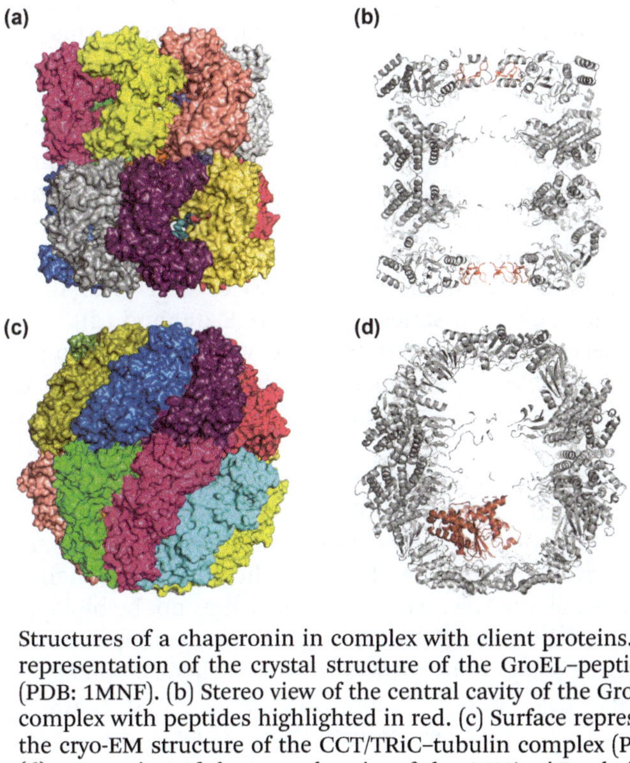

Figure 8.2 Structures of a chaperonin in complex with client proteins. (a) Surface representation of the crystal structure of the GroEL–peptide complex (PDB: 1MNF). (b) Stereo view of the central cavity of the GroEL–peptide complex with peptides highlighted in red. (c) Surface representation of the cryo-EM structure of the CCT/TRiC–tubulin complex (PDB: 7NVN). (d) Stereo view of the central cavity of the CCT/TRiC–tubulin complex with the tubulin molecule highlighted in red.

single conformation, as found in this crystal structure. As noted by the authors, several other structures had been published in which pieces of peptides from GroES or the N-terminus of GroEL had also bound to the inner surface of GroEL.[21] In any of these cases, the link between this peptide binding and the binding of a partially folded client is unclear. Additionally, as previously discussed by Helen Saibil, cryo-EM visualization of protein binding to GroEL shows that the density of these proteins is found deeper in the cavity than this peptide binding site,[22–24] suggesting that full-length proteins primarily exist elsewhere within the cavity. This difference does not mean that the peptide binding location located in the crystal structure is irrelevant, but unlike for many non-chaperone proteins, peptide binding sites for chaperones likely only tell one part of a complicated story that involves both binding and folding.

In 2011, a crystal structure of CCT/TRiC bound to a client was resolved by Muñoz *et al.* and it serves as an instructive contrast in issues to that of the previous GroEL structure.[25] In the case of CCT/TRiC, the authors did not set out to co-crystallize a client complex, but instead CCT/TRiC co-purified with a tubulin subunit bound in its cavity. The resulting crystals were of low resolution and only diffracted to 5.5 Å. As a result, the structure was low in detail and not all secondary structures of the CCT/TRiC could be placed unambiguously. The tubulin density, while being substantial, did not contain a clearly identifiable secondary structure. Whether this was due to the low resolution of the structure, the dynamics of the tubulin in potentially a partially folded state, other issues with the electron density, or a combination of these factors was not entirely clear. The possibility that the extra density was just an experimental artifact was excluded by confirmatory cryo-EM studies. The authors did not attempt to model a definitive protein sequence into the tubulin density and interpreted the density as representing a partially unfolded version of tubulin. Recently, two cryo-EM studies have reported near-atomic-resolution structures of CCT/TRiC in complex with tubulin in various folded states (see Figure 8.2(c) and (d)).[25,26] These structural complexes have revealed that the well-folded regions of substrate proteins are specifically stabilized by the interior wall of CCT/TRiC. In addition, the disordered C-terminal tail in the chamber can further facilitate folding.

The contrast of the client-bound crystal structures of GroEL and CCT/TRiC with recent cryo-EM structures of CCT/TRiC–tubulin complexes highlights some of the limitations of X-ray crystallography in studying many chaperone–client complexes. In the case of GroEL, the client was engineered to bind tightly to one site, but the resulting complex may not be biologically relevant. Meanwhile, in the case of CCT/TRiC, the single-snapshot view of crystallography made imaging a partially unfolded client challenging, while cryo-EM analysis may often be more suitable for addressing heterogenous chaperone–client complexes. As discussed below in the context of Spy, the problem of sample heterogeneity in crystallography can be circumventable, but it requires significant experimental investment.

8.4 Trigger Factor

Trigger factor is a ribosome-associated chaperone that plays important roles in the co- and post-translational folding and assembly process in bacteria.[27] The crystal structure of trigger factor on its own revealed a dragon-shaped architecture comprising three domains: an N-terminal domain, a peptidyl–prolyl isomerase domain and a C-terminal domain. The N-terminal domain contains a signature ribosomal recognition site motif.[28] The C-terminal domain, which forms a cradle-shaped structure, contains the major substrate binding sites.[29,30] The peptidyl–prolyl isomerase domain is homologous to the FK506-binding class of proline isomerization proteins. This domain acts as an auxiliary substrate-binding surface and is capable of catalyzing proline isomerization, a reaction that is at least in principle important for protein folding.[31] Insights regarding chaperone-mediated co-translational folding have come from the crystal structure of the N-terminal domain of trigger factor in complex with the 50S ribosome.[29,32,33] In these structural models, the trigger factor interacts with the ribosomal protein L23 and creates a dome over the ribosomal exit tunnel. Binding to the ribosome induces a conformational rearrangement in the N-terminal domain of the trigger factor, so that hydrophobic patches on the surface of the trigger factor face the ribosomal exit tunnel. This allows the trigger factor to associate with newly synthesized polypeptides immediately as they emerge from the ribosome and thus protect them from misfolding and aggregation. Crosslinking studies showed that elongating polypeptide chains associate with trigger factor in a directed manner: short chains encounter the hydrophobic region of trigger factor's N-terminal domain, longer chains appear to explore the entire interior of the C-terminal domain and the longest reach the peptidyl–prolyl isomerase domain.[34] Cryo-EM structures of trigger factor bound to translating ribosomes demonstrate that trigger factor –ribosome complex becomes less dynamic as nascent polypeptides elongate, although the density map corresponding to the polypeptide is elusive.[34,35] While these early structural studies supported the model by which trigger factor protects newly-synthesized polypeptides as they emerged from the ribosome, an *in vivo* study using selective ribosome profiling showed that trigger factor only associates with the translating ribosome that carries a nascent polypeptide chain of more than 100 amino acids.[36] This could ensure the modified enzymes, such as peptide deformylases and methionine aminopeptidases, have access to polypeptides prior to the binding of trigger factor.

In addition to its action on the co-translational folding process, it is also clear that the ribosome-free trigger factor in the cytosol can act as a chaperone in a posttranslational manner.[37] Saio *et al.* have determined high-quality NMR structural models of trigger factor–PhoA peptide complexes free in solution (see Figure 8.3(a)).[38] A detailed description of these results is outside of the scope of this chapter but is well discussed in Chapter 6. Suffice it to say these NMR structural models show that the trigger factor stabilizes

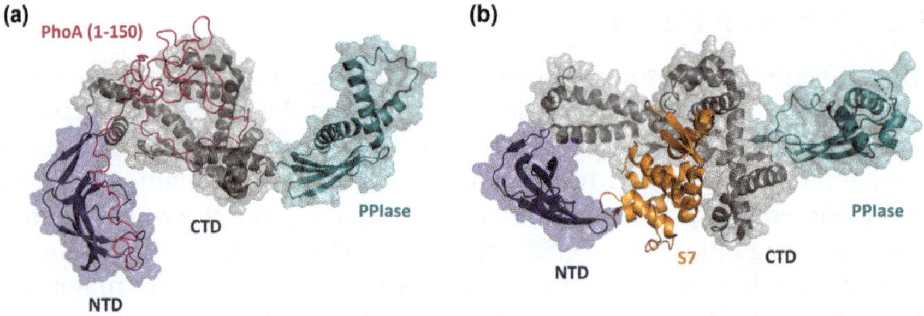

Figure 8.3 Different substrate binding modes of trigger factor. (a) NMR structure of the *E. coli* trigger factor–PhoA peptide (1–150) complex (PDB: 2MLY) revealed the interaction between the trigger factor and the fully unfolded substrate. The structure of the *E. coli* trigger factor is represented as cartoons and the individual domains are colored differently. N-terminal domain (NTD), dark blue; peptidyl–prolyl isomerase (PPIase) domain, teal and C-terminal domain (CTD), dark grey. The structure of the PhoA peptide is colored pink. For clarity, only one NMR model is shown in the figure. (b) Crystal structure of the *T. maritima* trigger factor–S7 complex (PDB: 3GTY) revealed the interaction between the trigger factor and the nearly folded substrate. The structure of the *T. maritima* trigger factor is presented in a similar fashion to the NMR structure of the trigger factor–PhoA complex. The structure of S7 is colored orange.

~100 aa-long PhoA peptides in a fully unfolded state primarily through interactions with multiple hydrophobic regions. It is hypothesized that by associating with multiple hydrophobic regions on PhoA peptides, trigger factor can effectively block inter-molecular contact and prevent protein aggregation. One limitation of this study is that trigger factor interacting partners were peptides in solution, and future studies will be needed to see how well the principles derived from this study apply to elongating peptide chains that emerge from the ribosome, or intact proteins post-translationally. In addition to its anti-aggregation activity as we just discussed, it is clear that trigger factor also plays more active roles in assisting protein folding.[39–42] For instance, single-molecule studies have shown that trigger factor could guide its client protein toward a productive folding pathway by populating a partially folded state, which is much less accessible in a spontaneous folding reaction.[41] Trigger factor has also been shown to exert its unfoldase activity by recognizing and unfolding a pre-existing folded polypeptide.[40] However, due to the lack of structural information on how the trigger factor associates with a (partially) folded protein, the structural mechanism by which trigger factor folds/unfolds its client protein is largely unknown.

Until now, only one structure of trigger factor in complex with an intact possible substrate protein is available. Martinez-Hackert *et al.* determined a co-crystal structure of trigger factor bound to the ribosomal protein S7 (see Figure 8.3(b)).[37] One obvious question is whether this represents an interaction with a genuine folding substrate or an interaction with part of the ribosome. Given the observation that many of trigger factor's *in vivo* substrates form oligomers or macromolecular assemblages, the authors argue that the crystal structure of the trigger factor–S7 complex may provide structural details regarding the trigger factor's roles in the assembly processes. The observation that trigger factor–bound S7 adopts a very near native conformation, with an RMSD of 0.65 Å as compared to ribosome-bound S7, makes these observations more relevant to how the trigger factor interacts with proteins late in their folding pathway rather than early during folding. The structural complex reveals that S7 sits in the trigger factor's cradle-shaped C-terminal domain, which is in agreement with previous studies showing that the trigger factor's substrate binding pocket is large enough to accommodate a small, folded protein or domain.[34] Second, while the structures of individual domains of S7–bound trigger factor are structurally very similar to those of apo trigger factor, substantial re-positioning of these individual domains occurs through flexible inter-domain linkers upon substrate binding. In addition, the interface of the binary complex revealed low shape complementarity, implying at least some potential for trigger factor to promiscuously bind to a number of client proteins. Curiously, the contacts between trigger factor and S7 in this crystal structure are primarily mediated by electrostatic forces rather than the hydrophobic contacts between the trigger factor and PhoA peptides observed using NMR[38] (see Figure 8.3). If the PhoA peptide interactions represent the contacts made with early folding intermediates and the S7 contacts those made with late folding intermediates, then trigger factor may be able to follow the same type of contact order progression as the chaperone Spy does with its substrate Im7 in which as the protein folds, different parts of the client bind to the chaperone. Indeed, trigger factor, similar to Spy, has recently been shown to both bind to proteins and allow protein folding while bound.[42] Pooling the results of interaction studies would suggest that trigger factor may have a heterogeneous substrate-binding surface using a mixture of hydrophilic and hydrophobic residues, which could enable the trigger factor to associate with both folded and unfolded proteins based on the characteristics of their solvent-exposed regions.[43] Similar characteristics of substrate binding surfaces have been shown in other chaperones, including Spy and GroEL. In both chaperone actions, switching from a hydrophobic binding mode to a hydrophilic binding mode has been shown to be an important feature for them to facilitate the folding of their client proteins.[7,44]

8.5 Spy

Spy is an *E. coli* periplasmic chaperone that offers an example of how crystallography can be used to determine client–chaperone interactions even when they are more dynamic. Spy is an ATP-independent chaperone that is heavily induced by stressors including ethanol and was found to protect the cell especially against tannin protein folding stress.[45] Crystal structures found Spy to be a small homodimer that forms a cradle shape, providing the first hint that proteins could go inside the cradle to be protected from stress.[45,46] Subsequent crosslinking experiments confirmed that the cradle was the site of protein binding.[47] ATP-independent chaperones are thought to bind tightly to their clients to prevent aggregation, but are not directly involved in the folding process. In contrast, Spy was found to both prevent protein aggregation and help proteins to fold.[45] Detailed kinetics experiments showed that client proteins would fold while bound to the cradle of Spy.[48,49]

To determine how this folding could be possible, one avenue of study would be to solve the ensemble of structures of the client bound to Spy's surface. This is a problem that would, on its surface, be amenable to NMR spectroscopy, given Spy and its clients' sizes, and these studies have also been performed.[50] However, the first approach taken was *via* crystallography.[51]

As described in previous sections, part of the challenge of using crystallography to determine the structures of clients interacting with GroEL and trigger factor was that crystallographers are incentivized to find conditions in which clients bind in a manner that is homogeneous and relatively undynamic, as this is the state that will most likely crystallize and show sufficient electron density for the client to model it accurately. While this approach proved useful for DnaK based on its mechanism, our current understanding of the chaperone function suggests that the single snapshot typically produced by crystallography will not be ideal.

Nevertheless, we set up crystallography experiments for Spy with a variant of its *in vivo* client Im7 that was destabilized and does not fold stably at room temperature[52] in an attempt to capture at least a partial snapshot of the complex. Conditions were found that produced highly reproducible and stable crystals that diffracted to high resolution and had clients bound at ratios that would suggest high enough occupancy to determine a co-structure, and these crystals did not form when the client was not present. However, the data that were collected produced maps that, perhaps predictably, were difficult to interpret. These maps showed large amounts of diffuse and low-intensity electron density within the cradle of Spy. It is not uncommon that ligands in crystal structures are poorly defined in initial maps and improve with rounds of model building and refinement, but this structure did not respond to such treatment. The introduction of a few client variants containing heavy atoms produced anomalous signals that were also spread across the cradle, often in multiple states. The data suggested that, as expected of the physiological case, the client was dynamic when bound to Spy and in multiple different conformations and therefore produced a weak electron density that was not strong enough to perform traditional crystal structure model building and refinement.

While this result was discouraging from the perspective of being able to quickly determine a co-structure, it did present an opportunity to dig deeper into chaperone–client interactions than a simple snapshot. Instead, we reasoned that the weak density held useful information on the conformational ensemble of client states while bound to the chaperone. However, the density by itself did not contain enough information to solve for the ensemble. We therefore developed an approach to localize the client using heavy atom derivatives. These heavy atom derivatives used the non-canonical amino acid *p*-iodophenylalanine, which was substituted in at a few positions on the client. Iodine exhibits a large anomalous signal that could be used to locate that amino acid specifically in three-dimensional space. Our early attempts suggested that multiple conformations were present in the crystal, as there were more peaks and they were more dispersed in the cradle than would have been possible with only one conformational state of the client.[51]

At this point, we departed from the traditional treatment of the data. If we had enough of these iodine anomalous signals giving specific positions of individual amino acids in multiple states, that could have been enough when combined with the weak electron density to determine the conformational ensemble of the client. There was no corollary in crystallography for this approach, but there was in NMR spectroscopy.[53,54] We therefore took known ensemble selection approaches from NMR spectroscopy and adapted them to use crystallography data.

Ensemble selections using NMR data follow a consistent theme with discrete steps: (1) a large pool of conformations should be computationally generated, often using an MD simulation. In our case, we performed a coarse-grained MD simulation of the client bound to the chaperone in the crystal. This MD simulation does not need to be perfectly accurate, it just needs to contain a subset of individual snapshots that can be selected using experimental data that would reflect the actual conformational ensemble. (2) Many small ensembles pulled from the large pool of conformations should be tested to see how well they back-predict the experimental data. If the algorithm is able to run to completion, it should produce the best small ensemble of conformations that can explain the data. (3) Validations should be run to test whether the small ensemble is appropriate and improves the fit to data as compared with the original large conformational pool. In our case, the most important validation was a 10-fold hold out cross-validation test, which showed that the ensemble selected using the anomalous signals and electron density was an improvement over the MD simulation alone and had predictive power. We named this overall strategy READ, for the residual anomalous and electron density data that it used.

The resulting ensemble contained 6 Im7 members, which, if arranged by RMSD to the native state, depicted the Im7 client folding while bound to Spy (see Figure 8.4). Most notably, the ensemble structure complemented mutational data we had recently published on mutations that improved Spy as a chaperone. Together, these mutations and structure validated the importance of the hydrophobic pocket on Spy's surface for client binding, as well as conformational dynamics in Spy to aid protein folding. The data also showed that as the client folded and its hydrophobics formed a protein core,

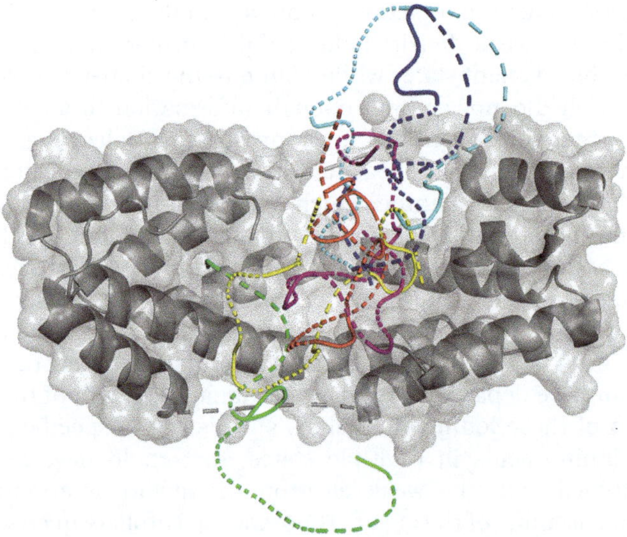

Figure 8.4 Structure of Spy-bound Im7$_{6–45}$ derived from the residual electron and anomalous density (READ) crystallography method (PDB: 5WNW).[51] The structure of Spy is represented as grey cartoons and the Spy-bound Im7$_{6–45}$ ensembles are represented as a multiple-colored cartoon.

the hydrophobic contacts between the client and the chaperone were replaced by electrostatic interactions. These data were complemented by other experiments in the lab performed shortly thereafter, which showed that the client binding and release cycle was dominated by this surface contact trading mechanism.[44]

This work drew criticism for its methods by Wang.[55] The criticisms were (1) "If a small anomalous peak observed in such a crystal were indeed to correspond to iodine with its peak height of only ∼10% that of the expected fully occupied iodine at the wavelength of 1.90749 Å (∼6.5 keV), only ∼10% of the substrate would be implied to be ordered, and the remaining 90% would be disordered", (2) that the anomalous signals were weak; weak enough that there were many noise peaks in the anomalous maps of similar size. Thus, it was not immediately apparent which peaks were client signals, and which were noise, with Wang claiming it entirely possible that we were fitting only noise. (3) Wang compared difference maps of different Spy datasets with different clients bound and found no detectable differences at the level of these maps, suggesting that either the amount of client bound was below detection or it was disordered enough to not impact the electron density.

We co-published a reply to Wang's commentary. Just using the initial data, we can conclude the following: (1) although Wang's 10% estimation of occupancy per position is accurate, he ignores the sum of multiple positions' occupancy towards the total occupancy of the client. (2) Although it was true that the anomalous peaks were weak and noisy, there were two issues with this claim. The first is that it was overstated by Wang in that he had over-modeled the Spy client and found additional conflicts as a result. The most important rebuttal came from the hold-out cross-validation test in the original manuscript that had already confirmed that the signal to noise ratio in the anomalous signals was sufficient to produce a predictive ensemble, eliminating the possibility that we were interpreting noise with no actual data. (3) The client map comparisons would be too insensitive given the large amount of disorder in the clients.

We were confident in our original data and its interpretation both due to the corroborating work of He *et al.*[50] and the cross-validation tests from the original work. That said, we did believe the level of noise was a limiting factor in the utility of READ and were looking to improve for future applications as well as to further confirm the initial findings on Spy.

The initial crystallography experiments had been performed using a relatively standard protein crystallography beamline at a synchrotron. Shortly after publication, we became aware of a new synchrotron beamline that was near commissioning that was engineered specifically to improve the signal to noise ratio of anomalous signals in the wavelength range of iodine, potentially by multiple orders of magnitude compared to our previous experiments.[56] This was an excellent opportunity to put our previous model and data to the test. We therefore re-crystallized and re-collected many of the anomalous variants at this new beamline. Multiple datasets were collected for each crystal to confirm the anomalous signals. We did not reconfirm every anomalous site due to the high stringency in these experiments in terms of reproducibility and crystal integrity; only a subset of the peptides produced multiple crystals and were able to withstand the rigorous data collection approach. For those three peptides we tested, the end result was much higher intensity anomalous signals for iodines, which clearly showed evidence of being in multiple states that were nearly identical to those we had discovered in our first set of experiments. Running the selection process again, we found that the higher-quality data produced nearly the same ensemble as well. Therefore, the new data confirmed our previous interpretations of the client–chaperone complex.[57]

Between the new protocol outlined for optimal data collection and publicly released code for the READ selection procedure,[58] all of the components are now available for others to use similar methods on other chaperone–client complexes. We suspect that there are probably many datasets that have been collected on chaperone–client crystals that displayed the same properties as those we used for the Spy–Im7 ensemble and would be ideal starting places for this procedure. Therefore, crystallography could still be an important player in determining the ensemble of chaperones bound to clients for years to come.

Although READ has advantages, it has distinct disadvantages as well. First, it is a time and resource-heavy experiment as it requires many crystals and a large amount of highly specialized beamtime. Second, the possibility exists that the amino acid substitutions necessitated by the READ approach could

Table 8.1 Crystal structures of chaperone–client complex.

	Hsp70:peptide	GroEL:peptide	TF:S7	Spy:Im7
PDB	1DKX	1MNF	3GTY	5WNW
Resolution	2 Å	3 Å	3.4 Å	1.8 Å
Key findings	Revealed the substrate recognition mechanism of Hsp70	Revealed the promiscuous binding mode of the peptide on the apical domain of GroEL	Demonstrated how trigger factor associates with its client protein in a near-native state	Used a new crystallography-based strategy, READ, to study a heterogeneous ensemble of chaperone–client complexes
Limitations	• The structure of the Hsp70–peptide complex may not be sufficient to reveal how Hsp70 recognizes a partially folded client protein.	• The link between peptide binding and the binding of partially unfolded protein is unclear. • Later experimental data suggested that partially unfolded protein is found deeper in the cavity than the peptide binding site that is revealed in this crystal structure.	• The physiological function of the 2 : 2 trigger factor–substrate complex revealed in the crystal structure is unclear.	• Time and resource-consuming. • Introducing mutations may interfere with the interactions with client proteins.

perturb the client ensemble either through intramolecular contacts within the client or intermolecular contacts with the chaperone. In the case of intermolecular contacts, some of the iodo-phenylalanine signals overlapped between different residues, suggesting that iodine could have its own binding preferences that could subtly or not-so-subtly shift the observed ensemble. This can be ruled out as a dominant factor by cross-validation tests to ensure back-prediction between differently-mutated clients (such as in the case of Spy) but is not something that can be known until well into the data collection process.

Alternate labeling strategies to iodo-phenylalanine could be tried, for example utilizing cysteines to link heavy metals. However, the nature of anomalous signals requiring large atoms that are not typically found in biology makes any mutation potentially problematic.

Similarly, the nature of crystallography has its limitations, including dependence on particular crystal conditions. For example, in our Spy structures, we observed several bound zinc atoms. Thus far, zinc binding does not appear to be functionally significant for Spy but could be vestigial activity based on related metal-binding proteins.[59] However, the crystal conditions we identified required zinc and these atoms were important crystal contacts. Could the zinc atoms or other crystal-packing perturb the relatively weak client binding in a way that is non-physiological? It is possible and would have to be evaluated on a case-by-case basis. Ultimately, READ is a powerful technique for visualizing the ensemble nature of client–chaperone contexts, but it also has drawbacks that could limit its usage in many cases.

8.6 Conclusions

The cases of Hsp70, GroEL, trigger factor and Spy offer differing lessons for the use of crystallography in determining client–chaperone co-structures (Table 8.1). In cases where traditional crystallography methods are used, the key is whether the physiological function of the chaperone will be adequately captured by a peptide in a single conformation. While this approach was informative for DnaK, GroEL and trigger factor offer cautionary tales. However, with the advent of READ used for Spy, crystallography has the potential to elucidate heterogeneous complexes with large conformational changes of the client. It is likely that several crystal conditions on major chaperones already exist that could be used as a starting place for READ. Therefore, it is possible to envision many future studies using this new technique to determine the ensembles of clients bound to chaperones.

Acknowledgements

This work was supported by the National Institute of General Medical Sciences grants R35GM142442 (SH). JCAB is supported by the Howard Hughes Medical Institute as an Investigator.

References

1. W. Neupert and M. Brunner, *Nat. Rev. Mol. Cell Biol.*, 2002, **3**, 555–565.
2. N. B. Nillegoda and B. Bukau, *Front. Mol. Biosci.*, 2015, **2**, 57.
3. S. Preissler and E. Deuerling, *Trends Biochem. Sci.*, 2012, **37**, 274–283.
4. X. Zhu, X. Zhao, W. F. Burkholder, A. Gragerov, C. M. Ogata, M. E. Gottesman and W. A. Hendrickson, *Science*, 1996, **272**, 1606–1614.
5. A. Gragerov, L. Zeng, X. Zhao, W. F. Burkholder and M. E. Gottesman, *Science*, 1996, **272**, 1606–1614.
6. S. Rüdiger, L. Germeroth, J. Schneider-Mergener and B. Bukau, *EMBO J.*, 1997, **16**, 1501–1507.
7. Y. E. Kim, M. S. Hipp, A. Bracher, M. Hayer-Hartl and F. U. Hartl, *Annu. Rev. Biochem.*, 2013, **82**, 323–355.
8. M. Zahn, N. Berthold, B. Kieslich, D. Knappe, R. Hoffmann and N. Strater, *J. Mol. Biol.*, 2013, **425**, 2463–2479.
9. R. Rosenzweig, A. Sekhar, J. Nagesh and L. E. Kay, *eLife*, 2017, **6**, e28030.
10. R. Kityk, J. Kopp, I. Sinning and M. P. Mayer, *Mol. Cell*, 2012, **48**, 863–874.
11. M. P. Mayer, S. Rüdiger and B. Bukau, *Biol. Chem.*, 2000, **381**, 877–885.
12. D. Schmid, A. Baici, H. Gehring and P. Christen, *Science*, 1994, **263**, 971–973.
13. T. Laufen, M. P. Mayer, C. Beisel, D. Klostermeier, A. Mogk, J. Reinstein and B. Bukau, *Proc. Natl. Acad. Sci. U. S. A.*, 1999, **96**, 5452–5457.
14. J. F. Swain, G. Dinler, R. Sivendran, D. L. Montgomery, M. Stotz and L. M. Gierasch, *Mol. Cell*, 2007, **26**, 27–39.
15. R. Kityk, J. Kopp and M. P. Mayer, *Mol. Cell*, 2018, **69**, 227–237.e224.
16. R. Rosenzweig, N. B. Nillegoda, M. P. Mayer and B. Bukau, *Nat. Rev. Mol. Cell Biol.*, 2019, **20**, 665–680.
17. M. Marcinowski, M. Holler, M. J. Feige, D. Baerend, D. C. Lamb and J. Buchner, *Nat. Struct. Mol. Biol.*, 2011, **18**, 150–158.
18. R. Schlecht, A. H. Erbse, B. Bukau and M. P. Mayer, *Nat. Struct. Mol. Biol.*, 2011, **18**, 345–351.
19. M. Y. Cheng, F. U. Hartl, J. Martint, R. A. Pollock, F. Kalousek, W. Neupertt, E. M. Hallberg, R. L. Hallberg and A. L. Horwich, *Nature*, 1989, **337**, 620–625.
20. L. Chen and P. B. Sigler, *Cell*, 1999, **99**, 757–768.
21. Z. Xu, A. L. Horwich and P. B. Sigler, *Nature*, 1997, **388**, 741–750.
22. D. K. Clare, P. J. Bakkes, H. van Heerikhuizen, S. M. van der Vies and H. R. Saibil, *Nature*, 2009, **457**, 107–110.
23. N. Elad, G. W. Farr, D. K. Clare, E. V. Orlova, A. L. Horwich and H. R. Saibil, *Mol. Cell*, 2007, **26**, 415–426.
24. H. Saibil, *Nat. Rev. Mol. Cell Biol.*, 2013, **14**, 630–642.
25. I. G. Muñoz, H. Yebenes, M. Zhou, P. Mesa, M. Serna, A. Y. Park, E. Bragado-Nilsson, A. Beloso, G. de Carcer, M. Malumbres, C. V. Robinson, J. M. Valpuesta and G. Montoya, *Nat. Struct. Mol. Biol.*, 2011, **18**, 14–19.

26. D. Gestaut, Y. Zhao, J. Park, B. Ma, A. Leitner, M. Collier, G. Pintilie, S. H. Roh, W. Chiu and J. Frydman, *Cell*, 2022, **185**, 4770–4787. e4720.
27. A. Hoffmann, B. Bukau and G. Kramer, *Biochim. Biophys. Acta*, 2010, **1803**, 650–661.
28. G. Kramer, T. Rauch, W. Rist, S. Vorderwulbecke, H. Patzelt, A. Schulze-Specking, N. Ban, E. Deuerling and B. Bukau, *Nature*, 2002, **419**, 171–174.
29. L. Ferbitz, T. Maier, H. Patzelt, B. Bukau, E. Deuerling and N. Ban, *Nature*, 2004, **431**, 590–596.
30. F. Merz, A. Hoffmann, A. Rutkowska, B. Zachmann-Brand, B. Bukau and E. Deuerling, *J. Biol. Chem.*, 2006, **281**, 31963–31971.
31. G. Kramer, H. Patzelt, T. Rauch, T. A. Kurz, S. Vorderwulbecke, B. Bukau and E. Deuerling, *J. Biol. Chem.*, 2004, **279**, 14165–14170.
32. D. Baram, E. Pyetan, A. Sittner, T. Auerbach-Nevo, A. Bashan and A. Yonath, *Proc. Natl. Acad. Sci. U. S. A.*, 2005, **102**, 12017–12022.
33. F. Schlunzen, D. N. Wilson, P. Tian, J. M. Harms, S. J. McInnes, H. A. Hansen, R. Albrecht, J. Buerger, S. M. Wilbanks and P. Fucini, *Structure*, 2005, **13**, 1685–1694.
34. F. Merz, D. Boehringer, C. Schaffitzel, S. Preissler, A. Hoffmann, T. Maier, A. Rutkowska, J. Lozza, N. Ban, B. Bukau and E. Deuerling, *EMBO J.*, 2008, **27**, 1622–1632.
35. J. Deeng, K. Y. Chan, E. O. van der Sluis, O. Berninghausen, W. Han, J. Gumbart, K. Schulten, B. Beatrix and R. Beckmann, *J. Mol. Biol.*, 2016, **428**, 3588–3602.
36. E. Oh, A. H. Becker, A. Sandikci, D. Huber, R. Chaba, F. Gloge, R. J. Nichols, A. Typas, C. A. Gross, G. Kramer, J. S. Weissman and B. Bukau, *Cell*, 2011, **147**, 1295–1308.
37. E. Martinez-Hackert and W. A. Hendrickson, *Cell*, 2009, **138**, 923–934.
38. T. Saio, X. Guan, P. Rossi, A. Economou and C. G. Kalodimos, *Science*, 2014, **344**, 1250494.
39. S. Haldar, R. Tapia-Rojo, E. C. Eckels, J. Valle-Orero and J. M. Fernandez, *Nat. Commun.*, 2017, **8**, 668.
40. A. Hoffmann, A. H. Becker, B. Zachmann-Brand, E. Deuerling, B. Bukau and G. Kramer, *Mol. Cell*, 2012, **48**, 63–74.
41. A. Mashaghi, G. Kramer, P. Bechtluft, B. Zachmann-Brand, A. J. Driessen, B. Bukau and S. J. Tans, *Nature*, 2013, **500**, 98–101.
42. K. Wu, T. C. Minshull, S. E. Radford, A. N. Calabrese and J. C. A. Bardwell, *Nat. Commun.*, 2022, **13**, 4126.
43. R. Mitra, K. Wu, C. Lee and J. C. A. Bardwell, *Annu. Rev. Biophys.*, 2022, **51**, 409–429.
44. P. Koldewey, F. Stull, S. Horowitz, R. Martin and J. C. A. Bardwell, *Cell*, 2016, **166**, 369–379.
45. S. Quan, P. Koldewey, T. Tapley, N. Kirsch, K. M. Ruane, J. Pfizenmaier, R. Shi, S. Hofmann, L. Foit, G. Ren, U. Jakob, Z. Xu, M. Cygler and J. C. Bardwell, *Nat. Struct. Mol. Biol.*, 2011, **18**, 262–269.

46. E. Kwon, D. Y. Kim, C. A. Gross, J. D. Gross and K. K. Kim, *Protein Sci.*, 2010, **19**, 2252–2259.
47. S. Quan, L. Wang, E. V. Petrotchenko, K. A. Makepeace, S. Horowitz, J. Yang, Y. Zhang, C. H. Borchers and J. C. Bardwell, *eLife*, 2014, **3**, e01584.
48. F. Stull, P. Koldewey, J. R. Humes, S. E. Radford and J. C. A. Bardwell, *Nat. Struct. Mol. Biol.*, 2016, **23**, 53–58.
49. K. Wu, F. Stull, C. Lee and J. C. A. Bardwell, *Nat. Commun.*, 2019, **10**, 4833.
50. L. He, T. Sharpe, A. Mazur and S. Hiller, *Sci. Adv.*, 2016, **2**, e1601625.
51. S. Horowitz, L. Salmon, P. Koldewey, L. S. Ahlstrom, R. Martin, S. Quan, P. V. Afonine, H. van den Bedem, L. Wang, Q. Xu, R. C. Trievel, C. L. Brooks and J. C. Bardwell, *Nat. Struct. Mol. Biol.*, 2016, **23**, 691–697.
52. C. L. Pashley, G. J. Morgan, A. P. Kalverda, G. S. Thompson, C. Kleanthous and S. E. Radford, *J. Mol. Biol.*, 2012, **416**, 300–318.
53. A. T. Frank, A. C. Stelzer, H. M. Al-Hashimi and I. Andricioaei, *Nucleic Acids Res.*, 2009, **37**, 3670–3679.
54. G. Nodet, L. Salmon, V. Ozenne, S. Meier, M. J. Jensen and M. Blackledge, *J. Am. Chem. Soc.*, 2009, **37**, 3670–3679.
55. J. Wang, *Nat. Struct. Mol. Biol.*, 2018, **25**, 989–990.
56. A. Wagner, R. Duman, K. Henderson and V. Mykhaylyk, *Acta Crystallogr., Sect. D: Struct. Biol.*, 2016, **72**, 430–439.
57. S. Rocchio, R. Duman, K. El Omari, V. Mykhaylyk, C. Orr, Z. Yan, L. Salmon, A. Wagner, J. C. A. Bardwell and S. Horowitz, *Acta Crystallogr., Sect. D: Struct. Biol.*, 2019, **75**, 1084–1095.
58. L. Salmon, L. S. Ahlstrom, J. C. A. Bardwell and S. Horowitz, *Methods Mol. Biol.*, 2018, **1764**, 491–504.
59. L. van der Weel, K. S. As, W. J. C. Dekker, L. van den Eijnden, W. van Helmond, C. Schiphorst, W. R. Hagen and P. L. Hagedoorn, *J. Inorg. Biochem.*, 2019, **192**, 98–106.

CHAPTER 9

Studying Molecular Chaperones and Their Client Interactions by Nanometer Distance Restraints from Electron Paramagnetic Resonance Spectroscopy

J. DRÖDEN AND M. DRESCHER*

Department of Chemistry and Konstanz Research School Chemical Biology, University of Konstanz, Universitätsstraße 10, 78457 Konstanz, Germany
*Email: malte.drescher@uni-konstanz.de

9.1 Introduction

The diverse family of molecular chaperones is a key player in the protein quality control system of the cell assisting in *de novo* protein folding and refolding to prevent misfolding and aggregation.[1–3] Understanding the underlying functions and mechanisms in the complex network of molecular chaperones is the focus of ongoing research.[4–11] Biophysical methods are of utmost importance to unravel the processes ensuring protein homeostasis.[12–18] Among them, electron paramagnetic resonance (EPR) spectroscopy in combination with site-directed spin-labeling (SDSL) is a powerful technique to gain insight into the structure and dynamics of molecular chaperones. SDSL-EPR is a virtually background-free technique facing no limitations in size, complexity or flexibility of the system under investigation.[19–21] Compared to *e.g.* Förster resonance energy transfer (FRET) experiments, required spin labels are small in size

New Developments in NMR No. 29
Biophysics of Molecular Chaperones: Function, Mechanisms and Client Protein Interactions
Edited by Sebastian Hiller, Maili Liu and Lichun He

Published by the Royal Society of Chemistry, www.rsc.org

and besides can be equally simplifying and simultaneously broadening the application. In contrast to single molecule techniques, EPR spectroscopy is able to characterize the entire structural ensemble of the sample providing the possibility to access the distribution of conformations quantitatively. Here, we present the state-of-the-art EPR methodology and focus on application to molecular chaperones to elucidate their structures, conformational changes and their client interactions.

EPR spectroscopy provides a broad range of experimental insights. Several EPR approaches are outlined in Chapter 10 by Shan and Qin. Here, we focus on distance restraints in the nanometer range obtained by EPR spectroscopy. While continuous wave EPR experiments provide dynamic and structural information, pulsed EPR methods emerged and are nowadays the leading EPR technique to obtain distance restraints. Several pulse sequences, among which the most widespread sequence is called double electron–electron resonance (DEER),[22,23] were developed to measure the distance between two unpaired electrons.

This chapter provides an introduction to spin labeling of proteins, followed by a description of continuous wave and pulsed EPR methods for distance determination. Here, the main focus lies on dead-time free four-pulse DEER experiments as a standard two-frequency technique to determine distances. Finally, recent research contributions investigating molecular chaperones and their clients by dipolar EPR spectroscopy are summarized.

9.2 Site-directed Spin-labeling

Most proteins are intrinsically diamagnetic and thus without further modification 'invisible' in the EPR experiment. SDSL pioneered by McConnell in 1965[24] and further developed by Hubbell and co-workers[25] is used to introduce non-native paramagnetic centers into proteins. Numerous publications already addressed the principle of SDSL in combination with EPR spectroscopy.[19–21,25–33]

SDSL requires the introduction of a site with unique reactivity at a selected position using site-directed mutagenesis. Most commonly, cysteine residues are used exploiting their low natural abundance and the reactivity of their sulfhydryl group. Subsequently, the protein can be modified with a paramagnetic reagent, the spin label, that gets covalently attached to the site of interest (see Scheme 9.1a).[19,25,26,28,30] Nevertheless, cysteine substitution mutagenesis and introduction of the spin label must not significantly perturb the protein structure, thermal stability, function and activity, which has to be assured by control experiments.[26,27,34]

9.2.1 Spin Labels Reactive to Cysteine Residues

Since the attachment of spin labels to cysteine residues is the most frequently used approach in SDSL, several spin labels exhibiting different chemical and spectroscopic properties have been developed. Thereby, the choice of spin label depends on the respective research question. However,

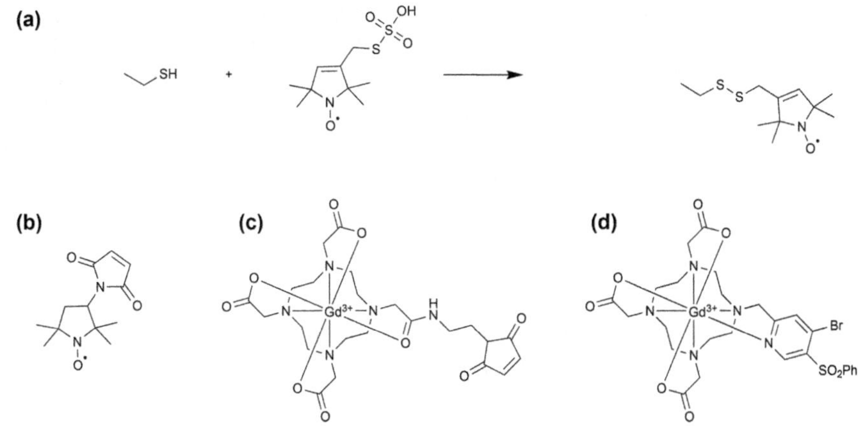

Scheme 9.1 (a) Site-directed spin labeling of cysteine residues with MTSSL. Chemical structures of other frequently used spin labels: (b) 3-maleimido-proxyl, (c) Gd(III)-DOTA-maleimide and (d) BrPSPy-DO3A-Gd(III).

the labeling procedure of cysteine residues is largely independent of the spin label used. Typically, the protein of interest is treated with a reducing agent, such as 1,4-dithiothreitol (DTT) or tris(2-carboxyethyl)phosphine (TCEP), to eliminate disulfide bridges between cysteine residues. After removal of the reducing agent, a typically six- to ten-fold molar excess of the spin label is added and incubation typically for at least one hour, up to 24 hours, is performed.[28] Here, the temperature is kept between 4 °C and room temperature. Subsequently, excess spin label is removed using for example size exclusion chromatography or dialysis.[28]

The most commonly used type of spin label is based on nitroxides. They consist of five- or six-membered rings containing an unpaired electron in the N–O moiety that is stabilized by four bulky groups in the α-position.[24] Here, the most popular one is the methanethiosulfonate spin label (MTSSL, see Scheme 9.1a).[35] It reacts readily, solely and quantitatively with sulfhydryl groups showing only small structural perturbations due to its small volume.[19,35] Besides attachment *via* thiosulfonate, maleimide or iodoacetamide moieties enable the covalent linkage of nitroxide-based spin labels to cysteine residues. A prominent example of a maleimide-linked spin label is 3-maleimido-proxyl (see Scheme 9.1b).[24]

Due to commercial availability, easy handling, rather low structural perturbation, and spectroscopic properties of nitroxides, they are perfectly suited for *in vitro* EPR studies. However, the limited steric shielding of the radical and especially the disulfide linkage of MTSSL are inadequate for in-cell experiments due to sensitivity to the reducing environment.[36–39] Here, other types of labels, such as transition metal ion-based spin labels, are promising candidates that could meet the demands of intracellular experiments.[40–44] Most prominently, Gd(III) ions are used as transition metal ions possessing the obligatory spin that is non-zero. The commercially available Gd(III)-DOTA-maleimide[45] or the recently developed BrPSPy-DO3A-Gd(III)[46] spin labels are excellent examples of such spin label types (see Scheme 9.1c and d). Again, these Gd(III)-based spin labels differ in the composition of the metal-chelating cage and the type of linker generating distinct chemical and spectroscopic properties. Alternatively, paramagnetic metal ions directly coordinating with the protein of interest can be used to perform EPR experiments, as it was successfully shown for Mn(II)–Mn(II) distance determination.[47]

The third class of spin labels includes trityl radicals. These bulky types of spin labels exhibit favorable spectroscopic properties, as well as higher bioresistance compared to nitroxides, arousing researchers' interest in further developments.[48–52] Here, they are mentioned only on account of completeness since chaperones have not been studied using this type of spin label yet.

9.3 Distance Restraints from Continuous Wave and Pulsed EPR Spectroscopy

Compared to nuclear magnetic resonance (NMR) spectroscopy, EPR spectroscopy enables the determination of distance restraints in the nanometer range. In contrast, usually only one restraint per measurement is obtained; however, the distances accessible in the EPR experiment are longer,[21] and also broad distance distributions are quantitatively accessible. EPR distance determination is not limited to the size and complexity of the system under study offering advantages compared to other biophysical techniques.[19–21,30,53]

Basically, in EPR spectroscopy, there exist two different main measuring modes: continuous wave (cw) and pulsed mode. Cw EPR spectra provide different primary parameters, such as side chain mobility, solvent accessibility or polarity of the immediate spin label environment.[26,27,30] The corresponding approaches are discussed in detail in Chapter 10 by Shan and Qin. Here, we focus on experimental approaches termed dipolar spectroscopy for determining inter-spin distances between at least two paramagnetic labels providing structural information, which give direct access to the structural ensemble of proteins and their complexes.[21] Mostly, doubly labeled proteins are used to obtain intra-molecular restraints; however, measuring inter-molecular distances between singly labeled proteins is also possible, as exploited for studying multimeric proteins.[54] In general, the sample is shock frozen in liquid nitrogen generating an overview of the conformational ensemble in a solution-like environment. The measurement itself is performed at cryogenic temperatures counteracting the transversal relaxation process and avoiding averaging of the dipolar coupling by molecular tumbling.[55] Whether measuring in cw or pulsed mode depends on the occurring distances (under typical conditions, 0.8–2 nm for cw and 1.7–8 nm for pulsed experiments).[56]

9.3.1 Distances from Continuous Wave EPR Experiments

Cw EPR experiments for determining distances are typically performed at X-band frequency (9.5 GHz) and optimally, a sample volume of 50 μL having a spin concentration of 50 μM is required.[57] Distances up to 2 nm can be estimated by cw experiments in the frozen state, ideally at a temperature of 120 K.[25,57–59] In this state, molecular tumbling is inhibited, and strong dipolar interactions corresponding to distances below 2 nm are reflected in the spectral shape. Often, a comparison of spectra of singly labeled proteins, in which intramolecular dipolar interactions are absent, is required to determine a dipolar broadening function required for analysis.[58] Distances and width of distance distributions can be extracted by spectral line shape analysis *via* simulation or line shape deconvolution methods.[56,58,59] In a less elaborated approach, the spin dipolar interactions can be estimated by the intensity ratio d_1/d (see Figure 9.1).[60] A ratio of less than 0.4 suggests the occurrence of distances below 2 nm.

9.3.2 Distances from Pulsed EPR Experiments

In analogy to NMR spectroscopy, also in EPR pulsed excitation can be applied. Various pulse sequences typically on a timescale of microseconds with nanosecond pulses have been developed; among them is dipolar spectroscopy detecting the dipolar interaction of coupled electron spins.[61] Among the most popular pulse sequences in dipolar spectroscopy are double-quantum coherence (DQC),[62] relaxation-induced dipolar modulation enhancement (RIDME)[63,64] and DEER,[22,23] also called pulsed electron–electron double resonance (PELDOR). Here, we focus on DEER, which is nowadays the most widely used EPR technique for distance determination.

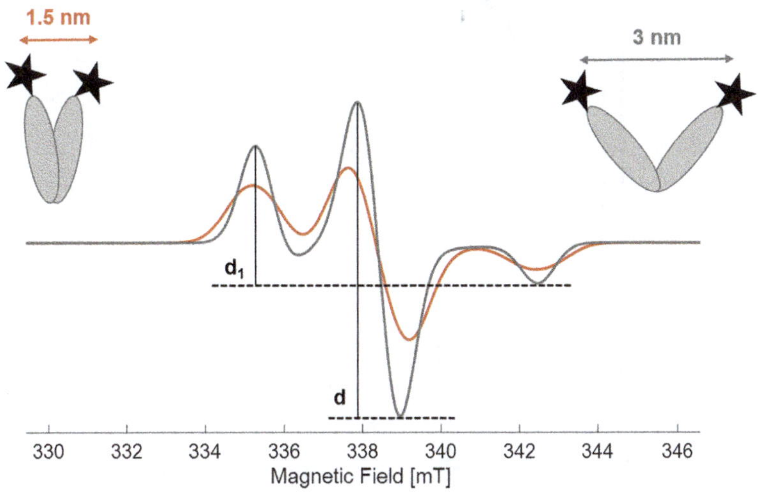

Figure 9.1 Distance determination *via* cw EPR spectroscopy. Simulated EPR X-band powder spectra of pairs of two nitroxides with inter-spin distances of 3 nm (grey) and 1.5 nm (red). The intensity ratio d_1/d is 0.57 for the former and 0.39 for the latter due to dipolar broadening. Simulation was performed using Matlab R2019b and the software package Easyspin 6.0.0-dev.33.[138]

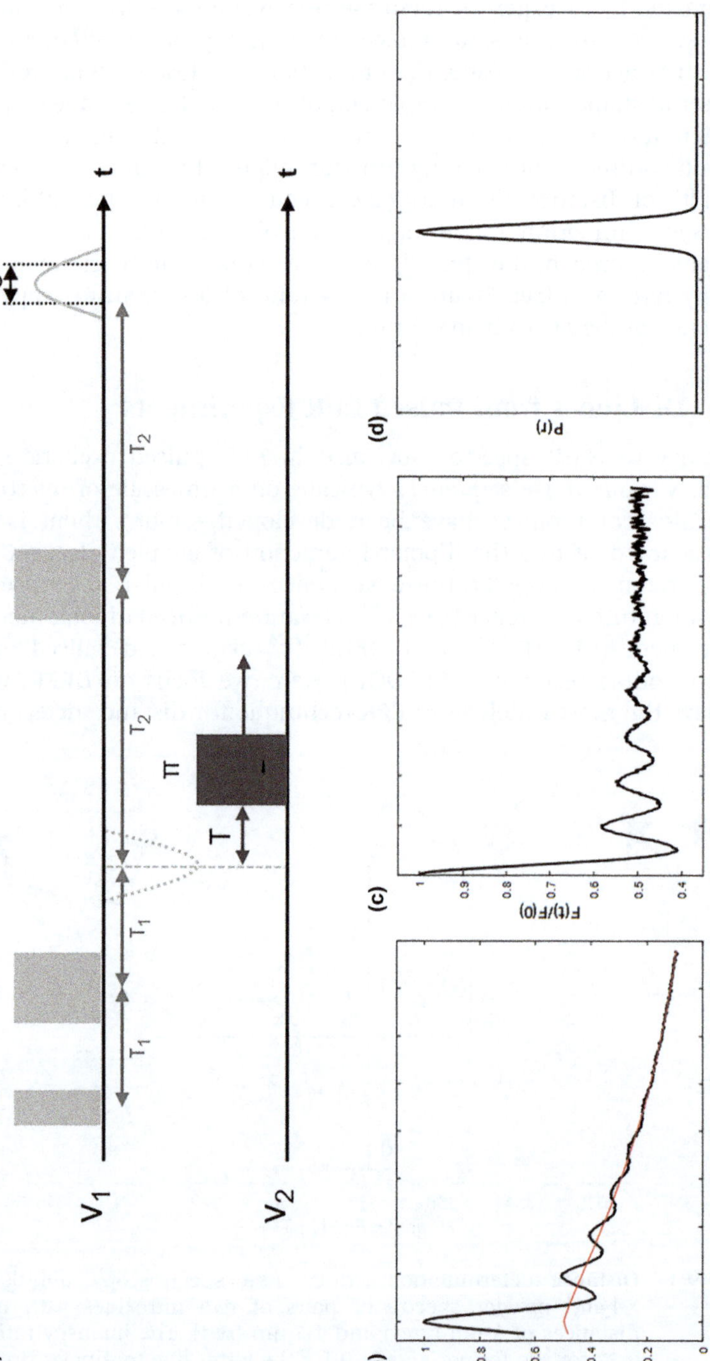
(a)
V1
V2
π/2
π
τ1
τ2
T
ig
t
(b)
V(t)/V(0)
Time [μs]
(c)
F(t)/F(0)
Time [μs]
(d)
P(r)
Distance [nm]

DEER yields a distribution of distances rather than a mean distance[30,55] accessing distances of 1.7 nm to 8 nm, in favorable cases up to 16 nm.[54] The upper limit of the accessible distance range can be increased by deuteration of the matrix,[55,65] *i.e.* the solution the protein resides, and by deuteration of the protein of interest itself.[66]

Initially, the three-pulse sequence was invented by Milov *et al.*[22,23] and further development resulted in a dead-time free four-pulse sequence (see Figure 9.2a).[67] It consists of a refocused primary echo sequence at the observer frequency ν_1 with inter-pulse delays τ_1 and τ_2 and an inversion pulse at the pump frequency ν_2 that is applied at variable times T.[55] The pump pulse addresses the electron spins B, thus changing the resonance frequency of the coupled A spins. This results in an echo intensity oscillating with the dipolar frequency of the coupled spins.[61] Besides the dipole–dipole interaction included in the signal, the echo is modulated by interactions with protons and deuterons.[53] However, at high frequencies like W-band (94 GHz) contributions from nuclear modulations are absent. Upon variation of the fixed inter-pulse delays and summation of all different echoes, unwanted modulations are suppressed.[53,55,68] The sensitivity of the DEER experiment could be enhanced upon measuring at higher frequencies, such as 34 GHz (Q-band) or 94 GHz (W-band) for Gd(III)–Gd(III) DEER.[55,61,68,69] Currently, the Q-band is the most popular frequency for DEER experiments on nitroxide spin labels.

In pulsed experiments, typical sample volumes are 10–25 μL;[28] however, due to the availability of high-power instruments and resonators, oversized samples with up to 60 μL are possible increasing the signal-to-noise ratio.[68] The ideal spin concentration depends on the expected distances; usually, spin concentrations of 50 μM are intended to achieve traces of good quality.[55] Recently, even nanomolar concentrations were reported to be feasible for DEER experiments approaching more physiologically relevant concentrations.[70,71] The measurement time typically varies between 1 and 12 h,[61] and depending on the sample quality and expected distance range, longer

Figure 9.2 Exemplary experimental DEER data of a nitroxide-biradical used as a molecular ruler. (a) Pulse sequence of the dead-time free four-pulse DEER experiment. The pulses at frequency ν_1 address the A spins, whereas the pulse at frequency ν_2 flips the B spins. The delay time T between the unobserved primary echo (grey dotted Gaussian-shaped curve) and π pump pulse is varied resulting in modulation of the refocused echo intensity $V(t)$. The integration gate (ig) is set to the full width at half height of the refocused echo (grey Gaussian-shaped curve) and integration yields the echo intensity $V(t)$. (b) DEER data of a nitroxide-biradical (black) that will be corrected for background contributions (red). The refocused echo intensity $V(t)$ is plotted against the varied time T of the pump pulse. (c) Form factor $F(t)$ obtained after background correction. (d) Distance distribution of the nitroxide-biradical obtained after Tikhonov regularization. Data analysis was performed using DeerAnalysis.[80] Exemplary DEER data are provided within the DeerAnalysis software package.

measurements of up to 24 h are required. Improvement is achieved upon more efficient excitation of spins realized by varying the shape of the pulses using novel arbitrary waveform generators (AWG).[72–75]

For DEER experiments, the optimal temperature for nitroxide spin labels is between 40 and 60 K and it is between 5 and 30 K for Gd(III)-based spin labels.[55,65] However, DEER measurements with rotation-restricted spiro-cyclohexyl-nitroxide spin labels at higher temperatures have been reported.[76] Moreover, the development of trityl-based spin labels featuring longer phase memory times guides future experiments towards more physiological temperatures.[21]

DEER measures the echo intensity $V(t)$ described as:

$$V(t) = F(t)B(t). \tag{9.1}$$

The form factor $F(t)$ reflects the intra-molecular dipolar interaction and the background signal $B(t)$ arises from inter-molecular interactions (see Figure 9.2b).[53,61] In order to calculate a distance distribution, the background contribution has to be removed which is not always trivial, especially if the measured trace is short.[53,61] The background function $B(t)$ for a homogenous distribution restricted to D dimensions is defined as an exponential function

$$B(t) = \exp\left(-kt^{\frac{D}{3}}\right) \tag{9.2}$$

with $D = 3$ for three-dimensional homogenous distribution and $D = 2$ for two-dimensional homogenous distribution. Preferably, the background is determined experimentally by measuring the respective singly labeled protein samples.[53,61,77] Upon background correction, the remaining form factor $F(t)$ appears as a damped oscillation containing the occurring dipolar frequencies (see Figure 9.2c). $F(t)$ is dominated by a single frequency that is inversely proportional to the cube of the mean distance r_{AB}.[53]

Transformation of the form factor results in the distance distribution (see Figure 9.2d) containing information on the mean distance, width and shape of the distribution.[55] However, this transformation is a moderately ill-posed problem and several algorithms were proposed to solve the problem.[53] The most used technique to overcome the ill-posed problem is Tikhonov regularization.[78,79] Here, an optimum regularization parameter α is defined, which is adjusted according to mathematical criteria.[53] The algorithms for background correction and extraction of distance distributions have been implemented in the open-source data analysis software package DeerAnalysis.[80] Later, neural network processing (DEERnet) was introduced in addition to traditional evaluation methods.[81] Recently, the comprehensive open-source Python package DeerLab has been introduced to the community enabling analysis of DEER data while processing the background contribution and the form factor at the same time.[82]

Finally, the calculated constraints can be used for modeling the protein structure, complex formation and conformational changes.[30] At the same time, the experimental results can be complemented by simulations. Most frequently, a rotamer approach is used to predict spin label positions based on a structural model of the protein, for example obtained by X-ray crystal structure analysis or theoretical approaches. Assuming this structural model is completely rigid, the rotamers of the label tethers and the resulting distance distribution are simulated.[53] Today, particular open-source software, such as MMM[83,84] or MtsslWizard,[85] are available enabling fast and easy predictions. It is also possible to perform full molecular dynamics simulations including the spin labels.[86]

9.4 Molecular Chaperones Studied by EPR Spectroscopy

In this section, recent work in the field of molecular chaperones studied by EPR spectroscopy in combination with SDSL is reviewed. In particular, we focus on results from dipolar spectroscopy, in contrast to the chapter by Shan and Qin that reports on label dynamics and rotational motions. Notably, in the outlined publications exclusively cysteine-based spin labeling was performed. Chaperone systems from different domains of life and their particular client interactions have been studied demonstrating the flexibility and broad application range of this labeling approach. In the following, important examples from the literature are grouped according to the chaperone system under investigation.

9.4.1 The Ribosome-associated Complex RAC

A recent contribution to chaperone research focused on the ribosome-associated complex (RAC) and aimed at understanding the RAC–ribosome interaction at a molecular level.[34] The stable heterodimer RAC is composed of Ssz1 and Zuo1, which comprises several domains (see Figure 9.3a) and solely interacts with the ribosome. Due to the lack of a complete high-resolution structure of the 3.3 MDa translation machinery in contact with the 107 kDa large RAC complex, a model having an α-helix as the middle domain in combination with the NMR structure of the four-helix bundle (4HB) fitted into the cryo-EM density of the RAC–ribosome complex was predicted (see Figure 9.3b).[87,88] Based on this working model, labeling positions within the helices of the 4HB and the middle domain of Zuo1 were chosen for obtaining EPR distance restraints (see Figure 9.3a). Numerous control experiments were performed to ensure the preservation of RAC's secondary structure, stability, *in vivo* functionality and ribosome binding capability upon introduction of cysteine residues and nitroxide spin labels. DEER traces that were measured in the presence of ribosomes did not differ from those measured for the RAC complex alone. The authors concluded

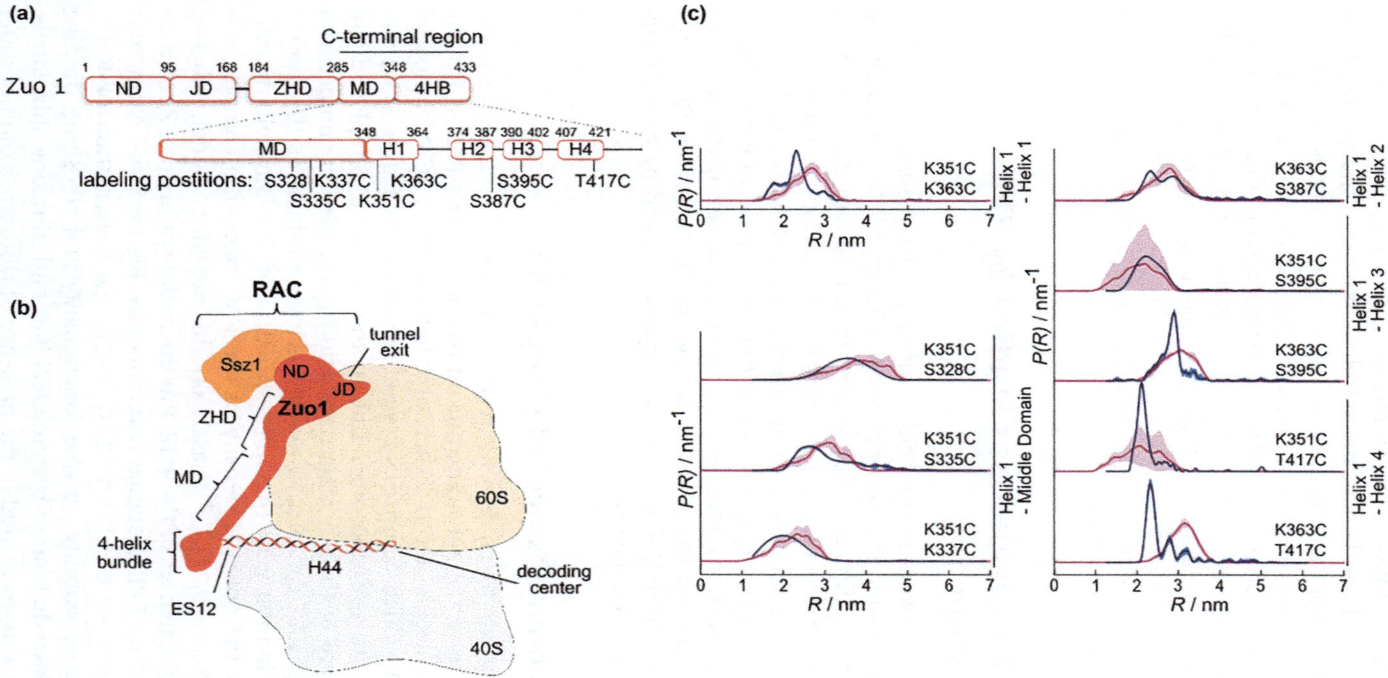

Figure 9.3 Structural investigation of the ribosome-associated complex RAC. (a) Domain organization of Zuo1: N-terminal domain (ND), J-domain (JD), Zuo1 homology domain (ZHD), middle domain (MD) and four-helix bundle (4HB). The engineered cysteine residues for spin labeling are indicated. (b) Model of the RAC–ribosome complex. (c) Experimental distance distributions of RAC (blue) compared to the predicted distance restraints based on the presented working model (magenta). Reproduced from ref. 34, https://doi.org/10.1038/s41598-021-87847-y, under the terms of the CC BY 4.0 license, http://creativecommons.org/licenses/by/4.0/.

that no conformational change in Zuo1 in the region monitored is taking place to control the chaperone activity. The proposed working model was supported by a set of DEER experiments (see Figure 9.3c). The experimental distance distributions of RAC match the simulated distributions based on the model using rotamer libraries. This study clearly emphasizes the power of SDSL-EPR spectroscopy being independent of the size and complexity of the system under study.

9.4.2 The Heat Shock Protein Hsp70

Similar to RAC's function, the eukaryotic heat shock protein 70 (Hsp70) and its bacterial analogue DnaK facilitate the correct folding of nascent polypeptide chains and denatured proteins.[89,90] It consists of two domains, the N-terminal nucleotide-binding domain (NBD) being responsible for ATP-binding and hydrolysis and the C-terminal substrate-binding domain (SBD), which comprises β-sheet topology for substrate binding and an α-helix acting as a moveable lid.[91–94] Binding of the substrate occurs *via* a short stretch of seven amino acid residues that is only present in unfolded proteins.[95,96] Using specifically designed small peptides modified with two MTSSL spin labels in combination with analyzing the dipolar broadening of low-temperature cw EPR spectra, Popp *et al.* were able to show that upon binding of DnaK, the inter-spin distances increase implying a stretching of the substrate. However, the conformation of the peptides is determined by the binding event itself rather than being dependent on the nucleotide present.[97]

The allosterically controlled Hsp70 cycle is characterized by large structural rearrangements.[93,98] Schlecht *et al.*[99] focused particularly on the lid movement having one MTSSL spin label placed in the lid and another in the β-sheet of the SBD. Low-temperature cw EPR measurements without ATP reported the lid to be closed, whereas in the presence of ATP, the spin labels are separated by more than 2 nm. Their EPR results complemented by disulfide-linking experiments indicated an equilibrium between the three conformational states 'lid closed', 'lid open' and 'intermediate-open', illustrating the conformational flexibility of the lid domain (see Figure 9.4).[99]

In a more recent study, the whole conformational ensemble of DnaK was mapped by Lai and co-workers using ion-mobility native mass spectrometry and DEER experiments.[100] In contrast to the previous study,[99] the authors unraveled the docking interfaces between the SBD and the NBD, as well as the position of the α-helical lid respective to the SBD or the NBD. Using four cysteine double mutants, the allosteric landscape of DnaK under different ligand-bound conditions was identified (see Figure 9.5), where DEER is exceptionally powerful in determining multiple, even sparsely populated conformations. They reported a domain-docked conformation (*D*) for ATP-bound DnaK, with the lid being either associated or dissociated from the NBD. For ADP and substrate-bound DnaK, many states were found to be populated; however, the NBD and the SBD are largely undocked. On the other hand, Lai *et al.* reported 90% of DnaK bound with ATP and substrate to

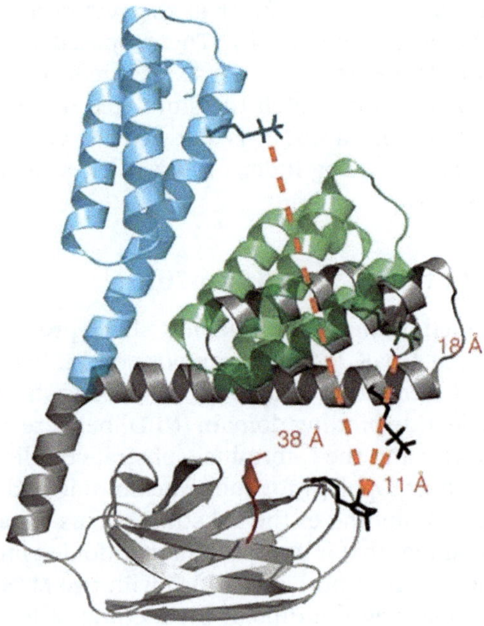

Figure 9.4 Cartoon representation of the SBD of DnaK demonstrating the three different lid conformations: 'lid closed' (grey), 'lid open' (blue) and 'intermediate-open' (green). MTSSL side chains are presented as sticks and the distances between the labels are indicated. Reproduced from ref. 99 with permission from Springer Nature, Copyright 2011.

populate a partially docked conformation (P) with the lid closed over the substrate that is distinct from the docked and the undocked state.

9.4.3 The Heat Shock Protein Hsp90

Another player in the protein quality control system is the eukaryotic heat shock protein 90 (Hsp90). It exists as a homodimer in which each monomer consists of three domains. The N-domain at the N-terminus comprising the ATP binding site is connected to the middle domain (M-domain) *via* a flexible linker and is followed by the C-domain, where dimerization takes place.[101–103] The ATP- and Mg-dependent Hsp90 cycle is coupled to global conformational changes[104] making EPR spectroscopy the most suitable technique to investigate these states. Giannoulis *et al.* recently performed Gd(III)–Gd(III) as well as Mn(II)–Mn(II) DEER experiments to clarify the conformations of yeast Hsp90 with molecular details.[47] The C-domains exhibit the same distance distributions at all stages of the cycle demonstrating their dimerization throughout (see Figure 9.6).

Subsequent Mn(II)–Mn(II) DEER measurements resolved the conformations of four different states: apo, pre- and post-hydrolysis, as well as the intermediate high-energy state (HES) (see Figure 9.7). In the absence of co-chaperones,

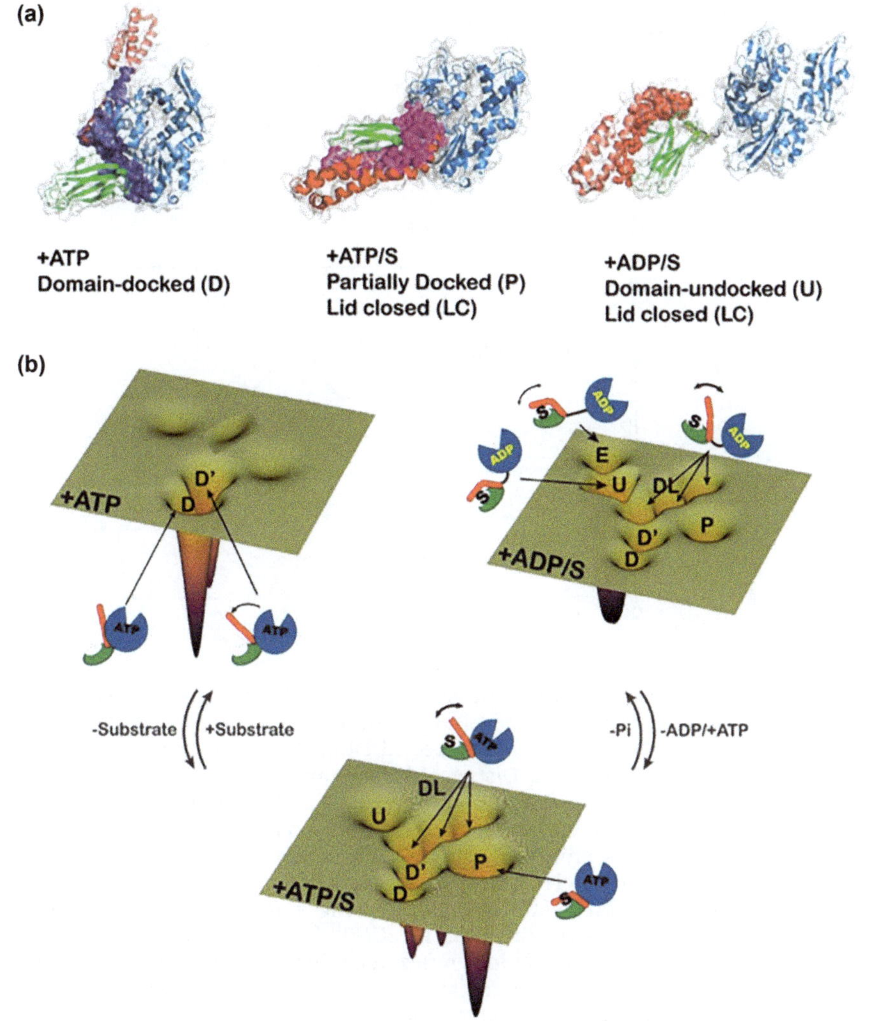

Figure 9.5 (a) Schematic representation of DnaK's conformations with the NBD in blue, β-sheet topology of the SBD in green and α-helical lid in red. In the presence of ATP and a substrate, DnaK is in the allosterically active state with an inter-domain interface formed between the NBD and the linker and between the α- and β-subdomain of the SBD (pink spheres). (b) Schematic allosteric energy landscapes of DnaK under different ligand-bound conditions. Each conformation detected in the experiments is represented, where the depth of each conformation is proportional to the relative abundance found by DEER. The respective conformations are D: docked, D′: domain docked with detached α-helical lid, U: undocked, DL: dynamic lid, E: extended and P: partially docked. Reproduced from ref. 100, https://doi.org/10.1074/jbc.M116.770404, under the terms of the CC BY 4.0 license, http://creativecommons.org/licenses/by/4.0/.

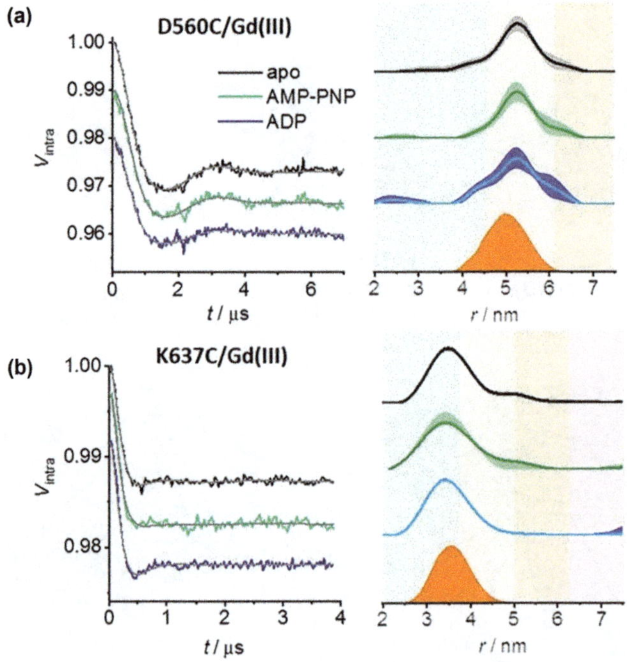

Figure 9.6 Gd(III)–Gd(III) DEER measurements on homo-dimeric yeast Hsp90 with spin labels placed in the C-domain at positions (a) D560C and (b) K637C. Background-corrected DEER time traces with fit (grey) are shown on the left. Distance distributions with respective confidence intervals are shown on the right. Both mutants were measured in the absence or in the presence of substrates reflecting the different states, apo (black), prehydrolysis (green) and posthydrolysis (blue) states, in the catalytic cycle of Hsp90. For comparison, expected distance distributions (orange) were modeled using a rotamer approach and the closed X-ray structure (PDB ID 2CG9[139]). Background-colored areas in the distance distributions reflect reliability regions with green denoting reliable shape, yellow reliable mean and width, orange reliable mean and red non-reliable area. Reproduced from ref. 30 with permission from PNAS, Copyright 2020.

the N-domains were found to be closed in the presence of ATP or ADP, whereas previous publications proposed an equilibrium between the open and closed ones.[105,106] Besides, the pre-hydrolysis conformational ensemble differs from those of the HES and post-hydrolysis state.

Hsp90 functions downstream of Hsp70 having a broad range of client proteins, among which is the intrinsically disordered protein (IDP) tau.[107] IDPs lack a defined three-dimensional structure, thus occupying a variety of conformations, which enormously complicates their biophysical investigation.[108] Nevertheless, intra- and inter-molecular DEER experiments in combination with an unconventional data analysis approach[109] were able to unravel the interaction mechanism of tau and Hsp90 at a molecular level.[110]

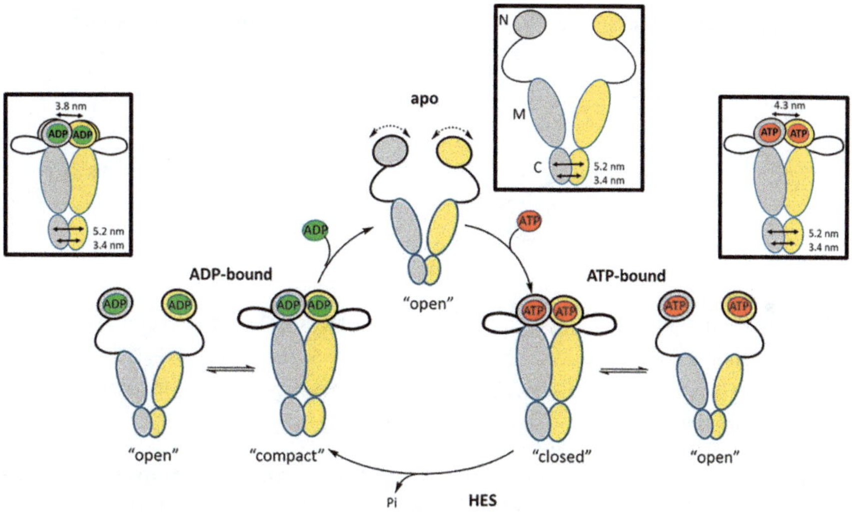

Figure 9.7 Conformational cycle of yeast Hsp90 as previously described in the literature.[105,106] The N-domains in the apo state can adopt a range of conformations (dashed arrows), while they are reported to be in equilibrium between open and closed conformations in the nucleotide-bound states. The insets show results from the DEER experiments displaying the maxima of the distance distributions. The N-domains were found to be closed in the presence of ATP or ADP. The C-domains are dimerized throughout the cycle displaying mean distances of 5.2 nm and 3.4 nm for probed mutants at all stages. Reproduced from ref. 47 with permission from PNAS, Copyright 2020.

Here, no quantitative distance distributions were calculated from the experimental DEER traces in order to avoid over-interpretation while studying the extremely broad conformational ensembles. Instead, the modulation depths of the traces were used to unravel the increase or decrease in spin–spin separation. This approach revealed tau adopting a dynamic paper-clip conformation in the absence of a chaperone as proposed earlier.[111,112] Upon interaction with Hsp90, the N-terminus of tau opens up uncovering the aggregation-prone repeat domain (see Figure 9.8). Strikingly, this unfolding promotes tau's self-aggregation and formation of small oligomers contradictory to the actual function of chaperones.

9.4.4 The SecB Chaperone

Besides chaperones that assist in *de novo* folding and refolding processes, there exist chaperones supporting the transport and translocation of translated, still unfolded proteins. A prominent example is the tetrameric *E. coli* chaperone SecB.[113] Its interaction with the unfolded substrates, B-chain of insulin and bovine pancreatic trypsin inhibitor (BPTI), was studied earlier using cw EPR spectroscopy.[114,115] In these studies, the cysteine-reactive spin labels attached to the substrates revealed the binding of SecB *via* a decrease

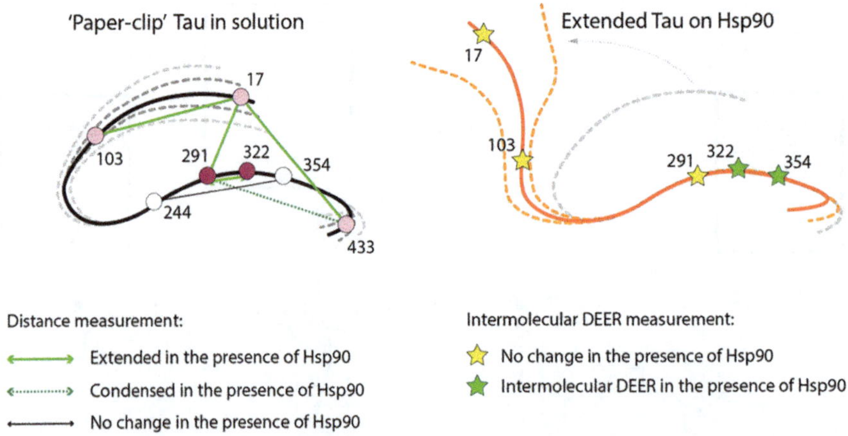

Figure 9.8 Schematic representation of tau in the absence (left) and in the presence of Hsp90 (right). Tau adopts a paper-clip conformation in the absence of Hsp90, whereas in the presence of Hsp90, the N-terminus opens up uncovering the aggregation-prone repeat domain of tau. Reproduced from ref. 110, https://doi.org/10.1126/sciadv.aax6999, under the terms of the CC BY 4.0 license, http://creativecommons.org/licenses/by/4.0/.

in the mobility of the spin label. Measurements at a low temperature (183 K) indicated a label separation of more than 2 nm confirming that substrate binding is achieved in an extended manner. Spin labeling of SecB in combination with analyzing dipolar broadening of cw spectra showed that in the absence of a substrate, the spin labels are separated by less than 2 nm, whereas upon addition of a substrate, they move apart from each other.[115] Unfortunately, the resolution of distances from cw EPR experiments is low, particularly the intensity ratio d_1/d (see Section 9.3.1) allows solely to claim whether spin labels are separated by less or more than 2 nm. In a follow-up study by Haimann *et al.*, the authors performed X-band DEER experiments using engineered cysteine double mutants of SecB to characterize conformational changes upon substrate binding in more detail.[116] Their results pointed towards a model, where the quaternary structure of SecB opens up in a scissor-like manner upon binding of a substrate (see Figure 9.9).

9.4.5 Other Types of Chaperones

DnaJ, the bacterial homologue of Hsp40, is an essential partner of DnaK stimulating ATPase activity in the DnaK cycle.[117] DnaJ consists of an N-terminal J domain important for DnaK binding and ATPase stimulation.[118] A polyproline motif links the J domain with the GF domain that is rich in glycine and phenylalanine residues and serves as a linker to the C-terminal domain.[119] Barends *et al.* gave insights into the structure and flexibility of DnaJ by performing X-ray crystallography, cross-linking

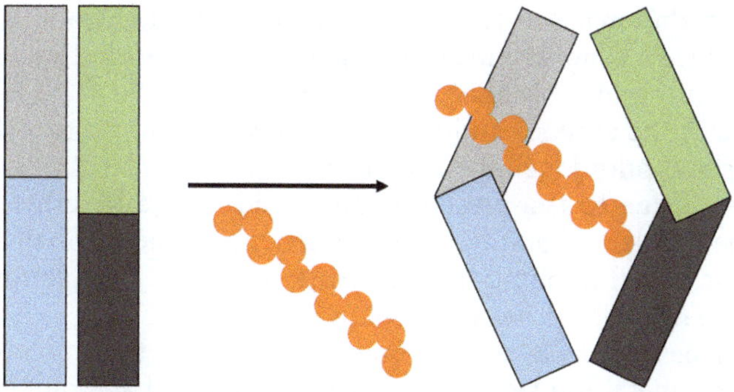

Figure 9.9 Schematic representation of the tetrameric SecB chaperone. Upon binding of a substrate (orange), the SecB structure opens up in a scissor-like manner.

and X-band DEER experiments.[120] Here, the authors report on an ordered GF domain as found in the crystallography experiments contrary to earlier findings of a rather disordered GF domain.[119,121] Their results were verified with DEER experiments exhibiting distances that are in accordance with the simulated distance distributions based on their crystal structure. In addition, DnaJ mutants probing the interface of the GF and C-terminal domain yielded broad distance distributions revealing DnaJ to be flexible in the junction of these domains.

Another contribution to the field of EPR spectroscopy is the study of the tetradecameric GroEL machinery that provides a protected area for protein folding.[54] Schmidt *et al.* combined sparse spin labeling and deuteration of protein and solvent with high-sensitivity Q-band DEER measurements to meet the requirements of the complex system and its expected distances. They were able to extend the dipolar evolution times to 80 μs allowing extraction of distances of up to 16 nm. The obtained distance distributions match the simulations based on the previously published crystal structure[122] emphasizing the potency of the described approach. This publication is considered a major breakthrough in the EPR community, nearly doubling the range of accessible distances forming the basis for studying other multimeric protein complexes and higher order oligomer systems.

The folding and assembly of the protein class of molybdenum-containing enzymes, such as the membrane-bound respiratory nitrate reductase complex NarGHI, is ensured by dedicated chaperones. Lorenzi *et al.* studied the structure and conformational dynamics of the chaperone NarJ by DEER spectroscopy and ion mobility mass spectrometry upon substrate interaction.[123] Having no high-resolution structure present, the chosen techniques give valuable structural information, as well as insights into client interactions. Cw EPR experiments initially revealed the interaction of NarJ with the substrate NarG. DEER experiments using engineered double cysteine mutants provided a broad distance distribution with three maxima in the absence of NarG. In comparison, the distance distribution is narrow exhibiting a unique maximum in the presence of a substrate. In a following work, Karthikeyan *et al.* introduced a novel nitroxide-based spin label selective for cysteine residues featuring bioresistance (M-TETPO).[124] Here, M-TETPO-labeled NarJ was studied in the absence and in the presence of a substrate yielding comparable results. Additionally, the first in-oocyte X-band DEER experiments were achieved due to the presence of nitroxide-shielding ethyl groups increasing the spin label stability in the cellular environment. Finally, M-TETPO is suited for in-cell application but further does not lose performance by changing the nitroxide-protecting groups.

Another type of spin label as presented in Section 9.2.1 is based on the paramagnetic metal ion Gd(III). Yagi *et al.* used the chaperone ERp29, ubiquitously expressed in the endoplasmic reticulum facilitating processing and transport of proteins, to characterize the performance of a novel cysteine-specific Gd(III)-based spin label.[40] DEER measurements in the W-band frequency on two different mutants revealed highly precise distance distributions of up to 6 nm. Moreover, the authors reported that the dimer interface of rat ERp29 is comparable with human ERp29, which was determined by crystallography.[125]

9.5 Outlook

In this chapter, we outlined ongoing research in the field of molecular chaperones studied by EPR spectroscopy. Interestingly, exclusively cysteine-specific spin labeling was performed. However, first attempts were made using other types of spin labels instead of the standard labels MTSSL and 3-maleimido-proxyl. Beyond labeling of cysteine residues, other strategies emerged based on the incorporation of non-canonical amino acids[32,126] in combination with copper-click chemistry,[127–130] strain-promoted azide-alkyne cycloaddition[131,132] or Suzuki–Miyaura coupling.[133] Even direct incorporation of a spin-labeled amino acid[134] was reported.

Clearly, the future of biological application of EPR spectroscopy points to performing experiments in the native environment of proteins.[39] First studies already proved the feasibility of the novel orthogonal labeling strategies combined with in-cell labeling[135] and in-cell EPR measurements.[136] Along with trityl-based spin labels enabling distance determination near physiological temperatures,[137] new perspectives for in-cell EPR spectroscopy in combination with SDSL are upcoming. Even though first experiments on molecular chaperones in model-cell systems were reported,[124] we expect more in-cell studies to come increasing the impact of SDSL-EPR spectroscopy in the field of molecular chaperones.

Abbreviations

AWG	Arbitrary waveform generator
BPTI	Bovine pancreatic trypsin inhibitor
Cw	Continuous wave

DEER	Double electron–electron resonance
DQC	Double quantum coherence
DTT	1,4-Dithiothreitol
EPR	Electron paramagnetic resonance
FRET	Förster resonance energy transfer
HES	High-energy state
Hsp70	Heat shock protein 70
Hsp90	Heat shock protein 90
IDP	Intrinsically disordered protein
NBD	Nucleotide-binding domain
NMR	Nuclear magnetic resonance
MTSSL	Methanethiosulfonate spin label
PELDOR	Pulsed electron–electron double resonance
RAC	Ribosome-associated complex
RIDME	Relaxation-induced dipolar modulation enhancement
SBD	Substrate-binding domain
SDSL	Site-directed spin labeling
TCEP	Tris(2-carboxyethyl)phosphine
4HB	Four-helix bundle

Acknowledgements

The authors would like to thank Mykhailo Azarkh for fruitful discussions and assistance in performing simulations. This work has received funding from the European Research Council (ERC) under the European Union's Horizon 2020 research and innovation programme (Grant Agreement number: 772027 – SPICE – ERC-2017-COG).

References

1. F. U. Hartl, A. Bracher and M. Hayer-Hartl, *Nature*, 2011, **475**, 324.
2. Y. E. Kim, M. S. Hipp, A. Bracher, M. Hayer-Hartl and F. Ulrich Hartl, *Annu. Rev. Biochem.*, 2013, **82**, 323.
3. J. Buchner, *J. Biol. Chem.*, 2019, **294**, 2074.
4. M. Santra, D. W. Farrell and K. A. Dill, *Proc. Natl. Acad. Sci. U. S. A.*, 2017, **114**, 2654.
5. I. Sučec, B. Bersch and P. Schanda, *Front. Mol. Biosci.*, 2021, **8**, 1017.
6. A. Barducci and P. De Los Rios, *Curr. Opin. Struct. Biol.*, 2015, **30**, 161.
7. S. Hiller, *Trends Biochem. Sci.*, 2019, **44**, 517.
8. K. Wu, F. Stull, C. Lee and J. C. A. Bardwell, *Nat. Commun.*, 2019, **10**(1), 4833.
9. G. Calloni, T. Chen, S. M. Schermann, H. C. Chang, P. Genevaux, F. Agostini, G. G. Tartaglia, M. Hayer-Hartl and F. U. Hartl, *Cell Rep.*, 2012, **1**, 251.

10. P. Genevaux, C. Georgopoulos and W. L. Kelley, *Mol. Microbiol.*, 2007, **66**, 840.
11. S. Wickner, T. L. L. Nguyen and O. Genest, *Annu. Rev. Microbiol.*, 2021, **75**, 719.
12. A. S. Wentink, N. B. Nillegoda, J. Feufel, G. Ubartaitė, C. P. Schneider, P. De Los Rios, J. Hennig, A. Barducci and B. Bukau, *Nature*, 2020, **587**, 483.
13. A. Pierro, E. Etienne, G. Gerbaud, B. Guigliarelli, S. Ciurli, V. Belle, B. Zambelli and E. Mileo, *Biomolecules*, 2020, **10**, 1062.
14. K. Weinhäupl, C. Lindau, A. Hessel, Y. Wang, C. Schütze, T. Jores, L. Melchionda, B. Schönfisch, H. Kalbacher, B. Bersch, D. Rapaport, M. Brennich, K. Lindorff-Larsen, N. Wiedemann and P. Schanda, *Cell*, 2018, **175**, 1365.
15. I. Sučec, Y. Wang, O. Dakhlaoui, K. Weinhäupl, T. Jores, D. Costa, A. Hessel, M. Brennich, D. Rapaport, K. Lindorff-Larsen, B. Bersch and P. Schanda, *Sci. Adv.*, 2020, **6**, eabd0263.
16. C. A. E. M. Spronk, S. Żerko, M. Górka, W. Koźmiński, B. Bardiaux, B. Zambelli, F. Musiani, M. Piccioli, P. Basak, F. C. Blum, R. C. Johnson, H. Hu, D. Scott Merrell, M. Maroney and S. Ciurli, *J. Biol. Inorg. Chem.*, 2018, **23**, 1309.
17. M. Alfano, G. Veronesi, F. Musiani, B. Zambelli, L. Signor, O. Proux, M. Rovezzi, S. Ciurli and C. Cavazza, *Chem. – Eur. J.*, 2019, **25**, 15351.
18. E. E. Boczek, J. Fürsch, M. L. Niedermeier, L. Jawerth, M. Jahnel, M. Ruer-Gruß, K. M. Kammer, P. Heid, L. Mediani, J. Wang, X. Yan, A. Pozniakovski, I. Poser, D. Mateju, L. Hubatsch, S. Carra, S. Alberti, A. A. Hyman and F. Stengel, *eLife*, 2021, **10**, e69377.
19. C. Altenbach, S. L. Flitsch, H. G. Khorana and W. L. Hubbell, *Biochemistry*, 1989, **28**, 7806.
20. W. L. Hubbell, D. S. Cafiso and C. Altenbach, *Nat. Struct. Biol.*, 2000, 7, 735–739.
21. W. L. Hubbell, C. J. López, C. Altenbach and Z. Yang, *Curr. Opin. Struct. Biol.*, 2013, **23**, 725.
22. A. D. Milov, A. B. Ponomarev and Y. D. Tsvetkov, *Chem. Phys. Lett.*, 1984, **110**, 67.
23. A. D. Milov and Y. D. Tsvetkov, *Appl. Magn. Reson.*, 1997, **12**, 495.
24. O. H. Griffith and H. M. McConnell, *Proc. Natl. Acad. Sci. U. S. A.*, 1966, **55**, 8.
25. W. L. Hubbell, H. S. Mchaourab, C. Altenbach and M. A. Lietzow, *Structure*, 1996, **4**, 779.
26. W. L. Hubbell and C. Altenbach, *Curr. Opin. Struct. Bio.*, 1994, **4**, 566.
27. W. L. Hubbell, A. Gross, R. Langen and M. A. Lietzow, *Curr. Opin. Struct. Biol.*, 1998, **8**, 649.
28. C. S. Klug and J. B. Feix, *Methods Cell Biol.*, 2008, **84**, 617.
29. J. P. Klare and H. J. Steinhoff, *Photosynth. Res.*, 2009, **102**, 377.
30. J. P. Klare, *Biol. Chem.*, 2013, **394**, 1281.

31. P. Roser, M. J. Schmidt, M. Drescher and D. Summerer, *Org. Biomol. Chem.*, 2016, **14**, 5468.
32. T. Braun, M. Drescher and D. Summerer, *Int. J. Mol. Sci.*, 2019, **20**, 373.
33. F. Torricella, A. Pierro, E. Mileo, V. Belle and A. Bonucci, *Biochim. Biophys. Acta, Proteins Proteomics*, 2021, **1869**, 140653.
34. S. J. Fries, T. S. Braun, C. Globisch, C. Peter, M. Drescher and E. Deuerling, *Sci. Rep.*, 2021, **11**, 8681.
35. L. J. Berliner, J. Grunwald, H. O. Hankovszky and K. Hideg, *Anal. Biochem.*, 1982, **119**, 450.
36. R. Igarashi, T. Sakai, H. Hara, T. Tenno, T. Tanaka, H. Tochio and M. Shirakawa, *J. Am. Chem. Soc.*, 2010, **132**, 8228.
37. M. Azarkh, O. Okle, P. Eyring, D. R. Dietrich and M. Drescher, *J. Magn. Reson.*, 2011, **212**, 450.
38. M. J. Lawless, A. Shimshi, T. F. Cunningham, M. N. Kinde, P. Tang and S. Saxena, *ChemPhysChem*, 2017, **18**, 1653.
39. A. Bonucci, O. Ouari, B. Guigliarelli and E. Mileo, *ChemBioChem*, 2019, **21**, 451.
40. H. Yagi, D. Banerjee, B. Graham, T. Huber, D. Goldfarb and G. Otting, *J. Am. Chem. Soc.*, 2011, **133**, 10418.
41. M. Qi, A. Groß, G. Jeschke, A. Godt and M. Drescher, *J. Am. Chem. Soc.*, 2014, **136**, 15366.
42. Y. Yang, F. Yang, Y.-J. Gong, J.-L. Chen, D. Goldfarb and X.-C. Su, *Angew. Chem., Int. Ed.*, 2017, **56**, 2914.
43. A. Feintuch, G. Otting and D. Goldfarb, *Methods Enzymol.*, 2015, **563**, 415.
44. Y. Yang, Y. J. Gong, A. Litvinov, H. K. Liu, F. Yang, X. C. Su and D. Goldfarb, *Phys. Chem. Chem. Phys.*, 2017, **19**, 26944.
45. A. Martorana, G. Bellapadrona, A. Feintuch, E. Di Gregorio, S. Aime and D. Goldfarb, *J. Am. Chem. Soc.*, 2014, **136**, 13458.
46. Y. Yang, F. Yang, Y. J. Gong, T. Bahrenberg, A. Feintuch, X. C. Su and D. Goldfarb, *J. Phys. Chem. Lett.*, 2018, **9**, 6119.
47. A. Giannoulis, A. Feintuch, Y. Barak, H. Mazal, S. Albeck, T. Unger, F. Yang, X.-C. Su and D. Goldfarb, *Proc. Natl. Acad. Sci. U. S. A.*, 2020, **117**, 395.
48. J. J. Jassoy, C. A. Heubach, T. Hett, F. Bernhard, F. R. Haege, G. Hagelueken, O. Schiemann, J. J. Jassoy, C. A. Heubach, T. Hett, F. Bernhard, F. R. Haege, G. Hagelueken and O. Schiemann, *Molecules*, 2019, **24**, 2735.
49. N. Fleck, C. Heubach, T. Hett, F. R. Haege, P. P. Bawol, H. Baltruschat and O. Schiemann, *Angew. Chem.*, 2020, **132**, 9854.
50. N. Fleck, C. Heubach, T. Hett, S. Spicher, S. Grimme and O. Schiemann, *Chem. – Eur. J.*, 2021, **27**, 5292.
51. Y. Yang, B.-B. Pan, X. Tan, F. Yanf, Y. Liu, X.-C. Su and D. Goldfarb, *J. Phys. Chem. Lett.*, 2020, **11**, 1141.
52. Z. Hasanbasri, K. Singewald, T. D. Gluth, B. Driesschaert and S. Saxena, *J. Phys. Chem. B*, 2021, **125**, 5265.

53. G. Jeschke and Y. Polyhach, *Phys. Chem. Chem. Phys.*, 2007, **9**, 1895.
54. T. Schmidt, M. A. Wälti, J. L. Baber, E. J. Hustedt and G. M. Clore, *Angew. Chem., Int. Ed.*, 2016, **55**, 15905.
55. G. Jeschke, *Annu. Rev. Phys. Chem.*, 2012, **63**, 419.
56. J. E. Banham, C. M. Baker, S. Ceola, I. J. Day, G. H. Grant, E. J. J. Groenen, C. T. Rodgers, G. Jeschke and C. R. Timmel, *J. Magn. Reson.*, 2008, **191**, 202.
57. M. Drescher and G. Jeschke, *EPR Spectroscopy: Applications in Chemistry and Biology*, Springer Berlin Heidelberg, Berlin, Heidelberg, 2012, vol. 321.
58. M. D. Rabenstein and Y.-K. Shint, *Proc. Natl. Acad. Sci. U. S. A.*, 1995, **92**, 8239.
59. H. J. Steinhoff, N. Radzwill, W. Thevis, V. Lenz, D. Brandenburg, A. Antson, G. Dodson and A. Wollmer, *Biophys. J.*, 1997, **73**, 3287.
60. S. J. Anthony-Cahill, P. A. Benfield, R. Fairman, Z. R. Wasserman, S. L. Brenner, W. F. Stafford, C. Altenbach, W. L. Hubbell and W. F. DeGrado, *Science*, 1992, **255**, 979.
61. D. Goldfarb, *J. Magn. Reson.*, 2019, **306**, 102.
62. P. P. Borbat and J. H. Freed, *Chem. Phys. Lett.*, 1999, **313**, 145.
63. S. Milikisyants, F. Scarpelli, M. G. Finiguerra, M. Ubbink and M. Huber, *J. Magn. Reson.*, 2009, **201**, 48.
64. S. Razzaghi, M. Qi, A. I. Nalepa, A. Godt, G. Jeschke, A. Savitsky and M. Yulikov, *J. Phys. Chem. Lett.*, 2014, **5**, 3975.
65. L. Garbuio, K. Zimmermann, D. Häussinger and M. Yulikov, *J. Magn. Reson.*, 2015, **259**, 163.
66. R. Ward, A. Bowman, E. Sozudogru, H. El-Mkami, T. Owen-Hughes and D. G. Norman, *J. Magn. Reson.*, 2010, **207**, 164.
67. M. Pannier, S. Veit, A. Godt, G. Jeschke and H. W. Spiess, *J. Magn. Reson.*, 2000, **142**, 331.
68. Y. Polyhach, E. Bordignon, R. Tschaggelar, S. Gandra, A. Godt and G. Jeschke, *Phys. Chem. Chem. Phys.*, 2012, **14**, 10762.
69. H. Ghimire, R. M. McCarrick, D. E. Budil and G. A. Lorigan, *Biochemistry*, 2009, **48**, 5782.
70. S. Kucher, C. Elsner, M. Safonova, S. Maffini and E. Bordignon, *J. Phys. Chem. Lett.*, 2021, **12**, 3679.
71. K. Ackermann, J. L. Wort and B. E. Bode, *J. Phys. Chem. B*, 2021, **125**, 5358.
72. T. Bahrenberg, Y. Rosenski, R. Carmieli, K. Zibzener, M. Qi, V. Frydman, A. Godt, D. Goldfarb and A. Feintuch, *J. Magn. Reson.*, 2017, **283**, 1.
73. T. F. Prisner, *J. Magn. Reson.*, 2019, **306**, 98.
74. P. E. Spindler, S. J. Glaser, T. E. Skinner and T. F. Prisner, *Angew. Chem., Int. Ed.*, 2013, **52**, 3425.
75. A. Scherer, S. Tischlik, S. Weickert, V. Wittmann and M. Drescher, *Magn. Reson*, 2020, **1**, 59.

76. V. Meyer, M. A. Swanson, L. J. Clouston, P. J. Boratyński, R. A. Stein, H. S. McHaourab, A. Rajca, S. S. Eaton and G. R. Eaton, *Biophys. J.*, 2015, **108**, 1213.
77. L. Fábregas Ibáñez and G. Jeschke, *Phys. Chem. Chem. Phys.*, 2020, **22**, 1855.
78. T. H. Edwards and S. Stoll, *J. Magn. Reson.*, 2018, **288**, 58.
79. Y.-W. Chiang, P. P. Borbat and J. H. Freed, *J. Magn. Reson.*, 2005, **177**, 184.
80. G. Jeschke, V. Chechik, P. Ionita, A. Godt, H. Zimmermann, J. Banham, C. R. Timmel, D. Hilger and H. Jung, *Appl. Magn. Reson.*, 2006, **30**, 473.
81. S. G. Worswick, J. A. Spencer, G. Jeschke and I. Kuprov, *Sci. Adv.*, 2018, **4**, eaat5218.
82. L. Fábregas Ibáñez, G. Jeschke and S. Stoll, *Magn. Reson.*, 2020, **1**, 209.
83. Y. Polyhach, E. Bordignon and G. Jeschke, *Phys. Chem. Chem. Phys.*, 2011, **13**, 2356.
84. G. Jeschke, *Protein Sci.*, 2018, **27**, 76.
85. G. Hagelueken, D. Abdullin and O. Schiemann, in *Methods in Enzymology*, Academic Press Inc., 2015, vol. 563, p. 595.
86. M. J. Schmidt, A. Fedoseev, D. Bücker, J. Borbas, C. Peter, M. Drescher and D. Summerer, *ACS Chem. Biol.*, 2015, **10**, 2764.
87. Y. Zhang, C. Ma, Y. Yuan, J. Zhu, N. Li, C. Chen, S. Wu, L. Yu, J. Lei and N. Gao, *Nat. Struct. Mol. Biol.*, 2014, **21**, 1042.
88. J. K. Ducett, F. C. Peterson, L. A. Hoover, A. J. Prunuske, B. F. Volkman and E. A. Craig, *J. Mol. Biol.*, 2013, **425**, 19.
89. F. U. Hartl and M. Hayer-Hartl, *Science*, 2002, **295**, 1852.
90. M. P. Mayer and B. Bukau, *Cell. Mol. Life Sci.*, 2005, **62**, 670.
91. K. M. Flaherty, C. DeLuca-Flaherty and D. B. McKay, *Nature*, 1990, **346**, 623.
92. C. J. Harrison, M. Hayer-Hartl, M. Di Liberto, F. U. Hartl and J. Kuriyan, *Science*, 1997, **276**, 431.
93. X. Zhu, X. Zhao, W. F. Burkholder, A. Gragerov, C. M. Ogata, M. E. Gottesman and W. A. Hendrickson, *Science*, 1996, **272**, 1606.
94. H. Wang, A. V. Kurochkin, Y. Pang, W. Hu, G. C. Flynn and E. R. P. Zuiderweg, *Biochemistry*, 1998, **37**, 7929.
95. A. Gragerov, L. Zeng, X. Zhao, W. Burkholder and M. E. Gottesman, *J. Mol. Biol.*, 1994, **235**, 848.
96. S. Rüdiger, L. Germeroth, J. Schneider-Mergener and B. Bukau, *EMBO J.*, 1997, **16**, 1501.
97. S. Popp, L. Packschies, N. Radzwill, K. P. Vogel, H. J. Steinhoff and J. Reinstein, *J. Mol. Biol.*, 2005, **347**, 1039.
98. M. P. Mayer, H. Schröder, S. Rüdiger, K. Paal, T. Laufen and B. Bukau, *Nat. Struct. Biol.*, 2000, **7**, 586.
99. R. Schlecht, A. H. Erbse, B. Bukau and M. P. Mayer, *Nat. Struct. Mol. Biol.*, 2011, **18**, 345.
100. A. L. Lai, E. M. Clerico, M. E. Blackburn, N. A. Patel, C. V. Robinson, P. P. Borbat, J. H. Freed and L. M. Gierasch, *J. Biol. Chem.*, 2017, **292**, 8773.
101. A. Bracher and F. U. Hartl, *Nat. Struct. Mol. Biol.*, 2006, **13**, 478.

102. K. Richter and J. Buchner, *Cell*, 2006, **127**, 251.
103. C. Prodromou, S. M. Roe, R. O'Brien, J. E. Ladbury, P. W. Piper and L. H. Pearl, *Cell*, 1997, **90**, 65.
104. A. K. Shiau, S. F. Harris, D. R. Southworth and D. A. Agard, *Cell*, 2006, **127**, 329.
105. M. Taipale, D. F. Jarosz and S. Lindquist, *Nat. Rev. Mol. Cell Biol.*, 2010, **11**, 515.
106. A. Schulze, G. Beliu, D. Helmerich, J. Schubert, L. Pearl, C. Prodromou and H. Neuweiler, *Nat. Chem. Biol.*, 2016, **12**, 628.
107. G. E. Karagöz, A. M. S. Duarte, E. Akoury, H. Ippel, J. Biernat, T. Morán Luengo, M. Radli, T. Didenko, B. A. Nordhues, D. B. Veprintsev, C. A. Dickey, E. Mandelkow, M. Zweckstetter, R. Boelens, T. Madl and S. G. D. Rüdiger, *Cell*, 2014, **156**, 963.
108. D. Eliezer, *Curr. Opin. Struct. Biol.*, 2009, **19**, 23.
109. D. Kurzbach, G. Platzer, T. C. Schwarz, M. A. Henen, R. Konrat and D. Hinderberger, *Biochemistry*, 2013, **52**, 5167.
110. S. Weickert, M. Wawrzyniuk, L. H. John, S. G. D. Rüdiger and M. Drescher, *Sci. Adv.*, 2020, **6**, eaax6999.
111. S. Jeganathan, M. Von Bergen, H. Brutlach, H. J. Steinhoff and E. Mandelkow, *Biochemistry*, 2006, **45**, 2283.
112. M. D. Mukrasch, S. Bibow, J. Korukottu, S. Jeganathan, J. Biernat, C. Griesinger, E. Mandelkow and M. Zweckstetter, *PLoS Biol.*, 2009, 7, 0399.
113. J. B. Weiss and P. J. BassfordJr., *J. Bacteriol.*, 1990, **172**, 3023.
114. V. G. Panse, P. Vogel, W. E. Trommer and R. Varadarajan, *J. Biol. Chem.*, 2000, **275**, 18698.
115. V. G. Panse, K. Beena, R. Philipp, W. E. Trommer, P. D. Vogel and R. Varadarajan, *J. Biol. Chem.*, 2001, **276**, 33681.
116. M. M. Haimann, Y. Akdogan, R. Philipp, R. Varadarajan, D. Hinderberger and W. E. Trommer, *Biol. Chem.*, 2011, **392**, 849.
117. J. P. Hendrick and F.-U. Hartl, *Annu. Rev. Biochem.*, 1993, **62**, 349.
118. D. Wall, M. Zylicz and C. Georgopoulos, *J. Biol. Chem.*, 1994, **269**, 5446.
119. M. Pellecchia, T. Szyperski, D. Wall, C. Georgopoulos and K. Wüthrich, *J. Mol. Biol.*, 1996, **260**, 236.
120. T. R. M. Barends, R. W. W. Brosi, A. Steinmetz, A. Scherer, E. Hartmann, J. Eschenbach, T. Lorenz, R. Seidel, R. L. Shoeman, S. Zimmermann, R. Bittl, I. Schlichting and J. Reinstein, *Acta Crystallogr., Sect. D: Biol. Crystallogr.*, 2013, **69**, 1540.
121. T. Szyperski, M. Pellecchia, D. Wall, C. Georgopoulos and K. Wüthrich, *Proc. Natl. Acad. Sci. U. S. A.*, 1994, **91**, 11343.
122. C. Bartolucci, D. Lamba, S. Grazulis, E. Manakova and H. Heumann, *J. Mol. Biol.*, 2005, **354**, 940.
123. M. Lorenzi, L. Sylvi, G. Gerbaud, E. Mileo, F. Halgand, A. Walburger, H. Vezin, V. Belle, B. Guigliarelli and A. Magalon, *PLoS One*, 2012, 7, e49523.

124. G. Karthikeyan, A. Bonucci, G. Casano, G. Gerbaud, S. Abel, V. Thomé, L. Kodjabachian, A. Magalon, B. Guigliarelli, V. Belle, O. Ouari and E. Mileo, *Angew. Chem., Int. Ed.*, 2018, **57**, 1366.
125. N. N. Barak, P. Neumann, M. Sevvana, M. Schutkowski, K. Naumann, M. Malešević, H. Reichardt, G. Fischer, M. T. Stubbs and D. M. Ferrari, *J. Mol. Biol.*, 2009, **385**, 1630.
126. M. R. Fleissner, E. M. Brustad, T. Kálai, C. Altenbach, D. Cascio, F. B. Peters, K. Hideg, S. Peuker, P. G. Schultz and W. L. Hubbell, *Proc. Natl. Acad. Sci. U. S. A.*, 2009, **106**, 21637.
127. S. Kucher, S. Korneev, S. Tyagi, R. Apfelbaum, D. Grohmann, E. A. Lemke, J. P. Klare, H.-J. Steinhoff and D. Klose, *J. Magn. Reson.*, 2016, **275**, 38.
128. P. Widder, F. Berner, D. Summerer and M. Drescher, *ACS Chem. Biol.*, 2019, **14**, 83.
129. T. Kálai, M. R. Fleissner, J. Jekő, W. L. Hubbell and K. Hideg, *Tetrahedron Lett.*, 2011, **52**, 2747.
130. E. H. Abdelkader, A. Feintuch, X. Yao, L. A. Adams, L. Aurelio, B. Graham, D. Goldfarb and G. Otting, *Chem. Commun.*, 2015, **51**, 15898.
131. A. Kugele, B. Silkenath, J. Langer, V. Wittmann and M. Drescher, *ChemBioChem*, 2019, **20**, 2479.
132. T. Plass, S. Milles, C. Koehler, C. Schultz and E. A. Lemke, *Angew. Chem., Int. Ed.*, 2011, **50**, 3878.
133. A. Kugele, T. Braun, P. Widder, L. Williams, M. J. Schmidt, D. Summerer and M. Drescher, *Chem. Commun.*, 2019, **55**, 1923.
134. M. J. Schmidt, J. Borbas, M. Drescher and D. Summerer, *J. Am. Chem. Soc.*, 2014, **3**, 1.
135. S. Kucher, S. Korneev, J. P. Klare, D. Klose and H.-J. Rgen Steinhoff, *Phys. Chem. Chem. Phys.*, 2020, **22**, 13358.
136. P. Widder, J. Schuck, D. Summerer and M. Drescher, *Phys. Chem. Chem. Phys.*, 2020, **22**, 4875.
137. O. Krumkacheva and E. Bagryanskaya, *J. Magn. Reson.*, 2017, **280**, 117.
138. S. Stoll and A. Schweiger, *J. Magn. Reson.*, 2006, **178**, 42.
139. M. M. U. Ali, S. Mark Roe, C. K. Vaughan, P. Meyer, B. Panaretou, P. W. Piper, C. Prodromou and L. H. Pearl, *Nature*, 2006, **440**, 1013.

CHAPTER 10

EPR Studies of Chaperone Interactions and Dynamics

A. SIEGEL,[a,†] J. SINGH,[b,†] P. Z. QIN*[b,c] AND S. O. SHAN*[a]

[a] Division of Chemistry and Chemical Engineering, California Institute of Technology, 1200 E. California Blvd, CA 91125, USA; [b] Department of Chemistry, University of Southern California, 3430 S. Vermont Ave, CA 90089, USA; [c] Department of Biological Sciences, University of Southern California, 3430 S. Vermont Ave, CA 90089, USA
*Emails: pzq@usc.edu; sshan@caltech.edu

10.1 Introduction

Generation and maintenance of protein homeostasis require proper folding, assembly, localization, and quality control of all proteins in the cell. A diverse set of molecular chaperones have evolved to carry out these different tasks.[1,2] Some chaperones act co-translationally on the ribosome to capture and direct the folding or localization of nascent proteins during their synthesis. Others, such as the Hsp70 system, act as central hubs in the cell that interact with numerous newly synthesized proteins released from the ribosome and assist in their downstream biogenesis pathways including folding, membrane targeting, and degradation.[3] Chaperonins such as GroEL act on client protein refractory to the Hsp70 system by sequestering client proteins in a cage structure, which provides a protected environment conducive to their folding.[4] Analogously, the Hsp90 chaperone system provides a dedicated machinery to assist the folding of clients that cannot fold spontaneously,[5] such as kinases and glucocorticoid receptors. In addition,

[†] Equal contribution.

New Developments in NMR No. 29
Biophysics of Molecular Chaperones: Function, Mechanisms and Client Protein Interactions
Edited by Sebastian Hiller, Maili Liu and Lichun He

Published by the Royal Society of Chemistry, www.rsc.org

numerous small chaperones evolved to protect newly synthesized membrane and organellar proteins from aggregation in the cytosol and to escort them to target destinations in the cell. When protein aggregation occurs under stress conditions, the Hsp103 family of chaperones act as disaggregases that cooperate with Hsp70 to remodel and refold aggregated proteins.[6,7] Their actions are assisted by small heatshock proteins (sHsps), which serve as proteostatic buffers that are activated by elevated proteostatic stress and bind to non-native states of proteins at early stages of misfolding to prevent the accumulation of irreversible aggregates.[8] Together, the diverse activities of the different chaperone systems and their hierarchical organization in the cell provide a robust network that maintains homeostasis of the proteome.

The unique features of molecular chaperones present conceptual and experimental challenges in understanding the principles and mechanisms of their action. For example, how do chaperones recognize their client proteins? With a few exceptions, most chaperones evolved to interact with a broad spectrum of client proteins and do not bind to specific clients *via* recognition of a well-defined sequence motif or three-dimensional surface. Instead, chaperones recognize the physicochemical properties of protein sequences and conformational states of the client. By necessity, they avoid binding to folded proteins and must instead interact with clients in non-native conformations, including the unfolded, partially folded, misfolded, or aggregated states, which do not present a well-defined interaction interface for recognition. Instead, emerging data show that chaperones form 'fuzzy' complexes with client proteins, in which multiple contact sites and weak interactions are sampled.[9] This makes it challenging to apply many well-established structural tools that were developed for studies of systems with well-defined stoichiometry, structure, and interactions.

While high-resolution structural techniques remain important for providing structures of molecular chaperones and generating mechanistic hypotheses, site-directed spin labeling coupled to electron paramagnetic resonance (SDSL–EPR) provides a complementary approach in chaperone research, especially in disordered and aggregated systems that often complicate alternative techniques. The information from EPR studies is particularly powerful when combined with biochemical data and the information from complementary biophysical techniques such as NMR and fluorescence spectroscopy. In this chapter, we will begin by reviewing the theory of continuous wave EPR and its utility in providing site-specific information on the local backbone mobility of proteins. We will then discuss how SDSL–EPR has been employed to address different classes of questions pertaining to the mechanism of molecular chaperones.

10.2 Overview of Continuous-wave EPR: Theory and Experimental Considerations

SDSL–EPR monitors site-specifically attached stable radicals (*i.e.*, species containing unpaired electrons) using EPR spectroscopy and provides structural (*e.g.*, distance constraints) and dynamic (*e.g.*, motions at the labeling site) information on the parent molecule.[10] It is applicable to high-molecular-weight heterogeneous systems under physiological conditions and thus is not hampered by issues encountered in other conventional structural biology techniques such as NMR (*e.g.*, difficult to apply to large complexes) and X-ray crystallography (*e.g.*, crystal growth; packing artifacts).[10] SDSL–EPR has been established as a powerful tool for studying lipid bilayers and membranes,[11,12] proteins,[10,13–17] and nucleic acids.[18,19] It is particularly suitable for investigating biomolecules that function in inhomogeneous environments, such as membrane-associated proteins[10,13] and enzymes at abiotic interfaces,[20] or within large heterogeneous complexes such as chaperone networks.

The most commonly used approach in SDSL studies of proteins is to substitute a unique cysteine at the desired labeling site and attach a nitroxide spin label, such as methanethiosulfonate (MTSL), to the cysteine.[10] The R1 label, which has the nitroxide pyrroline ring attached to the cysteine *via* a disulfide bond (see Figure 10.1(A)),[21] is the most widely used spin label for protein studies and will be the focus of the discussions here (see Figure 10.1(B)). However, we note that a large number of nitroxide variants have been reported for protein studies. SDSL methods have also been developed for other types of labels such as TAM,[22,23] Cu^{2+},[24] and Gd^{3+},[25] and using non-cysteine-based chemistry (*e.g.*, click chemistry with unnatural amino acids).[26] This chapter will focus on the SDSL of chaperones using continuous-wave EPR (CW-EPR); studies using pulsed-EPR are discussed in Chapter 9 by Dröden and Drescher: Studying Molecular Chaperones and Their Client Interactions by Nanometre Distance Restraints from Electron Paramagnetic Resonance Spectroscopy.

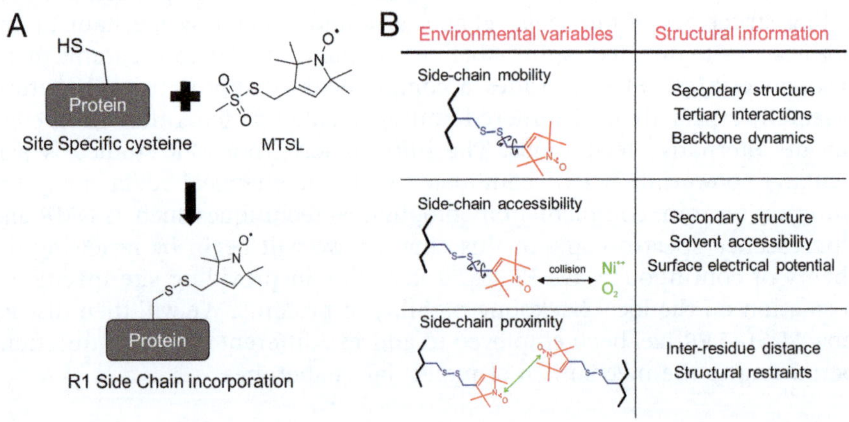

Figure 10.1 General approach and applications of site-directed spin labeling. (A) Reaction of MTSL with an engineered cysteine residue, for site-specific incorporation of the R1 side chain. (B) Structural applications of SDSL discussed in this review: site-specific information on side-chain mobility, side-chain accessibility, and side-chain proximity.

10.2.1 Theory of CW-EPR

A brief description of the basic physical concepts that are most relevant to SDSL studies of chaperones using CW-EPR is presented here. For a more in-depth understanding of the theory relevant to SDSL–EPR, the readers are referred to several excellent classical books and review articles.[27–31]

EPR spectroscopy uses microwave radiation to manipulate and detect interactions between radicals and an external magnetic field. The nitroxide spin label contains an unpaired electron with a spin quantum number $S=1/2$. In the absence of an external magnetic field, the spin states of a free electron are degenerate. In the presence of a static magnetic field (B_0), the electrons experience the Zeeman interaction, where the electron occupies one of two states, parallel or antiparallel to the static magnetic field (see Figure 10.2). The energy difference between these two states depends on the magnetic field strength and can be described as:

$$\Delta E = g\beta_e B_0 \tag{10.1}$$

where g is the electron spin g-factor, β_e is the Bohr magneton, and B_0 is the strength of the magnetic field. In CW-EPR, the sample is irradiated continuously with electromagnetic radiation in the microwave frequency range. At resonance, the electron absorbs energy from the external electromagnetic radiation. The resonance condition is defined by:

$$\Delta E = g\beta_e B_0 = h\nu \tag{10.2}$$

where h is the Planck constant and ν is the frequency of the electromagnetic radiation. When the resonance condition is met, electrons can be excited

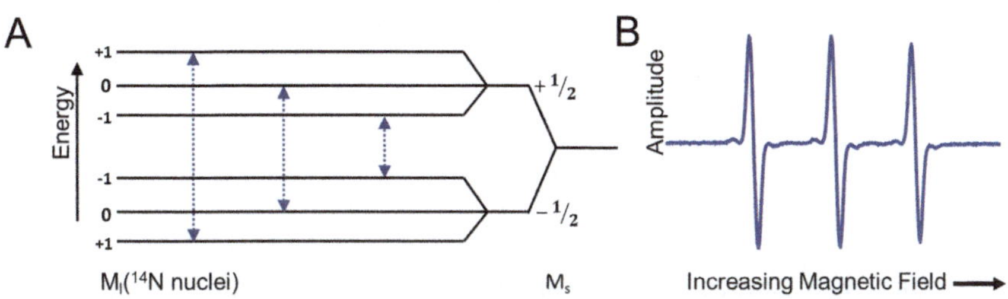

Figure 10.2 Physical basis of the CW-EPR spectral line-shape of a nitroxide label. (A) Electron spin energy level diagram. In the presence of an applied magnetic field, the electron energy levels separate according to the spin quantum number S_z (the Zeeman effect) and the nuclear spin quantum number M_I (the hyperfine interaction). (B) A representative EPR spectrum of tempol nitroxide, obtained by varying the applied magnetic field strength at a fixed microwave frequency. The three peaks correspond to the three transitions indicated by the dotted arrows in panel (A). Reproduced from ref. 18 with permission from Elsevier, Copyright 2008.

from a low energy state ($S_z = -1/2$) to a higher energy state ($S_z = +1/2$), and absorption of microwave radiation occurs. This is the basis of the CW-EPR measurement.

In addition to the Zeeman interaction, the unpaired electron also experiences a small local magnetic field produced by nearby nuclei. This is known as the hyperfine interaction and may counter or enhance the externally applied field. In a nitroxide, the dominant hyperfine interaction is between the unpaired electron and the ^{14}N nucleus of the nitroxyl group. The $I = 1$ ^{14}N nucleus splits the nitroxide EPR signal into $2I + 1 = 3$ lines, each corresponding to a ^{14}N nuclear spin state ($m_I = -1$, 0, $+1$) (see Figure 10.2(A)). As the nuclear spin must be conserved in a quantum mechanics allowed transition, the resonance condition is modified to include the hyperfine interaction (A):

$$h\nu = g\beta_e B_0 + m_I A \tag{10.3}$$

This gives rise to three observed transitions (absorption) in a ^{14}N nitroxide spectrum (see Figure 10.2(A)).[32]

Conventionally, the CW-EPR spectrum is measured by sweeping the external magnetic field (B_0) at a fixed microwave frequency. A spectrum is generally obtained using source modulation and the lock-in amplification technique, which increases the signal-to-noise ratio. The resulting spectrum is the first derivative of the absorption spectrum (see Figure 10.2(B)).

In addition to the short-range nuclear-electron hyperfine interaction, electrons can also experience an electron–electron dipolar interaction. For the dipolar interaction between two spins A and B, an additional term E_{dip} is applied to the electron energy level, formulated as:

$$E_{dip} = D_{dip} \cdot S_z^A \cdot S_z^B \cdot \frac{(1 - 3\cos^2\theta)}{r_{AB}{}^3} \tag{10.4}$$

and

$$D_{dip} = \frac{\mu_0 g_A g_B \beta_e{}^2}{4\pi} = h\nu_{dip} \tag{10.5}$$

where r_{AB} is the distance between spin A and spin B, and θ is the angle between the distance vector r_{AB} and the external magnetic field B_0. The vacuum permeability is μ_0 and when $g_A = g_B = 2$, the dipolar frequency $\nu_{dip} = 51\,921$ (MHz) (Å^3). E_{dip} represents the dipolar splitting and is inversely proportional to the r^3 distance and θ angle orientation between the two labels. The electron–electron dipolar interaction can be regarded as a perturbation to the energy levels of the non-interacting spin and therefore may affect the observed CW-EPR spectrum, generally resulting in "broadening" of the spectral lineshape. Spectral broadening due to static dipolar interaction or distance-dependent relaxation can be used to perform CW-based inter-spin distance measurements in double-labeled samples between 8 and 25 Å. An in-depth discussion of CW-based inter-spin distance measurement can be found in the literature.[21,33] At even shorter distances, the electrons also experience an exchange interaction which can further broaden the spectra, which may complicate the distance analysis.[34] At inter-spin distances exceeding 25 Å, dipolar broadening of the CW-EPR spectra becomes negligible compared to other mechanisms, and pulse techniques, such as DEER (PELDOR) or DQC, are needed to measure the dipolar interactions. For discussions of the use of DEER in chaperone studies, the readers are referred to Chapter 9 by Dröden and Drescher.

10.2.2 EPR as a Site-specific Probe for Protein Structure and Interaction

10.2.2.1 Mobility of a Single Nitroxide Label

With a nitroxide, both g and A are 3×3 tensors with unequal tensor components (*i.e.*, anisotropic). The electron spin energy levels are dictated by the averaging of the g and A tensors, which result from rotational (re-orientation) dynamics of the nitroxide. The observed CW-EPR spectrum measures the transitions between the electron spin energy levels, which reports on the overall rotational dynamics of the electron spin.

For a nitroxide covalently attached to a macromolecule, the overall rotational motion of the label dictates the observed CW-EPR spectral lineshape. The overall label dynamics, characterized by an overall rotational correlation time τ, is the combined result of three types of motions (see Figure 10.3(A)): the rotational motion of the entire macromolecule (characterized by a rotational correlation time τ_R), the torsional rotation about bonds that connect the nitroxide ring to the macromolecule (τ_i), and the motion of the segment of the macromolecule at or near the labeling point (τ_B).[18,21] The τ_R motion is uniform at all labeling sites and can be used to deduce information regarding the entire molecule (*e.g.*, dimerization). The τ_i and τ_B motions are specific to the labeling site and are the primary sources of information on the local environment of a biomolecule.

The most commonly used microwave frequency is at X-band (~9.5 GHz) and resonance occurs at $B_0 \approx 3399$ Gauss (0.34 T). The influence of nitroxide dynamics on the CW-EPR lineshape can be classified into three regimes.[28,35] In the fast motion limit corresponding to $\tau < 0.1$ ns, averaging of the g and A tensors is nearly complete, and the nitroxide spectrum shows three sharp lines (see Figure 10.3(A)). At the rigid limit ($\tau > 50$ ns), the label adopts a static spatial distribution with respect to the microwave frequency, resulting in a spectrum that shows the sum of lineshapes from all possible values of the g and A tensors (see Figure 10.3(A)). At the intermediate "slow-motion" regime, averaging of the g and A tensors is incomplete and varies depending on the overall label motion, resulting in clear variations in the observed X-band CW-EPR lineshape (see Figure 10.3(A)) that provide rich information about the parent macromolecule.

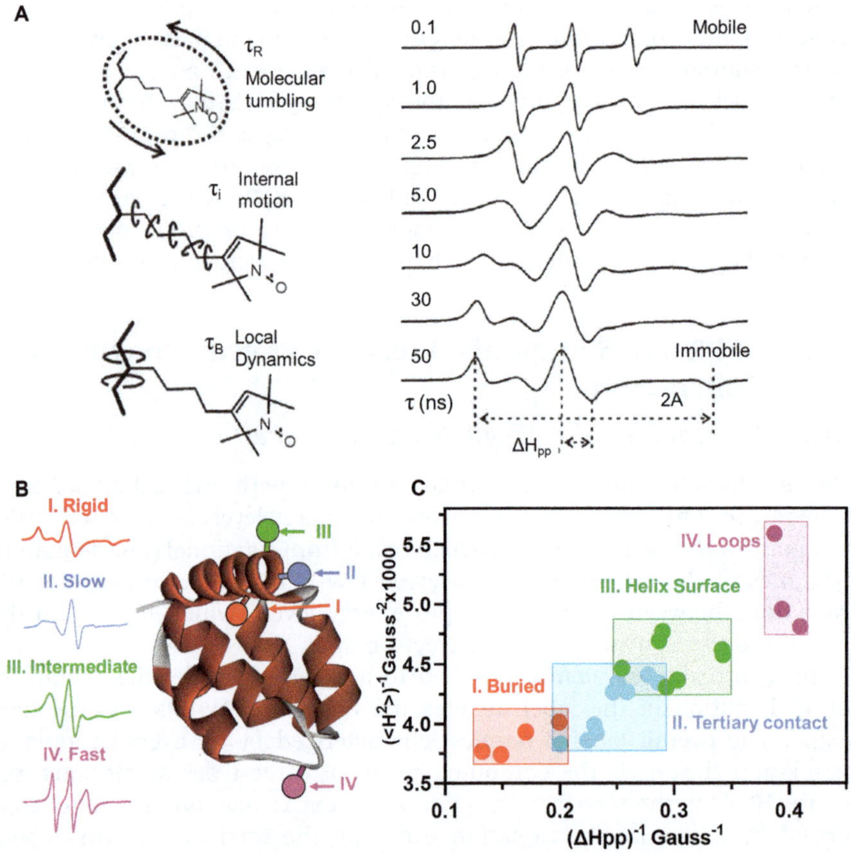

Figure 10.3 Nitroxide mobility as a site-specific probe of protein structure and dynamics. (A) The relationship between nitroxide dynamics and the CW-EPR line shape. Left: The three modes of motion that contribute to nitroxide dynamics; right: simulated X-band EPR spectra of nitroxides undergoing isotropic rotation at different rotational correlation ties (τ), with the semi-quantitative line shape parameters, central linewidth (ΔH) and apparent hyperfine splitting (2A), shown below. (B) Influence of the label environment on nitroxide motion and the CW-EPR line shape. Spheres represent nitroxide labels experiencing different regimes of mobility, due to the structure and dynamics of the attachment position in the model protein. (C) Mobility plot showcasing the reciprocal second moment $(\langle H_2 \rangle)^{-1}$ *versus* the reciprocal central linewidth $(\Delta H)^{-1}$ of 22 nitroxide label sites in the T4 lysozyme. The spectral parameters allow for the general classification of protein structure and dynamics. Reproduced from ref. 18 and 36 with permission from Elsevier, Copyright 2008.

A large number of works have shown that the X-band CW-EPR spectrum of the R1 label varies depending on the local protein environment. Most noticeably, Hubbell and coworkers have classified R1 lineshapes with respect to protein structure and dynamics (see Figure 10.3(B) and (C)).[10,21] The studies identified characteristic lineshape groups that categorize "mobility" of R1 labels[21,36] and developed semi-quantitative parameters to extract R1 "mobility" from an observed spectrum.[10,21] These include the central linewidth (ΔH; see Figure 10.3(A)), the apparent hyperfine splitting (2A; see Figure 10.3(A)), and the second moment $\langle H_2 \rangle$ measuring the overall breadth of the spectrum, with a high-mobility label giving a "narrow" spectrum with a small ΔH, small 2A, and low $\langle H_2 \rangle$.[15] Both the lineshape groups and the semi-quantitative mobility parameters were shown to correlate with structural elements at the R1 site, such as sites that are buried in the protein interior, surface exposed, on a flexible loop, or in tertiary contact (see Figure 10.3(B) and (C)).[21,36] Furthermore, by scanning the R1 label, one amino acid at a time, along a segment of the protein primary sequence, one can observe periodic variations in R1 mobility that reveal the secondary structure within the folded protein (*e.g.*, α-helix and β-sheet).[21,10,13] This has been a powerful tool for the study of protein structure, folding, and interaction and has been extensively applied in investigations of molecular chaperones.

These principles of R1 mobility analysis can be extended to studies of protein network interactions. Nitroxide dynamics can be altered at protein interaction interfaces upon binding and can reveal the specific residues involved in complex formation. Zhang *et al.* utilized EPR spectral changes to show how two proteins in the bacterial signal recognition particle (SRP) and SRP receptor (SR), interact to form an early intermediate complex followed by rearrangement to a stable complex (see Figure 10.4(A)).[37] The spin label at SR I236C showed intermediate mobility as an unbound protein. Formation of the early intermediate caused minor spectral broadening

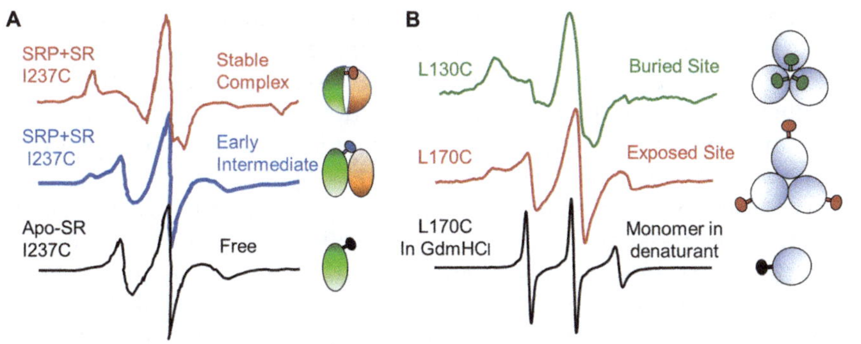

Figure 10.4 Nitroxide mobility analysis in studies of protein network interactions. (A) Interactions of bacterial signal recognition particle (SRP) and SRP receptor (SR) detected by SDSL–EPR. The EPR spectra are shown for the R1 label at SR I237C in Apo-SR (black), the early intermediate (blue), and the stable complex (red), with an adjacent cartoon representation of the protein labeling site on the right. (B) Studies on the architecture of $Lhcb_5$ aggregates by SDSL–EPR. The EPR spectra are shown for the R1 label in $Lhcb_5$-L130C aggregate (green), $Lhcb_5$-L170C aggregate (red), and solubilized $Lhcb_5$-L170C monomer (black), with an adjacent cartoon representation of labeling sites on the right.[37,38]

(see Figure 10.4(A), blue), whereas formation of the stable complex caused immobilization of the R1 label, as shown by large changes in the central linewidth and apparent hyperfine splitting of the CW-EPR spectrum (see Figure 10.4(A), red). This allowed the researchers to distinguish SR I237 as a site specifically involved in stable complex assembly. As another useful extension, R1 mobility can be used to define the architecture of molecular aggregates in chaperone research (see Figure 10.4(B)). Nguyen *et al.* utilized SDSL to map the structure of protein aggregates formed by $Lhcb_5$, which comprise the light-harvesting complex.[38] EPR spectra revealed that $Lhcb_5$-L129C is a buried site within the LHCP aggregate based on the highly immobilized spectrum (see Figure 10.4(B), green). In comparison, $Lhcb_5$-L170C displays substantially higher mobility and represents a surface-exposed site on the aggregate (see Figure 10.4(B), red). Control experiments with LHCP solubilized using a chemical denaturant yielded a narrow spectrum representative of solubilized monomers with unrestricted nitroxide motion (see Figure 10.4(B), black). Thus, analysis of R1 sidechain mobility is a unique tool for probing site-specific information within a complex macromolecular environment and is further discussed throughout this review.

In addition to semi-quantitative mobility analysis, one can simulate the entire CW-EPR spectrum. For example, the Freed group has developed a macroscopic-order microscopic-disorder (MOMD) model that describes the motion of the nitroxide label under the constraint of an orienting potential.[39] The MOMD model allows the iterative fitting of an experimental spectrum. Fitting yields an effective correlation time (τ) and an order parameter (S, describing the restriction in the angular motion) that describe the partially ordered diffusive motion of the nitroxide label with respect to the target macromolecule. Compared to the semi-quantitative lineshape analysis (*i.e.*, ΔH, 2A), spectral simulations yield quantitative descriptions of the label motion based on a given physical model, although it is known that certain parameters (*e.g.*, τ and S in MOMD simulations) may be correlated to a degree and not uniquely determined. A popular software package for the simulation and analysis of EPR spectra is EasySpin.[40]

To deduce information about the parent molecule, it is essential to connect the spectral simulations with the specific system under investigation and with the results obtained from complementary measurements. For example, the Hubbell group has carried out in-depth studies of the R1 label attached to the T4 lysozyme using a combination of spectral simulations, X-ray crystallography, and substitution of nitroxide functional groups.[21] The results show that at solvent-exposed helical surface sites, the R1 spectra can be described by a one-component spin population with a single dynamic mode, and thus, the R1 motions report on the variations in the nanosecond dynamics of the protein backbone.[21,41,42]

The observed CW-EPR spectrum may also contain "multiple components" as the spin label populations experience different local environments. A multicomponent spectrum can be treated as a linear combination of individual component spectra. One can carry out spectral decomposition to obtain the relative percentages of each component using individual component spectra that are either measured or constructed theoretically.[21] Studies on R1 labels have revealed that a multi-component spectrum can arise from heterogeneous local protein conformations or from discrete R1 conformers with different configurations of the bonds connecting the pyrroline ring to the protein.[21] In the cases where the individual components can be connected with distinct protein states, the spectral decomposition analysis can give valuable information such as the equilibrium between the protein states.[21,43,44]

10.2.2.2 Solvent Accessibility Measured from a Single Nitroxide Label

Another parameter obtainable by CW-EPR is the relative solvent accessibility of a spin labeled side chain.[21] The principle of solvent accessibility measurement is based on assessing the collision frequency, W_{ex}, between the nitroxide label and a fast-relaxing paramagnetic reagent (R_{ex}), such as O_2, nickel(II) ethylenediamine-*N*,*N*-diacetic acid (NiEDDA), and chromium(III) oxalate (CROX). Such collision shortens the lattice relaxation time (T_1) and the spin–spin relaxation time (T_2) of the label in a fashion that is proportional to W_{ex}. As such, measuring the dependence of nitroxide label T_1 and/or T_2 on R_{ex} can provide a direct assessment of W_{ex}.

Nitroxide solvent accessibility has been measured using both CW- and pulse-EPR techniques.[21] A facile CW-EPR method to measure solvent accessibility is power saturation.[45] In this approach, the peak amplitude in the CW-EPR spectra is recorded as a function of varying microwave power (P) to obtain a $P_{1/2}$ parameter, the power at which the EPR signal amplitude is reduced to 50% of its maximal unsaturated value. $P_{1/2}$ is inversely proportional to the product of T_1 and T_2, whose perturbations by the paramagnetic reagent are proportional to W_{ex}. Similar to the mobility analysis, solvent accessibility measurements are combined with nitroxide scanning to assess the folding and/or interaction at a specific site on the protein. Specifically, variations of solvent accessibility along a primary sequence can yield periodic patterns that identify (or distinguish) α-helices (periodicity of 3.6), β-sheets (periodicity of 2), or loops (no apparent periodicity).[21,10,13] The classic example of scanning residues 127–134 of the T4 lysozyme (see Figure 10.5(A)) reveals α-helical periodicity, due to the higher collision frequency of the solvent-exposed residues (Figure 10.5(B)).

Finally, the properties of the paramagnetic reagent can be manipulated to gain information on interactions between the target protein and its environment.[21,10,13] For example, O_2 and NiEDDA generally give very similar solvent accessibility patterns for a folded protein in an aqueous solution. However, for proteins interacting with membranes or lipids, O_2 and NiEDDA may show different patterns, as O_2 has high solubility in the non-polar lipid phase whereas NiEDDA has low solubility. A combined analysis of O_2 and NiEDDA can thus yield information on the relative positioning of a protein segment with respect to the membrane.[21,46,47]

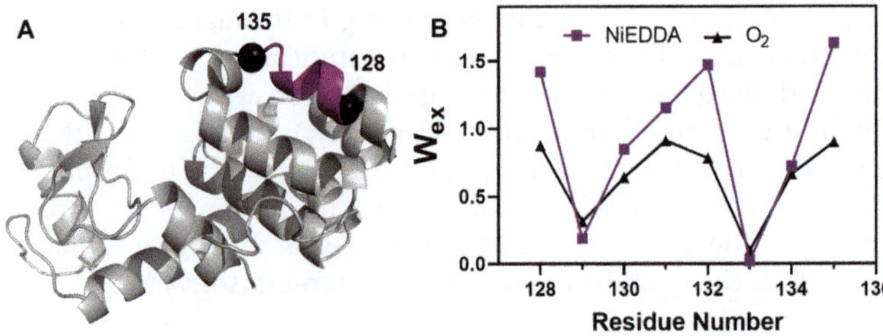

Figure 10.5 Nitroxide scanning to gauge side chain solvent accessibility measured by power saturation. (A) Cartoon model of the T4 lysozyme with an α-helix (128–135) subject to a nitroxide scan colored in purple. The start and end residues of the scan are indicated by black spheres. (B) Plot of nitroxide collision frequency (W_{ex}) *versus* R1 label position with paramagnetic relaxation agent O_2 (black triangles) and NiEDDA (purple squares). The increase in W_{ex} indicates increased solvent accessibility of the R1 nitroxide labels along the α-helix. Reproduced from ref. 45 with permission from Elsevier, Copyright 2005.

10.2.2.3 Distance-dependent Dipolar Coupling Between a Pair of Nitroxide Labels

Distances measured *via* nitroxide labels and CW-EPR have been successfully used to provide direct structural constraints for investigating protein structure and conformational changes,[14,21] although these studies have been constrained by the upper limit of ∼25 Å in the measurable distance (see section 10.2.1). The developments of pulsed techniques have drastically expanded the capability of SDSL–EPR for structural biology, for which readers are referred to Chapter 9 by Dröden and Drescher.

When the distance between a pair of R1 labels is 8–20 Å, the electron–electron dipolar interaction results in measurable broadening of the X-band CW-EPR spectrum compared to that of the noninteracting spins.[21,48] Comparison of the lineshape of the double-labeled spectrum to the sum of the two single-labeled spectra yields a broadening function ΔΔH, the difference in the ΔH values between the two spectra, from which the distance between the two spins can be extracted using deconvolution or convolution methods. High-quality labeled protein samples and high-quality spectra must be measured to avoid broadening artifacts. Uncertainties in the measured interspin distance are likely due to R1 rotamer orientations and produce an error distribution range of approximately ±3 Å. In a classic demonstration of CW-EPR-based distance measurements with the T4 lysozyme,[49] a pair of spin labels 10 Å apart causes significant broadening ($\Delta\Delta H \approx 10$ Gauss) compared to the sum of the single-labeled control spectra (Figure 10.6, blue), whereas an R1 pair 25 Å apart shows negligible broadening ($\Delta\Delta H \approx 0.3$ Gauss). Analogous double labeling schemes can be utilized to give structural constraints to chaperones, client proteins, or their complexes.

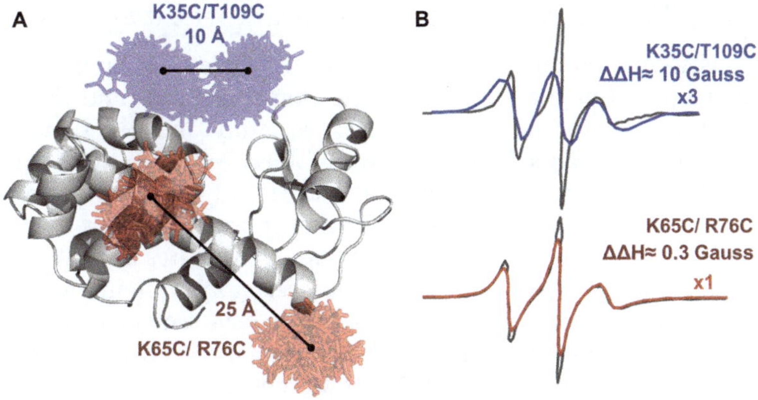

Figure 10.6 Distance-dependent dipolar coupling between a pair of nitroxide labels. (A) The structure of the T4 lysozyme with R1 rotamers modeled *via* the spin label modeling program ALLNOX,[93,94] which predicted average distances of $\sim 10 \pm 3$ Å for the K35C/T109C pair (blue) and $\sim 25 \pm 3$ Å for the K65C/R76C pair. (B) Dipolar broadened spectrum of double-labeled K35C/T109C (blue) and K65C/R76C (red) overlaid with the sum of their respective single-labeled spectra (grey). The changes in the central linewidth ($\Delta\Delta H$) due to dipolar broadening and the spectra scaling factor are displayed above their respective spectrum. Reproduced from ref. 49 with permission from American Chemical Society, Copyright 1997.

10.3 SDSL–EPR Studies of Molecular Chaperones: Structure, Dynamics, and Interactions

10.3.1 Defining Chaperone–Client Interactions

Understanding how chaperones interact with client proteins is of central importance to dissecting their mechanisms and cellular roles. However, chaperone–client interactions differ from well-defined protein–protein interactions, as discussed earlier in the Introduction. The dynamic and non-specific nature of chaperone–client interactions presents multiple challenges for their investigation, and high-resolution structural information on chaperone–client complexes remains scarce.

Direct protein interactions alter the local environment and backbone rigidity of sites labeled with nitroxide spin labels, making SDSL–EPR a valuable tool for the detection of chaperone–client interactions. SDSL–EPR can be used to provide site-specific binding information that complements observations from other structural and biochemical techniques. For particularly challenging chaperone–client systems characterized by disorder, heterogeneity, and a high molecular weight that complicate other structural approaches, SDSL–EPR provides a robust method to detect and define chaperone–client interactions.

10.3.1.1 The Bacterial Secretory Pathway (SecA and SecB)

Post-translational protein export in bacteria is carried out by the secretory system composed of the homotetrameric chaperone SecB and the ATPase SecA.[50] SecB captures newly synthesized secretory and outer membrane proteins and maintains them in the unfolded, translocation-competent state. SecB interacts with SecA, which is localized at the plasma membrane and uses its ATPase cycle to translocate nascent proteins across the SecYEG translocation machinery. While high-resolution structures of SecA and SecB have been available, how SecB recognizes secretory proteins and how it interacts with and hands off client proteins to SecA remained outstanding questions that were addressed by a series of elegant SDSL–EPR studies.

In the study of Crane *et al.* in 2005, 41 solvent-exposed residues covering 43% of the SecB surface were mutated to cysteines and labeled with the nitroxide spin probe MTSL without affecting the activity.[51] EPR lineshape analysis showed that many spin labels are highly mobile in free SecB but become immobile upon SecA binding, as reflected in the broadening of ΔH and the appearance of hyperfine splitting that led to an overall increase in the spectral breadth (2A). In some cases, quantitative fitting of the spectral splitting further yielded the order parameter (S) that describes the fraction of immobile spin-labels and the overall correlation time (τ) that describes the motion of the spin label in each state. Collectively, the sites at which EPR spin labels undergo changes in mobility mapped to a large and contiguous surface on SecB were interpreted as the interaction interface for SecA, based on which the SecB–SecA interaction was modeled (see Figure 10.7(A), red).[51]

Using a slightly expanded set of spin labels on SecB covering 49% of its surface, SecB's interaction with a model client protein, galactose-binding protein (GBP), was also investigated by EPR.[52] As client binding by SecB is dynamic and reversible, and released clients partition between folding or rebinding to SecB, EPR studies were carried out under experimental conditions that modulate client folding. For GBP, refolding is calcium-dependent and slowed using a Ca^{2+} chelator, whereas the addition of Ca^{2+} favors client folding. Comparison of the EPR spectra under these conditions showed that SecB only binds to client proteins in the unfolded or non-native states, whereas only minor spectral changes were observed when calcium was present to favor folding.

The client interaction sites on SecB were mapped based on the immobilization of spin labels under conditions that favor client binding. Of the 44 spin-labeled sites tested, 29 showed modest or significant immobilization upon client binding and were interpreted to be client interaction sites (see Figure 10.7(A), blue). Importantly, the same set of spin labels undergo immobilization when these studies were extended to additional Sec pathway substrates,[53] suggesting that a variety of secretory proteins use the same client interaction surfaces on SecB. In addition, sites where the cysteine mutation and spin labeling disrupted client binding also suggested important sites of contact with clients. Together, these experiments revealed a

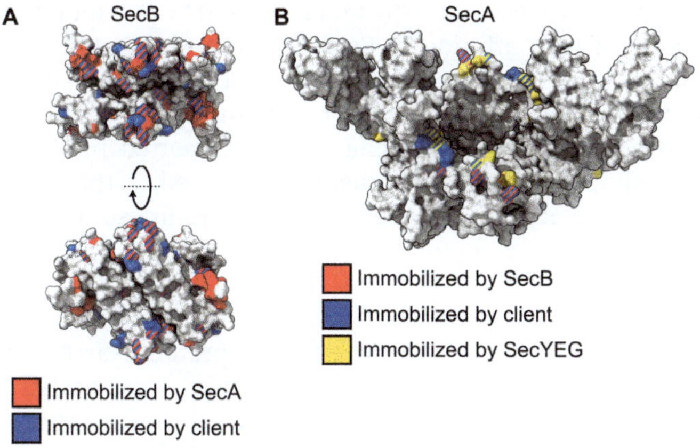

Figure 10.7 Interaction sites of SecA and SecB defined by EPR-SDSL mobility analysis. (A) Structure of SecB (PDB: 1FX3)[96] with residues immobilized upon binding of SecA (red) or the proGBP client (blue) colored on the surface. (B) Structure of SecA (PDB: 2FSF)[95] with residues immobilized upon the binding of SecB (red), proGBP client (blue), or SecYEG (yellow) colored on the surface. Together this shows the docking interface between SecA and SecB, as well as the overlapping sites that mediate the docking of various interaction partners. Data are from Crane *et al.* 2005, Crane *et al.* 2006, Cooper *et al.* 2008, and Lilly *et al.* 2009.[51–54]

more extensive set of client binding regions on the SecB homotetramer surface than predicted by the crystal structure. This indicates that client binding on SecB is not limited to a specific interaction groove, but rather, client proteins form promiscuous contacts with sites throughout the SecB surface following a handful of potential pathways that wrap around the chaperone. The large number of potential client interaction sites on SecB was proposed to enable its high client affinity (K_d's of 5–30 nM) despite the lack of client specificity. Furthermore, the flexibility of client interaction sites on SecB was suggested to contribute to a concerted mechanism of client handoff from SecB to SecA, in which SecA displaces some client interactions with SecB but retains others such that client proteins are never completely dissociated from SecB.

Using a similar approach, SDSL–EPR was used to identify the interaction sites of SecA with its various interaction partners, including SecB, client protein, and SecYEG.[54] As with SecB, only unfolded client proteins generated mobility changes for spin-labeled SecA. The presence of a signal sequence on client proteins made no difference, indicating that SecA interacts with the mature region of the protein. Furthermore, each interaction partner has distinct but partially overlapping interaction sites on SecA (see Figure 10.7(B)). For example, SecYEG induces spin probe immobilization in a pattern similar to those observed with substrates and SecB. Despite this apparent overlap, biochemical data showed that the binding of various SecA interaction partners is not mutually exclusive, and SecA bound to the SecYEG translocon has a high affinity for SecB.[55] These observations suggest conformational rearrangements and adjustment of interaction surfaces as SecA interacts with its successive binding partners during the protein export cycle. The presence of conformational changes is supported by the observation that several spin labels on SecA display increased, rather than reduced, mobility upon interaction with SecB, SecYEG, and liposomes.

10.3.1.2 NarJ

NarJ is a dedicated chaperone for the metalloprotein NarG that forms the membrane-associated nitrate reductase complex. NarJ assists in the assembly of the NarGHI complex at the membrane, and its absence leads to premature membrane anchoring of NarGH and failure to assemble the correctly folded NarGHI complex. In Lorenzi *et al.* 2011, the NarJ binding site was mapped using nitroxide spin probes incorporated across its solvent-accessible surface.[56] Upon binding of a peptide derived from NarG, the spin probe at E118C in a hydrophobic cavity of NarJ undergoes a large shift in the EPR spectra, with a highly immobile component accounting for roughly 89% of the population compared to ~50% in free NarJ. This, together with DEER experiments, showed that NarG binds to NarJ's hydrophobic cavity and reduces the conformational sampling of NarJ. This model was corroborated *in vivo* by Karthikeyan *et al.* 2017, who used a novel nitroxide spin probe with a much longer lifetime in living cells, M-TETPO.[57] By injecting M-TETPO labeled NarJ into living eukaryotic oocytes, the client interaction of NarJ was studied by monitoring the NarG-induced changes in the EPR spectra. Similar client-induced EPR spectral changes were observed in oocytes and *in vitro*, demonstrating that the binding interaction between NarJ and the NarG peptide remains the same in the *in vivo* environment.

10.3.1.3 Nucleic Acid Chaperones

In addition to chaperones that handle protein clients, some nucleic acid chaperones evolved to assist with the folding of RNA. ORF1p is an RNA chaperone that binds to single-stranded RNA and is critical for L1 retrotransposition, the process by which the long interspersed nuclear element-1 is transcribed, retrotranslated, and reinserted into a different part of the genome. ORF1p consists of a coiled-coil domain that mediates its trimerization, a middle domain consisting of an RNA recognition motif (RRM), and a C-terminal domain (CTD) that is also implicated in RNA binding. Despite the crystal structures of ORF1p, its interactions with RNA and how this assists with retrotransposition remained open questions.

In the study of Januszyk *et al.* in 2010, on alanine scanning mutagenesis, a contiguous RNA binding surface mapped in the RRM and CTD was identified using alanine scanning mutagenesis.[58] To directly monitor client interactions, 27 single cysteines were engineered throughout the putative

RNA binding region. Even in apo ORF1p, the spectra for most spin labels are multi-component, indicating a mixture of dynamic and motionally restricted states in slow exchange. Upon RNA binding, multiple spin labels in the buried core of the RRM show a significantly increased immobile population. As these sites are not solvent accessible, these spectral changes must reflect RNA-induced conformational changes favoring a more structured state of the RRM. Additional RNA-induced spectral changes, together with biochemical data, mapped the RNA binding sites on ORF1p to the β3 and β4 strands of RRM and the α1 helix of the CTD. It was proposed that these distinct RNA binding sites could be positioned adjacent to one another in a model of the ORF1p trimer and thus form a large contiguous surface for high-affinity RNA binding.

10.3.2 Client Conformational Remodeling

Chaperones are conformationally selective: they preferentially recognize client proteins in non-native states while minimizing interactions with correctly folded proteins.[59] Whether and how different chaperones induce a more folding-competent conformation of client proteins through their binding interactions remains an outstanding question. SDSL–EPR offers a valuable method to address this question. The mobility of the nitroxide spin label is often by itself an excellent measure of the foldedness of the client. Additionally, the signal amplitude from a spin label is greatly impacted by dipolar relaxation when two spin probes are within 8–20 Å of one another, as discussed earlier. This is particularly useful for detecting close tertiary contacts in client proteins, as well as secondary structures when spin labels are placed at the i and $i+2$ residues in β-sheets, or at the i and $i+4$ residues in α-helices.

10.3.2.1 Small Heat Shock Proteins (sHsps)

sHsps are an abundant class of ATP-independent chaperones characterized by their small size (12–43 kDa) and a core α-crystallin domain flanked by variable N- and C-terminal domains. sHsps stabilize partially folded or unfolded clients to prevent their further misfolding or aggregation, often prior to refolding attempts by ATP-dependent chaperone systems. The importance of sHsps in the maintenance of proteostasis in physiology and pathology is reviewed elsewhere.[8,60–62]

Using the well-characterized T4 lysozyme (T4L) as a model client, SDSL–EPR was used to detect structural changes at various sites of T4L upon engagement with an sHsp, αA-crystallin.[63] While spin labels are highly mobile in free T4L, binding with αA-crystallin induced significant broadening of the EPR spectra and the appearance of an immobile population that reflects T4L bound to the chaperone. Using the amplitude of the immobile component as a measure of the degree of chaperone binding, quantitative binding isotherms were constructed for a series of T4L mutants in which folding was systematically destabilized by introducing 'cavities' in the folded core. These data showed that the binding of αA-crystallin to T4L mutants correlated with the degree of destabilization in the latter, indicating that αA-crystallin specifically recognizes non-native states. This conclusion was supported by the observation that higher temperatures, which destabilize folding, also lead to increased chaperone binding to T4L.

The ability of sHsps to remodel client conformation was directly demonstrated by Claxton *et al.* 2007 using a combination of fluorescence and EPR spectroscopy.[64] Two spectroscopic probes, consisting of a pair of nitroxide spin labels or a bimane–Trp pair, were introduced in T4L at sites that fingerprint the tertiary fold. Trp quenches bimane fluorescence at distances <15 Å. Analogously, spin probes undergo dipolar coupling in the distance range of 8–20 Å. Five pairs of probes were used to monitor the folding of the N- and C-terminal domains of T4L as well as the proximity between the two domains. In addition, pairs of spin probes were incorporated at the i and $i+4$ residues to monitor helix folding (see Figure 10.8(A)). In all cases, quenching of bimane fluorescence by Trp was observed in folded T4L but abolished upon binding to αA-crystallin and another sHsp, αB-D3. Analogously, spin–spin coupling in folded T4L was largely abolished upon chaperone binding, indicating the loss of both secondary and tertiary structures (see Figure 10.8(B)). Finally, EPR lineshape analysis showed that αA-crystallin induced immobilization of spin probes in the C-terminal but not the N-terminal domain of T4L and that probes in the C-terminal domain are protected by αA-crystallin from NiEDDA. These results suggest an asymmetry of recognition, in which sHsps preferentially bind to the more hydrophobic C-terminal domain of T4L.

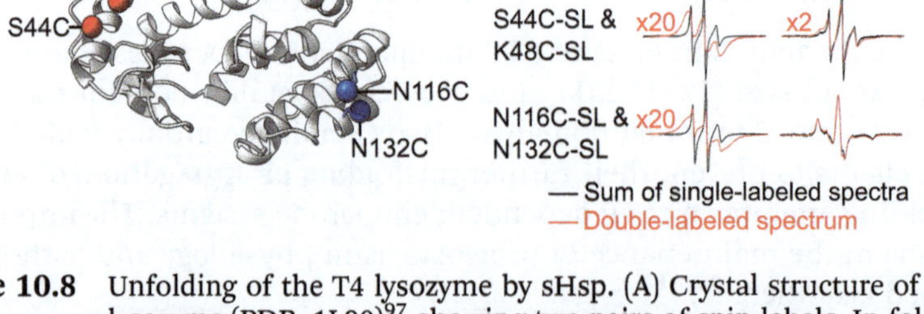

Figure 10.8 Unfolding of the T4 lysozyme by sHsp. (A) Crystal structure of the T4 lysozyme (PDB: 1L90)[97] showing two pairs of spin labels. In folded T4 lysozyme, the S44CR1/K48CR1 pair (red) and the N116CR1/N132CR1 pair (blue) are 6.3 Å and 4.7 Å apart, respectively. (B) In free T4 lysozyme (left), the EPR spectra from double-labeled samples (red) show strong spin–spin coupling and are 20-fold broadened relative to the sum of the single-labeled samples (black). In the presence of α-crystallin (right), spin–spin coupling is abolished in the double-labeled samples (red), indicating a loss of secondary and tertiary structure in the T4 lysozyme. Reproduced from ref. 64 with permission from Elsevier, Copyright 2008.

10.3.2.2 GroEL

The Hsp60 family of chaperonins is required for the folding of a subset of the proteome that contains complex tertiary folds rich in β-sheet content. The chaperone cycle for the best-studied member of this class, GroEL, has been extensively studied.[4] GroEL forms a tetradecamer assembly composed of two heptameric rings, each forming a cage within which client proteins are captured and protected from aggregation with the assistance of the co-chaperone GroES.[65] ATP binding induces a conformational change in GroEL to expose hydrophilic surfaces in the interior of the cage and thus generates a protected environment conducive to client folding. The first step and pre-requisite of this assisted folding cycle is the ability of GroEL to capture unfolded client proteins, which occurs prior to GroES and ATP binding.

The ability of GroEL to select specific client conformation was probed using SDSL–EPR by Persson *et al.* in 1999.[66] Nitroxide spin probes were introduced into a GroEL substrate, HCAII, and GroEL binding was detected by EPR lineshape analysis. Intriguingly, GroEL induced increased mobility in the spin labels, and significant binding was only detected at high temperatures and assisted by low doses of GdmHCl, conditions that favor protein unfolding. These observations suggest that GroEL binds to client proteins in the unfolded or molten globule state. In the study of Owenius *et al.* in 2009, the effect of GroEL on the client folding status was more directly tested using a pair of nitroxide spin probes in the core β-sheet of HCAII.[67] The spin labels are 8.8 Å apart in the folded structure, and their dipolar coupling was readily detected in free HCAII. When the temperature was increased to 50 °C, HCAII underwent extensive aggregation. In contrast, HCAII remained soluble in the presence of GroEL, indicating its binding and protection by the chaperone. Importantly, GroEL–bound HCAII does not display significant spin–spin coupling between the nitroxide probes, indicating unfolding of its structural core. These results support the model that GroEL can function as an unfoldase at the early stages of its chaperone cycle before ATP/GroES binding, through binding-induced unfolding of the client protein.

10.3.3 Conformational Changes of the Chaperone

Molecular chaperones undergo controlled client binding and release during their chaperone cycle. On the one hand, chaperones must effectively bind and protect aggregation-prone client proteins in the unfolded, misfolded, or aggregated state. On the other hand, the role of most chaperones is not to trap their client proteins but to assist in their refolding, localization, or degradation, all of which require controlled client release. These opposing functional requirements are resolved *via* conformational changes in the chaperone itself, which drive the chaperone cycle and provide points of regulation by nucleotide, cochaperones, and other biological signals.[68] SDSL–EPR has been successfully used in many studies to probe conformational changes in the chaperone.

10.3.3.1 Hsp70

The Hsp70 family of proteins comprise a central hub in the cellular chaperone network that provides the first layer of protection for newly synthesized proteins released from the ribosome. Hsp70s consist of an N-terminal nucleotide-binding domain (NBD) that binds and hydrolyzes ATP and a C-terminal substrate binding domain (SBD) that mediates client binding. Early work using peptide substrates showed that hydrophobic peptides bind to a conserved hydrophobic cleft in the beta-sheet domain of SBD (SBDβ). Client binding is regulated by cochaperone-driven ATPase cycles in the NBD: in the ATP-state, an α-helical lid domain in SBD (SBDα) docks on the NBD, generating an open SBD with rapid client binding and dissociation kinetics. Upon ATP hydrolysis (activated by J-proteins), the SBDα lid closes on SBDβ, generating a closed SBD that binds to clients more tightly. This model, however, raised questions as to how Hsp70 accommodates client proteins of much larger size within its substrate binding cleft.

In the study of Schlecht *et al.* in 2010, the mechanics of SBD movements during client recognition of the bacterial Hsp70 DnaK were studied.[69] The motion of the SBDα lid was detected *via* spin–spin relaxation between pairs of nitroxide probes incorporated in SBDα and SBDβ and compared between DnaK bound with clients of different sizes and in different nucleotide states (see Figure 10.9(A)). These experiments showed that the SBD cycles between lid-open and lid-closed states in apo-DnaK, whereas ATP binding drives it to a wide-open conformation (see Figure 10.9(B)). As expected, DnaK adopts a predominantly lid-closed conformation in the absence of ATP when bound to the peptide substrate. In contrast, the EPR spectrum of DnaK showed no spin–spin interaction in the presence of a folded client protein, σ^{32} and had identical lineshapes to the spectrum of the ATP-bound lid-open conformation. These results indicate that the binding of natively folded protein substrates forces the DnaK SBD to adopt an open conformation.

Additional dynamics in the Hsp70 SBD during client binding were studied with BiP, the major Hsp70 homologue in the ER. Yang *et al.* in 2016 reported a new crystal structure of ATP-bound BiP in which the SBD appears occluded, with the $L_{1,2}$ loop between the first and second β-strands in SBDβ swinging towards the $L_{3,4}$ loop in an orientation that precludes binding to the putative peptide binding pocket.[70] The movement of the $L_{1,2}$ loop in response to changes in the nucleotide state was probed using nitroxide spin labels incorporated in the vicinity of this loop. EPR line shape analysis showed that the spin labels displayed increased mobility upon ATP binding, supporting ATP-induced dynamics in the $L_{1,2}$ loop that opens up the peptide binding groove. Fluorescence quenching experiments corroborated this conclusion and further showed that a J-protein cochaperone for BiP, ERdj3, opens up the $L_{1,2}$ loop. Finally, mutation of three conserved glycine residues in the β1-$L_{1,2}$ region abolished the ATP-induced loop mobility; the same mutations weakened peptide binding to BiP 30–99 fold and severely

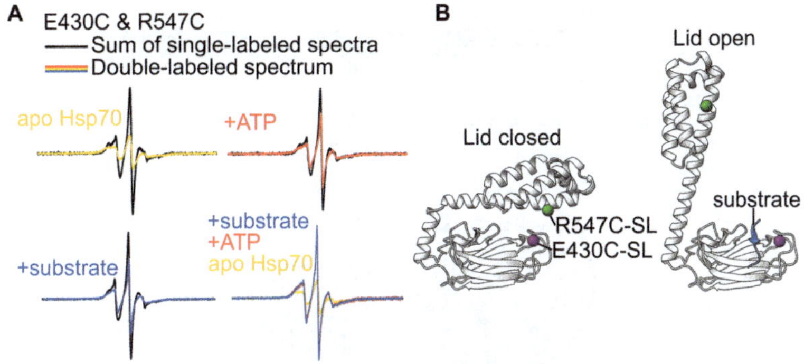

Figure 10.9 Hsp70 lid opening and closing detected by SDSL–EPR. (A) EPR spectra of Hsp70's SBD double-labeled with spin probes on the helical lid (R547C-R1) and beta subdomains (E430C-R1) of the SBD. In the absence of a nucleotide or a substrate (top left), the labeled sites are <10 Å apart, resulting in strong spin–spin coupling in the spectra from the double-labeled sample (yellow) relative to the sum of the single-labeled spectra (black). Addition of ATP (red) or a substrate (blue) reduced the spin–spin coupling, consistent with the nitroxide pair separated by >20 Å. The spectra of a double-labeled sample under all three conditions are overlayed for comparison (bottom right). (B) Model of the Hsp70 lid opening consistent with the EPR-derived distances. The spin labels at R547C (green) and E430C (purple) are <10 Å apart in the crystal structure of the 'lid closed' conformation (PDB: 1DKX; left).[98] Addition of ATP or a substrate (blue) drives the Hsp70 SBD to a dynamic 'lid open' conformation (modeled; right). Reproduced from ref. 69 with permission from Springer Nature, Copyright 2011.

impaired the client folding activity of BiP, demonstrating the importance of the $L_{1,2}$ movement in BiP function.

10.3.3.2 *Small Heat Shock Proteins*

sHsps are a ubiquitous class of chaperones that help maintain cellular proteostasis in response to stress (see Section 10.3.2.1). In the absence of nucleotide binding and hydrolysis cycles that propel conformational changes in ATP-dependent chaperones, sHsps evolved alternative mechanisms for the regulation of their chaperone cycle. Recent work suggests a general mechanism for their regulation, whereby sHsps assemble into inactive, high-order oligomers in the absence of stress. Under stress conditions, phosphorylation of sHsps drives the dissociation of oligomers into more active dimers. The high molecular weight species, compositional heterogeneity, and intrinsic disorder of sHsps have complicated studies of their structure and interactions.

In the study of McDonald *et al.* in 2011, cysteine scanning mutagenesis combined with SDSL–EPR was used to identify the oligomeric and dimeric interfaces of Hsp27.[71] Five clusters of residues in the N-terminal domain of

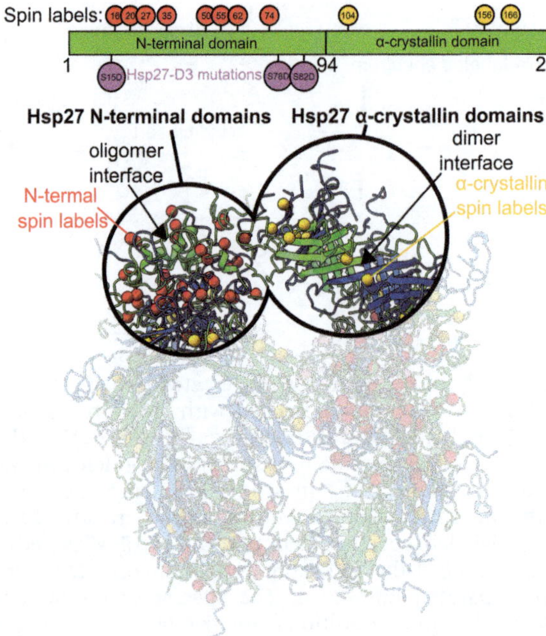

Figure 10.10 Dimeric and oligomeric interfaces of Hsp27. Top: Domain diagram showing spin-labeled positions in the N-terminal and α-crystallin domains of Hsp27 and the three Ser-to-Asp mutants in Hsp27-D3. Lower panel: Hsp27 subunits (blue and green) are shown in the structure of an oligomeric Hsp27 24mer (PDB: 6DV5; bottom).[99] Spin labels placed throughout the N-terminal domain (red) are immobile in WT Hsp27 but became mobile and had 10–20 fold higher solvent accessibility in the Hsp27-D3 mutant that remains a dimer. Spin labels placed in the α-crystallin domain (yellow) remain immobile in both WT Hsp27 and Hsp27-D3. Reproduced from ref. 71 with permission from American Chemical Society, Copyright 2012.

Hsp27 preceding the α-crystallin domain were replaced with single cysteines and labeled with nitroxide spin probes (see Figure 10.10). The EPR spectral lineshapes were compared between wildtype Hsp27, which exists predominantly in the oligomeric state, and a phosphomimetic mutant of Hsp27 (Hsp27-D3), which prevents oligomer formation but maintains the dimeric complex. The results showed that many spin probes in the N-terminal domain are highly immobile in the Hsp27 oligomer but become mobile in the Hsp27-D3 dimer. Solvent accessibility of the spin probes to NiEDDA was also 10–20 fold higher in Hsp27-D3 compared to wildtype Hsp27. In contrast, the mobility of spin labels in the α-crystallin domain of Hsp27 was unaffected by the D3 mutation. These results provide evidence that the N-terminal domain of Hsp27 mediates oligomer assembly and is released upon dissociation into dimers. Importantly, the increased spin probe mobility and solvent accessibility of the N-domain were reversed upon substrate binding, suggesting that this domain is also involved in client binding. Together, these results support a model for Hsp27 activation in which part of its client binding interface is sequestered by interaction with one another in the oligomer and that phosphorylation-induced oligomer disassembly exposes these sites to capture and protect client proteins.

10.3.3.3 High Temperature Requirement A (HtrA)

HtrA, also known as DegP, is a dual-function protein in the bacterial periplasm, a space devoid of ATP as an energy source. HtrA has two biochemical activities that operate at different temperatures: it functions predominantly as an ATP-independent chaperone at normal temperatures but switches on its proteolytic activity at high temperatures. The structural basis for this temperature-regulated functional switch was investigated by Kim *et al.* in 2008 using a combination of EPR and fluorescence spectroscopy.[72] Pairs of nitroxide spin labels were incorporated into the proteolytic active site and an α-helical lid (H_L) flanking the catalytic site of HtrA, and the inter-spin distances were measured based on dipolar coupling between the spin probes. At room temperature, strong spin–spin coupling was observed and yielded an inter-spin distance of ~12 Å between the active site and the H_L loop. However, 5 min of heat treatment abolished the inter-spin dipolar coupling, indicating that H_L moved away from the catalytic site of HtrA. Opening the H_L loop at elevated temperatures was further supported by fluorescence quenching experiments. Moreover, SDSL–EPR experiments using nitroxide spin probes engineered into other structural elements of HtrA demonstrated that the protein retains its overall tertiary and secondary structures at high temperatures. Together, these experiments suggest a model in which the proteolytic activity of HtrA is activated by high-temperature-induced local motion in the H_L loop, leading to higher substrate accessibility to the proteolytic active site.

10.3.3.4 Signal Recognition Particle (SRP)

SRP is a chaperone-like targeting factor that directs co-translational protein transport to the bacterial plasma membrane or the eukaryotic ER membrane. SRP recognizes and protects hydrophobic signal sequences or transmembrane domains (TMDs) on the nascent polypeptide emerging from the ribosome and delivers them to protein translocation sites at the target membrane *via* interaction with the SRP receptor (SR). Targeting is driven by a dimerization-activated GTPase cycle in which two homologous GTPases, one in SRP and one in SR, undergo a sequential series of conformational changes upon their dimerization that culminate in their reciprocal GTPase activation. The first step in this cycle is the formation of a low-affinity early SRP–SR intermediate complex, which then undergoes a slow, GTP-dependent rearrangement to form a stable closed complex (see Figure 10.11(A)). The early intermediate is essential for the initiation of protein targeting, and its specific stabilization by ribosomes translating

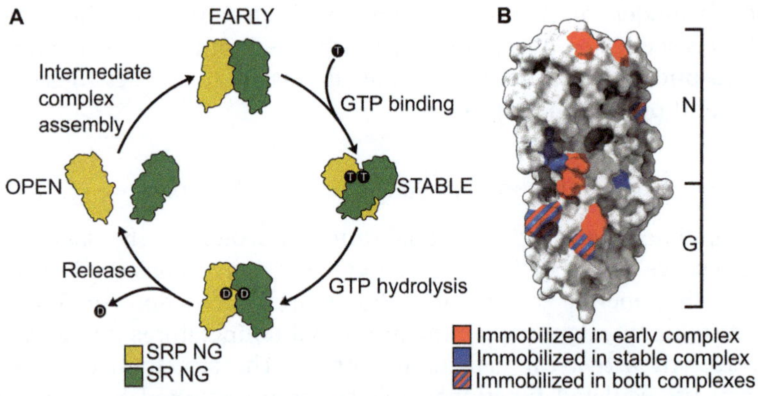

Figure 10.11 SDSL–EPR map differences in the SRP–SR binding interface between the early intermediate and stable complexes. (A) The SRP–SR interaction cycle. Free SRP and SR rapidly associate to form an 'early' intermediate independently of GTP. The complex undergoes a GTP-dependent rearrangement to form a tighter 'stable' complex, from which GTP hydrolysis is activated and triggers SRP/SR dissociation and nucleotide exchange. (B) Structure of the *E. coli* SR (PDB: 6N9B)[100] with residues that undergo mobility changes upon formation of the early intermediate or the stable complex colored in red and blue, respectively.[37]

SRP-dependent proteins provides a key mechanism to ensure the fidelity of the pathway. However, the weak affinity and transient nature of the early intermediate, as well as the sizes of the SRP (100 kDa) and SR (55 kDa), present significant challenges for most structural methods.

In the study of Zhang *et al.* in 2011, the interaction and conformation of the early intermediate were studied by a combination of SDSL–EPR and fluorescence spectroscopy.[37] To map the binding interface between SRP and SR, nitroxide spin labels were placed throughout the surface of the SR GTPase domain, comprising an α-helical N-domain and a G-domain that pack against each other. Comparison of the EPR spectra of spin probes in free SR *versus* those in the SRP·SR complex, biochemically locked at the early intermediate stage or in the stable complex, showed that the former lacks interactions between the G-domains that characterize the stable complex (see Figures 10.4(A) and 10.11(B)). Instead, spin labels in the N-domain of SR as well as the N–G domain interface were immobilized in the early intermediate (see Figure 10.11(B)), indicating that SRP and SR initially contact one another *via* their N-domains. These results were supported by time-resolved Förster resonance energy transfer (FRET) measurements, which showed a high FRET population in the early intermediate for pairs of FRET probes in the N-domains of SRP and SR, which vanished upon transition to the stable complex. The opposite was observed for FRET pairs incorporated in the G-domains. Together with molecular docking and mutational analyses, these data suggest a model in which SRP–SR interaction initiates with electrostatically driven interactions between their N-domains, followed by intramolecular rearrangements at the N–G interface of both GTPases to enable a close approach between their G-domains leading to the GTP-stabilized closed complex (see Figure 10.11(A)). This model was subsequently corroborated by the cryo-EM structure of the early intermediate in complex with the translating ribosome.[73]

10.3.3.5 cpSRP43

cpSRP43 is an ATP-independent chaperone that protects the light-harvesting chlorophyll *a*, *b*-binding proteins (LHCPs), which comprise the light-harvesting antenna complex, during their transit through the chloroplast stroma *en route* to the thylakoid membrane. As with most membrane protein chaperones, client interaction of cpSRP43 must be spatially regulated: it needs to capture client proteins rapidly and effectively in the aqueous phase to prevent their aggregation, but also to promptly release them at the target membrane. Biochemical and NMR studies showed that cpSRP43's chaperone activity towards LHCPs is activated by an interaction partner in the stroma, cpSRP54, which drives cpSRP43 into an active closed conformation. At the thylakoid membrane, cpSRP43 further interacts with the stromal domain of the Alb3 translocase, which induces cpSRP43 into a less active conformation and drives the release of LHCP at the site of its membrane insertion.[74]

To understand the activation mechanism of cpSRP43, the nature of the substrate- and activator-induced conformational changes in cpSRP43 was probed using complementary biophysical techniques.[75] 9 single cysteines were engineered throughout the substrate binding domain (SBD) of cpSRP43, consisting of an N-terminal chromodomain (CD1), four ankyrin repeat motifs, and a bridging helix that caps the C-terminus (see Figure 10.12). These cysteines were labeled either with a nitroxide spin probe or with the ^{19}F-BTFA probe, and the EPR and ^{19}F-NMR spectra were compared in the absence and in the presence of cpSRP54. EPR lineshape analysis showed that the spin labels in ankyrin repeat motifs 2–4 (Anks 2–4) become less mobile upon binding of the cpSRP54 peptide or the substrate LHCP, whereas the probes in CD1 and Ank1 were unperturbed by cpSRP54 or LHCP binding (see Figure 10.12). These results indicate increased backbone mobility and disorder in the C-terminal half of the ankyrin repeat domain in the open conformation. In addition, ^{19}F-NMR directly detected the open and closed conformation of cpSRP43 as distinct cross peaks and showed that the open state peaks are broadened and more sensitive to denaturants compared to those for the closed state, especially towards the C-terminus of the ankyrin repeat domain. Together with the results of H/D exchange experiments, these results demonstrate that activation of cpSRP43 occurs *via* a disorder-to-order transition in its ankyrin repeat domain, in which cooperative stable packing between neighboring ankyrin repeat motifs is

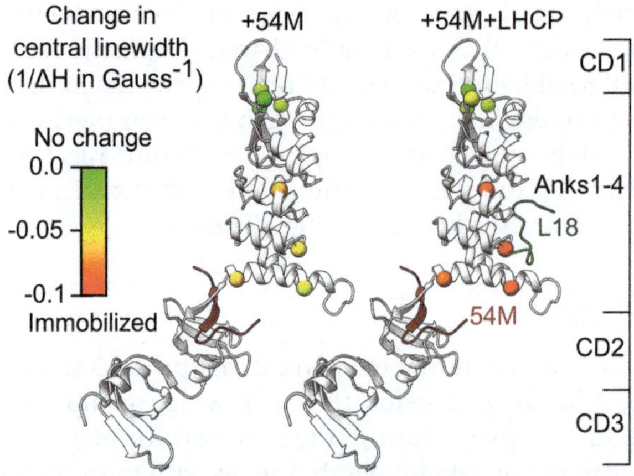

Figure 10.12 EPR mobility changes show a disorder-to-order transition upon activation of the chaperone cpSRP43. A composite structure of cpSRP43 (PDBs: 3DEP, 3UI2, 5E4W)[101–103] is shown with spin-labeled sites represented as spheres. Spin-labeled sites are colored by the relative changes to ΔH upon addition of the activator (54M peptide) (left) and the substrate LHCP (right), as indicated. Both ligands induce immobilization of the C-terminal ankyrin repeat motifs (Anks 2–4) but little change to chromodomain 1 (CD1).[75]

required to generate a contiguous contact surface for effective capture and protection of the LHCPs.

10.3.4 Molecular Architecture of Protein Aggregates

Stress conditions and genetic mutations can overwhelm the proteostatic capacity of the cellular chaperone network, leading to the accumulation of toxic protein aggregates.[76] Protein aggregation is linked to numerous neurodegenerative and other diseases rooted in protein misfolding.[77–80] Many of these diseases are characterized by the deposition of fibrillar amyloid plaques, often composed of specific proteins such as the amyloid beta peptide (Aβ) associated with Alzheimer's disease, α-synuclein associated Parkinson's disease, polyglutamine (polyQ) repeat proteins associated with Huntington's disease, and the islet amyloid polypeptide (IAPP) associated with pancreatic β-cell failure.

X-ray diffraction and crystallography studies demonstrated that the amyloid fibrils formed by various proteins share a highly stable core, composed of a pair of β-sheets that form an interdigitated 'steric zipper' and run perpendicular to the fibril axis.[81,82] Despite this fundamental similarity, fibril structures can vary depending on whether the β-strands are parallel or anti-parallel within the sheet, whether the sheets pack with the same or different surfaces, and whether they are oriented parallel or antiparallel to one another. In addition, the mechanism by which cellular proteins are converted into fibrils and the source of their toxicity remain unclear.

Amyloid fibril formation is a multi-step process that begins with the formation of small, amorphous, on- or off-pathway oligomers, followed by their conversion to a nucleus that propagates to form protofibrils and mature fibrils.[79,83–86] Emerging data suggest that early oligomers can be more cytotoxic compared to mature fibrils. Finally, most stress conditions lead to the rapid accumulation of amorphous protein aggregates that also disrupt the proteostasis network in the cell. The heterogeneity and large size of protein aggregates and their lack of well-defined structures pose enormous challenges for many of the well-developed tools and techniques in structural biology.

SDSL–EPR provides an excellent approach to overcoming many of these barriers, providing residue-specific information on the organization and architecture of protein aggregates as well as the sequences that drive their formation. When a protein of interest is labeled with a single nitroxide spin probe and mixed with an excess of unlabeled protein during aggregate formation, EPR lineshape yields information on whether the site of labeling lies at the solvent-accessible surface *versus* the buried core of the aggregate. When protein aggregates are prepared entirely from singly spin-labeled proteins, dipolar coupling and spin exchange occur between nitroxide probes that are adjacent to one another in the aggregate, providing information on its quaternary structure. Finally, a pair of spin probes at the i, $i+4$ positions in α-helices or i, $i+2$ positions in β-strands can assess the presence or absence of secondary structures in the aggregate. These approaches have been used to study the structure of an increasing number of amyloid-forming proteins and have also proved effective in defining the architecture of amorphous protein aggregates.

10.3.4.1 Aβ

Neuronal accumulation of amyloid plaques composed of the Aβ peptide is a hall mark of Alzheimer's disease. Both Aβ40 and Aβ42 peptides are generated from the cleavage of the amyloid precursor protein by γ-secretase. Intriguingly, while Aβ40 is more abundant, Aβ42 dominates the amyloid plaques, indicating its higher aggregation propensity. Despite numerous studies, the structure of Aβ42 fibrils has remained controversial, in part due to polymorphism of the fibril (see next paragraph). A combination of solid-state NMR, H/D exchange, and mutagenesis studies did not generate sufficient structural constraints for high-resolution modeling. To address this issue, Gu *et al.* in 2015 incorporated spin labels at 22 sites in Aβ42 and monitored strong spin-exchange interactions in the resulting amyloid fibril,[87] which occurs for inter-spin distances of <5.5 Å. Strong spin exchange was observed for most of the spin probes, supporting a parallel in-register β-sheet structure in the fibril. Quantitative analysis from spectral simulations revealed four sets of residues that undergo strong spin exchange interactions, which map to the β_1 and β_2 strands also observed in the Aβ40 fibril structure and two additional β-strands at the C-terminus of Aβ42.

The EPR-derived constraints were used for structural modeling with Rosetta and generated a molecular model for Aβ42 protofilament consisting of four parallel in-register β-sheets. The C-terminal β-turn-β motif in the Aβ42 fibril is absent in Aβ40 fibrils, suggesting a model to explain the higher aggregation propensity of Aβ42 and its importance in pathogenesis.

A major challenge in the study of amyloid fibrils is their polymorphism, whose underlying structural basis and relationship to pathology remain to be determined. Wang *et al.* in 2019 addressed this question by comparing the molecular arrangement of the Aβ42 peptide in fibrils grown under quiescent *versus* agitated conditions, which leads to the formation of elongated *versus* short fibrils, respectively.[88] SDSL–EPR studies of these two types of fibrils showed that their secondary structures are similar. Nevertheless, fibrils generated under agitation show higher inter-spin exchange frequency compared to those grown under quiescent conditions, suggesting a more tightly packed core in the former. These results also show that tuning inter-strand side chain packing interactions could play a role in fibril polymorphism.

Many studies have found that early oligomers that precede fibril formation are more cytotoxic than the amyloid fibrils themselves and are increasingly recognized as the most pathogenic species. Gu *et al.* in 2012 studied the molecular architecture of a toxic Aβ42 oligomer, termed the globulomer, using SDSL–EPR.[89] Nitroxide spin probes were introduced at 14 positions throughout Aβ42. Spin-labeled Aβ42 was mixed with excess unlabeled Aβ42 during oligomer formation, which minimizes inter-spin interactions to allow for mobility analysis. In addition, globulomers formed solely from spin-labeled Aβ42 allowed measurement of inter-spin distances *via* dipolar coupling and provided restraints on the molecular organization of Aβ42 molecules in the globulomer. The data showed that individual Aβ42 molecules are arranged in parallel within the globulomers, with the C-terminal region forming a tightly packed core with intermolecular distances of ∼12 Å. This distance also argued against the presence of parallel in-register β-sheet structures in the globulomer, indicating that the in-register β-sheet structures observed in the mature fibril require a rearrangement of their intermolecular packing at a later stage, such as during or after the nucleation event.

10.3.4.2 Tau

Filamentous aggregates of the microtubule-binding protein tau constitute another hallmark of Alzheimer's disease and other neurodegenerative diseases, with early aggregates and oligomers of tau suspected as the neurotoxic species. Tau is a client of the chaperone Hsp90, which induces oligomerization of tau and initiates its proteasomal degradation, but blocks the formation of filamentous aggregates. To understand the molecular mechanism by which Hsp90 regulates tau oligomerization, a combination of CW-EPR and DEER measurements was used to probe the conformational changes of tau during Hsp90-induced oligomerization.[104] Intramolecular DEER distance measurements of double spin-labeled tau supported a 'paper-clip' model proposed from previous FRET and NMR measurements, in which the N- and C-termini are close to one another, and the tau repeat domain (TauRD) is in a more compact conformation than would be found in a purely random coil (see Figure 10.13, left panel). EPR lineshape analysis showed that Hsp90 induced local restrictions in the mobility of spin labels at residues 291 and 322 in TauRD, supporting this region as the Hsp90 binding site (see Figure 10.13, right panel). In contrast, spin labels near the N-terminus of tau increased in mobility upon Hsp90 addition, suggesting a conformational change that opens up the paper-clip fold (see Figure 10.13, right panel). This model is further supported by DEER measurements, which observed pronounced separation between pairs of spin probes labeled at the N-termini and TauRD as well as between those at the N- and C-termini of Tau upon Hsp90 binding. On the other hand, intermolecular DEER measurements to map the contact sites in the tau oligomer showed that the oligomer interface is also located in TauRD. Together, these results suggest that Hsp90 induces tau to adopt a more 'open' conformation in which the N-terminus detaches from TauRD and from the C-terminus, thus opening up the paper-clip structure and exposing the TauRD region to undergo subsequent oligomerization.

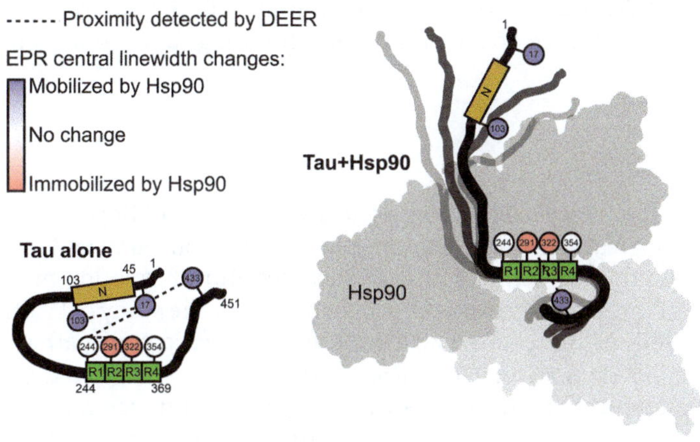

Figure 10.13 Model of tau conformations in the absence (left) and in the presence (right) of Hsp90. Hsp90 decreased the mobility of spin labels in the repeat domains (291, 322 in red) while increasing the mobility of spin labels in the N-terminal domain and at the C-terminus (17, 103, 433 in blue). DEER measurements show that pairs of spin labels at the N-terminus and repeat domains of tau are closer in the absence of Hsp90 (dotted lines in the left panel; 17–103, 17–433, 17–244, 244–291) and separated upon addition of Hsp90, whereas the C-terminus of tau is closer to the repeat domains upon Hsp90 binding (dotted line in the right panel: 291–433). Together, this supports tau adopting a compact 'paper-clip' conformation by itself (left) but a more open conformation upon chaperone binding (right). Reproduced from ref. 104, https://doi.org/10.1126/sciadv.aax6999, under the terms of the CC BY 4.0 license, http://creativecommons.org/licenses/by/4.0/.

10.3.4.3 PrP

Neuronal accumulation of protein deposits formed by the prion protein PrP is strongly associated with transmissible spongiform encephalopathies, commonly referred to as prion disorders. PrP aggregation results from the conversion of the normal cellular prion protein, PrP^{C}, to an abnormal conformer, PrP^{SC}, which forms a self-perpetuating infectious agent. While high-resolution structures of the PrP^{C} monomer are available, PrP^{SC} displays distinct biophysical properties compared to PrP^{C} despite sharing the same covalent structure. In the study of Cobb *et al.* in 2007, spin probes were incorporated at 47 sites in human PrP(90–271),[90] a fragment strongly associated with PrP infectivity and that contains all the mutations associated with hereditary prion disease. Sites of contact between neighboring PrP molecules within the amyloid were mapped based on inter-spin interactions from dipolar coupling, which provide distance constraints in the 8–20 Å range, and spin exchange, which occurs at even shorter distances. The results showed that, contrary to previous theoretical models, the amyloid core is formed by C-terminal residues 159–219 in PrP. The strong inter-spin exchange between probes in this region is consistent with the parallel, in-register organization of stacked β-strands and excludes an anti-parallel arrangement. This core structure showed high stability and remains tightly packed in the presence of 2 M GdmHCl, supporting this region as the driver for amyloidosis.

10.3.4.4 Het-s

In addition to disease-causing aggregates, some amyloid fibrils contribute to distinct biological functions. An example of a functional amyloid is the Het-s protein from the filamentous fungus, in which fibril formation prevents the exchange of cytoplasm between distinct species. In the study of Wälti *et al.* in 2016, the role of the chaperone GroEL in the aggregation pathway of the prion-forming domain of Het-s was studied using a combination of biophysical methods.[91] Solution NMR and negative stain electron microscopy showed that GroEL accelerates the fibrillation of Het-s, with the fibril densely decorated with GroEL at a regular spacing of 199 Å. CW-EPR measurements, using Het-s labeled with a spin probe near the N-terminus of the prion domain, showed that the spin label is highly mobile in the Het-s monomer, with a rotational correlation time of $\tau_c \sim 0.5$ ns. Fibrillation or the initial binding of GroEL induced a less mobile population in the EPR spectra, with $\tau_c \approx 3$ ns. However, the formation of Het-s fibril in the presence of GroEL introduced a population in which the nitroxide spin probe lost essentially all mobility ($\tau_c \approx 199$ ns), which was attributed to a direct interaction between the spin-labeled position on Het-s and the apical face of GroEL. Interestingly, the immobile population and the more flexible species ($\tau_c \approx 3$ ns) were equally populated, consistent with the ratio of outer to inner Het-s filaments. This strongly suggests that GroEL coats the outer filaments of Het-s at nearly all positions, whereas the inner filaments do not contact GroEL.

10.3.4.5 LHCPs

LHCPs are a superfamily of conserved and highly abundant membrane proteins in green algae and land plants. They comprise the light-harvesting antenna complex at the thylakoid membrane, which absorbs sunlight and regulates the transfer of excitation energy to photosystem II for oxygenic photosynthesis. Analogous to most organellar membrane proteins, the biogenesis of LHCPs poses multiple challenges to cellular protein homeostasis. LHCPs are synthesized in the cytosol, imported into the chloroplast, and delivered by the chloroplast signal recognition particle (cpSRP) pathway to the Alb3 translocase, which mediates their insertion into the thylakoid membrane. LHCPs contain three transmembrane domains (TMDs) and are highly prone to aggregation in aqueous environments; therefore their transit through the stroma requires effective protection by cpSRP. As described in Section 10.3.3.5, the cpSRP43 subunit of cpSRP is responsible for this chaperone activity. It specifically recognizes a conserved L18 motif in the TM2-TM3 loop of LHCP and forms promiscuous hydrophobic contacts with all the TMDs.[92] Despite the lack of an ATPase domain, cpSRP43 also displays robust disaggregase activity towards the LHCPs, indicating that the energy from chaperone–client binding can drive effective reversal of protein aggregation.

To understand the molecular basis of cpSRP43's disaggregase activity, Nguyen *et al.* in 2013 probed the molecular architecture of the LHCP aggregate using a combination of SDSL–EPR and thiol-specific modification.[38] 30 single cysteines were engineered throughout LHCP, and their solvent accessibility in the LHCP aggregate was probed by NEM modification. In addition, each cysteine was labeled with the nitroxide spin probe, whose local backbone mobility in the LHCP aggregate was measured by EPR lineshape analyses. The results showed that spin probes at the N-terminal loop and in the L18 recognition motif of LHCP were highly mobile, whereas the probes in all the TMDs, the TM1-TM2 loop, and the C-terminus of LHCP are strongly immobilized in the aggregate (see Figure 10.14(A)). These results are in agreement with the NEM accessibility analyses and together show that the N-terminus and L18 motif of LHCP are exposed and highly accessible on the surface of the aggregate, whereas the hydrophobic TMDs pack against one another inside the aggregate to minimize their solvent exposure (see Figure 10.14(B)). This molecular architecture of the LHCP aggregate suggests a two-step mechanism for ATP-independent disaggregase activity that was further supported by biochemical experiments: cpSRP43 first recognizes and coats the aggregate surface *via* high-affinity interaction with the L18 sequence; this is followed by 'invasion' into the aggregate core as the chaperone establishes additional contacts with the TMDs and competes with their packing interactions, leading to disruption of the aggregate core and disassembly of the aggregate.

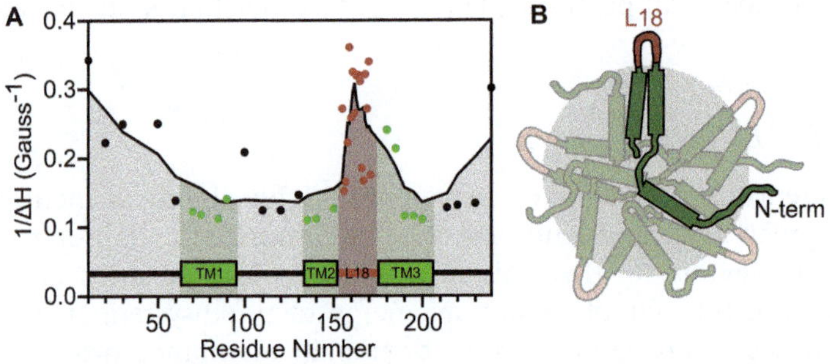

Figure 10.14 Structure of an aggregate formed by the light-harvesting chlorophyll-binding protein (LHCP). (A) Rolling average of the inverse central linewidths for spin labels throughout LHCP, with spin labels in the TMDs in green, L18 motif in red, and inter-TMD loops in black. (B) Model of an LHCP aggregate based on the mobility of spin labels throughout LHCP in the aggregates. The N-terminus and L18 motif are exposed on the surface, whereas the TMDs are buried in the aggregate core. cpSRP43's specific recognition of the surface-exposed L18 motif facilitates recognition of the aggregate to initiate disaggregation. Reproduced from ref. 38, https://doi.org/10.1074/jbc.M113.462812, under the terms of the CC BY 4.0 license, http://creativecommons.org/licenses/by/4.0/.

10.4 Conclusions

Site-directed spin-labeling and EPR can be used to study chaperone systems that are challenging or intractable to other higher-resolution structural techniques. The relative simplicity of obtaining and analyzing EPR spectra from spin-labeled proteins also makes it an excellent complementary technique to extend observations from other techniques such as crystallography, FRET, and NMR. Reciprocally, molecular interpretation of EPR data to derive mechanistic insight often needs to leverage complementary information from biochemistry and other biophysical methods. Continuing advances in pulsed-EPR techniques have also made EPR studies of proteins more quantitative across longer distances, which can provide valuable information on large-scale conformational changes to chaperones and their clients (see Chapter 9 by Dröden and Drescher on DEER). As more structures of well-folded proteins are solved each year, the study of large protein complexes and intrinsically disordered proteins become new frontiers in structural biology. EPR is uniquely suited to explore these frontiers and will remain a useful tool to provide information on the mechanism of action of molecular chaperones.

Acknowledgements

We gratefully acknowledge support from grants DOE.SC0020661 and NIH R35 GM136321 to S. S., NSF MCB-1716423, NSF MCB-1818107, and NIH GM124413 to P. Z. Q., and NIH T32-GM118289 to J. S.

References

1. F. U. Hartl, A. Bracher and M. Hayer-Hartl, *Nature*, 2011, **475**, 324–332.
2. Y. E. Kim, M. S. Hipp, A. Bracher, M. Hayer-Hartl and F. Ulrich Hartl, *Annu. Rev. Biochem.*, 2013, **82**, 323–355.
3. R. Rosenzweig, N. B. Nillegoda, M. P. Mayer and B. Bukau, *Nat. Rev. Mol. Cell Biol.*, 2019, **20**, 665–680.
4. M. Hayer-Hartl, A. Bracher and F. U. Hartl, *Trends Biochem. Sci.*, 2016, **41**, 62–76.
5. F. H. Schopf, M. M. Biebl and J. Buchner, *Nat. Rev. Mol. Cell Biol.*, 2017, **18**, 345–360.
6. E. C. Duran, C. L. Weaver and A. L. Lucius, *Front. Mol. Biosci.*, 2017, **4**, 54.
7. S. M. Doyle and S. Wickner, *Trends Biochem. Sci.*, 2009, **34**, 40–48.
8. K. Reinle, A. Mogk and B. Bukau, *J. Mol. Biol.*, 2021, 167157.
9. S. Hiller, *Trends Biochem. Sci.*, 2019, **44**, 517–527.
10. W. L. Hubbell and C. Altenbach, *Curr. Opin. Struct. Biol.*, 1994, **4**, 566–573.
11. W. L. Hubbell and H. M. McConnell, *J. Am. Chem. Soc.*, 1971, **93**, 314–26.
12. J. H. K. Rozenfeld, E. L. Duarte, T. R. Oliveira and M. T. Lamy, *Biophys. Rev.*, 2017, **9**, 633–647.
13. W. L. Hubbell, A. Gross, R. Langen and M. A. Leitzow, *Curr. Opin. Struct. Biol.*, 1998, **8**, 649–656.
14. W. L. Hubbell, D. S. Cafiso and C. Altenbach, *Nat. Struct. Biol.*, 2000, 7, 735–739.
15. G. E. Fanucci and D. S. Cafiso, *Curr. Opin. Struct. Biol.*, 2006, **16**, 644–653.
16. H. Li, Y. Pan, Z. Yang, J. Rao and B. Chen, *Trends Food Sci. Technol.*, 2021, **109**, 37–50.
17. F. Torricella, A. Pierro, E. Mileo, V. Belle and A. Bonucci, *Biochim. Biophys. Acta, Proteins Proteomics*, 2021, 140653.
18. G. Z. Sowa and P. Z. Qin, in *Progress in Nucleic Acid Research and Molecular Biology*, 2008, vol. 82, pp. 147–197.
19. I. Krstić, B. Endeward, D. Margraf, A. Marko and T. Prisner, in *EPR Spectroscopy, Applications in Chemistry and Biology*, ed. M. Drescher and G. Jeschke, Springer, Berlin/Heidelberg, 2012, vol. 321, pp. 159–198.
20. Y. Pan, H. Li, J. Farmakes, F. Xiao, B. Chen, S. Ma and Z. Yang, *J. Am. Chem. Soc.*, 2018, **140**, 16032–16036.
21. C. Altenbach, C. J. López, K. Hideg and W. L. Hubbell, in *Methods in Enzymology*, ed. P. Z. Qin and K. Warncke, Academic Press, 2015, vol. 564, pp. 59–100.
22. Z. Yang, M. D. Bridges, C. J. López, O. Y. Rogozhnikova, D. V. Trukhin, E. K. Brooks, V. Tormyshev, H. J. Halpern and W. L. Hubbell, *J. Magn. Reson.*, 2016, **269**, 50–54.

23. E. G. Bagryanskaya, O. A. Krumkacheva, M. V. Fedin and S. R. A. Marque, in *Methods in Enzymology*, ed. P. Z. Qin and K. Warncke, Academic Press, 2015, vol. 563, pp. 365–396.
24. A. Gamble Jarvi, X. Bogetti, K. Singewald, S. Ghosh and S. Saxena, *Acc. Chem. Res.*, 2021, **54**, 1481–1491.
25. A. Feintuch, G. Otting and D. Goldfarb, in *Methods in Enzymology*, ed. P. Z. Qin and K. Warncke, Academic Press, 2015, vol. 563, pp. 415–457.
26. T. Braun, M. Drescher and D. Summerer, *Int. J. Mol. Sci.*, 2019, **20**, 373.
27. *Spin Labeling: Theory and Applications*, ed. L. J. Berliner, Academic Press, New York, 1976.
28. D. Marsh, *Mol. Biol., Biochem. Biophys.*, 1981, **31**, 51–142.
29. C. P. Poole Jr., *Electron Spin Resonance: A Comprehensive Treatise on Experimental Techniques*, Dover Publications, 1982.
30. *Distance Measurements in Biological Systems by EPR*, ed. G. R. Eaton, S. S. Eaton and L. J. Berliner, Kluwer, New York, NY, 2000, vol. 19.
31. A. Schweiger and G. Jeschke, *Principles of pulse electron paramagnetic resonance*, Oxford University Press, Oxford, 2001.
32. D. Marsh, *Spin-Label Electron Paramagnetic Resonance Spectroscopy*, CRC Press, 2019.
33. J. A. Cooke and L. J. Brown, in *Protein folding, misfolding, and disease*, Springer, 2011, pp. 73–96.
34. G. L. Closs, M. D. Forbes and P. Piotrowiak, *J. Am. Chem. Soc.*, 1992, **114**, 3285–3294.
35. H. O. Griffith and P. C. Jost, in *Spin labeling theory and application*, ed. L. J. Berliner, Academic Press, New York, 1976, pp. 453–423.
36. H. S. Mchaourab, M. A. Lietzow, K. Hideg and W. L. Hubbell, *Biochemistry*, 1996, **35**, 7692–7704.
37. X. Zhang, V. Q. Lam, Y. Mou, T. Kimura, J. Chung, S. Chandrasekar, J. R. Winkler, S. L. Mayo and S. Shan, *Proc. Natl. Acad. Sci. U. S. A.*, 2011, **108**, 6450–6455.
38. T. X. Nguyen, P. Jaru-Ampornpan, V. Q. Lam, P. Cao, S. Piszkiewicz, S. Hess and S. Shan, *J. Biol. Chem.*, 2013, **288**, 13420–13430.
39. D. E. Budil, S. Lee, S. Saxena and J. H. Freed, *J. Magn. Reson., Ser. A*, 1996, **120**, 155–189.
40. S. Stoll and A. Schweiger, *J. Magn. Reson.*, 2006, **178**, 42–55.
41. L. Columbus and W. L. Hubbell, *Trends Biochem. Sci.*, 2002, **27**, 288–295.
42. L. Columbus and W. L. Hubbell, *Biochemistry*, 2004, **43**, 7273–7287.
43. M. D. Bridges, K. Hideg and W. L. Hubbell, *Appl. Magn. Reson.*, 2010, **37**, 363–390.
44. L. Yu, W. Wang, S. Ling, S. Liu, L. Xiao, Y. Xin, C. Lai, Y. Xiong, L. Zhang and C. Tian, *Sci. Rep.*, 2015, **5**, 1–11.
45. C. Altenbach, W. Froncisz, R. Hemker, H. Mchaourab and W. L. Hubbell, *Biophys. J.*, 2005, **89**, 2103–2112.

46. C. Altenbach, D. A. Greenhalgh, H. G. Khorana and W. L. Hubbell, *Proc. Natl. Acad. Sci. U. S. A.*, 1994, **91**, 1667.
47. K. J. Oh, H. Zhan, C. Cui, K. Hideg, R. J. Collier and W. L. Hubbell, *Science*, 1996, **273**, 810–812.
48. M. D. Rabenstein and Y.-K. Shin, *Proc. Natl. Acad. Sci. U. S. A.*, 1995, **92**, 8239–8243.
49. H. S. Mchaourab, K. J. Oh, C. J. Fang and W. L. Hubbell, *Biochemistry*, 1997, **36**, 307–316.
50. A. Tsirigotaki, J. De Geyter, N. Šoštarić, A. Economou and S. Karamanou, *Nat. Rev. Microbiol.*, 2017, **15**, 21–36.
51. J. M. Crane, C. Mao, A. A. Lilly, V. F. Smith, Y. Suo, W. L. Hubbell and L. L. Randall, *J. Mol. Biol.*, 2005, **353**, 295–307.
52. J. M. Crane, Y. Suo, A. A. Lilly, C. Mao, W. L. Hubbell and L. L. Randall, *J. Mol. Biol.*, 2006, **363**, 63–74.
53. A. A. Lilly, J. M. Crane and L. L. Randall, *Protein Sci.*, 2009, **18**, 1860–1868.
54. D. B. Cooper, V. F. Smith, J. M. Crane, H. C. Roth, A. A. Lilly and L. L. Randall, *J. Mol. Biol.*, 2008, **382**, 74–87.
55. F.-U. Hartl, S. Lecker, E. Schiebel, J. P. Hendrick and W. Wickner, *Cell*, 1990, **63**, 269–279.
56. M. Lorenzi, L. Sylvi, G. Gerbaud, E. Mileo, F. Halgand, A. Walburger, H. Vezin, V. Belle, B. Guigliarelli and A. Magalon, *PLoS One*, 2012, 7, e49523.
57. G. Karthikeyan, A. Bonucci, G. Casano, G. Gerbaud, S. Abel, V. Thomé, L. Kodjabachian, A. Magalon, B. Guigliarelli, V. Belle, O. Ouari and E. Mileo, *Angew. Chem., Int. Ed.*, 2018, **57**, 1366–1370.
58. K. Januszyk, M. R. Fleissner, L. Atchabahian, F.-K. Shieh, C. Altenbach, S. L. Martin, F. Guo, W. L. Hubbell and R. T. Clubb, *Protein Sci.*, 2011, **20**, 1231–1243.
59. S. Hiller, *J. Biomol. NMR*, 2021, **75**, 1–8.
60. T. M. Treweek, S. Meehan, H. Ecroyd and J. A. Carver, *Cell. Mol. Life Sci.*, 2015, **72**, 429–451.
61. U. Jakob, M. Gaestel, K. Engel and J. Buchner, *J. Biol. Chem.*, 1993, **268**, 1517–1520.
62. M. Haslbeck, T. Franzmann, D. Weinfurtner and J. Buchner, *Nat. Struct. Mol. Biol.*, 2005, **12**, 842–846.
63. H. S. Mchaourab, E. K. Dodson and H. A. Koteiche, *J. Biol. Chem.*, 2002, **277**, 40557–40566.
64. D. P. Claxton, P. Zou and H. S. Mchaourab, *J. Mol. Biol.*, 2008, **375**, 1026–1039.
65. H. R. Saibil, W. A. Fenton, D. K. Clare and A. L. Horwich, *J. Mol. Biol.*, 2013, **425**, 1476–1487.
66. M. Persson, P. Hammarström, M. Lindgren, B.-H. Jonsson, M. Svensson and U. Carlsson, *Biochemistry*, 1999, **38**, 432–441.
67. R. Owenius, A. Jarl, B.-H. Jonsson, U. Carlsson and P. Hammarström, *J. Chem. Biol.*, 2010, **3**, 127–139.

68. M. P. Mayer, *Mol. Cell*, 2010, **39**, 321–331.
69. R. Schlecht, A. H. Erbse, B. Bukau and M. P. Mayer, *Nat. Struct. Mol. Biol.*, 2011, **18**, 345–351.
70. J. Yang, Y. Zong, J. Su, H. Li, H. Zhu, L. Columbus, L. Zhou and Q. Liu, *Nat. Commun.*, 2017, **8**, 1201.
71. E. T. McDonald, M. Bortolus, H. A. Koteiche and H. S. Mchaourab, *Biochemistry*, 2012, **51**, 1257–1268.
72. D. Y. Kim, E. Kwon, Y.-K. Shin, D.-H. Kweon and K. K. Kim, *J. Mol. Biol.*, 2008, **377**, 410–420.
73. L. F. Estrozi, D. Boehringer, S. Shan, N. Ban and C. Schaffitzel, *Nat. Struct. Mol. Biol.*, 2011, **18**, 88–90.
74. F.-C. Liang, G. Kroon, C. Z. McAvoy, C. Chi, P. E. Wright and S. Shan, *Proc. Natl. Acad. Sci. U. S. A.*, 2016, **113**, E1615–E1624.
75. A. Siegel, C. Z. McAvoy, V. Lam, F.-C. Liang, G. Kroon, E. Miaou, P. Griffin, P. E. Wright and S.-O. Shan, *J. Mol. Biol.*, 2020, **432**, 166708.
76. F. U. Hartl and M. Hayer-Hartl, *Nat. Struct. Mol. Biol.*, 2009, **16**, 574–581.
77. V. Kumar, N. Sami, T. Kashav, A. Islam, F. Ahmad and Md. I. Hassan, *Eur. J. Med. Chem.*, 2016, **124**, 1105–1120.
78. F. Chiti and C. M. Dobson, *Annu. Rev. Biochem.*, 2017, **86**, 27–68.
79. T. P. J. Knowles, M. Vendruscolo and C. M. Dobson, *Nat. Rev. Mol. Cell Biol.*, 2014, **15**, 384–396.
80. F. U. Hartl, *Annu. Rev. Biochem.*, 2017, **86**, 21–26.
81. T. Lührs, C. Ritter, M. Adrian, D. Riek-Loher, B. Bohrmann, H. Döbeli, D. Schubert and R. Riek, *Proc. Natl. Acad. Sci. U. S. A.*, 2005, **102**, 17342–17347.
82. M. Sunde, L. C. Serpell, M. Bartlam, P. E. Fraser, M. B. Pepys and C. C. F. Blake, *J. Mol. Biol.*, 1997, **273**, 729–739.
83. S. I. A. Cohen, S. Linse, L. M. Luheshi, E. Hellstrand, D. A. White, L. Rajah, D. E. Otzen, M. Vendruscolo, C. M. Dobson and T. P. J. Knowles, *Proc. Natl. Acad. Sci. U. S. A.*, 2013, **110**, 9758–9763.
84. D. J. Selkoe, *Nature*, 2003, **426**, 900–904.
85. C. Haass and D. J. Selkoe, *Nat. Rev. Mol. Cell Biol.*, 2007, **8**, 101–112.
86. R. Kayed, E. Head, J. L. Thompson, T. M. McIntire, S. C. Milton, C. W. Cotman and C. G. Glabe, *Science*, 2003, **300**, 486–489.
87. L. Gu, J. Tran, L. Jiang and Z. Guo, *J. Struct. Biol.*, 2016, **194**, 61–67.
88. H. Wang, L. Duo, F. Hsu, C. Xue, Y. K. Lee and Z. Guo, *Sci. Rep.*, 2020, **10**, 5720.
89. L. Gu, C. Liu and Z. Guo, *J. Biol. Chem.*, 2013, **288**, 18673–18683.
90. N. J. Cobb, F. D. Sönnichsen, H. Mchaourab and W. K. Surewicz, *Proc. Natl. Acad. Sci. U. S. A.*, 2007, **104**, 18946–18951.
91. M. A. Wälti, T. Schmidt, D. T. Murray, H. Wang, J. E. Hinshaw and G. M. Clore, *Proc. Natl. Acad. Sci. U. S. A.*, 2017, **114**, 9104–9109.
92. C. Z. McAvoy, A. Siegel, S. Piszkiewicz, E. Miaou, M. Yu, T. Nguyen, A. Moradian, M. J. Sweredoski, S. Hess and S. Shan, *J. Biol. Chem.*, 2018, **293**, 8861–8873.

93. Y. Ding, V. Kathiresan, X. Zhang, I. S. Haworth and P. Z. Qin, *J. Phys. Chem. A*, 2019, **123**, 3592–3598.
94. K. N. Beasley, B. T. Sutch, M. H. Ma'mon, R. Langen, P. Z. Qin and I. S. Haworth, *Methods Enzymol.*, 2015, **563**, 569–593.
95. Y. Papanikolau, M. Papadovasilaki, R. B. G. Ravelli, A. A. McCarthy, S. Cusack, A. Economou and K. Petratos, *J. Mol. Biol.*, 2007, **366**, 1545–1557.
96. Z. Xu, J. D. Knafels and K. Yoshino, *Nat. Struct. Biol.*, 2000, **7**, 1172–1177.
97. A. E. Eriksson, W. A. Baase and B. W. Matthews, *J. Mol. Biol.*, 1993, **229**, 747–769.
98. X. Zhu, X. Zhao, W. F. Burkholder, A. Gragerov, C. M. Ogata, M. E. Gottesman and W. A. Hendrickson, *Science*, 1996, **272**, 1606–1614.
99. L. Nappi, A. H. Aguda, N. A. Nakouzi, B. Lelj-Garolla, E. Beraldi, N. Lallous, M. Thi, S. Moore, L. Fazli, D. Battsogt, S. Stief, F. Ban, N. T. Nguyen, N. Saxena, E. Dueva, F. Zhang, T. Yamazaki, A. Zoubeidi, A. Cherkasov, G. D. Brayer and M. Gleave, *J. Clin. Invest.*, 2020, **130**, 699–714.
100. C. Faoro and S. F. Ataide, *J. Struct. Biol.*, 2019, **208**, 107387.
101. K. F. Stengel, I. Holdermann, P. Cain, C. Robinson, K. Wild and I. Sinning, *Science*, 2008, **321**, 253–256.
102. I. Holdermann, N. H. Meyer, A. Round, K. Wild, M. Sattler and I. Sinning, *Nat. Struct. Mol. Biol.*, 2012, **19**, 260–263.
103. A. Horn, J. Hennig, Y. L. Ahmed, G. Stier, K. Wild, M. Sattler and I. Sinning, *Nat. Commun.*, 2015, **6**, 8875.
104. S. Weickert, M. Wawrzyniuk, L. H. John, S. G. D. Rüdiger and M. Drescher, *Sci. Adv.*, 2020, **6**, eaax6999.

CHAPTER 11

Probing Single Chaperone Substrates

F. WRUCK,[a,b] M. J. AVELLANEDA,[a,c] M. M. NAQVI,[a,d] E. J. KOERS,[a,e] K. TILL,[a] L. GROSS,[a] F. MOAYED,[a] A. ROLAND,[a] L. W. H. J. HELING,[f,g] A. MASHAGHI[f,g] AND S. J. TANS*[a,h]

[a] AMOLF, Science Park 104, 1098 XG Amsterdam, The Netherlands; [b] LUMICKS BV, Amsterdam, The Netherlands; [c] Institute of Science and Technology Austria, Am Campus 1, Klosterneuburg 3400, Austria; [d] Department of Pharmacology, University of Cambridge, CB2 1PD Cambridge, UK; [e] Centre of Membrane Proteins and Receptors (COMPARE), University of Birmingham and University of Nottingham, Midlands, UK; [f] Medical Systems Biophysics and Bioengineering, Leiden Academic Centre for Drug Research, Faculty of Science, Leiden University, 2333CC, Leiden, The Netherlands; [g] Centre for Interdisciplinary Genome Research, Faculty of Science, Leiden University, Einsteinweg 55, 2333CC Leiden, The Netherlands; [h] Bionanoscience Department of Delft University of Technology and Kavli Institute of Nanoscience Delft, 2629HZ Delft, The Netherlands
*Email: S.Tans@amolf.nl

11.1 Introduction

Elucidating how cells control the conformational state of their proteins is one of the major problems in biology.[1,2] Indeed, most non-infectious diseases are thought to be rooted in the formation of erroneous protein structures, a process exacerbated by cellular stress, ageing, and environmental and genetic factors.[3] It has been well established that molecular chaperones are key to efficient protein folding.[4,5] Important progress has

New Developments in NMR No. 29
Biophysics of Molecular Chaperones: Function, Mechanisms and Client Protein Interactions
Edited by Sebastian Hiller, Maili Liu and Lichun He

Published by the Royal Society of Chemistry, www.rsc.org

also been achieved in determining chaperone structures, as well as their conformational changes, biochemistry, and interactions with co-factors. However, we know far less about the underlying molecular mechanisms at the level of substrate conformational changes.[6–8] Yet, during their lifetime, amino acid chains undergo highly diverse conformational transitions that are decisive to success and failure: they can collapse into compact yet dynamic states,[9] fold progressively during their synthesis,[10] become partially unfolded to enable signaling pathways,[11] and get incorporated into protein complexes,[12] protein aggregates,[13] and stress granules,[14] while also being removed from these higher-order structures.[15,16] Though many of these processes have been investigated extensively, it has been difficult to study protein conformation changes directly.

This lack of progress is understandable. Conformational changes of fully folded proteins, such as those associated with allosteric transitions, typically involve a few key states that can be controlled externally and hence can be studied using averaging structural methods such as cryo-EM and crystallography.[17] Conversely, structural states of unfolded and partially folded protein substrates are typified by large conformational ensembles, dynamics from microseconds to minutes, and heterogeneous interactions with chaperones and their ATP-driven cycles.[18,19] These factors pose major experimental detection challenges. In recent years, however, crucial technical advances have emerged that promise to address them more directly. Structural approaches including cryo-EM, NMR, and hydrogen exchange mass spectrometry can increasingly resolve marginally stable complexes, interaction sites and surfaces, and rapid movements within polypeptide backbones, while single-molecule FRET has been pivotal in identifying structural heterogeneity, among many more effects crucial to understanding how protein conformations are affected by chaperones.[8,20–24]

This chapter aims to describe recent advances in probing chaperone function using single-molecule manipulation methods including atomic force microscopy (AFM), magnetic tweezers, and optical tweezers, noting that most studies employed the latter (see Figure 11.1). We thus focus on the substrate rather than the chaperone dynamics and on mechanical probing approaches rather than fluorescence-based techniques. Since its inception, single-molecule manipulation has been pivotal in revealing many important phenomena, ranging from RNA polymerase translocation dynamics to kinesin stepping and multi-step protein folding pathways.[25–28] What this approach lacks in direct structural information, it makes up for in real-time quantification of movements and stability. In recent years, it has been increasingly applied to elucidating protein state control. This first view on the conformational transitions of chaperone substrates has revealed many unexpected effects, ranging from the stabilization of folding intermediates[29–31] to the promotion of folding under tension,[32] the strengthening of polypeptide collapse,[33] and the forceful translocation of looped polypeptides.[34] Hence, now is a good moment to take stock of this new direction and assess how it is altering our understanding of chaperone function and protein state control.

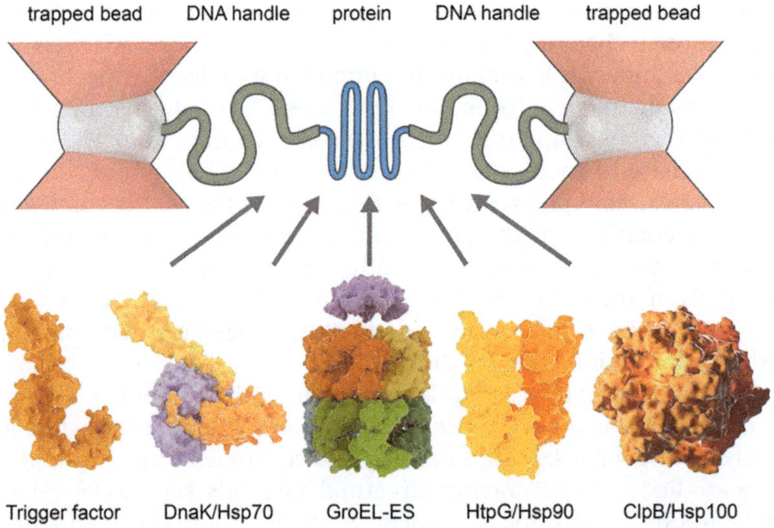

Figure 11.1 The substrate manipulation approach to studying chaperone function. Protein substrates are held in place by tethering them to DNA handles, which in turn are attached to micron-sized beads that are trapped in focused laser beams. By moving the laser beams, one can mechanically perturb the tethered protein while monitoring its dimensions and internal tension or force in time. In turn, this allows direct probing of diverse phenomena, and how they are induced or altered by chaperones that are free in solution, including unfolding and refolding transitions, intermediate folded states, and their mechanical stability, and gradual collapse processes and refolding probabilities. The use of tandem-repeat protein constructs allows the study of aggregation between them. This approach is increasingly combined with single-molecule fluorescence detection, which can for instance report on the binding and unbinding of single chaperone molecules.

The outline of this chapter will be as follows: after providing background on the optical tweezers technique, we follow proteins from their initial synthesis by ribosomes, to the interactions with the major *E. coli* chaperones including trigger factor, DnaK, HtpG, and GroEL–ES, and then turn to aggregation prevention by the small heat shock proteins and disaggregation by ClpB (see Figure 11.1), and conclude with an outlook where we discuss new challenges and questions for the coming years.

11.2 The Optical Tweezers Approach

Optical tweezers have found many applications in physics and the life sciences[35–37] ever since Arthur Ashkin's first description of optical trapping of dielectric particles in liquid.[38] Using near-infrared lasers to form gradient force traps made it possible to indirectly apply pN-scale forces on large molecules, nm-sized motor enzymes and ribozymes (*e.g.* ribosomes), internal parts of cells, and whole living cells in their native environment without causing optical damage.[39,40] The use of optical traps in single-molecule biophysics has greatly enhanced our understanding of a wide range of molecular motors found within the cell.[41] Optical trap force spectroscopy has enabled researchers to carry out precise measurements of the minuscule forces and displacements that govern many biological processes at the single molecule level.[42]

A simple optical trap is formed by focusing a Gaussian beam to create a diffraction-limited spot using a lens with a high numerical aperture such as an inverted microscope objective.[36] Optical trapping of micron-sized dielectric particles near the focal point of such a gradient trap is easiest explained in terms of conservation of momentum.[43] As light impinges on a transparent spherical particle with a higher refractive index than the surrounding medium, some of the light is scattered, but most of the light passes through the sphere. As illustrated in Figure 11.2a, the combined momentum flux change of the Gaussian beam passing through the sphere results in an overall force experienced by the bead pointing towards the region with the greatest light

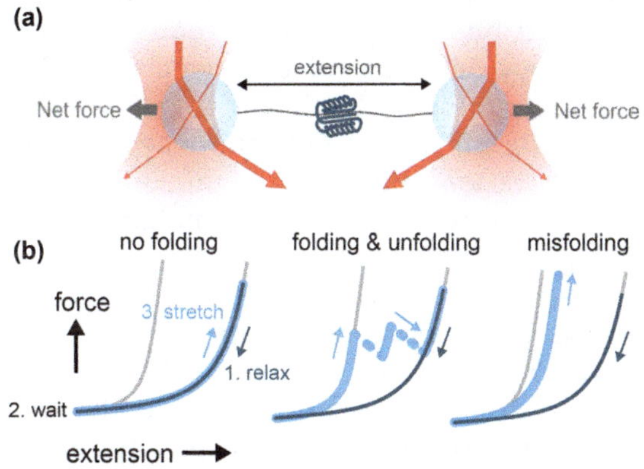

Figure 11.2 The principle of optical trapping and the protein force spectroscopy approach. (a) Simplified ray diagram depicting the net force experienced by two tethered beads in two optical traps due to refraction of the trapping laser at the surface of the beads (red arrows). Key measured quantities are the extension or distance between the beads, and the net force acting on the molecular construct bridging the two beads. (b) Schematic of resulting force spectroscopy data for a slow folding protein, as it either does not refold (left), does fold and then unfolds again *via* a folding intermediate (middle), or misfolds and then does not unfold during relax-wait-stretch cycles (right). Gray lines indicate the theoretical forces and extensions when stretching or relaxing a compact folded state and an extended unfolded state, using the worm-like chain (WLC) model. Hence, one can detect diverse changes that chaperones can potentially exert. They may change the probability of refolding or induce it against applied forces, promote intermediate folds identified as data following additional WLC curves, but also suppress folded states including misfolds, and alter the stability of intermediate folded states as identified by their unfolding force.

intensity, the center of the focal point. A particle can be trapped near the focal point if this so-called gradient force overcomes the scattering force due to light scattered by the sphere. If unperturbed, the sphere will be trapped at a slight offset along the optical axis from the focal point, where the gradient force balances the scattering force.

Typically, two such optical traps are used in optical tweezers experiments, formed within the focal plane of a single inverted microscope objective, where one trap can be moved or steered relative to the other. Through off-the-shelve surface modifications, one can 'tether' individual biomolecules (or biomolecular assemblies) of interest between two such optically tapped beads. By increasing the distance between the tethered optically trapped beads one can stretch the tethered molecule until at some point the trapped beads get pulled out of the center of their traps, imparting a light-momentum change-induced restoring force on the tethered assembly akin to a spring under tension, following Hooke's law for a narrow range of forces for up to hundreds of pN in force, dependent on the trapping laser intensity as well as the bead size.

Other microscopy techniques can be combined with optical tweezers, such as multi-color confocal microscopy and/or stimulated emission depletion fluorescence microscopy, allowing for correlated force–fluorescence spectroscopy with super resolution.[44,45] Although more difficult to realize, other single-molecule techniques, such as AFM and magnetic tweezers, have also been combined with various fluorescence microscopy techniques, while specific care needs to be taken to enable simultaneous correlated force–fluorescence measurements.[46,47]

Optical tweezers are ideal for exerting pN-level forces on individual molecules or molecular assemblies, while simultaneously detecting length changes on the order of nanometers and force changes as small as 0.1 pN, while filtering out noise due to Brownian motion. Both the spatial and force resolution increase with applied tension. Some custom instruments have even been able to achieve Angstrom-level resolution at high applied forces.[48] This sensitivity enables for instance the detection of conformational changes of individually tethered proteins during folding and unfolding transitions. By tuning the applied forces one can study proteins in partially folded and misfolded states, which is difficult to do by other means (see Figure 11.2b). Single-molecule techniques like optical tweezers have proven essential for the study of protein folding, misfolding, and synthesis,[49–54] which are asynchronous processes that are difficult to observe using ensemble methods. Optical tweezers have been used at the single-molecule level to observe the stepping of motor proteins,[55–58] DNA–protein complexes,[59] the unfolding and refolding of RNA molecules and proteins[60,61] and to provide information on the translation machinery by reporting on the strength of the interaction between the ribosome and mRNA,[62] its translocation along a short hairpin-forming mRNA molecule,[63] as well as the release of an arrested nascent chain.[64] In a similar vein, and of specific interest in this chapter, optical tweezers can provide unique insight into the effects of chaperones such as trigger factor (TF) and Hsp70 (DnaK) on the folding, misfolding, and aggregation of their substrate proteins.[31,65]

Optical tweezers studies also come with several challenges and considerations. A key initial difficulty is to tether the substrate protein to beads, which is typically done using dsDNA handles (see Figure 11.1). DNA handles and proteins can be attached using cysteine chemistry, enzymatic reactions, and the ligation of long DNA strands to short DNA oligos.[67–72] Performing this step with high enough yields and preserving the native folded state in non-trivial, specifically for proteins exhibiting limited solubility and stability. For attaching the DNA handles to beads one typically employs biotin-streptavidin or DIG-anti-DIG, which can be formed *in situ* within the optical tweezers instrument. One may note that the attached DNA linkers reduce the protein chain entropy, and hence can affect folding in principle. However, this entropy reduction is small compared to the other relevant energies, in particular when folding occurs in the relaxed state at zero tension (see Figure 11.2b, middle). Applied forces effectively tilt the protein's free-energy landscape.[66] This does not affect folding at zero tension, as force is then merely used to probe the folded state after folding (see Figure 11.2b, middle). Landscape tilting does affect quantitative reconstructions of folding energy landscapes of (typically smaller) proteins that fold when under tension. DNA linkers may in principle also interfere sterically, for instance with chaperone binding, as will be discussed more in the GroEL section. High forces can reduce the average tether lifetime and lead to melting of the DNA handles whilst small forces may effectively drown measured displacements in Brownian noise. Hence, most measurements are typically carried out between *ca.* 2 and 60 pN. Depending on the protein substrate of interest, folding pathways may be complex and heterogeneous, in particular when interacting with chaperones. Obtaining data of high quality and quantity is thus challenging and may benefit from automation.

11.3 Co-translational Folding

11.3.1 Nascent Chains at the Ribosome

In recent years it has become increasingly clear that the ribosome plays an active role in the folding of proteins during synthesis while also facilitating the interaction with other chaperones, such as trigger factors in *E. coli*. Although mostly treated separately, the line drawn between protein synthesis and chaperone action is becoming increasingly blurred. Here single-molecule studies in particular have uncovered some of these intriguing mechanisms. Studying mRNA translation at the single molecule level is challenging, partially due to the highly dynamic nature of the process, but also because of the large number of components and interactions involved. System complexity is a specific challenge in single-molecule investigations, as a malfunction in one component can lead to the failure of the experiment. Indeed, ribosomes are among the most elaborate molecular systems that have been studied at the single-molecule level. More specifically, the ribosome consists of 2 subunits (termed large and small), which are made up of 3 long rRNA molecules and more than 50 ribosomal proteins in prokaryotes, 4 rRNA, and 79–80 proteins in eukaryotes. This large (*ca.* 20–30 nm) molecular assembly translates mRNA with the aid of aminoacylated tRNAs and several translation initiation, elongation, and release factors. Individual amino acids are linked together within the peptidyl transferase center, from there the newly minted (nascent) polypeptide chain traverses a *ca.* 10 nm long and *ca.* 1–2 nm wide ribosomal tunnel, cutting through the large ribosomal subunit. Emerging at the ribosomal exit vestibule, the polypeptide chain can fold into its native structure. The key enabling factor for single-molecule experiments has been the availability of customizable *in vitro* transcription/translation systems from commercial sources, and the development of specific assays. For nascent chain folding studies, the latter typically involves attaching a (several) kilobase-long DNA handles to one of the ribosomal proteins, and another one to the N-terminus of the stalled nascent chain,[73,74] while active translation can be measured in real time using handles that are attached to the ribosome and the mRNA,[75] the ribosome and the nascent chain[73] or just the mRNA on its own.[63] The experiments usually involve intentional stalling of the ribosome at specific sites and/or the incorporation of artificially labelled amino acids.

Single-molecule studies of translation have yielded crucial insights into this highly dynamic and heterogenous process. For instance, recent single-molecule studies have shown how protein folding can take place during the synthesis of the polypeptide chain.[73,76] This cotranslational folding was shown to be impacted by the close vicinity of the ribosome itself, since the bulk of the ribosome is made up of negatively charged rRNA resulting in electrostatic interactions between the charged amino acids of the nascent polypeptide chain and the rRNA, causing a delay in the folding of the complete newly synthesized protein.[74] Structural studies have shown that some proteins can already form secondary structures and even tertiary structures deep within the ribosomal tunnel.[77,78] In the case of the small zinc finger domain ADR1a a recent single molecule study utilizing optical tweezers and correlated smFRET has demonstrated that folding can in fact be accelerated within the confines of the ribosomal exit tunnel.[79] Strikingly, the tunnel itself is acting like a folding-accelerating chaperone in this case. Such mechanical manipulation experiments allow the nascent chain to be unfolded, while keeping the rest of the system intact, and refolding to be followed in real time upon relaxation of mechanical tension, which is difficult to achieve by traditional refolding experiments that employ chemical denaturants and monitor the mean refolding progress within the population.

The rate of synthesis of proteins is also modulated by the electrostatic interactions between the nascent chain traversing the tunnel and the tunnel (mostly rRNA) itself. Clusters of positively charged amino acids arginine and lysine, and to a lesser extent histidine, can in fact significantly slow down the rate of synthesis.[80,81] The incorporation of multiple prolines in a row can also lead to temporary stalling of mRNA translation, which can be rescued by proline-specific elongation factors (EFP) in bacteria for instance.[82] Rare codons and mRNA secondary structures also play a role in modulating translation.[83,84] In contrast, forces applied to the nascent polypeptide chain during translation can increase the rate of synthesis by lowering the transition state free energy barrier to peptide bond formation.[85] The simple act of folding on the ribosome exerts a pulling force on the order of up to 10 pN.[86]

The protein synthesis rate plays an important role in cotranslational folding. Slowed and stalled translation gives the N-terminal and already tunnel-traversed segments of the nascent polypeptide valuable time to fold, potentially preventing aggregation with other partially translated proteins within the cytosol in polysomes for instance.[87] The inherent heterogeneity of protein synthesis, where cotranslational folding appears to be tightly coupled to the rate of translation, makes it a difficult system to study using bulk methods. Single-molecule methods such as optical tweezers are increasingly enabling assays that may soon allow us to follow the dynamics and mechanisms of the involved processes in exquisite detail and will help in disentangling the intricate interplay of the many different components and processes involved, especially when it comes to unravelling the cotranslational role of chaperones. See Figure 11.3.

11.3.2 Trigger Factor

As soon as they exit the ribosomal tunnel, nascent chains can interact with a plethora of proteins that could affect their folding pathway and dynamics, which optical tweezers can uniquely address. In bacteria, the chaperone trigger factor (TF) is directly associated with the ribosome, at the tunnel exit, and hence it is typically the first chaperone that nascent chains interact with.[88,89] Discovered in 1987,[90] TF is not essential as deletion of the TF encoding gene *tig* seems to be compensated by enhanced action of the chaperones DnaK and GroEL.[89] However, a combined deletion of TF and DnaK was found to be lethal above ~30 °C causing misfolding and aggregation of several hundred cytosolic proteins.[88] Despite sharing a lot of overlapping functions with DnaK and GroEL, the ATP-independent TF appears to engage with protein chains in a unique way, which leaves many open questions about how TF affects the folding process. Over the last decades, many techniques ranging from structural to single-molecule studies have been employed to decipher the function of TF.

The 48 kDa TF protein consists of three domains, the N-terminal, C-terminal, and the PPIase domain, namely because this domain displays catalytic activity as peptidyl–prolyl *cis*/*trans* isomerase.[89] Together these domains adopt a dragon-shaped structure, where the N-terminal domain forms the "tail", the PPIase domain the "head" and the C-terminal domain a central body, constituting nearly half of the TF molecule and residing between the N-terminal and PPIase domains with two protruding "arms".[88] TF binds to the ribosome *via* its N-terminal domain, using the ribosomal exit

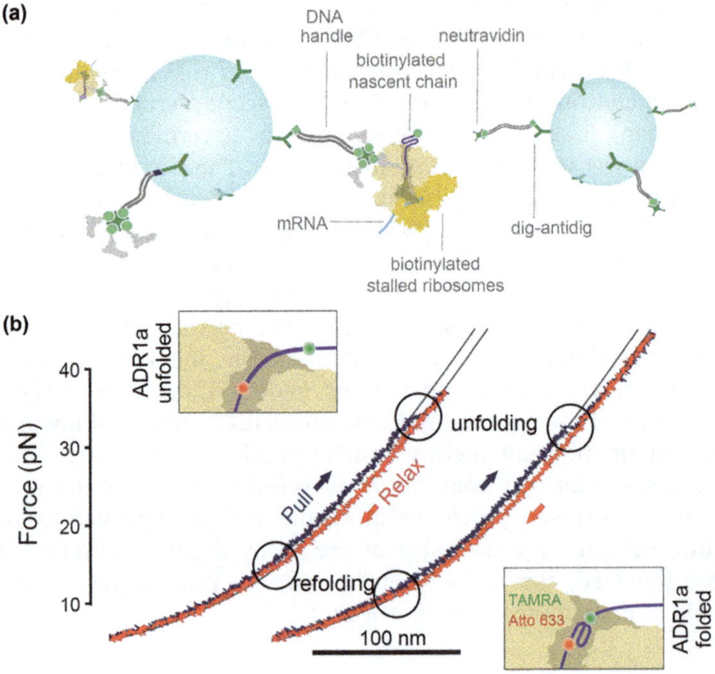

Figure 11.3 Optical tweezers study of nascent chain folding at the ribosome. (a) Schematic diagram indicating the key components and linking strategy to enable nascent chain manipulation. In the strategy indicated here, biotinylated ribosomes are first linked to beads *via* DNA handles, biotin-neutravidin, and dig-antidig linkages, and then induced to translate a nascent chain, which is biotinylated at the N-terminus and stalled using the SecM strong stalling sequence. After entry into the optical tweezer's apparatus, the final link with a second bead, *via* another DNA handle is made, hence allowing the force spectroscopy approach. (b) Two stretching and relaxation data sets on ADR1a, which is shown to unfold and refold. It showed that refolding is accelerated, when ADR1a was positioned inside the ribosomal tunnel, which is notable given the important steric constraints that the tunnel imposes.[79] By incorporating fluorescent probes into the nascent chain during translocation, simultaneous FRET detection of these unfolding and refolding events was achieved. Data reproduced from ref. 79, https://doi.org/10.1038/s42003-021-02055-8, under the terms of the CC BY 4.0 license, http://creativecommons.org/licenses/by/4.0/.

site proteins L23 and L29 as major docking site.[91] Ribosome binding stabilizes TF in an open confirmation, creating a cradle-like cavity between the ribosomal surface and the N-terminal tail, and the C-terminal arms of TF, which is about 40 Å deep and 35 Å wide.[92,93] The open cavity is thought to be linked to the function of TF, which is incompletely understood however.[88] For long it has been thought that the open cradle-shaped cavity provides a protective environment for emerging nascent chains, shielding them from unwanted interactions with their hydrophobic segments as well as with other cellular components, thereby preventing aggregation and protein degradation.[94–96] Thus, TF was believed to fulfill a classic holdase function that stabilizes unfolded states.[97]

However, single-molecule studies revealed several other functions, including the ability to bind and stabilize partially folded states and promote correct folding.[31,98] The first studies focused on fully synthesized protein substrates, away from the ribosome. In cells, TF is present in a two-to three-fold excess relative to ribosomes, suggesting that it may also play a role when not bound to the ribosome.[88] In one study, optical tweezers were used to study the impact of TF on folding pathways.[31] In data from repeated relax-wait-stretch cycles, unfolding and folding transitions are observed as increases and decreases in the extension, respectively, noting that chain segments that are extended by tension can be orders of magnitude longer than folded ones (see Figure 11.2). The maltose-binding protein MBP was found to show a distinctive (un)folding pathway, which was significantly altered in the presence of TF. Partially folded states were "visited" more frequently, for longer, and resisted higher forces, suggesting that TF directly binds the folded part of the protein chain, not just the unfolded chain.[31] At zero force, the protein chain continued to fold, observable by reductions in measured extension, and hence was not irreversibly stabilized by TF. Within tandem-repeated MBP constructs, TF increased the probability of each MBP copy to adopt the native core structure, while decreasing the probability to engage with neighboring MBP copies to form aggregated structures. Together, these data indicated that by binding partially folded structures, TF shields them from non-native interactions with distant sites along the protein sequence[31,94] (see Figure 11.4). Notably, a subsequent magnetic tweezer study showed that the interaction of TF depends on the extension of the substrate chain, as modulated by mechanical tension.[98] Here, the experimental procedure differs slightly, and for instance lets extended and unfolded proteins rapidly jump to a particular force, after which folding is monitored as extension decreases in time. Specifically, for the small globular protein L, it was found that the folding probability for an intermediate force regime (5–9 pN) is increased by up to ~40% in the presence of TF. At zero force the chaperone appeared to hinder the refolding transition, fulfilling potentially a stabilizing function.[98]

Taken together, these single-molecule results provided insights into how TF reshapes the folding free energy landscape (see Figure 11.4). The stabilization of partially folded states indicates the formation of energy minima in between the unfolded and fully folded states, or a deepening of existing ones. Such minima may in fact decelerate folding within individual substrate repeats. But at the same time, the 'spatial separation' of different substrate repeats by TF raises kinetic barriers towards aggregated states, as revealed by the tandem MBP repeats.[94] Increased folding rates mediated by TF may also be caused by an overall lowering of the folding energy barrier, which we here refer to as folding acceleration, and hence also

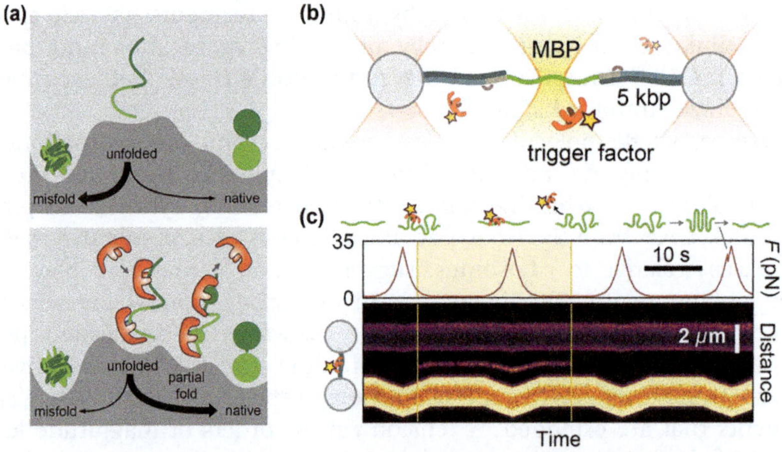

Figure 11.4 Trigger factor mechanisms revealed by substrate manipulation studies. (a) Cartoon of how folding landscapes can be reshaped by trigger factor, observed by optical tweezers study of single-domain, and engineered multi-domain substrates.[31] Top: Without a chaperone, interactions between domains lead to efficient misfolding. Bottom: Trigger factor binds and stabilizes not only the unfolded chain, but also intermediate states with stable tertiary structure, which corresponds to a formation of energy valleys, or a deepening of existing ones. Owing to the resulting protection against interactions between domains, the folding barrier to misfolds is increased in height, thus limiting misfolding between domains. Hence, native folding is promoted indirectly by limiting misfolding pathways. Trigger factor can also be seen as setting a length scale for native folds, as its binding effectively promotes local over distant intra-chain interactions. Reproduced from ref. 31, with permission from Springer Nature, Copyright 2013. (b) Schematic of combined optical tweezers and fluorescence study of chaperone–substrate interactions. A key technical challenge is to attach long DNA handles efficiently and stably, which is required to increase the distance between the substrate and the bead surface, which is typically highly fluorescent and thus perturbs fluorescence detection at the substrate location.[106] (c) Corresponding kymograph during stretching and relaxation cycles showing a single fluorescently labelled trigger factor binding and unbinding to the tethered MBP protein (thin line in the center).[106] The data are consistent with MBP refolding less frequently when the trigger factor is bound.[31] Here MBP does refold when unbound, as seen by the unfolding feature at the end of the force data in time.[106] Reproduced from ref. 106, https://doi.org/10.1038/s42004-020-0267-4, under the terms of the CC BY 4.0 license, http://creativecommons.org/licenses/by/4.0/.

impacts single substrates in the absence of aggregation. Folding energy barriers can be effectively reduced by lowering the entropy of the unfolded chain, which may be achieved by multiple contact sites on the chaperone surface, even as contacting must remain dynamic and thus preserve entropy to continue folding, yielding what has been termed a 'fuzzy complex'.[8,99] The inner surface of the TF cradle exposes hydrophilic and hydrophobic residues and continuous hydrophobic patches, which may enable TF to interact in this fashion, with substrates of diverse compositions and sizes.[96,100] Stabilization of partially folded states cannot be explained by binding to unfolded substrate segments, however. TF may instead interact with patches on the folded substrate surface, which is exposed when partially folded, and not accessible when fully folded. This idea is supported by the observation that TF does not interact with fully folded substrates in optical tweezers experiments.[31] Additionally, simulations showed that TF employs the flexible arms and polypeptide loop at its tips to interact with its substrates forming "touching" and "hugging" complexes in a dynamic way.[100]

The first single-molecule results on TF roles at the ribosome have been reported recently.[95] A newly synthesized unfolded nascent chain segment of the multidomain protein elongation factor G was found to denature an already synthesized, and folded domain. While the ribosome alone did not protect against this denaturation, TF did, presumably by limiting interactions between the different domains. TF was also found to prevent interdomain misfolding and as a result speeds up folding, in line with the 'spatial separation' model discussed above (see Figure 11.4). The data also showed intermediate states with altered molecular extension in the N-terminal G-domain folding pathway in the presence of TF. Considering the flexibility of TF it was suggested that these altered molecular extensions stem from the binding of the nascent chain to several sites within the TFs inner surface, which would reduce the extension of the protein while keeping it largely unfolded and entropy lowered, thus reducing inter-domain misfolding and promoting domain folding. It was also hypothesized that by forming multiple contacts spaced apart along the client protein chain, entropy is reduced, which in turn may also facilitate subsequent folding.[95]

These novel single-molecule findings thus extend the picture that TF acts as an unfolded chain holdase that decelerates folding. NMR relaxation experiments had indeed shown binding to several distinct regions within TF's inner surface, suggesting TF keeps substrate proteins in an extended, unfolded conformation.[96] Earlier crosslinking studies had indicated that growing nascent chains initially follow a predefined, domain wise, path through the entire interior of TF in an unfolded conformation, which indicated that TF can act as a holdase.[89] Another NMR study even indicated an unfoldase role of TF. By using all its substrate binding sites, TF may be able to unfold transiently formed structures.[92,96] However, since TF cannot use ATP, it is assumed that its unfolding activity is restricted by the intrinsic thermodynamic stability of the substrate.[92] The ability of TF to form dimers has been highly debated.[92,101,102] The dimers may represent inactive storage, as it occludes substrate binding and prevents promiscuous binding of TF.[102–104] TF dimerization potentially could also impose a substrate selection filter, with only high-affinity clients able to bind. Depending on size, folding state, and amino acid composition, the emerging nascent chains increase the affinity of TF for ribosomes by about 2 to 30-fold ($K_D \sim 40$–700 nM).[94,103,105]

In conclusion, single-molecule optical tweezer studies have added a new dimension to the bulk-biochemical, NMR, and structural insights, owing to its ability to follow large conformational changes and folding pathway dynamics. Since TF acts in a dynamic reaction cycle governed by translation, with a half-life of TF-ribosome nascent chain complexes of about 15 to 50 s, the detailed dynamics of how TF engages nascent chains are difficult to unveil.[92,94,95,101,102] Optical tweezers combined with fluorescence detection of chaperone binding will be a useful tool to address these questions.

11.4 Post-translational Folding

11.4.1 The Hsp70 System

The 70 kDa heat shock proteins, termed Hsp70 in mammalian cells and DnaK in bacteria, are a ubiquitous family of chaperone systems that are highly conserved across species and thought to mainly interact post-translationally, though co-translational engagement has been reported in certain cases.[107] They are upregulated in response to temperature and other types of environmental stress, yet also play key roles in normal physiology, as revealed by bulk experimental techniques.[108] Hsp70 chaperones are important to *de novo* folding and many other cellular processes, though the mechanisms of action remain incompletely understood. They aid in the translocation of proteins across membranes, regulate the activity of several key enzymes, and can disaggregate certain protein aggregates. Its numerous vital physiological roles are emphasized by the involvement of Hsp70 in many human pathologies ranging from cancer to protein aggregation diseases.[109–111] DnaK is composed of an N-terminal nucleotide-binding domain and a C-terminal substrate-binding domain that contains a moveable 30-amino acid lid. DnaK–peptide binding is regulated by the ATP cycle: when ATP is bound the chaperone exists primarily in its open conformation. Substrate association and dissociation rates are high and affinity is low. Unfolded peptide segments can bind in the substrate binding cleft, which stimulates ATP hydrolysis, in synergy with the co-chaperones (DnaJ and GrpE in *E. coli*). In the resulting ADP-bound state, the lid is predominantly closed. The association and dissociation rates are then significantly lower, resulting in a highaffinity.[112,113]

Bulk measurement techniques, however, are less suited to elucidate folding dynamics and intermediate folded states. Such intermediate states typically occur only transiently, in a non-synchronized manner within the population, and cannot readily be stabilized to study them, as for instance is possible with unfolded states using chemical denaturants or temperature. These issues leave key questions at the core of DnaK functioning openly. It has long been thought that DnaK interacts exclusively with unfolded substrate segments. But can DnaK also affect substrates at later stages of folding, as for instance observed for trigger factor in the previous section? Does it mainly protect against aggregation or do more? These questions motivated several force-spectroscopy studies at the single-molecule level.[29,30,114]

However, the involvement of co-chaperones and the ATP hydrolysis cycle posed additional difficulties. The latter is in fact more of a challenge in single-molecule studies, as any binding or conformational event causes heterogeneity in the temporal dynamics at the molecular level, which rapidly become impossible to disentangle, whereas such events are typically averaged out or invisible in bulk assays.

Hence, the single-molecule investigations followed a bottom-up approach, using different model substrates.[29,30,114] One study[30] focused on the maltose binding protein (MBP) constructs also employed in the trigger factor investigation: a single MBP (1MBP) and a tandem-repeat construct (4MBP). In the absence of DnaK, mechanically unfolding 4MBP with optical tweezers and subsequent relaxation to low forces in order to attempt folding, rather yielded misfolding. The latter was seen by subsequent stretching of the construct, which often showed compact structures that either failed to unfold 4MBP (tight misfold) or unfolded in steps larger than expected for one MBP repeat (weak misfold), indicating that small aggregates had formed. Addition of the DnaK chaperone system (DnaK, DnaJ, and GrpE) and ATP buffer showed that the unfolding and relaxation cycles rarely led to tight misfolds, while weak misfolds still formed, and native-like refolding of single MBP repeats increased. These results were consistent with bulk studies, which showed that the DnaK system can promote folding by limiting aggregation.

To probe underlying mechanisms while limiting temporal heterogeneity, DnaK was studied in distinct nucleotide states, in the absence of co-chaperones, acting on the single MBP construct.[30] In the ADP state, unfolded MBP showed a very low refolding rate, consistent with bulk and single-molecule FRET studies,[115] which showed that DnaK–ADP can stabilize unfolded states. More notably, DnaK–ADP was found to also stabilize near native refolded MBP substrates. These compact structures could not be unfolded when stretching up to 65 pN, which is the technical maximum force of the assay (see Figure 11.5a and b). This DnaK-induced stabilization was not observed for fully folded native states of MBP, or for DnaK in the ATP-bound and APO states. Another single-molecule study, which used AFM for the mechanical manipulation of poly-ubiquitin substrates, interestingly also found that DnaK–ADP could stabilize a folding intermediate.[29] Here, the folding intermediate rather appeared to be a molten globule state without a stable tertiary structure. The interactions thus effectively stabilized unfolded conformations, consistent with the MBP study. Consistently, both studies thus revealed that DnaK can bind and stabilize compact substrate conformations (see Figure 11.5c and d), which cannot be explained by the canonical model, and shows that DnaK acts along the folding pathway rather than only at the beginning.

The observed stabilization of folding intermediates raises the question of which part of DnaK is responsible. Truncation of the DnaK lid was found to abolish this stabilization, directly showing its importance in this interaction. Conversely, mutating the groove (DnaK-V426F), which is known to reduce

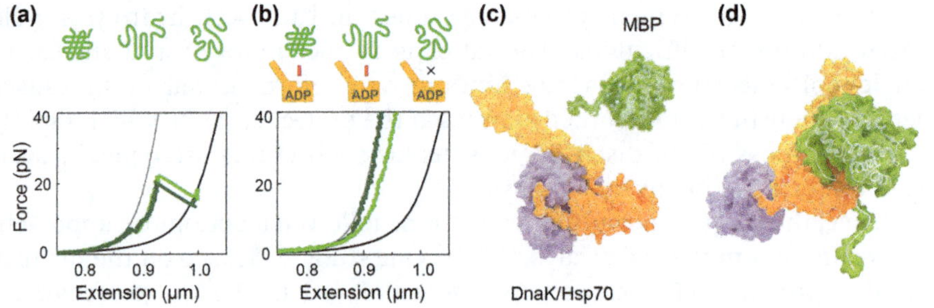

Figure 11.5 Stabilization of near-native states by DnaK. (a) Typical stretching curves of maltose binding protein after refolding, showing a fully folded state that unfolds, first into a partially folded core state, and subsequently into the fully unfolded state.[30] (b) In the presence of DnaK and ADP, the refolded states can be stabilized, as shown by the lack of unfolding.[30] (c) Structural representation of DnaK in the open conformation and near native MBP. (d) Impression of near-native MBP stabilized by bound DnaK in the ADP state, when it is predominantly in a conformation with the lid domain (yellow) in the closed conformation. Data on DnaK mutants show that both the lid domain and the peptide binding groove in the orange domain play a role in the stabilization.[30] Data reproduced from ref. 30 with permission from Springer Nature, Copyright 2016.

peptide affinity by over 10-fold,[116] did allow for stabilization, though after exposing the substrate for longer periods of time. Moreover, stabilization appeared less strong, as the unfolding forces, while higher than without chaperones, were lower than for *WT* DnaK and occurred *via* folding intermediates. Consistent with this weaker peptide affinity, MBP refolding was re-established, in contrast to *WT* and lid truncated DnaK. Testing the 4MBP substrates indicated that the groove mutant was still able to suppress tight aggregation and aid in limiting the stability of weak misfolds, consistent with the idea that the lid interacts with substrates directly, rather than merely closing the groove. The DnaK lid and groove do act cooperatively, however, as the groove mutation did limit the efficiency of stabilization (see Figure 11.5d). It should also be noted that stabilization against applied force differs from thermodynamic stabilization yet does reveal direct interactions between DnaK and the surface of the near-native folded structure.

DnaJ is another key part of the DnaK system. DnaJ is a dimer of two 40 kDa subunits that can bind and stabilize unfolded proteins, while also interacting directly with DnaK. DnaJ alone was found to suppress MBP and polyubiquitin refolding,[29,30] consistent with its known holdase role. The polyubiquitin study showed that DnaJ recognizes substrates in a force-dependent manner. The measured suppression of substrate refolding became more efficient for larger forces, maximizing at about 160 pN, and then decreased again for higher forces, indicating that DnaJ recognizes a cryptic sequence in extended substrates. One AFM single-molecule study however reported DnaJ rather promoted the refolding of ubiquitin repeats within poly-ubiquitin constructs.[114] However, DnaJ at the same time limited a misfolding that likely involved interactions between neighboring ubiquitin repeats. Hence, ubiquitin repeats bound by DnaJ may be prevented from aggregating with surrounding repeats, hence allowing the latter to form their native structure. DnaK and DnaJ together were more capable of suppressing misfolding and promoting native folding than either chaperone alone,[29,30,114] consistent with their joint presence *in vivo*. This also means that DnaJ is not dominant in its stabilization of unfolded states, which may indicate that DnaK can promote DnaJ release.[117] However, refolding rates of single ubiquitin and MBP were not shown to be increased by the DnaK system, which could mean that their main role is to prevent aggregation or may also reflect the comparatively high autonomous refolding efficiencies of these specific substrates. The most optimal polyubiquitin refolding was observed for a 1 : 2 ratio of DnaJ over DnaK.[114]

Overall, the single-molecule investigations have opened a hidden layer of substrate interactions and conformational interplay and produced fundamental extensions of the Hsp70 canonical model. The findings also suggest many new open questions and lines of study. For instance, the structural basis of Hsp70 in complex with folding intermediates remains elusive, which could be addressed by NMR. Hsp70 is also known to be a potent disaggregase, as studied recently.[118] The disaggregation, with the underlying entropic pulling mechanism,[119] and folding promotion roles of Hsp70 appear tenuous, and how they relate remains to be elucidated. Stabilization of folding intermediates by Hsp70 may be important to this issue. Hsp70 could aid disaggregation by competing for sticky patches on the surfaces of incompletely or misfolded substrates, which can become exposed in the dissolution process driven by Brownian motion but also tend to aggregate. Hsp70 could promote folding by transiently capturing and releasing key folding intermediates, driven by ATP hydrolysis, and one can speculate about other mechanisms. More generally, it is unclear whether Hsp70 can lower folding barriers, rather than only raising barriers to aggregated states. These are important open topics for which single-molecule methods are ideally suited. The same holds for elucidating the role of Hsp70 during translation at the ribosome, its dynamic interactions with other chaperones including GroEL, Hsp90, and the disaggregase Hsp100, and its crucial role in many regulatory pathways and cellular malfunction.

11.4.2 GroEL–GroES

Many proteins fail to fold autonomously, with the folding landscape driving them to aggregated, non-native states.[120,121] Chaperones are broadly considered to assist the folding of proteins by shielding them from aggregation interactions.[122] Strikingly, it has remained unresolved whether and how chaperones can also lower folding barriers and accelerate the folding process even in the absence of aggregation. Elucidating this question is difficult. Folding barriers are typically considered kinetic in nature and perhaps also for that reason poorly understood for proteins of typical sizes. This more direct folding assistance route would also suppress aggregation, by limiting the time spent in unfolded, aggregation-prone states.

The very notion that protein folding can be assisted at all was introduced with the discovery of the barrel-shaped chaperonin complexes,[123] with GroEL/GroES from *Escherichia coli* is the most well-studied example. It is known to confine proteins within its unique cylindrical structure, which provides an environment that is conducive to folding. The GroEL structure consists of two heptameric rings of 14 identical 57 kDa subunits, with a large cavity. GroES is a small ring-shaped oligomer made of seven 10 kDa size subunits, that caps the GroEL cavity in ATP dependent manner.[124] Numerous biochemical[125] and structural studies provided a wealth of information on the GroEL–GroES reaction cycle.[126] Substrates initially bind the hydrophobic apical domains of GroEL and binding of ATP and GroES then release the protein into the cavity where it folds in isolation. ATP binding to the other ring then release GroES and the folded protein from the cavity by intra and inter-rings allosteric signaling.[127] However, the core folding mechanism of GroEL remains debated.[128] A key issue is whether GroEL acts as a passive protein aggregation inhibitor[129] or whether it actively accelerates folding, as discussed above, and if so what mechanism is employed. The notable GroEL closed cavity has suggested that unfolded substrates are sterically confined, which can lower their configurational entropy and hence lower the energy barrier, or alternatively GroEL could promote folding by actively unfolding misfolded conformations.[130,131] Moreover, various questions regarding the functional GroEL–GroES ATP cycles such as the substrate-accepting state of GroEL, the folding competent state, and the precise role of GroES are unanswered.

Recently, optical tweezers were used for the first time to study folding by GroEL–GroES at the single molecule level[132] (see Figure 11.6). Force spectroscopy measurements were performed on wt-MBP and a slow folding dm-MBP variant, in which they were repeatedly unfolded by stretching them at the N- and C-termini, followed by relaxation to zero force to attempt folding during typically 5 seconds, and finally stretching again to assess the new folded state. These experiments indeed showed clear increases in the folding rate with GroEL–GroES present. The effect was particularly strong for dm-MBP, whose core refolding frequency went from less than 10% to about 90% of repeatedly performed relax-wait-stretch cycles. Such an acceleration of folding had thus far not been observed for GroEL or other chaperones and single protein substrates in the absence of aggregation interactions.

To elucidate underlying mechanisms, an approach of freezing the chaperone in its main nucleotide states was used, as was described for DnaK in the previous section. Surprisingly, GroEL in the ADP and ATP-bound states, but without GroES, was also able to accelerate folding, for wt-MBP, dm-MBP as well as for rhodanese protein. The real-time monitoring of optical tweezers proved useful in understanding the underlying mechanism: even before arriving in the relaxed state, the tethered substrate already started to gradually compact as the force was decreasing (see Figure 11.6b). Hence, the

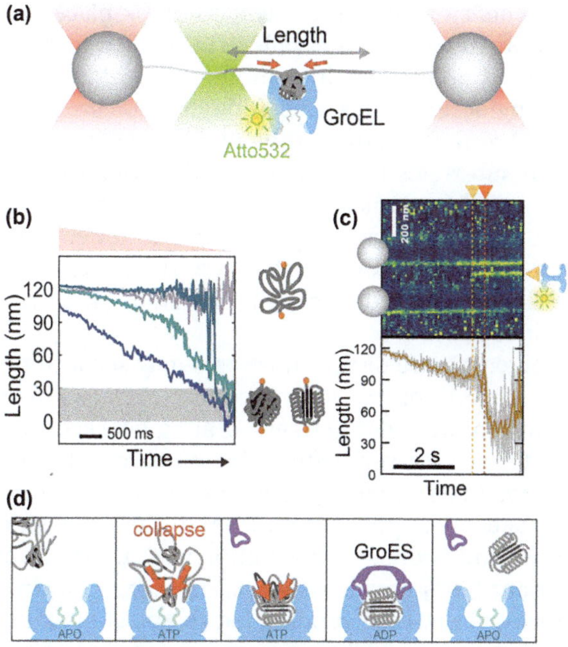

Figure 11.6 GroEL-mediated acceleration of protein folding by enhancing the collapse.[33] (a) Schematic of joint tweezers-fluorescence setup, which visualizes GroEL binding, while monitoring the length of the tethered protein substrate. Red arrows indicate detected forces that drive the collapse and folding of the substrate within the GroEL cavity. (b) Corresponding contour length in time, showing both gradual and stepwise decreases that indicate collapse and folding triggered by GroEL, for three different dm-MBP molecules with GroEL and ADP. Gray trace: data without GroEL. (c) Kymograph of confocal fluorescence, which scans across the beads and tethered protein. Individual yellow pixels result from freely diffusing GroEL molecules, continuous lines from left to right indicate fluorescence from the bead surface, and the line from the middle to the end, marked by yellow triangles, indicates GroEL binding to the tethered substrate. Corresponding contour length plot of MBP (below) shows the folding step (see red triangle) occurring after GroEL binding (yellow triangle). (d) Model resulting from the optical tweezers data. Unfolded substrates initially bind the apical domains at the GroEL rim, while the unbound substrate segments are drawn into the GroEL cavity by forces that strengthen its general collapse and folding. GroES binding to the apical rim displaces the bound substrate segments, which are then free to collapse and complete folding. ATP drives the substrate and GroES binding and release cycle. Data is reproduced from ref. 33, https://doi.org/10.1126/sciadv.abl6293, under the terms of the CC BY 4.0 license, https://creativecommons.org/licenses/by/4.0/.

"collapse" of the substrate proteins, which can be discerned without chaperones but in a much weaker fashion, was enhanced by GroEL. Moreover, the stronger the collapse, the higher the subsequent probability of folding.

Control experiments showed that the collapse took place within the GroEL cavity, as truncating the C-terminal tails of GroEL led to decreased collapse and folding. Using fluorescence in combination with tweezers, it could be shown that folding occurred just after a single fluorescently labelled GroEL chaperone bound to the substrate, thus providing direct evidence of accelerated folding of substrates in complexes with GroEL (see Figure 11.6c). In addition, the roles of the apical domains became clearer by using small peptides that bind it competitively, and hence lower substrate affinity. Consistent with acceleration occurring inside the cavity rather than on the apical rim, folding acceleration was still observed. These results were notable given previous work:

First, the tethered nature of the substrates may seem to preclude optical tweezers' manipulation of substrates within a closed GroEL–ES cavity. However, the data showed that hermetic sealing between GroEL and GroES is not required for acceleration within the cavity, and hence that the DNA handles or the protein terminal segments can be sandwiched between GroEL and GroES, consistent with previous observations of polypeptides positioned in between GroEL and GroES.[133] This notion was supported by experiments on a GroEL variant (SR1) that locks GroES in place in a single binding reaction and showed that GroES binding led to apical domain release of the substrate, refolding, and tight stabilization – consistent with the substrate constrained within a closed cavity. In bulk, SR-1 in the ADP state did not show substrate release and folding.[125] This lack of refolding may be due to GroES binding inefficiently to ADP-saturated SR-1 when the substrate is already bound to the apical ring, or that full release of the protein and subsequent folding then fails. Encapsulation can be forced for certain substrates with ADP though without full release and folding.[134] Hence, even though ADP will support GroES binding to GroEL-APO or SR1, if there is substrate protein on the ring first, GroES may not efficiently bind over (large) substrate proteins, or not properly release the protein.

Second, folding acceleration by GroEL alone had not been observed before. Yet, previous work is not inconsistent, as different states and transitions are detected. Specifically, the substrates refolded by GroEL alone were near-native and still complexed with GroEL, shown by increased unfolding forces as for DnaK–ADP in the previous section, rather than fully native and released, which is needed for detection in bulk assays.[24,125,131,135] Indeed, the averaged refolding signal in the latter typically evolves over minutes,[125,135,136] whereas single molecules are observed to fold within seconds of contact. Single-molecule FRET is closer to the optical tweezers data, as it also allows for the detection of sub-populations that follow different pathways. For instance, the ATP and ADP states of GroEL alone consistently showed bimodal distributions of FRET efficiencies for labelled substrates.[136] On the other hand, single-molecule FRET cannot follow GroEL interaction and folding events in time, nor easily distinguish between partially folded and collapsed states, which is key to showing accelerated formation of near-native states.

Third, the observed acceleration mechanism differs fundamentally from previous models (see Figure 11.6d). Enhancing the collapse brings residues together, which must contact to fold, and hence lower folding barriers. At a microscopic level, the hydrophobic effect is thought to drive polypeptide collapse, and it has been shown that it can be altered by nearby surfaces, and it would be of interest to study such water-related phenomena in the context of chaperones.[137–140] Interestingly, substrates in complex with other chaperones including Spy and Skp have also been found in compact and dynamic molten globule states using NMR,[8,141] suggesting that collapse enhancement could be a more general chaperone feature. For GroEL–GroES, various other mechanisms have been proposed. Lin *et al.*[23] observed a compact and expanded intermediate state of Rubisco using ensemble FRET, in line with GroEL promoting folding by unfolding misfolds. Compact intermediates were also observed by Chakraborty *et al.*[131] dm-MBP in bulk assays, as well as eight-fold enhanced folding rates in the presence of GroEL–GroES, explained as GroEL–GroES lowering the folding barrier by reducing the entropy of the unfolded substrate. Other data instead suggested that aggregation suppression by GroEL is central to observed increased dm-MBP refolding,[142] while experiments at very low (100 pM) dm-MBP concentrations that preclude aggregation still showed acceleration.[143] An active folding role of the GroEL cavity was also confirmed by Weaver *et al.*[144] who studied another stringent substrate PepQ, using fluorescence and cryo-electron microscopy. It was also found that the amphiphilic GroEL C-termini tails inside the cavity directly interact with the PepQ folding intermediate deep inside the GroEL cage.[23] Using single-molecule FRET, Hofmann et. al[145] reported no change in the folding rates of the N and L regions of rhodanese with and without GroEL–GroES, while the C domain folded two orders slowly in the presence of GroEL–GroES. However, Priya *et al.* reported a higher folding yield of the freeze–thaw rhodanese in the presence of GroEL–GroES, due to the unfolding of the misfolded proteins.[135] These and other folding mechanisms are not mutually exclusive and could act in conjunction.

The dual-barrel nature of GroEL and resulting dual GroES binding sites are also notable structurally. By fluorescently labelling the slow ATP hydrolyzing D398A GroEL variant, Takei *et al.*[146] used single-molecule total internal reflection microscopy to localize GFP proteins at GroEL–GroES sites. The authors demonstrated the formation of symmetric or football complexes[147] and observed the simultaneous folding of two GFP molecules within two GroEL–GroES complexes. The data showed that the football complexes were not short-lived intermediates and were observed in the presence of both denatured and folded substrates. Whereas dual color fluorescence correlation spectroscopy measurements[148] showed that these complexes were appreciably detected only in the presence of unfolded substrates while in the presence of folded or in the absence of proteins they were rarely observed. Noshiro *et al.*[149] used high-speed AFM imaging and also confirmed the formation of symmetric complexes. Interestingly, the authors reported that such complexes are not short-lived intermediates but have lifetimes ranging from 1.5 to 2.6 seconds and their frequency only moderately depends on the protein conditions.

In summary, single-molecule studies have provided a new view of how polypeptide chains can be impacted by GroEL–GroES. They indicate that true folding catalysts exist, with catalysis being enabled by strengthening the collapse. The ability to monitor substrate conformational changes inside the GroEL cavity opens up the possibility to study other substrates, GroEL–GroES dynamics, the role of GroEL in co-translational folding, as well as the human homolog TRiC/CCT. This optical tweezers approach may also be used to elucidate the role of collapse modulation in protein phase separation and more broadly throughout the protein quality control machinery.

11.4.3 Hsp90

Heat shock protein 90 (Hsp90) is a highly conserved ATP-dependent chaperone that is essential in eukaryotic cells. It is a highly abundant protein involved in protein folding that often works in tandem with Hsp70[150] to promote protein folding efficiency.[151] In contrast to Hsp70, Hsp90 serves a narrower clientele which includes 60% of the protein kinases, 30% of ubiquitin ligases, and 7% of transcription factors. Hsp90 often shows important functional interplay with Hsp70 and typically recognizes highly unstable proteins.[152] For example, in a study of FANCA (Fancomi anemia group A) mutants it was shown that mutants that preferably bind Hsp90 over Hsp70 show a less severe phenotype than mutants binding Hsp70.[153] Hsp90 is important for protein homeostasis, for example during heat shock but also plays an important role during cell growth and profileration[154] which resulted in the development of Hsp90 inhibitors as anticancer agents.[155] Despite these important cellular roles, not much is known about the conformational changes that Hsp90 imparts on its substrates, indicating a need for single-molecule investigations.

Hsp90 consists of three domains – an ATP binding domain, a client and co-chaperone binding domain, and a dimerization domain. The folding of Hsp90 was used as a model for the folding of multi-domain proteins and studied by manipulation with optical tweezers.[156] The study described the folding kinetics of monomeric yeast Hsp90 (Hsp82). They found that the dimerization domain folds in line with the classical two-state model of folding of a small protein. In contrast, the client binding and ATP binding domain show intra- and inter-domain misfolds slowing down the folding of Hsp90 significantly. An optical tweezers study of the same group in 2018 revealed similar folding rates of bacterial Hsp90 (HtpG) and yeast Hsp90, while the ER paralog Grp94 folding is slower and has altered highly stable charged linker which provides a more rigid interface between the ATP binding domain and the client/co-chaperone binding domain. These notable findings on the role of domain interactions during the folding of this important chaperone raise key questions about its function.[157]

Hsp90–client interactions have proven to be notoriously enigmatic, even as recent models provide fresh structural insight. Clients were found to be unstructured for tau,[158] semi-unfolded for Cdk4,[159] or in the near-native folding state for the Glucocorticoid Receptor,[160,161] highlighting different modi operandi of Hsp90. An optical tweezer study of the effect of bacterial and human Hsp90 on clients shines more light on Hsp90 molecular function.[162] In contrast to previous mentioned studies, optical tweezers were used to study how the client conformation is affected, by interactions with Hsp90. Compared to the chaperone–substrate systems discussed in the other sections, a specific challenge here is the typically disordered and structurally heterogenous nature of the Hsp90 clients. This not only renders the biochemistry of DNA handle and bead attachments far more inefficient and complex, but also introduces the need to consider conformational ensembles rather than the classic chaperone task of achieving distinct folded states. Studies were performed mainly on Luciferase, and its interaction with the bacterial Hsp90 (HtpG), as well as on the wild-type receptor ligand binding domain of the human glucocorticoid receptor, and its interactions with human Hsp90. During the relaxation of mechanically unfolded client chains, it was found that Hsp90 can promote a gradual contraction of the client chain, in an ATP-stimulated manner and despite counteracting forces. Interestingly, the Hsp90-induced gradual contractions are reminiscent of the collapse enhancement seen for GroEL in the previous section, though the dependence on ATP hydrolysis also sets it apart. Polypeptide collapse itself is thought to be key to autonomous protein folding.[163–165] The Hsp90 interactions were also shown to suppress misfolding and aggregation, which typically involve contacts between more distance residues. The experimental findings thus indicate a model where Hsp90 stimulates local intra-chain interactions within unstable clients and suppresses distant ones. Such promotion of local interactions has also been proposed as a mechanism of action for trigger factor[31,100] and DnaK,[30] as discussed in earlier sections of this chapter. In contrast to these chaperone systems, however, the local folds induced by Hsp90 can be significantly smaller (tens or residues or less). Another difference with trigger factor is the reliance of Hsp90 on the ATP hydrolysis cycle,[100] though the precise role of the latter in substrate compaction is still unclear. It also remains an open question what the structural basis is of these Hsp90-induced conformers, how heterogeneous and dynamic they are, and how and where they interact on the Hsp90 surface. The role of these conformational effects at the single-molecule level on the interplay between Hsp90 and Hsp70 is also a major new chapter that remains unresolved.

Probing the dynamics of proteins that are conformationally unstable and heterogeneous, and closer to being disordered, as well as their interactions with the protein quality control machinery, is one of the next challenges for single-molecule protein manipulation. Overcoming the technological hurdles is relevant, given the many interesting and fundamental open questions we have about these protein systems, and their importance in various medical conditions.

11.5 Controlling Protein Aggregation

11.5.1 Small Heat Shock Proteins

Small Heat shock proteins (sHsps) are a highly diverse family of molecular chaperones. They are characterized by an α-crystallin domain (ACD) of 90 amino acids, flanked by N- and C-terminal extensions.[166,167] During stress conditions, sHsps are the first line of defense. They co-aggregate with misfolding proteins to prevent the formation of highly stable aggregates.[168,169] When the period of stress has passed, sHsp-associated protein substrates can be re-solubilized and refolded by ATP-dependent disaggregation chaperones Hsp70 and Hsp100.[170,171]

Among the various species of sHsps, manipulation studies of protein substrates at the single-molecule level have been performed on Hsp42, which is one of the two sHsps of yeast.[169] In this study, a construct of four maltose-binding proteins arranged in tandem (4MBP) was unfolded and allowed to refold in the presence and in the absence of Hsp42, using optical tweezers. As shown in the previous sections on trigger factor and DnaK, the 4MBP construct effectively provides high local protein concentration, and can mimic the formation of mini aggregates. Mechanically stretching natively folded 4MBP yields a gradual unfolding transition followed by four distinct unfolding events. The former corresponds to the C-terminal α-helixes unfolding and detaching from the core MBP structure, and the latter to the unfolding of the four MBP core structures, as characterized by a change in contour length of 92 nm and an unfolding force of 25 pN, on average. In the absence of a chaperone, upon relaxation to low force, the exposed internal segments of different MBP repeats find a chance to interact with each other and form non-native contacts between them. The subsequent stretching pulls show two kinds of aggregated structures: (1) compact structures that withstand forces over the experimental maximum of 65 pN, and are termed "tight aggregates", and (2) compact structures that do unfold, but then release a chain segment that exceeds the length one MBP core structure, referred to as "weak aggregates".

In the presence of the Hsp42 chaperone, protein aggregation within the 4MBP construct was found to be partially suppressed. Specifically, the tight aggregates that were abundant in the absence of chaperones were no longer detected, even as weak aggregates were still observed. In addition, a substantial increase in the formation of native-like core structures was detected. The latter is notable, as current models in which sHSPs suppress aggregation by binding and stabilizing non-folded protein segments would rather suppress any tertiary structure formation. To further investigate the nature of these native-like core structures, the authors probed a single maltose binding protein construct (sMBP). The results showed that core structures formed with Hsp42 present are less force-resistant than the native core state, with a mean unfolding force of ∼15 pN. These data suggest that Hsp42 directly interacts with near-native partially folded structures of its substrate. Subsequent hydrogen exchange mass spectrometry confirmed the ability of Hsp42 to stabilize near-native substrate states. Through its interactions, Hsp42 could interfere with native intramolecular substrate contacts, and hence prevent conversion to the fully native state, yield a lower resistance to forced unfolding, and limit aggregation by physically separating sequestered protein monomers.[169]

Another chaperone that was observed to bind near-native structures is Hsp33.[172] Like sHsps, Hsp33 is an ATP-independent chaperone that suppresses aggregation. However, it is redox-regulated and undergoes major conformational changes upon exposure to oxidative stress. Active Hsp33 has intrinsically disordered regions that provide a binding site to unfolding substrate proteins.[173] Optical tweezers studies on the same 4MBP construct in the presence of an increasing amount of (active) Hsp33 showed that a significant part of the 4MBP chain now remained unstructured. These data indicated that Hsp33 binds extended polypeptide chains and suppresses both aggregation between MBP repeats and folding transitions within repeats. Interestingly, they also showed significant increases in the relative proportion of MBP cores over aggregates during refolding, which indicated a form of selectivity for native-like structures. To disentangle aggregation and refolding effects, the authors probed sMBP constructs in the presence of Hsp33. It showed a decreased refolding rate, consistent with binding and stabilizing unfolded conformations. The sMBP and 4MBP data could be described using a statistical physics model that systematically defines the ensemble of states of folding and Hsp33 binding that the 4MBP system can be in. It indicated that native folding could be promoted effectively, by limiting the options of unfolded segments to aggregate, rather than promoting folding directly, as for instance observed for GroEL–ES in a previous section of this chapter. In addition, the data also showed that the native-like core-structures were smaller in size than the native core fold. One may speculate that these intermediate states are stabilized by the competitive binding of the intrinsically disordered regions of Hsp33.

A recurring theme hence is that these anti-aggregation chaperones not only bind unfolded, but also near-native conformations, manifested either by reduced stability against forced unfolding or a reduced size of near-native structure. These capabilities allow the chaperones to interact early along the unfolding pathway and could protect them from further unfolding, which may be a more efficient way to limit aggregation but could also accelerate refolding of the native structure once the cellular stress is relieved. The findings raise questions on the elementary differences and similarities within the large family of sHSPs.[174] One may speculate that different HSPs selectively act on certain types of substrates based on the unfolding pathway and early unfolding intermediates of the latter. The ability to probe substrate conformations in a wide range, from unfolded state to partial fold and multi-protein mini aggregates will be a useful tool to dissect their modes of action.

11.5.2 The ClpB Disaggregase

Beyond assisting protein folding and preventing protein aggregation, some chaperones have evolved the important ability to rescue and reactivate proteins trapped within aggregates.[13] In metazoans, this task is performed by Hsp70, Hsp110 and a complex network of J-proteins.[175] While their disaggregation mechanism remains largely unknown, recent studies suggest that it relies on entropic pulling.[107,118,176–178] In bacteria, protein disaggregation is mediated by a highly specialized chaperone, ClpB. This donut-shaped protein belongs to the AAA+ class of translocases and consists of an N-domain that binds its substrates, a middle domain that regulates its ATP activity through interaction with DnaK, and a C-terminal domain featuring its catalytic activity.[179] Regulation of ClpB activity is crucial for cell homeostasis, as evidenced by the fatal toxicity of unrepressed mutants,[180] yet the details of the interaction between DnaK and the ClpB middle domain are not fully understood. Recently, single-molecule FRET was used to reveal that the middle domain dynamics were remarkably fast, allowing precise tuning of ClpB activity.[181]

Despite the central role of ClpB in protein homeostasis within bacteria, its underlying mechanism has remained controversial until very recently. The processivity of other structurally related AAA+ translocases, such as ClpX and ClpA, was demonstrated a decade ago using optical tweezers,[182–184] but could not be established for ClpB. In fact, a series of studies using stop-flow suggested that ClpB and its yeast homologous Hsp104 were non-processive and could only translocate a couple of substrate amino acids at a time before disengaging.[185,186] Instead, it was proposed that ClpB could function similarly to the metazoan machinery through entropic pulling or Brownian ratcheting, also consistent with structural studies.[187] Remarkably, bulk studies have also shown that ClpB can process aggregates flanked by stable, folded proteins.[188] The authors suggested that the chaperone can accommodate several polypeptide chains within its central pore simultaneously – despite the lack of supporting structural evidence – and hence process the substrate by threading a polypeptide loop. At the same time, a recent single-molecule AFM study showed that the ring can open and close dynamically,[189] indicating that ClpB can wrap itself around internal segments of polypeptide chains, without the need for open ends. This finding offers an alternative explanation for the bulk data without the need to invoke multiple-chain translocation.

Unambiguous evidence for both processivity and loop extrusion was recently obtained using a combination of optical tweezers and single-particle fluorescence tracking, at a resolution below the diffraction limit.[44] In the study, a substrate protein (maltose binding protein or MBP) tethered between two trapped beads was first unfolded and then relaxed to a low force that prevented refolding. Of note, and in contrast to previous optical tweezers studies on translocases, the substrate did not present any free ends in the chosen configuration. This state can be argued to resemble a

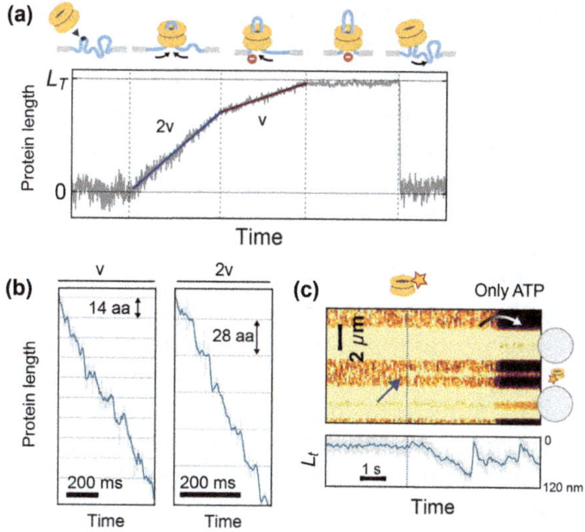

Figure 11.7 Processive substrate translocation by the disaggregase ClpB. (a) Length of the looped and translocated part of the protein chain, as determined by measured decreases in overall distance between the substrate N- and C-termini.[34] After ClpB binding, the translocated length increases continuously until the entire chain is translocated, showing processive action. The speed halves at a certain moment, consistent with a switch from translocating two arms of the loop to one arm when ClpB encounters the DNA handle on the other arm, until ClpB also encounters the second DNA handle and arrests fully. (b) High-resolution data showing that translocation occurs in steps, with the step size being two-fold smaller when the translocation speed halves.[34] These data are consistent with both arms translocating simultaneously through the central ClpB pore. (c) Fluorescence kymograph showing fluorescent ClpB binding (blue arrow), whose position can be used to determine the translocation of each arm independently[34] (not shown). Medium is switched from with labelled ClpB to without (only ATP) to reduce background. L_t indicates translocated length. Data is reproduced from ref. 34 with permission from Springer Nature, Copyright 2020.

native ClpB substrate, as the termini of an aggregated protein are more likely to be buried inside the aggregate – thus unavailable, like in the tethered substrate –, while internal segments are exposed on the surface. The addition of ClpB and ATP to the tethered substrate resulted in gradual contractions and sudden releases in the substrate length, indicative of processive protein translocation (see Figure 11.7a).

How can ClpB translocate substrates that do not present free ends? While the chaperone could transiently open in order to engage on an internal polypeptide segment, translocation would then not produce contractions in the used optical tweezers setup (as an analogy, a ring sliding along a string held by both ends would not generate any tension or pulling). To decipher the translocation topology, the authors implemented fluorescence detection together with force spectroscopy measurements. By imaging the position of a single fluorescently labelled and bound ClpB, one could measure its distance from both termini of the extended substrate – and hence deduce the length of the translocated part of the substrate (see Figure 11.7b). These dynamic data revealed that ClpB can indeed thread a polypeptide loop, like threading a string through the eye of a needle, and proceeds to translocate the loop in a highly processive fashion, without pausing until the end of the MBP substrate. The data also show that two chains can be accommodated in the central pore. Notably, ClpB was seen to switch between translocating either of the two chains or both simultaneously (see Figure 11.7a).

The observed polypeptide loop extrusion may be a universal strategy in protein processing systems, as recent studies suggest handling of multiple substrate chains by other translocases like p97 or Vps4.[190,191] Dissecting the complex dynamics that underlie this process is challenging, and optical tweezers constitute a key tool for this aim. In the ClpB study, measurements at high spatial and temporal resolution showed individual translocation steps.[44] Remarkably, when only one chain was translocated, the step size was 14 residues, and it doubled to 28 residues when two chains were translocated simultaneously (see Figure 11.7b). Given the structural evidence that each ClpB subunit can displace the substrate by 2 aminoacids,[192] the findings suggested a model in which all six ClpB subunits fire one after another in rapid succession (too fast to be resolved as individual steps), followed by a pause, after which a new firing sequence produces another observable step. A recent study used single-molecule FRET to explore the ultrafast dynamics of the pore loops on a ClpB subunit, and the findings pointed to a Brownian-ratchet rather than a hand-over-hand mechanism.[193] Furthermore, given that translocation can take place against remarkably high forces of up to 50 pN,[44] it is a puzzling question how ClpB can maintain such a strong grip on both chains during translocation, while allowing them to slide pass each other. Structural studies have shown that nine to ten pore loops inside the ClpB channel – which mediate substrate contact and translocation – are constantly in contact with the substrate (a single polypeptide chain), which explains the tight grip of ClpB on a single chain.[192–194]

How the presence of two chains inside the pore affects chaperone–substrate contact arrangement and coordination remains an important open question. Further structural studies using dedicated substrates (like circular peptides[191] or unfolded polypeptides flanked by stable domains[188]) will be important in elucidating these issues. Optical tweezers experiments using mixtures of monomers with impairing mutations could also shed light on the coordination between the subunits and their interaction with the substrate, as has been shown for other AAA+ translocases.[195–197]

These results provide a more complete picture of the disaggregating mechanism of ClpB. Interestingly, ClpB and its closely related yeast homolog Hsp104 exhibit different disaggregation efficiencies depending on the nature of the aggregates, amyloids in particular.[198] It is still unclear whether this difference can be solely explained by substrate specificity and recognition (mediated by the N-terminal, which has diverged between both chaperones[199]), or rather the underlying disaggregation mechanism being radically different depending on the topology of the aggregated substrate. Optical tweezers studies of disaggregation using other substrates (such as prion or disordered proteins[200,201]) can provide useful insights, as it is viable to observe the real-time action of ClpB on individual aggregates.[44] Harnessing the disaggregation power of ClpB and Hsp104 to treat human disorders in which uncontrolled protein aggregation seems to be involved (such as Parkinson's or Alzheimer's diseases) is a possibility recently explored,[202] yet much more research is needed to fill the knowledge gaps.

11.6 Outlook

In the past decade, optical tweezers and other manipulation methods have begun to contribute a direct view of how cells control protein conformations. Unlike many other methods, the detection principle is essentially independent of the protein state, which has resulted in surprise findings ranging from the promotion of intermediate folds to the modulation of collapse strength, and the processive translocation of polypeptide loops. Hence, a picture emerges of far more diverse conformational steering than previously reported, with chaperones targeting a host of protein states and transitions throughout the folding landscape.

Within this diversity, common features are also emerging. Specifically, most chaperone systems appear to interact not only with unfolded protein chain segments but also with segments exhibiting tertiary structure – and hence also bind protein surfaces rather than extended polypeptides or molten globules only. This feature has been observed for most studied systems, including trigger factor, Dnak/Hsp70, HtpG/Hsp90, GroEL–ES, and the small heat shock proteins, with a notable exception for the secretion pathway chaperone SecB.[203] Such interactions with tertiary structures could allow chaperones to remain associated during folding, protect sticky patches on the surfaces of partially folded states against aggregation, and could potentially also lower folding barriers. Its structural basis also remains an open question. We surmise that these exposed sticky surface patches, which ultimately become buried in the native states, are areas of chaperone contact.[31,204] Another emerging question is whether and how chaperones achieve direct folding acceleration, beyond the indirect promotion of folding *via* aggregation suppression. Folding acceleration was observed for substrates within the ribosomal tunnel and GroEL, which raises the question of whether a cavity, the hydrophobic effect, and the water structure play key roles.[137,139] The involved polypeptide collapse in GroEL offers a notable link to NMR observations of diverse chaperones interacting with molten-globule states.[8] Indeed, collapse modulation could be a more general mechanism to control protein states, including for instance phase-separated states.[205]

The foundational importance of protein states indicates various other urgent questions that are opened by these manifestations of conformational control. Nascent chain folding is one practically uncharted territory. While early models suggested that most chaperones act post-translationally, and hence decreased the relevance of this issue, recent work increasingly indicates ubiquitous co-translational functions for many chaperones.[206] Moreover, the selective ribosomal profiling method that is powerful in revealing them also points to protein–protein assembly as a major co-translational process.[207] Indeed, recent work revealed over 800 proteins in human cells that homo-dimerize during their joint translation, establishing co-translational assembly as a ubiquitous phenomenon.[208] These observations raise crucial questions on the conformational basis, the relation between folding and assembly, and the role of chaperones, which can be uniquely addressed with single-molecule manipulation. On a more basic level, it generally remains unclear how substrate conformational changes are correlated with (co)chaperone binding events and the ATP cycle. It was possible to show how GroEL binds first and folding occurs later. Various alternatives may be exploited, however. Chaperones may stabilize spontaneously formed folded structures, first bind and then actively unfold folded or misfolded states or engage in a temporal interplay with other chaperone systems, as suggested for Hsp70 and Hsp90.[150,209] The access to substrate conformational changes and real-time detection of (co)chaperone binding and unbinding will be important to disentangle these critical interdependencies.

Addressing these and many other questions will be an intriguing challenge for the coming decade. It also invites a deeper integration of different techniques, with the aim to connect insights at the structural, dynamic, and cellular levels. Doing so has the potential to overcome the inherent limitations of different approaches, and to reveal yet-unknown manifestations of protein control within cells.

References

1. J. Tyedmers, A. Mogk and B. Bukau, Cellular strategies for controlling protein aggregation, *Nat. Rev. Mol. Cell Biol.*, 2010, **11**, 777–788.
2. D. Balchin, M. Hayer-Hartl and F. U. Hartl, In vivo aspects of protein folding and quality control, *Science*, 2016, **353**, 13.
3. J. S. Valastyan and S. Lindquist, Mechanisms of protein-folding diseases at a glance, *Dis. Models Mech.*, 2014, **7**, 9–14.
4. Y. E. Kim, M. S. Hipp, A. Bracher, M. Hayer-Hartl and F. U. Hartl, Molecular chaperone functions in protein folding and proteostasis, *Annu. Rev. Biochem.*, 2013, **82**, 323–355.
5. M. P. Mayer, Gymnastics of Molecular Chaperones, *Mol. Cell*, 2010, **39**, 321–331.
6. M. J. Avellaneda, E. J. Koers, M. M. Naqvi and S. J. Tans, The chaperone toolbox at the single-molecule level: From clamping to confining, *Protein Sci.*, 2017, **26**, 1291–1302.

7. A. Mashaghi, G. Kramer, D. C. Lamb, M. P. Mayer and S. J. Tans, Chaperone action at the single-molecule level, *Chem. Rev.*, 2014, **114**, 660–676.
8. S. Hiller, Chaperone-Bound Clients: The Importance of Being Dynamic, *Trends Biochem. Sci.*, 2019, **44**, 517–527.
9. G. J. Haran, How, when and why proteins collapse: the relation to folding, *Curr. Opin. Struct. Biol.*, 2012, **22**, 14–20.
10. M. Thommen, W. Holtkamp and M. V. Rodnina, Co-translational protein folding: progress and methods, *Curr. Opin. Struct. Biol.*, 2016, **42**, 83–89.
11. M. Taipale, D. F. Jarosz and S. Lindquist, HSP90 at the hub of protein homeostasis: emerging mechanistic insights, *Nat. Rev. Mol. Cell Biol.*, 2010, **11**, 515–528.
12. J. N. Wells, L. T. Bergendahl and J. A. Marsh, Co-translational assembly of protein complexes, *Biochem. Soc. Trans.*, 2015, **43**, 1221–1226.
13. K. Liberek, A. Lewandowska and S. Ziętkiewicz, Chaperones in control of protein disaggregation, *EMBO J.*, 2008, **27**, 328–335.
14. Y. Shin and C. P. Brangwynne, Liquid phase condensation in cell physiology and disease, *Science*, 2017, **357**, DOI: 10.1126/science.aaf4382.
15. R. Sousa, Structural mechanisms of chaperone mediated protein disaggregation, *Front. Mol. Biosci.*, 2014, **1**, DOI: 10.3389/fmolb.2014.00012.
16. H. Meyer and C. C. Weihl, The VCP/p97 system at a glance: connecting cellular function to disease pathogenesis, *J. Cell Sci.*, 2014, **127**, 3877–3883.
17. R. Zahn, S. Perrett and A. R. Fersht, Conformational states bound by the molecular chaperones GroEL and SecB: A hidden unfolding (annealing) activity, *J. Mol. Biol.*, 1996, **261**, 43–61.
18. C. M. Dobson, Protein folding and misfolding, *Nature*, 2003, **426**, 884–890.
19. M. Mickler, M. Hessling, C. Ratzke, J. Buchner and T. Hugel, The large conformational changes of Hsp90 are only weakly coupled to ATP hydrolysis, *Nat. Struct. Mol. Biol.*, 2009, **16**, 281–286.
20. L. D. Cabrita, *et al.*, A structural ensemble of a ribosome-nascent chain complex during cotranslational protein folding, *Nat. Struct. Mol. Biol.*, 2016, **23**, 278–285.
21. D. H. Chen, *et al.*, Visualizing GroEL/ES in the act of encapsulating a folding protein, *Cell*, 2013, **153**, 1354–1365.
22. B. M. Burmann, *et al.*, Regulation of alpha-synuclein by chaperones in mammalian cells, *Nature*, 2020, **577**, 127–132.
23. Z. Lin and H. S. Rye, Expansion and compression of a protein folding intermediate by GroEL, *Mol. Cell*, 2004, **16**, 23–34.
24. F. Georgescauld, *et al.*, GroEL/ES Chaperonin Modulates the Mechanism and Accelerates the Rate of TIM-Barrel Domain Folding, *Cell*, 2014, **157**, 922–934.
25. C. J. Bustamante, Y. R. Chemla, S. Liu and M. D. Wang, Optical tweezers in single-molecule biophysics, *Nat. Rev. Methods Primers*, 2021, **1**, DOI: 10.1038/s43586-021-00021-6.

26. C. Cecconi, E. A. Shank, C. Bustamante and S. Marqusee, Direct observation of the three-state folding of a single protein molecule, *Science*, 2005, **309**, 2057–2060.
27. J. P. Junker, F. Ziegler and M. Rief, Ligand-dependent equilibrium fluctuations of single calmodulin molecules, *Science*, 2009, **323**, 633–637.
28. M. M. Naqvi, *et al.*, Single-molecule folding mechanisms of the apo- and Mg(2+)-bound states of human neuronal calcium sensor-1, *Biophys. J.*, 2015, **109**, 113–123.
29. J. Perales-Calvo, D. Giganti, G. Stirnemann and S. Garcia-Manyes, The force-dependent mechanism of DnaK-mediated mechanical folding, *Sci. Adv.*, 2018, **4**, eaaq0243.
30. A. Mashaghi, *et al.*, Alternative modes of client binding enable functional plasticity of Hsp70, *Nature*, 2016, **539**, 448–451.
31. A. Mashaghi, *et al.*, Reshaping of the conformational search of a protein by the chaperone trigger factor, *Nature*, 2013, **500**, 98–101.
32. S. Haldar, R. Tapia-Rojo, E. C. Eckels, J. Valle-Orero and J. M. Fernandez, Trigger factor chaperone acts as a mechanical foldase, *Nat. Commun.*, 2017, **8**, DOI: 10.1038/s41467-017-00771-6.
33. M. M. Naqvi, *et al.*, Protein chain collapse modulation and folding stimulation by GroEL–ES, *Sci. Adv.*, 2022, **8**, eabl6293.
34. M. J. Avellaneda, *et al.*, Processive extrusion of polypeptide loops by a Hsp100 disaggregase, *Nature*, 2020, **74**, DOI: 10.1038/s41586-020-2017-2.
35. A. Ashkin, Optical trapping and manipulation of neutral particles using lasers, *Proc. Natl. Acad. Sci. U. S. A.*, 1997, **94**, 4853–4860.
36. K. Svoboda and S. M. Block, Biological Applications of Optical Forces, *Annu. Rev. Biophys. Biomol. Struct.*, 1994, **23**, 247–285.
37. C. J. Bustamante, Y. R. Chemla, S. Liu and M. D. Wang, Optical tweezers in single-molecule biophysics, *Nat. Rev. Methods Primers*, 2021, **1**, 25.
38. A. Ashkin, J. M. Dziedzic, J. E. Bjorkholm and S. Chu, Observation of a single-beam gradient force optical trap for dielectric particles, *Opt. Lett.*, 1986, **11**, 288–290.
39. A. D. Mehta, M. Rief, J. A. Spudich, D. A. Smith and R. M. Simmons, Single-Molecule Biomechanics with Optical Methods, *Science*, 1999, **283**, 1689–1695.
40. M. Ozkan, M. Wang, C. Ozkan, R. Flynn and S. Esener, Optical Manipulation of Objects and Biological Cells in Microfluidic Devices, *Biomed. Microdevices*, 2003, **5**, 61–67.
41. J. R. Moffitt, Y. R. Chemla, S. B. Smith and C. Bustamante, Recent advances in optical tweezers, *Annu. Rev. Biochem.*, 2008, **77**, 205–228.
42. C. Bustamante, Y. R. Chemla, N. R. Forde and D. Izhaky, Mechanical processes in biochemistry, *Annu. Rev. Biochem.*, 2004, **73**, 705–748.
43. D. Gao, *et al.*, Optical manipulation from the microscale to the nanoscale: fundamentals, advances and prospects, *Light: Sci. Appl.*, 2017, **6**, e17039.

44. M. J. Avellaneda, *et al.*, Processive extrusion of polypeptide loops by a Hsp100 disaggregase, *Nature*, 2020, **578**, 317–320.
45. I. Heller, *et al.*, STED nanoscopy combined with optical tweezers reveals protein dynamics on densely covered DNA, *Nat. Methods*, 2013, **10**, 910–916.
46. R. Kassies, *et al.*, Combined AFM and confocal fluorescence microscope for applications in bio-nanotechnology, *J. Microsc.*, 2005, **217**, 109–116.
47. F. E. Kemmerich, *et al.*, Simultaneous Single-Molecule Force and Fluorescence Sampling of DNA Nanostructure Conformations Using Magnetic Tweezers, *Nano Lett.*, 2016, **16**, 381–386.
48. C. Bustamante, Y. R. Chemla and J. R. Moffitt, High-resolution dual-trap optical tweezers with differential detection: instrument design, *Cold Spring Harb. Protoc.*, 2009, **2009**, pdb.ip73.
49. E. Rhoades, E. Gussakovsky and G. Haran, Watching proteins fold one molecule at a time, *Proc. Natl. Acad. Sci. U. S. A.*, 2003, **100**, 3197–3202.
50. B. Schuler and W. A. Eaton, Protein folding studied by single-molecule FRET, *Curr. Opin. Struct. Biol.*, 2008, **18**, 16–26.
51. A. N. Kapanidis and T. Strick, Biology, one molecule at a time, *Trends Biochem. Sci.*, 2009, **34**, 234–243.
52. P. O. Heidarsson, *et al.*, Direct single-molecule observation of calcium-dependent misfolding in human neuronal calcium sensor-1, *Proc. Natl. Acad. Sci. U. S. A.*, 2014, **111**, 13069–13074.
53. H. Yu, *et al.*, Direct observation of multiple misfolding pathways in a single prion protein molecule, *Proc. Natl. Acad. Sci. U. S. A.*, 2012, **109**, 5283–5288.
54. A. Mashaghi, S. Mashaghi and S. J. Tans, Misfolding of Luciferase at the Single-Molecule Level, *Angew. Chem.*, DOI: 10.1002/anie.2014055662014.
55. K. Svoboda, C. F. Schmidt, B. J. Schnapp and S. M. Block, Direct observation of kinesin stepping by optical trapping interferometry, *Nature*, 1993, **365**, 721–727.
56. C. L. Asbury, A. N. Fehr and S. M. Block, Kinesin moves by an asymmetric hand-over-hand mechanism, *Science*, 2003, **302**, 2130–2134.
57. J. W. Shaevitz, E. A. Abbondanzieri, R. Landick and S. M. Block, Backtracking by single RNA polymerase molecules observed at near-base-pair resolution, *Nature*, 2003, **426**, 684–687.
58. E. A. Abbondanzieri, W. J. Greenleaf, J. W. Shaevitz, R. Landick and S. M. Block, Direct observation of base-pair stepping by RNA polymerase, *Nature*, 2005, **438**, 460–465.
59. W. Grange, *et al.*, VirE2: a unique ssDNA-compacting molecular machine, *PLoS Biol.*, 2008, **6**, e44.
60. J. Liphardt, B. Onoa, S. B. Smith, I. J. Tinoco and C. Bustamante, Reversible unfolding of single RNA molecules by mechanical force, *Science*, 2001, **292**, 733–737.
61. C. Bustamante, L. Alexander, K. Maciuba and C. M. Kaiser, Single-Molecule Studies of Protein Folding with Optical Tweezers, *Annu. Rev. Biochem.*, 2020, **89**, 443–470.

62. S. Uemura, *et al.*, Peptide bond formation destabilizes Shine-Dalgarno interaction on the ribosome, *Nature*, 2007, **446**, 454–457.
63. J. D. Wen, *et al.*, Following translation by single ribosomes one codon at a time, *Nature*, 2008, **452**, 598–603.
64. D. H. Goldman, *et al.*, Ribosome. Mechanical force releases nascent chain-mediated ribosome arrest in vitro and in vivo, *Science*, 2015, **348**, 457–460.
65. A. Mashaghi, *et al.*, Alternative modes of client binding enable functional plasticity of Hsp70, *Nature*, 2016, 1–16.
66. W. J. Greenleaf, M. T. Woodside and S. M. Block, High-resolution, single-molecule measurements of biomolecular motion, *Annu. Rev. Biophys. Biomol. Struct.*, 2007, **36**, 171–190.
67. M. J. Avellaneda, E. J. Koers, D. P. Minde, V. Sunderlikova and S. J. Tans, Simultaneous sensing and imaging of individual biomolecular complexes enabled by modular DNA–protein coupling, *Commun. Chem.*, 2020, **3**, 20.
68. Y. Hao, C. Canavan, S. S. Taylor and R. A. Maillard, Integrated Method to Attach DNA Handles and Functionally Select Proteins to Study Folding and Protein-Ligand Interactions with Optical Tweezers, *Sci. Rep.*, 2017, **7**, 10843.
69. F. Moayed, A. Mashaghi and S. J. Tans, A polypeptide-DNA hybrid with selective linking capability applied to single molecule nano-mechanical measurements using optical tweezers, *PLoS One*, 2013, **8**, e54440.
70. C. Cecconi, E. A. Shank, F. W. Dahlquist, S. Marqusee and C. Bustamante, Protein-DNA chimeras for single molecule mechanical folding studies with the optical tweezers, *Eur. Biophys. J.*, 2008, **37**, 729–738.
71. M. Jahn, J. Buchner, T. Hugel and M. Rief, Folding and assembly of the large molecular machine Hsp90 studied in single-molecule experiments, *Proc. Natl. Acad. Sci. U. S. A.*, 2016, **113**, 1232–1237.
72. Z. Ganim and M. Rief, Mechanically switching single-molecule fluorescence of GFP by unfolding and refolding, *Proc. Natl. Acad. Sci. U. S. A.*, 2017, 201704937.
73. F. Wruck, A. Katranidis, K. H. Nierhaus, G. Büldt and M. Hegner, Translation and folding of single proteins in real time, *Proc. Natl. Acad. Sci. U. S. A.*, 2017, **114**, E4399–E4407.
74. C. M. Kaiser, D. H. Goldman, J. D. Chodera, I. Tinoco Jr. and C. Bustamante, The ribosome modulates nascent protein folding, *Science*, 2011, **334**, 1723–1727.
75. T. Liu, *et al.*, Direct measurement of the mechanical work during translocation by the ribosome, *eLife*, 2014, **3**, e03406.
76. W. Holtkamp, *et al.*, Cotranslational protein folding on the ribosome monitored in real time, *Science*, 2015, **350**, 1104–1107.
77. S. Bhushan, *et al.*, α-Helical nascent polypeptide chains visualized within distinct regions of the ribosomal exit tunnel, *Nat. Struct. Mol. Biol.*, 2010, **17**, 313–317.

78. O. B. Nilsson, *et al.*, Cotranslational Protein Folding inside the Ribosome Exit Tunnel, *Cell Rep.*, 2015, **12**, 1533–1540.
79. F. Wruck, *et al.*, The ribosome modulates folding inside the ribosomal exit tunnel, *Commun. Biol.*, 2021, **4**, 523.
80. J. Lu and C. Deutsch, Electrostatics in the ribosomal tunnel modulate chain elongation rates, *J. Mol. Biol.*, 2008, **384**, 73–86.
81. C. A. Charneski and L. D. Hurst, Positively charged residues are the major determinants of ribosomal velocity, *PLoS Biol.*, 2013, **11**, e1001508.
82. P. Huter, *et al.*, Structural Basis for Polyproline-Mediated Ribosome Stalling and Rescue by the Translation Elongation Factor EF-P, *Mol. Cell*, 2017, **68**, 515–527.e516.
83. T. E. Quax, N. J. Claassens, D. Soll and J. van der Oost, Codon Bias as a Means to Fine-Tune Gene Expression, *Mol. Cell*, 2015, **59**, 149–161.
84. Y. Mao, H. Liu, Y. Liu and S. Tao, Deciphering the rules by which dynamics of mRNA secondary structure affect translation efficiency in Saccharomyces cerevisiae, *Nucleic Acids Res.*, 2014, **42**, 4813–4822.
85. B. Fritch, *et al.*, Origins of the Mechanochemical Coupling of Peptide Bond Formation to Protein Synthesis, *J. Am. Chem. Soc.*, 2018, **140**, 5077–5087.
86. S. E. Leininger, F. Trovato, D. A. Nissley and E. P. O'Brien, Domain topology, stability, and translation speed determine mechanical force generation on the ribosome, *Proc. Natl. Acad. Sci. U. S. A.*, 2019, **116**, 5523–5532.
87. K. C. Stein and J. Frydman, The stop-and-go traffic regulating protein biogenesis: How translation kinetics controls proteostasis, *J. Biol. Chem.*, 2019, **294**, 2076–2084.
88. A. Hoffmann, B. Bukau and G. Kramer, Structure and function of the molecular chaperone Trigger Factor, *Biochim. Biophys. Acta*, 2010, **1803**, 650–661.
89. F. Merz, *et al.*, Molecular mechanism and structure of Trigger Factor bound to the translating ribosome, *EMBO J.*, 2008, **27**, 1622–1632.
90. E. Crooke and W. Wickner, Trigger factor: a soluble protein that folds pro-OmpA into a membrane-assembly-competent form, *Proc. Natl. Acad. Sci. U. S. A.*, 1987, **84**, 5216–5220.
91. D. Baram, *et al.*, Structure of trigger factor binding domain in biologically homologous complex with eubacterial ribosome reveals its chaperone action, *Proc. Natl. Acad. Sci. U. S. A.*, 2005, **102**, 12017–12022.
92. C. M. Kaiser, *et al.*, Real-time observation of trigger factor function on translating ribosomes, *Nature*, 2006, **444**, 455–460.
93. L. Ferbitz, *et al.*, Trigger factor in complex with the ribosome forms a molecular cradle for nascent proteins, *Nature*, 2004, **431**, 590–596.
94. F. Wruck, *et al.*, Protein Folding Mediated by Trigger Factor and Hsp70: New Insights from Single-Molecule Approaches, *J. Mol. Biol.*, 2018, **430**, 438–449.

95. K. Liu, K. Maciuba and C. M. Kaiser, The Ribosome Cooperates with a Chaperone to Guide Multi-domain Protein Folding, *Mol. Cell*, 2019, **74**, 310–319.e317.
96. T. Saio, X. Guan, P. Rossi, A. Economou and C. G. Kalodimos, Structural basis for protein antiaggregation activity of the trigger factor chaperone, *Science*, 2014, **344**, 1250494.
97. A. Raine, M. Lovmar, J. Wikberg and M. Ehrenberg, Trigger factor binding to ribosomes with nascent peptide chains of varying lengths and sequences, *J. Biol. Chem.*, 2006, **281**, 28033–28038.
98. S. Haldar, R. Tapia-Rojo, E. C. Eckels, J. Valle-Orero and J. M. Fernandez, Trigger factor chaperone acts as a mechanical foldase, *Nat. Commun.*, 2017, **8**, 668.
99. D. Fan, *et al.*, Exploring the roles of substrate-binding surface of the chaperone site in the chaperone activity of trigger factor, *FASEB J.*, 2018, fj201701576.
100. K. Singhal, J. Vreede, A. Mashaghi, S. J. Tans and P. G. Bolhuis, The Trigger Factor Chaperone Encapsulates and Stabilizes Partial Folds of Substrate Proteins, *PLoS Comput. Biol.*, 2015, **11**, e1004444.
101. R. Maier, B. Eckert, C. Scholz, H. Lilie and F. X. Schmid, Interaction of trigger factor with the ribosome, *J. Mol. Biol.*, 2003, **326**, 585–592.
102. H. Patzelt, *et al.*, Three-state equilibrium of Escherichia coli trigger factor, *Biol. Chem.*, 2002, **383**, 1611–1619.
103. L. Morgado, B. M. Burmann, T. Sharpe, A. Mazur and S. Hiller, The dynamic dimer structure of the chaperone Trigger Factor, *Nat. Commun.*, 2017, **8**, 1992.
104. S. K. Lakshmipathy, *et al.*, Identification of nascent chain interaction sites on trigger factor, *J. Biol. Chem.*, 2007, **282**, 12186–12193.
105. A. Rutkowska, *et al.*, Dynamics of trigger factor interaction with translating ribosomes, *J. Biol. Chem.*, 2008, **283**, 4124–4132.
106. M. J. Avellaneda, E. J. Koers, D. P. Minde, V. Sunderlikova and S. J. Tans, Simultaneous sensing and imaging of individual biomolecular complexes enabled by modular DNA–protein coupling, *Commun. Chem.*, 2020, **3**, 20.
107. R. Rosenzweig, N. B. Nillegoda, M. P. Mayer and B. Bukau, The Hsp70 chaperone network, *Nat. Rev. Mol. Cell Biol.*, 2019, **20**, 665–680.
108. G. Calloni, *et al.*, DnaK functions as a central hub in the E. coli chaperone network, *Cell Rep.*, 2012, **1**, 251–264.
109. M. P. Mayer and B. Bukau, Hsp70 chaperones: cellular functions and molecular mechanism, *Cell. Mol. Life Sci.*, 2005, **62**, 670–684.
110. H. H. Kampinga and E. A. Craig, The HSP70 chaperone machinery: J proteins as drivers of functional specificity, *Nat. Rev. Mol. Cell Biol.*, 2010, **11**, 579–592.
111. M. E. Murphy, The HSP70 family and cancer, *Carcinogenesis*, 2013, **34**, 1181–1188.

112. R. Kityk, J. Kopp, I. Sinning and M. P. Mayer, Structure and dynamics of the ATP-bound open conformation of Hsp70 chaperones, *Mol. Cell*, 2012, **48**, 863–874.
113. M. Sikor, K. Mapa, L. V. von Voithenberg, D. Mokranjac and D. C. Lamb, Real-time observation of the conformational dynamics of mitochondrial Hsp70 by spFRET, *EMBO J.*, 2013, **32**, 1639–1649.
114. J. M. Nunes, M. Mayer-Hartl, F. U. Hartl and D. J. Muller, Action of the Hsp70 chaperone system observed with single proteins, *Nat. Commun.*, 2015, **6**, 6307.
115. R. Kellner, *et al.*, Single-molecule spectroscopy reveals chaperone-mediated expansion of substrate protein, *Proc. Natl. Acad. Sci. U. S. A.*, 2014, **111**, 13355–13360.
116. M. P. Mayer, *et al.*, Multistep mechanism of substrate binding determines chaperone activity of Hsp70, *Nat. Struct. Biol.*, 2000, **7**, 586–593.
117. J. Gamer, *et al.*, A cycle of binding and release of the DnaK, DnaJ and GrpE chaperones regulates activity of the Escherichia coli heat shock transcription factor sigma32, *EMBO J.*, 1996, **15**, 607–617.
118. A. S. Wentink, *et al.*, Molecular dissection of amyloid disaggregation by human HSP70, *Nature*, 2020, **587**, 483–488.
119. P. De Los Rios, A. Ben-Zvi, O. Slutsky, A. Azem and P. Goloubinoff, Hsp70 chaperones accelerate protein translocation and the unfolding of stable protein aggregates by entropic pulling, *Proc. Natl. Acad. Sci. U. S. A.*, 2006, **103**, 6166–6171.
120. A. L. Horwich, K. B. Low, W. A. Fenton, I. N. Hirshfield and K. Furtak, Folding in-Vivo of Bacterial Cytoplasmic Proteins - Role of Groel, *Cell*, 1993, **74**, 909–917.
121. C. M. Dobson, Protein folding and misfolding, *Nature*, 2003, **426**, 884–890.
122. F. U. Hartl and M. Hayer-Hartl, Protein folding - Molecular chaperones in the cytosol: from nascent chain to folded protein, *Science*, 2002, **295**, 1852–1858.
123. A. L. Horwich and W. A. Fenton, Chaperonin-assisted protein folding: a chronologue, *Q. Rev. Biophys.*, 2020, **53**, e4.
124. H. R. Saibil, W. A. Fenton, D. K. Clare and A. L. Horwich, Structure and allostery of the chaperonin GroEL, *J. Mol. Biol.*, 2013, **425**, 1476–1487.
125. H. S. Rye, *et al.*, Distinct actions of cis and trans-ATP within the double ring of the chaperonin GroEL, *Nature*, 1997, **388**, 792–798.
126. P. B. Sigler, *et al.*, Structure and function in GroEL-mediated protein folding, *Annu. Rev. Biochem.*, 1998, **67**, 581–608.
127. H. S. Rye, *et al.*, GroEL–GroES cycling: ATP and nonnative polypeptide direct alternation of folding-active rings, *Cell*, 1999, **97**, 325–338.
128. M. Hayer-Hartl, A. Bracher and F. U. Hartl, The GroEL–GroES Chaperonin Machine: A Nano-Cage for Protein Folding, *Trends Biochem. Sci.*, 2016, **41**, 62–76.
129. A. L. Horwich, A. C. Apetri and W. A. Fenton, The GroEL/GroES cis cavity as a passive anti-aggregation device, *FEBS Lett.*, 2009, **583**, 2654–2662.

130. M. J. Todd, G. H. Lorimer and D. Thirumalai, Chaperonin-facilitated protein folding: optimization of rate and yield by an iterative annealing mechanism, *Proc. Natl. Acad. Sci. U. S. A.*, 1996, **93**, 4030–4035.
131. K. Chakraborty, *et al.*, Chaperonin-catalyzed rescue of kinetically trapped states in protein folding, *Cell*, 2010, **142**, 112–122.
132. M. M. Naqvi, *et al.*, Protein chain collapse modulation and folding stimulation by GroEL–ES, *Sci. Adv.*, 2022, **8**, eabl6293.
133. F. Motojima and M. Yoshida, Polypeptide in the chaperonin cage partly protrudes out and then folds inside or escapes outside, *EMBO J.*, 2010, **29**, 4008–4019.
134. C. Chaudhry, *et al.*, Role of the gamma-phosphate of ATP in triggering protein folding by GroEL–GroES: function, structure and energetics, *EMBO J.*, 2003, **22**, 4877–4887.
135. S. Priya, *et al.*, GroEL and CCT are catalytic unfoldases mediating out-of-cage polypeptide refolding without ATP, *Proc. Natl. Acad. Sci. U. S. A.*, 2013, **110**, 7199–7204.
136. S. Sharma, *et al.*, Monitoring protein conformation along the pathway of chaperonin-assisted folding, *Cell*, 2008, **133**, 142–153.
137. L. Hua, R. Zangi and B. J. Berne, Hydrophobic Interactions and Dewetting between Plates with Hydrophobic and Hydrophilic Domains, *J. Phys. Chem. C*, 2009, **113**, 5244–5253.
138. J. A. Riback, *et al.*, Innovative scattering analysis shows that hydrophobic disordered proteins are expanded in water, *Science*, 2017, **358**, 238–241.
139. J. L. England and V. S. Pande, Potential for modulation of the hydrophobic effect inside chaperonins, *Biophys. J.*, 2008, **95**, 3391–3399.
140. B. Schuler, A. Soranno, H. Hofmann and D. J. Nettels, Single-molecule FRET spectroscopy and the polymer physics of unfolded and intrinsically disordered proteins, *Annu. Rev. Biophys.*, 2016, **45**, 207–231.
141. J. Thoma, B. M. Burmann, S. Hiller and D. J. Muller, Impact of holdase chaperones Skp and SurA on the folding of beta-barrel outer-membrane proteins, *Nat. Struct. Mol. Biol.*, 2015, **22**, 795–802.
142. N. K. Tyagi, W. A. Fenton, A. A. Deniz and A. L. Horwich, Double mutant MBP refolds at same rate in free solution as inside the GroEL/GroES chaperonin chamber when aggregation in free solution is prevented, *FEBS Lett.*, 2011, **585**, 1969–1972.
143. A. J. Gupta, S. Haldar, G. Milicic, F. U. Hartl and M. Hayer-Hartl, Active cage mechanism of chaperonin-assisted protein folding demonstrated at single-molecule level, *J. Mol. Biol.*, 2014, **426**, 2739–2754.
144. J. Weaver, *et al.*, GroEL actively stimulates folding of the endogenous substrate protein PepQ, *Nat. Commun.*, 2017, **8**, 15934.
145. H. Hofmann, *et al.*, Single-molecule spectroscopy of protein folding in a chaperonin cage, *Proc. Natl. Acad. Sci. U. S. A.*, 2010, **107**, 11793–11798.
146. Y. Takei, R. Iizuka, T. Ueno and T. Funatsu, Single-molecule observation of protein folding in symmetric GroEL-(GroES)2 complexes, *J. Biol. Chem.*, 2012, **287**, 41118–41125.

147. X. Ye and G. H. Lorimer, Substrate protein switches GroE chaperonins from asymmetric to symmetric cycling by catalyzing nucleotide exchange, *Proc. Natl. Acad. Sci. U. S. A.*, 2013, **110**, E4289–E4297.
148. S. Haldar, *et al.*, Chaperonin-Assisted Protein Folding: Relative Population of Asymmetric and Symmetric GroEL:GroES Complexes, *J. Mol. Biol.*, 2015, **427**, 2244–2255.
149. D. Noshiro and T. Ando, Substrate protein dependence of GroEL–GroES interaction cycle revealed by high-speed atomic force microscopy imaging, *Philos. Trans. R. Soc. London, Ser. B*, 2018, **373**, 20170180.
150. E. Kirschke, D. Goswami, D. Southworth, P. R. Griffin and D. A. Agard, Glucocorticoid receptor function regulated by coordinated action of the Hsp90 and Hsp70 chaperone cycles, *Cell*, 2014, **157**, 1685–1697.
151. T. Moran Luengo, R. Kityk, M. P. Mayer and S. G. D. Rudiger, Hsp90 Breaks the Deadlock of the Hsp70 Chaperone System, *Mol. Cell*, 2018, **70**, 545–552.e549.
152. M. Taipale, *et al.*, Quantitative analysis of HSP90-client interactions reveals principles of substrate recognition, *Cell*, 2012, **150**, 987–1001.
153. G. I. Karras, *et al.*, HSP90 Shapes the Consequences of Human Genetic Variation, *Cell*, 2017, **168**, 856–866.e812.
154. L. Whitesell and S. L. Lindquist, HSP90 and the chaperoning of cancer, *Nat. Rev. Cancer*, 2005, **5**, 761–772.
155. X. Yan and L. Yajun, Recent Advances in the Discovery of Novel HSP90 Inhibitors: An Update from 2014, *Curr. Drug Targets*, 2020, **21**, 302–317.
156. M. Jahn, J. Buchner, T. Hugel and M. Rief, Folding and assembly of the large molecular machine Hsp90 studied in single-molecule experiments, *Proc. Natl. Acad. Sci. U. S. A.*, 2016, **113**, 1232–1237.
157. M. Jahn, *et al.*, Folding and Domain Interactions of Three Orthologs of Hsp90 Studied by Single-Molecule Force Spectroscopy, *Structure*, 2018, **26**, 96–105.e104.
158. G. E. Karagöz, *et al.*, Hsp90-Tau complex reveals molecular basis for specificity in chaperone action, *Cell*, 2014, **156**, 963–974.
159. K. A. Verba, *et al.*, Atomic structure of Hsp90-Cdc37-Cdk4 reveals that Hsp90 traps and stabilizes an unfolded kinase, *Science*, 2016, **352**, 1542–1547.
160. C. M. Noddings, R. Y.-R. Wang, J. L. Johnson and D. A. Agard, Structure of Hsp90–p23–GR reveals the Hsp90 client-remodelling mechanism, *Nature*, 2021, **601**, 465–469.
161. R. Y.-R. Wang, *et al.*, GR chaperone cycle mechanism revealed by cryo-EM: inactivation of GR by GR:Hsp90:Hsp70:Hop client-loading complex, *bioRxiv*, 2020, preprint, DOI: 10.1101/2020.11.05.370247.
162. A. Mashaghi, *et al.*, Direct observation of Hsp90-induced compaction in a protein chain, *Cell Rep.*, 2022, **41**, DOI: 10.1016/j.celrep.2022.111734.
163. K. A. Dill, Theory for the folding and stability of globular proteins, *Biochemistry*, 1985, **24**, 1501–1509.

164. P. O. Heidarsson, *et al.*, A highly compliant protein native state with a spontaneous-like mechanical unfolding pathway, *J. Am. Chem. Soc.*, 2012, **134**, 17068–17075.
165. P. S. Kim and R. L. Baldwin, Specific intermediates in the folding reactions of small proteins and the mechanism of protein folding, *Annu. Rev. Biochem.*, 1982, **51**, 459–489.
166. T. R. Alderson, *et al.*, Local unfolding of the HSP27 monomer regulates chaperone activity, *Nat. Commun.*, 2019, **10**, 1068.
167. E. Basha, H. O'Neill and E. Vierling, Small heat shock proteins and alpha-crystallins: dynamic proteins with flexible functions, *Trends Biochem. Sci.*, 2012, **37**, 106–117.
168. C. L. Johnston, *et al.*, Single-molecule fluorescence-based approach reveals novel mechanistic insights into human small heat shock protein chaperone function, *J. Biol. Chem.*, 2021, **296**, 100161.
169. S. Ungelenk, *et al.*, Small heat shock proteins sequester misfolding proteins in near-native conformation for cellular protection and efficient refolding, *Nat. Commun.*, 2016, 7, 13673.
170. A. G. Cashikar, M. Duennwald and S. L. Lindquist, A chaperone pathway in protein disaggregation. Hsp26 alters the nature of protein aggregates to facilitate reactivation by Hsp104, *J. Biol. Chem.*, 2005, **280**, 23869–23875.
171. A. Mogk, E. Deuerling, S. Vorderwulbecke, E. Vierling and B. Bukau, Small heat shock proteins, ClpB and the DnaK system form a functional triade in reversing protein aggregation, *Mol. Microbiol.*, 2003, **50**, 585–595.
172. F. Moayed, *et al.*, The Anti-Aggregation Holdase Hsp33 Promotes the Formation of Folded Protein Structures, *Biophys. J.*, 2020, **118**, 85–95.
173. D. Reichmann, *et al.*, Order out of disorder: working cycle of an intrinsically unfolded chaperone, *Cell*, 2012, **148**, 947–957.
174. A. Mogk, C. Ruger-Herreros and B. Bukau, Cellular Functions and Mechanisms of Action of Small Heat Shock Proteins, *Annu. Rev. Microbiol.*, 2019, **73**, 89–110.
175. A. Mogk, E. Kummer and B. Bukau, Cooperation of Hsp70 and Hsp100 chaperone machines in protein disaggregation, *Front. Mol. Biosci.*, 2015, **2**, 22.
176. O. Faust, *et al.*, HSP40 proteins use class-specific regulation to drive HSP70 functional diversity, *Nature*, 2020, **587**, 489–494.
177. J. Tittelmeier, *et al.*, The HSP110/HSP70 disaggregation system generates spreading-competent toxic α-synuclein species, *EMBO J.*, 2020, **39**, e103954.
178. N. B. Nillegoda, *et al.*, Crucial HSP70 co-chaperone complex unlocks metazoan protein disaggregation, *Nature*, 2015, **524**, 247–251.
179. S. M. Doyle and S. Wickner, Hsp104 and ClpB: protein disaggregating machines, *Trends Biochem. Sci.*, 2009, **34**, 40–48.
180. Y. Oguchi, *et al.*, A tightly regulated molecular toggle controls AAA+ disaggregase, *Nat. Struct. Mol. Biol.*, 2012, **19**, 1338–1346.

181. H. Mazal, *et al.*, Tunable microsecond dynamics of an allosteric switch regulate the activity of a AAA+ disaggregation machine, *Nat. Commun.*, 2019, **10**, 1–12.
182. M. E. Aubin-Tam, A. O. Olivares, R. T. Sauer, T. A. Baker and M. J. Lang, Single-molecule protein unfolding and translocation by an ATP-fueled proteolytic machine, *Cell*, 2011, **145**, 257–267.
183. R. A. Maillard, *et al.*, ClpX(P) generates mechanical force to unfold and translocate its protein substrates, *Cell*, 2011, **145**, 459–469.
184. A. O. Olivares, A. R. Nager, O. Iosefson, R. T. Sauer and T. A. Baker, Mechanochemical basis of protein degradation by a double-ring AAA+ machine, *Nat. Struct. Mol. Biol.*, 2014, **21**, 871–875.
185. C. L. Durie, *et al.*, Hsp104 and Potentiated Variants Can Operate as Distinct Nonprocessive Translocases, *Biophys. J.*, 2019, **116**, 1856–1872.
186. T. Li, *et al.*, Escherichia coli ClpB is a non-processive polypeptide translocase, *Biochem. J.*, 2015, **470**, 39–52.
187. S. N. Gates, *et al.*, Ratchet-like polypeptide translocation mechanism of the AAA+ disaggregase Hsp104, *Science*, 2017, **357**, 273–279.
188. T. Haslberger, *et al.*, Protein disaggregation by the AAA+ chaperone ClpB involves partial threading of looped polypeptide segments, *Nat. Struct. Mol. Biol.*, 2008, **15**, 641–650.
189. T. Uchihashi, *et al.*, Dynamic structural states of ClpB involved in its disaggregation function, *Nat. Commun.*, 20182021, **9**, 2147.
190. J. van den Boom, *et al.*, Targeted substrate loop insertion by VCP/p97 during PP1 complex disassembly, *Nat. Struct. Mol. Biol.*, 1–8.
191. H. Han, *et al.*, Structure of Vps4 with circular peptides and implications for translocation of two polypeptide chains by AAA+ ATPases, *eLife*, 2019, **8**, e44071.
192. C. Deville, K. Franke, A. Mogk, B. Bukau and H. R. Saibil, Two-step activation mechanism of the ClpB disaggregase for sequential substrate threading by the main ATPase motor, *Cell Rep.*, 2019, **27**, 3433–3446.e3434.
193. H. Mazal, M. Iljina, I. Riven and G. Haran, Ultrafast pore-loop dynamics in a AAA+ machine point to a Brownian-ratchet mechanism for protein translocation, *Sci. Adv.*, 2021, **7**, eabg4674.
194. A. N. Rizo, *et al.*, Structural basis for substrate gripping and translocation by the ClpB AAA+ disaggregase, *Nat. Commun.*, 2019, **10**, 1–12.
195. K. L. Zuromski, R. T. Sauer and T. A. Baker, Modular and coordinated activity of AAA+ active sites in the double-ring ClpA unfoldase of the ClpAP protease, *Proc. Natl. Acad. Sci. U. S. A.*, 2020, **117**, 25455–25463.
196. O. Iosefson, A. O. Olivares, T. A. Baker and R. T. Sauer, Dissection of axial-pore loop function during unfolding and translocation by a AAA+ proteolytic machine, *Cell Rep.*, 2015, **12**, 1032–1041.
197. P. Rodriguez-Aliaga, L. Ramirez, F. Kim, C. Bustamante and A. Martin, Substrate-translocating loops regulate mechanochemical coupling and power production in AAA+ protease ClpXP, *Nat. Struct. Mol. Biol.*, 2016, **23**, 974–981.

198. J. Shorter and S. Lindquist, Hsp104 catalyzes formation and elimination of self-replicating Sup35 prion conformers, *Science*, 2004, **304**, 1793–1797.
199. D. M. Johnston, M. Miot, J. R. Hoskins, S. Wickner and S. M. Doyle, Substrate discrimination by ClpB and Hsp104, *Front. Mol. Biosci.*, 2017, **4**, 36.
200. K. Neupane, A. Solanki, I. Sosova, M. Belov and M. T. Woodside, Diverse metastable structures formed by small oligomers of α-synuclein probed by force spectroscopy, *PLoS One*, 2014, **9**, e86495.
201. H. Yu, *et al.*, Direct observation of multiple misfolding pathways in a single prion protein molecule, *Proc. Natl. Acad. Sci. U. S. A.*, 2012, **109**, 5283–5288.
202. S. Vashist, M. Cushman and J. Shorter, Applying Hsp104 to protein-misfolding disorders, *Biochem. Cell Biol.*, 2010, **88**, 1–13.
203. P. Bechtluft, *et al.*, Direct observation of chaperone-induced changes in a protein folding pathway, *Science*, 2007, **318**, 1458–1461.
204. K. Singhal, J. Vreede, A. Mashaghi, S. J. Tans and P. G. Bolhuis, The Trigger Factor Chaperone Encapsulates and Stabilizes Partial Folds of Substrate Proteins, *PLoS Comput. Biol.*, 2015, **11**, e1004444.
205. L. P. Bergeron-Sandoval, N. Safaee and S. W. Michnick, Mechanisms and Consequences of Macromolecular Phase Separation, *Cell*, 2016, **165**, 1067–1079.
206. G. Kramer, D. Boehringer, N. Ban and B. Bukau, The ribosome as a platform for co-translational processing, folding and targeting of newly synthesized proteins, *Nat. Struct. Mol. Biol.*, 2009, **16**, 589–597.
207. Y. W. Shieh, *et al.*, Operon structure and cotranslational subunit association direct protein assembly in bacteria, *Science*, 2015, **350**, 678–680.
208. M. Bertolini, *et al.*, Interactions between nascent proteins translated by adjacent ribosomes drive homomer assembly, *Science*, 2021, **371**, 57–64.
209. T. O. Street, L. A. Lavery and D. A. Agard, Substrate binding drives large-scale conformational changes in the Hsp90 molecular chaperone, *Mol. Cell*, 2011, **42**, 96–105.

CHAPTER 12

Integrative Methods to Investigate Chaperones in Regulating Protein Phase Separation and Aggregation

D. LI,*[a,b] S. ZHANG*[c] AND C. LIU*[c]

[a] Bio-X Institutes, Key Laboratory for the Genetics of Developmental and Neuropsychiatric Disorders, Ministry of Education, Shanghai Jiao Tong University, Shanghai 200030, China; [b] Zhangjiang Institute for Advanced Study, Shanghai Jiao Tong University, Shanghai 200240, China; [c] Interdisciplinary Research Center on Biology and Chemistry, Shanghai Institute of Organic Chemistry, Chinese Academy of Sciences, Shanghai 201210, China
*Emails: lidan2017@sjtu.edu.cn; zhangshengnan@sioc.ac.cn; liulab@sioc.ac.cn

12.1 Emerging Roles of Chaperones in Protein Phase Separation and Aggregation

12.1.1 Chaperones in Regulation of Protein Phase Separation

Molecular chaperones have been intensively studied for their functions in assisting innate protein synthesis and folding, and preventing protein mislocalization, misfolding, and abnormal aggregation in cells.[1] In the last decade, many different proteins were found to undergo liquid–liquid phase separation (LLPS) which drives the formation of different types of

New Developments in NMR No. 29
Biophysics of Molecular Chaperones: Function, Mechanisms and Client Protein Interactions
Edited by Sebastian Hiller, Maili Liu and Lichun He

Published by the Royal Society of Chemistry, www.rsc.org

membraneless organelles (MLOs) including stress granules (SGs), P body, paraspeckle, and Caji body.[2–4] Notably, the MLOs such as SGs contain hundreds of different proteins and RNAs which are highly concentrated in a confined area, thus bear extremely high pressure for abnormal protein aggregation.[5,6] Importantly, many canonical molecular chaperones including Hsp90, Hsp70, Hsp60, and Hsp40 were found to exist in different MLOs,[6,7] indicating that molecular chaperones may play an important role in maintaining proper protein conformation and activity in highly dynamic liquid-like state within MLOs. Indeed, recent studies showed that Hsp70 is important in maintaining the highly dynamic nature of TDP-43 nuclear bodies and SGs.[8,9] Whereas, genetically depletion or pharmaceutically inhibition of Hsp70 severely impairs the dynamics of MLOs, and results in liquid-to-solid phase transition and pathological amyloid protein aggregation associated with neurodegenerative diseases.[8,9] Thus, investigation of how different molecular chaperones regulate protein LLPS is of importance not only for revealing the molecular mechanism underlying the maintenance of the MLOs dynamic property and function, but also for understanding how dysregulation of protein LLPS may contribute to the pathological process in human diseases.

12.1.1.1 Chaperones in Preventing Protein LLPS

In 2018, several groups reported that a nuclear import receptor karyopherin-β2 (Kapβ2, also called as TNPO1) may act as a potent molecular chaperone for preventing protein LLPS of FUS both *in vitro* and in cells (see Figure 12.1(a)).[10–13] Moreover, Kapβ2 can even reverse FUS LLPS by dissolving the preformed FUS liquid-like droplets.[13] Further structural studies revealed that Kapβ2 tightly interacts with the C-terminal 26-residue proline-tyrosine nuclear localization signal (PY-NLS).[10,13] Interestingly, on top of this specific interaction, Kapβ2 also weakly binds to multiple segments across different domains of FUS including low complexity (LC) domain, Arg-Gly-Gly (RGG) motif, and RNA recognition motif (RRM) which all contribute to mediating FUS LLPS.[10] Thus, this promiscuously weak interaction enables Kapβ2 to compete and weaken the inter-molecular interaction of FUS, thus prevents FUS LLPS. Mutation of FUS PY-NLS identified in familial ALS patients disrupts the binding of Kapβ2 to FUS and promotes FUS aggregation in cells.[14,15] These studies provide molecular understanding on how a molecular chaperone tethers its client protein and prevents its LLPS.

More recently, the small heat shock protein Hsp27 was identified to exhibit a potent activity in inhibiting protein LLPS of FUS *in vitro* and FUS recruitment into SGs in cells.[16] Similar to Kapβ2, Hsp27 can also dissolve the liquid-like droplets formed by FUS (see Figure 12.1(b)). Interestingly, Hsp27 uses its N-terminal domain (NTD) to weakly bind to the entire FUS LC domain with a slight binding preference for Ser residues in LC. The NMR paramagnetic relaxation enhancement (PRE) experiment further showed that binding of Hsp27 can dramatically diminish the long-range and

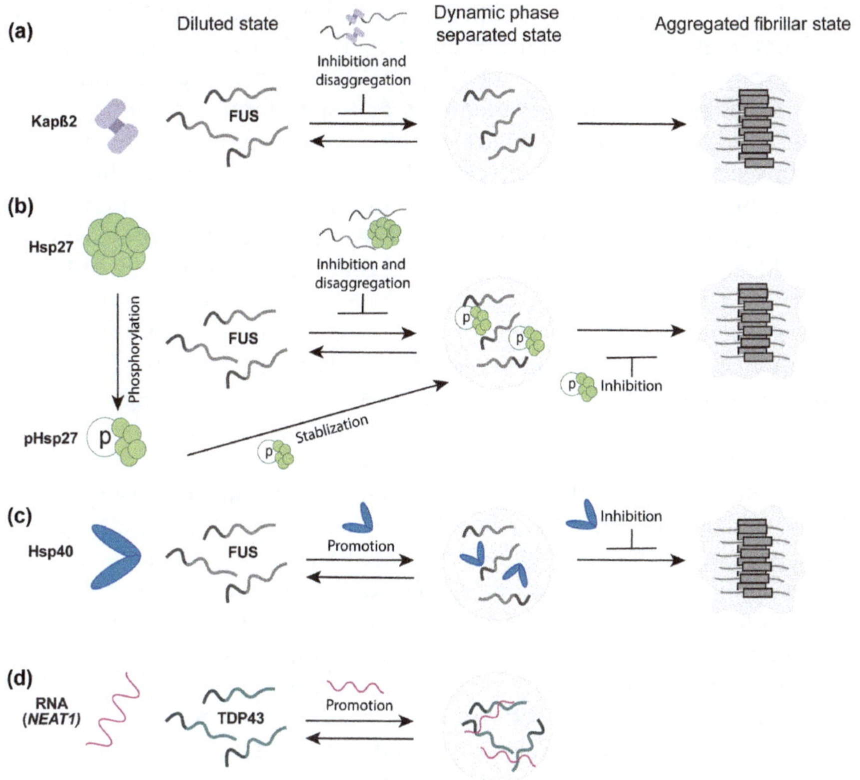

Figure 12.1 Chaperones in the regulation of protein phase separation. (a) Kapβ2 acts as a potent molecular chaperone for preventing and even reversing protein LLPS of FUS. (b) Hsp27 can inhibit and dissolve protein LLPS of FUS. In addition, stress-induced phosphorylation of Hsp27 negatively regulates its chaperone activity against FUS LC LLPS, promoting the partition of Hsp27 into the liquid-like droplet of FUS LC to stabilize protein LLPS and prevent its further fibrillation. (c) Hsp40 promotes the LLPS of FUS and prevents its further solidification and amyloid aggregation. (d) lncRNA *NEAT1* acts as a molecular chaperone to promote the LLPS of TDP43.

transient inter-molecular interaction of FUS LC which are important in mediating FUS LC LLPS. More interestingly, the chaperone activity of Hsp27 for inhibiting FUS LC LLPS can be negatively regulated by stress-induced phosphorylation at the NTD of Hsp27.[16] This suggests that the inhibitory activity of molecular chaperone for protein LLPS can be dynamically regulated in cells.

12.1.1.2 Chaperones in Maintaining Protein LLPS

Mounting evidences demonstrate that molecular chaperones such as Hsp70, Hsp40, HspB8, and Hsp27 can co-phase separate with client proteins (*e.g.*, TDP-43, FUS) and, more importantly, maintain the liquid-like property of clients in condensates and prevent their liquid-to-solid phase transition and aggregation (see Figure 12.1(c)).[8,9,16–18] Surprisingly, recently study showed that several different members of Hsp40 family including human Hdj1 and Hdj2, yeast Sis1 and Yjd1 exhibit a high capability to undergo LLPS by themselves.[17] They can be recruited into different MLOs including SGs and ubiquitin-rich nuclear bodies, implying their potential distinct roles in mediating different MLOs formation and regulating protein homeostasis in MLOs. Mechanistic study further revealed that Hdj1 binds the client FUS *via* its flexible G/F-rich region to promote FUS LLPS and employs its client-binding C-terminal domains (CTD) to capture the amyloidogenic LC of FUS to prevent its further solidification and amyloid aggregation.[17] Therefore, chaperones may use different domains to target different regions of the client protein for co-phase separating with the client and maintaining the highly-dynamic liquid-like property of the co-phase separated system.

In addition to Hsp40, other molecular chaperones including Hsp70, Hsp27, and HspB8 were also found to be involved in maintaining their client proteins in dynamic phase separated state.[8,9,18] For instance, mass spectrometry and cellular studies identified Hsp70 as the key player in chaperoning TDP-43 in TDP-43 nuclear bodies. Genetic depletion or pharmacological inhibition of Hsp70 severely impaired the dynamic nature of the TDP-43 nuclear bodies.[8,9] Further structural investigation showed that Hsp70 may directly capture the highly aggregation-prone region of TDP-43 to prevent its aggregation in the dynamic and liquid phase.[8] Two small heat shock proteins Hsp27 and HspB8 were also found to co-phase separate with FUS and TDP-43, and efficiently inhibit the amyloid aggregation of these two clients.[16,18] Altogether, these studies suggest that different chaperones may be commonly involved in keeping the dynamic properties of their client proteins in the liquid-like phase separated state as well as maintaining the protein homeostasis and biological function of different MLOs.

12.1.1.3 RNAs as Molecular Chaperones for Regulating Protein LLPS

Different MLOs including SGs, paraspeckles and nuclear bodies contain not only hundreds of different proteins but also different types of RNAs.[19–21] A recent study reports the long noncoding RNA (lncRNA) *SLERT* may act as a molecular chaperone to directly bind to the RecA domain of DDX21 and fine-tune the dynamic property and liquidity of protein condensate formed by DDX21 in nucleolus which is essential for regulating ribosomal RNA production.[22] This observation expands the concept of molecular chaperone from protein to RNA in modulating protein homeostasis in the phase separated state. In another case, the lncRNA *NEAT1* which is significantly un-regulated upon cellular stress was found to promote LLPS of TDP-43 *in vitro* and mediate the stress-induced assembly of TDP-43 nuclear bodies in cells (see Figure 12.1(d)). While, the ALS disease-causing mutation D169G of TDP-43 which impairs TDP-43 binding to *NEAT1* results in the formation of pathological aggregation of TDP-43 in cytoplasm.[23] Thus, RNA may play important roles in chaperoning protein in the phase separated state and different types of RNA exhibit different or even opposite activities in modulating protein LLPS in MLOs.

12.1.1.4 Different Chaperones Orchestrates Protein Homeostasis in Liquid-phase Separated State

Different canonical chaperones including Hsp40, Hsp60, Hsp27, Hsp70 as well as different types of RNA molecules are involved in the dynamical assembly of different MLOs.[24,25] Moreover, accumulating evidences showed that different molecular chaperones may organize into a functional chaperone network to keep the protein homeostasis in different MLOs.[24] Different MLOs contain different combination of proteins and RNAs with distinct physiochemical properties, they may require different subset of molecular chaperones. Even for the same MLO such as SGs, at different (dis)assemble stages, the chaperone network may be temporally changed and reorganized in response to the different duration and intensity of the internal or external stress.[24] However, our mechanistic knowledge on how chaperone network is spatiotemporally organized in different MLOs is still in its infancy. It's urgent to systematically investigate the chaperone networks in maintaining protein homeostasis of MLOs by using integrative methods which will be discussed in the next section, as well as developing new methods such as high-resolution imaging and proximity labelling techniques.

12.1.2 Chaperones in Modulating Protein Aggregation

Abnormal protein aggregation and fibrillation have been commonly observed in a variety of neurodegenerative diseases (ND) such as Alzheimer's disease (AD), Parkinson's disease (PD), and Amyotrophic lateral sclerosis (ALS).[26–29] Once formed under diseased condition, the pathological amyloid fibrils may spread in diseased brains *via* cell-to-cell transmission, induce neuroinflammation, destruct the balance of cellular protein homeostasis, and eventually cause cell death.[30,31] Molecular chaperones play an essential role in safeguarding these amyloid proteins from pathological aggregation. Numerous chaperones including Hsp40, Hsp70, Hsp27, Hsp104, and Valosin-containing protein (VCP)/p97 were found to exhibit potent chaperone activities in preventing or even reversing protein amyloid aggregation (see Figure 12.2).[32] Importantly, many genetic mutations identified from familial ALS and frontotemporal dementia (FTD) patients are localized on these molecular chaperones such as Hsp27 and VCP, which may impair the chaperone activity and facilitate protein pathological aggregation under diseased condition.[33,34] This provides strong genetic evidence supporting the important role of chaperones in preventing pathological protein aggregation related

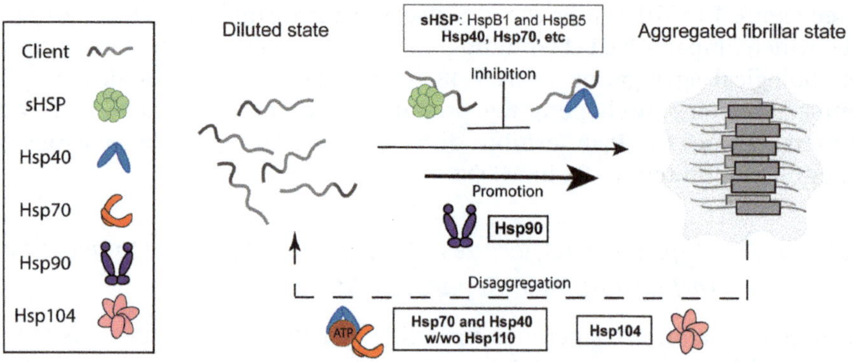

Figure 12.2 Chaperones in modulating protein aggregation. Numerous chaperones including sHSP, Hsp40, Hsp70, and others exhibit potent chaperone activities in preventing protein amyloid aggregation. While, Hsp90 can directly bind to Tau with the assistance of co-chaperones, which facilitates Tau fibrillation. In addition, several chaperones, such as Hsp104, Hsp110, Hsp70, and Hsp40 may work alone or together to dissolve amyloid fibrils.

to ND. Interestingly, some chaperones such as Hsp90 and its co-chaperone can promote amyloid fibrillation (see Figure 12.2), suggesting their deleterious effects in modulating amyloid protein aggregation.[35] Thus, different chaperones play distinct and even opposing roles in regulating amyloid proteins under different physiological or physiological conditions.

12.1.2.1 Chaperones to Prevent Protein Aggregation

Many different canonical and noncanonical molecular chaperones were identified to exhibit potent activities in preventing pathological aggregation of a variety of different amyloid proteins including Tau, α-syn, Abeta, prion, and IAPP.[36–40] The small heat shock proteins (sHSPs) including Hsp27 and HspB5 are intensively studied for their chaperone activity in preventing amyloid aggregation of Tau and α-syn.[41–43] Intriguingly, the anti-aggregation activity of sHSPs is fine-tuned by its NTD phosphorylation.[16,43,44] Under normal condition, Hsp27 and HspB5 forms heterogenous multimers (20–40 mer) with a moderate chaperone activity for inhibiting α-syn aggregation.[45] Upon acute or chronic stress, the NTD of Hsp27 can be phosphorylated which results in the dissociation of the multimer into low-molecular weight oligomer and releasing its client binding motif from the interior of the multimer.[16] Meanwhile, the chaperone activity of Hsp27 against α-syn aggregation is elevated dramatically.[46] Thus, the chaperone activity of Hsp27 is coupled by its conformational transition and self-assembly which can be regulated by external stimulus.

Systematical characterization of different molecular chaperones for binding to α-syn revealed that a wide spectrum of canonical and noncanonical chaperones including Hsp90, Hsc70, Hsp40, and Hsp27 can directly bind α-syn for inhibiting its amyloid aggregation.[36,47] Importantly, different chaperones can target different regions of α-syn. For example, Hsp90, Hsp27, and Hsc70 can bind to the N-terminal of α-syn, while Hdj1 binds to the C-terminal of α-syn. Moreover, mixing different chaperones together may dramatically increase the inhibitory activity for α-syn.[36] Similar phenomena were also observed for characterizing molecular chaperones of Tau. Different chaperones including Hsp90, Hsc70, Hsp27, and Hsp40 bind to different regions of Tau and display distinct anti-aggregation activity of Tau.[48–51] Interestingly, a recent study found that the enzyme nicotinamide mononucleotide adenylyltransferase (NMNAT) may act as a noncanonical chaperone and uses its enzymatic pocket to capture phosphorylated Tau (pTau) by specifically binding to its disease-related phosphorylation sites of the microtubule-binding domains.[37] Moreover, NMNAT efficiently inhibits amyloid fibrillation of pTau *in vitro* and prevents its aggregation and pathology in a fly model of Tauopathy.[37] Altogether, different chaperones may form an efficient chaperone network for binding the different regions or distinct post-translational modified species of the client amyloid proteins (*e.g.*, α-syn, Tau, and pTau) and cooperatively prevent their pathological amyloid aggregation in cells.

12.1.2.2 Chaperones to Reverse Protein Aggregation

Hsp104 from yeast has long been investigated for its potent chaperone activity in reversing preformed amyloid fibrils.[52] This disaggregation activity was believed to be essential in processing yeast prions for functional propagation.[53] Strikingly, Hsp104 exhibits a strong activity in dissolving fibrils formed by a wide spectrum of different human amyloid proteins including α-syn, Tau, and TDP-43.[54] Structural study revealed that Hsp104 features hexameric ring complexes with a spiral architecture.[55] Notably, mutagenesis screening revealed that single-point mutation of Hsp104 including A503V and Y507C may dramatically boost the chaperone activity of Hsp104 to disaggregate fibrils,[54] indicating a potential for engineering Hsp104 as super disaggregation machinery for clearing up the pathological protein fibrils (see Figure 12.2).

Unlike yeast Hsp104 which can dissolve amyloid fibrils on its own, mammalian Hsp110 requires assist of Hsp70 and Hsp40 for disaggregating amyloid fibrils.[56] Interestingly, recent works found that Hsp70 and Hsp40 are able to jointly disassemble preformed fibrils of α-syn in the presence or absence of Hsp110.[56,57] In this scenario, Hsp40 may firstly capture the flexible C-terminal of α-syn which is flanking on the fibril surface, and then recruit Hsp70 and Hsp110 for pulling α-syn fibril apart.[57] Thus, this disaggregase system contains multiple chaperones which transiently interact with each other, and cooperatively capture and remodel the aggregated fibrillar form of α-syn (see Figure 12.2). Compared to the traditional two-component system (chaperone–client), this disaggregation system is more complicated and requires integrating multiple biophysical approaches to study the structural basis of the complicated interplay between chaperones and the insoluble form of client proteins, which will be further discussed in the method description part.

12.1.2.3 *Chaperones to Promote Protein Aggregation and Accumulation*

Rather than preventing or dissolving protein aggregates, molecular chaperones can surprisingly promote protein aggregation and prevent them from clearance.[35,58] The most studied case is Hsp90 which can directly bind to Tau and facilitate its pathological fibrillation (see Figure 12.2).[35] A structural study combining solution NMR and small-angle X-ray scattering (SAXS) revealed that Hsp90 binds to a board region of Tau in the microtubule-binding domain including the key amyloidogenic region by forming a relatively weak but largely extended (over 100 Å long) interface.[51] The Hsp90-binding region of Tau is distinct from that of Hsp70 binding. Moreover, the Hsp90 co-chaperone – Aha1 can dramatically enhance the Hsp90-facilitated Tau fibrillation.[35] While another co-chaperone FKBP51 is found to interact with Hsp90 for synergizing to impair Tau clearance.[58] This result demonstrated that with the help of different co-chaperones, Hsp90 may impair Tau homeostasis by promoting Tau aggregation and accumulation. However, how different co-chaperones modulate Hsp90 activity and why the Tau binding pattern of Hsp90 results in promoting but not preventing Tau fibrillation remain poorly understood.

In summary, different molecular chaperones exhibit a broad spectrum of activities to regulate protein aggregation by targeting different regions of the client protein. Under different cellular circumstances, different subsets of molecular chaperones may be recruited and activated to fulfil their functions in either preventing or promoting protein aggregation. With the accumulation of our knowledge on how chaperone network modulates pathological amyloid protein, we could start to think of developing small molecule-based modulators to fine-tune the activity of special chaperone networks for preventing amyloid aggregation and pathology of amyloid proteins.

12.2 Methods to Investigate Chaperones Interacting with Client Proteins in Different States

Protein phase separation and aggregation feature protein conformational change, dynamic (dis)assembly, and transition between different states including the diluted state, the dynamic liquid-like phase separated state, and the more solidified amyloid aggregated state.[2,24,59] The same proteins may exhibit largely distinct physiochemical and structural properties in different states. Therefore, structural and mechanistic investigation of the interplay between chaperones and the client proteins in different states is highly challenging and requires a collection of different and integrative biophysical and biochemical approaches. In the following, we will introduce different methods used to study how different chaperones recognize their client proteins in the three different states.

12.2.1 Methods for Exploring the Interaction of Chaperones with Client Proteins in the Diluted State

12.2.1.1 Solution NMR Spectroscopy

Intensive studies demonstrated that the intrinsic disorder region (IDR) rather than the structured region of proteins has a high propensity for mediating protein phase separation and amyloid aggregation.[2,60] Indeed, several pathological amyloid proteins including α-syn, Tau, IAPP, and Abeta are intrinsic disordered proteins (IDPs). Compared to other structural biology methods including X-ray crystallography and cryo-EM, solution NMR is the most powerful approach to investigate the structure and molecular interaction of IDPs. Indeed, solution NMR has been widely used to characterize the conformation of IDPs in the diluted state as well as their interplay with different chaperones.[8,10,61,62] Among the different NMR-based methods, the two-dimensional (2D) ^{1}H–^{15}N heteronuclear single quantum coherence (HSQC) spectrum is a simple fingerprint spectrum that could be used to map the interface of the chaperone and IDPs (see Figure 12.3). *In vitro* titration of unlabelled chaperone to the ^{15}N-labelled client may identify the residues within the binding interface of clients to the chaperone by monitoring the chemical shift perturbations or intensity changes of each residue of client protein upon chaperone titration. For instance, titration of different chaperones including Hsp90, Hsp27, Hsp40, and Hsp60 to ^{15}N-labelled monomeric α-syn aqueous solution revealed that α-syn employs different regions to bind different chaperones (*e.g.*, the C-terminal of α-syn for Hsp40 binding, the N-terminal region for Hsp27 and Hsp90 binding).[36,47] Conversely, titration of the unlabelled client protein to the ^{15}N-labelled chaperone may determine the key residues & regions used by chaperones for capturing clients. However, as for the high molecular-weight chaperones (*e.g.*, Hsp90, Hsp60, and Hsp104) and large client proteins (*e.g.*, Tau and TDP-43), too many signals and slow tumbling of large proteins cause severe signal overlap and linewidth broadening of the conventional HSQC spectrum. Then, transverse relaxation-optimized spectroscopy (TROSY) at high magnetic field strengths, in combination with various isotope labelling techniques, such as deuteration, side-chain labelling of Leu, Ile, and Phe, were used to reduce the signal overlap and achieve the assignment and interactions of proteins with molecular masses up to 100 kDa.[63] By using selectively isoleucine-labelled Hsp90 and methyl-TROSY, the N-terminal and middle domain of Hsp90 (170 kDa for Hsp90 dimer) were identified to bind Tau.[51] Nevertheless, this powerful approach helps to reveal the structural basis of the interplay between many chaperone–client pairs including Tau–Hsp90 and transthyretin–Hsp90. Moreover, by using TROSY and selective ^{13}C-methyl labelling of Ala, Met, Ile, Leu, Val, and Thr residues, the solution structure and dynamics features of an Hsp40 in complex with an unfolded client protein were determined by NMR.[64] The atomic complex structure reveals that Hsp40 recognizes the unfolded client by using its

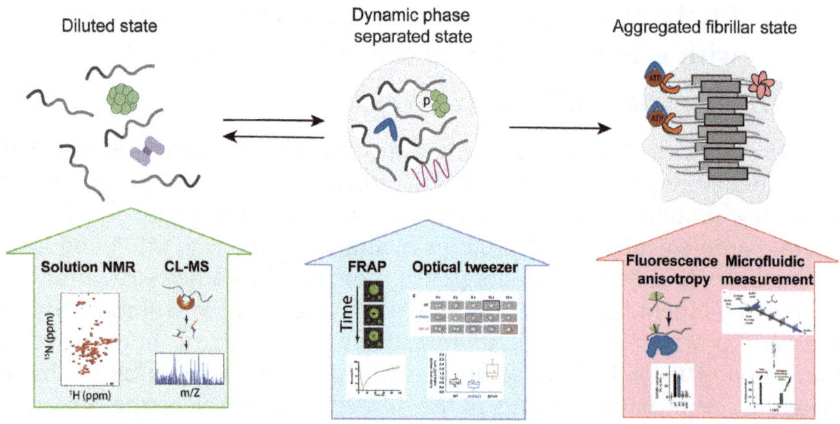

Figure 12.3 Methods to investigate the interaction between chaperones and clients in three different states. Solution NMR is the most powerful approach for investigating the structure and molecular interaction between chaperones and clients in diluted states. CL–MS is another approach to identifying the interfaces of the chaperone–client pair. In the dynamics-separated state, FRAP and optical tweezer experiments are widely used to monitor the dynamic interaction between chaperones and clients. Different protein-binding assays, *e.g.*, fluorescence anisotropy and microfluidic platform, are used to study the binding processes between chaperones and amyloid fibrils. The two figures for optical tweezers were reproduced from ref. 78, https://www.nature.com/articles/s41467-019-09902-7, under the terms of the CC BY 4.0 license, http://creativecommons.org/licenses/by/4.0/. The bottom figure for fluorescence anisotropy was reproduced from ref. 56 with permission from Springer Nature, Copyright 2020. The two figures for microfluidic measurement were reproduced from ref. 87 with permission from American Chemical Society, Copyright 2016.

multivalent binding sites (up to four) to dynamically interact with the 7 strong and 8 weak binding sites of the client protein. Competitive binding of the C-tail of Hsp70 to the client binding sites of Hsp40 displaces the unfolded client. This complex structure greatly expands our knowledge of how Hsp40 exerts its chaperone activity with Hsp70.

Moreover, In-cell NMR, which directly delivers the *in vitro* purified ^{15}N-labelled proteins into the cell and records the HSQC spectra under different cellular treatments,[65–67] provides a powerful approach to investigate the interaction between the client protein and chaperones in cells. A recent study by Hiller and co-workers explored the interplay between α-syn and different chaperones including Hsp90 and Hsc70 in cells, and revealed these chaperones may bind the N-terminal and Tyr39-containing region of α-syn in cells to stabilize it from binding to the mitochondria membrane and aggregation.[47]

Paramagnetic NMR-based method such as paramagnetic relaxation enhancement (PRE) is powerful and extensively used for probing the transient and long-range interactions between proteins.[68] *S*-(1-Oxyl-2,2,5,5 tetramethyl-2,5-dihydro-1*H*-pyrrol-3-yl) methyl methanesulfonothioate (MTSL) is a commonly used cysteine-specific spin label, which can provide distance information up to ~25 Å with an MTSL label.[69] One study used PRE to investigate the binding of MTSL-labelled Tau proteins to Hsp90, in which three different positions located in the Hsp90-binding surface of Tau were independently labelled and caused paramagnetic signal broadening of ^{15}N–Hsp90, defining a large interface on Hsp90 for Tau binding.[70] Another study showed that the binding of Hsp27 to MTSL-labelled FUS-LC restored the global paramagnetic signal broadening, implying that Hsp27 binds to FUS LC and diminishes the self-association of FUS LC.[16] In general, the PRE experiment is sensitive in detecting the long range and transient interactions of low populated components and is a powerful tool for studying the weak interactions between the chaperone and amyloid clients. Altogether, solution NMR spectroscopy provides an unparalleled method to investigate the structural basis of the interaction between different chaperones and client proteins, especially for the weak interaction between chaperone and IDPs in the diluted state.

12.2.1.2 Cross-linking Mass Spectrometry (CL–MS)

The transient interaction between chaperones and clients can be captured by covalently linking the two proteins together. The cross-linked regions from the chaperone–client interface with spatial proximity can be further identified by mass spectrometry.[18,49] Different types of cross-linkers were developed to cross-link different types of residues together with a range of the linking distance. For example, bis(sulfosuccinimidyl)suberate (BS^3), which is a primary amine-reactive crosslinker, can covalently link paired lysine residues in spatial proximity with a C_α–C_α distance $<$24 Å. While, for linking lysine residues to the negatively charged residues including aspartic acid and glutamine acid, the cross-linkers 1-ethyl-3-(3-dimethylaminopropyl) carbodiimide (EDC) and sulfo-*N*-hydroxysulfosuccinimide (sulfo-NHS) can be used. EDC directly conjugates carboxylates (–COOH) to primary amines (–NH_2) between two cross-linked residues, and sulfo-NHS is further introduced for stabilization.

CL–MS has been used to successfully identify the interfaces of several chaperone–client pairs (see Figure 12.3). For example, one study used BS^3 to cross-link pTau and NMNAT for monitoring the protein interface and identified several lysine residues around the enzymatic pocket of NMNAT for binding pTau.[37] This observation well explains why the enzymatic substrate of NMNAT may diminish the binding and chaperone activity of NMNAT to pTau. Another study used the bifunctional reagent disuccinimidyl dibutyricurea (DSBU) which primarily cross-links lysine residues with a C_α–C_α distance as far as 35 Å. They identified that the chaperone SurA can capture its client – the outer membrane protein (OMP) by using its core domain.[71]

12.2.1.3 Cryo-EM and Other Integrative Methods

Since the resolution revolution of cryo-EM technology in 2013,[72,73] cryo-EM has been emerging as one of the most powerful methods to determine the structures of large protein complexes.[55,72,74] However, due to the high conformational flexibility of clients, most of which are IDPs, and the relatively weak and transient interaction between the chaperones and client proteins, it remains challenging to determine the cryo-EM structures of chaperones in complex with full-length client proteins. Whereas, cryo-EM structures of several important chaperones including Hsp104 and Hsp90 as well as the complex structures of chaperones with the corresponding short binding segments were determined.[55,75,76] These cryo-EM studies provide important structural information for understanding how different types of chaperones recognize their client proteins. For example, Hsp104 captures and translocates the clients by undergoing a ratchet-like conformational change fuelled by ATP hydrolysis and employs a highly conserved pore-loop Tyr residues to thread the unfolded polypeptides through the internal channel for clients unfolding.[55]

In addition to NMR, cryo-EM, and CL–MS, other methods including small angle X-ray scattering (SAXS), molecular dynamics (MD) simulation, and single molecular approach were also used to investigate the structural basis of interaction between chaperones and clients in the diluted state. A successful case to combine different integrative methods together is the structural characterization of the interplay between Hsp90 and Tau.[51] NMR titration was performed for the first time to map the interacting residues between Hsp90 and Tau, which reveals that Hsp90 utilizes its extended binding interface consisting of the hydrophobic and negatively charged residues from the N-terminal and middle domain to bind to a broad region within the microtubule-binding domain of Tau including the two aggregation-prone segments (^{275}VQIIN^{279} and ^{306}VQIVY^{310}). SAXS was then used to explore the conformational change of Hsp90 upon binding to Tau. Later, the same group combined NMR titration and PRE to map the interaction between Hsp90, Tau, and a co-chaperone – PPIase FKBP51.[70] Another remarkable study combined NMR titration, SAXS, X-ray crystallography, and other biophysical approaches to reveal kapβ2 as a noncanonical chaperone binding to its client protein FUS by both specifically capturing its C-terminal PY-NLS region and weakly binding to multiple regions across full-length FUS.[10]

12.2.2 Methods to Study the Interaction of Chaperones with Client Proteins in Dynamic Phase Separated State

As introduced in the first section, chaperones are widely involved in the biogenesis of different MLOs and play important roles in maintaining protein homeostasis of MLOs and the dynamic nature of proteins in the phase-separated state.[24] Thus, it's of great value to investigate at the atomic level how chaperones recognize their client proteins in the dynamic phase separated state. However, the biophysical properties of proteins in the dynamic phase separated state is largely different from those in the diluted state. The proteins in the phase-separated state are condensed with a high protein density, and the protein conformation is highly dynamic, especially for the IDPs.[77] Weak and multivalent inter- and intra-molecular interaction govern the formation and maintenance of protein phase separated state, and the protein–protein interaction is highly diverse with high heterogeneity.[2,3,59] Thus, classic biophysical approaches are difficult for exploring the structural and mechanistic details of the interaction between chaperones and client proteins in the phase-separated state. The atomic view of the interplay between chaperones and client proteins in the liquid-like protein condensate is still absent. Nevertheless, a couple of different methods including optical spectroscopy, single molecule method, and CL–MS have been used to probe the interaction and activity of chaperones for maintaining clients in the phase-separated state (see Figure 12.3).

12.2.2.1 *Optical Spectroscopy*

Fluorescence recovery after photobleaching (FRAP) experiment has long been used to monitor the mobility and interior dynamics of proteins in the liquid-like droplet.[5] The rapid fluorescence recovery rate of fluorescence-labelled proteins implies the dynamic nature of the targeted protein in droplets. The time-course FRAP is powerful to monitor the liquid-to-solid phase transition and solidification of the proteins in droplets (see Figure 12.3), which may lead to the disruption of protein dynamic phase separation and formation of pathological protein aggregates. In this scenario, a gradual decrease in the recovery rate and delay of the recovery time of the photo-bleached protein in the droplet can be detected. Importantly, by performing the time-course FRAP assay, several studies showed that different chaperones including Hsp27, HspB8, Hsp40, and Hsp70 can co-phase separate with their clients, maintain the high fluorescence recovery rate of the client proteins in the liquid-like state, and efficiently prevent their clients against liquid-to-solid phase transition both in the liquid-like droplet *in vitro* and the protein condensates in cells.[8,9,16–18] The FRAP data provided strong evidence to support that chaperone plays a key role in maintaining the dynamic nature of client proteins in the phase-separated state.

12.2.2.2 *Single Molecule Methods*

Several single molecule-based approaches including the optical tweezer experiment and single molecular FRET (smFRET) were recently applied to investigate the interplay between the chaperone and client in the dynamic phase-separated state.[18,78,79] Optical tweezer is a powerful approach to determining the fusion rate of protein droplets by trapping two protein droplets in two optical traps and moving them into contact for triggering droplet fusion. By calculating the fusion rate, one can evaluate the material property (*e.g.*, viscosity, protein diffusion rate) of the liquid-like droplet formed by the target proteins (see Figure 12.3). A previous study used this approach to demonstrate that the single-point disease-related mutation (D214V) of hnRNPA1 can decrease the fusion rate of hnRNPA1 droplets and facilitate the liquid-to-solid phase transition of hnRNPA1.[78] Recently, the optical tweezer experiment was used to measure the fusion property of droplets formed by FUS alone and FUS mixed with HspB8. Consistent with the FRAP results, optical tweezer data showed HspB8 can keep the high fusion rate of FUS droplets, strengthening the essential role of HspB8 in chaperoning FUS in the highly dynamic liquid-like phase-separated state.[18]

The smFRET is sensitive to probe the conformational change of proteins at the single-molecule level. This approach was recently applied to study the conformational dynamics of Tau in the liquid-like droplet.[79] In this study, several FRET pairs were designed and installed in the different regions of Tau. By monitoring the smFRET signal and comparing the FRET efficiency histograms of Tau in the diluted state and the phase-separated state, the authors show that, upon forming liquid-like droplets, Tau undergoes conformational changes to form a more expanded conformation which may further facilitate aberrant liquid-to-solid phase transition of Tau in the liquid-like droplet. This demonstrated the power of smFRET in monitoring the protein conformation in the dynamic phase-separated state. However, how previously identified different chaperones of Tau such as Hsp90, Hsp27, and Hsp70 influence the conformation of Tau in the liquid-like droplet remains to be investigated.

12.2.2.3 CL–MS and Other Methods

As mentioned above, CL–MS is a powerful method to probe the interaction of chaperones and clients in the diluted state. Notably, it has recently been applied to map the intra- and inter-molecular interaction of protein complexes in the phase-separated state.[18] This study examined the interaction between HspB8 and FUS by using the crosslinker disuccinimidyl suberate (DSS) with the spacer arm of 11.4 Å, which can cross-link the lysine residues from the same protein or two interacting proteins in spatial proximity. Interestingly, in the diluted state, the CL–MS only identified very limited cross-linking peptides between HspB8 and FUS. Whereas, the cross-linked peptides were dramatically increased once HspB8 and FUS co-phase separated in the dynamic phase separated state. This demonstrated that HspB8 may form condensate-specific interaction with FUS in the liquid droplet which may enable HspB8 chaperone to efficiently prevent FUS solidification and aggregation in the highly dynamic liquid-like droplet.

Other biophysical and computational methods including solution NMR,[61] fluorescence correlation spectroscopy,[80] nanoscale localized Infrared spectroscopy,[10] and MD simulation[81] have been successfully applied to investigate the overall protein conformation, secondary structure changes of proteins, and protein dynamics of different protein samples in the liquid-like droplets. Unfortunately, these powerful methods are not yet used to study the interplay between chaperones and clients in the liquid-like droplet according to the published works. As shown by the CL–MS results of interaction between HspB8 and FUS,[18] chaperones may adopt distinct conformations and binding patterns to chaperone their clients in the liquid-like state compared to that in the diluted state. Thus, it is worth exploring the molecular mechanism and structural basis of chaperone–client interaction within the liquid-like droplet by combining the abovementioned as well as other newly developed integrative methods.

12.2.3 Methods to Explore the Interaction of Chaperones with Client Proteins in the Aggregated Fibrillar State

Unlike the highly dynamic liquid-like phase-separated state, the aggregated fibrillar state of client proteins is relatively stable and features well-defined fibrillar architectures.[74,82] Proteins can enter the aggregated state by either self-assembling directly from the diluted state or undergoing the liquid-to-solid phase transition from the highly dynamic phase-separated state.[2,3,27,28,59] Many pathological proteins including Tau, α-syn, Abeta, and FUS can form pathological amyloid fibrils which are closely associated with a variety of devastating human diseases such as AD, PD, and ALS.[27,83–85] Different chaperones were found to modulate pathological liquid-to-solid phase transition and amyloid fibrillation of client proteins as well as interact with their clients in the aggregated fibrillar state.[16,24,56] Thus, developing methods to investigate the structural basis and molecular mechanism underlying how chaperones interact with clients in the aggregate state is of great importance.

12.2.3.1 Biochemical Methods

Different protein-binding assays (*e.g.*, fluorescence anisotropy, quartz crystal microbalance assay, surface plasmon resonance (SPR), and sedimentation assay) have been successfully conducted to measure the binding of chaperones and amyloid fibrils formed by different clients.[56,86] For example, Bukau and co-workers used fluorescence anisotropy to measure the binding affinity of DNAJB1 to α-syn monomer and fibrils (see Figure 12.3). Alexa Fluor 488-labelled DNAJB1 were prepared and titrated to α-syn monomer and fibril samples, respectively. Then, the binding curve of DNAJB1 to different α-syn species was drawn based on the measurement of steady-state fluorescence anisotropy and microscale thermophoresis. Surprisingly, the results revealed that DNAJB1 preferentially binds to the fibrillar state of α-syn with a 200-fold higher binding affinity compared to that of monomeric α-syn.[56] Despite that the underlying mechanism of this extremely high selective binding of DNAJB1 to fibrillar rather than monomeric α-syn client remains poorly understood, this result strongly suggests chaperones may play an important yet not fully-appreciated role in capturing and modulating the aggregated states of amyloid proteins.

12.2.3.2 Microfluidic Measurements

The microfluidic platform was recently used to study the physical and chemical properties underlying the binding processes between chaperones and amyloid clients, which monitors the micron-scale mass transport simultaneously in space and time for microfluidic diffusional sizing.[87] This approach uses a laminar flow within a microfluidic channel and creates a well-defined interface for the analyte and a running buffer (see Figure 12.3). Measurements are conducted by observing the diffusive spreading of the components of the analyte solution as they travel downstream along the channel. Global analysis of the combined space and time data enables the hydrodynamics radii of individual components within a complex mixture of different species and quantifies their relative concentrations. Recently, Knowles and co-workers used the microfluidic platform to probe the thermodynamics and kinetics of the binding between the chaperone αB-crystallin (also called HspB5) and α-syn amyloid fibrils.[88] They found that the high nanomolar binding of α-syn fibrils to HspB5 is driven by entropy rather than enthalpy, and the significant entropic gain originates from the disassembly of the oligomeric chaperones. These results suggest that oligomeric chaperones function as an entropic buffer system, in which chaperones are stored as inactive multimers and would release the smaller oligomeric active forms to chaperone clients by disassembling. The same group further combined microfluidic measurement and chemical kinetics to probe the disaggregation of α-syn fibrils by the Hsp70 family.[57] They found that α-syn fibrils could be completely disaggregated to monomers only when all the three chaperone components including Hsc70 and its co-chaperones DnaJB1 and Apg2, and ATP are present. In addition, they observed that α-syn monomers are removed directly from the fibril ends with little contribution from intermediate fibril fragmentation steps. All these findings extend our mechanistic understanding of how chaperones bind amyloid fibrils for disaggregation.

12.2.3.3 Other Methods

In the past ten years, cryo-EM and solid-state NMR have been widely used to determine the molecular structures of amyloid fibrils formed by different pathological proteins.[74,84,89] However, the structure of amyloid fibrils in complexes with chaperones has been lacking so far. Since emerging evidence demonstrated that the chaperones may tightly bind to amyloid fibrils, it will be very interesting to find the tightly interacted chaperone–fibril complex for investigating the structural details of the complex by using cryo-EM and/or solid-state NMR. One of the critical steps towards structure determination will be to prepare the homogenous and stable chaperone–fibril complex sample for structural characterization.

12.2.4 Summary

In this chapter, we summarized the emerging roles of molecular chaperones in regulating the client protein dynamic LLPS as well as pathological aggregation. Furthermore, we introduced different methods used for investigating

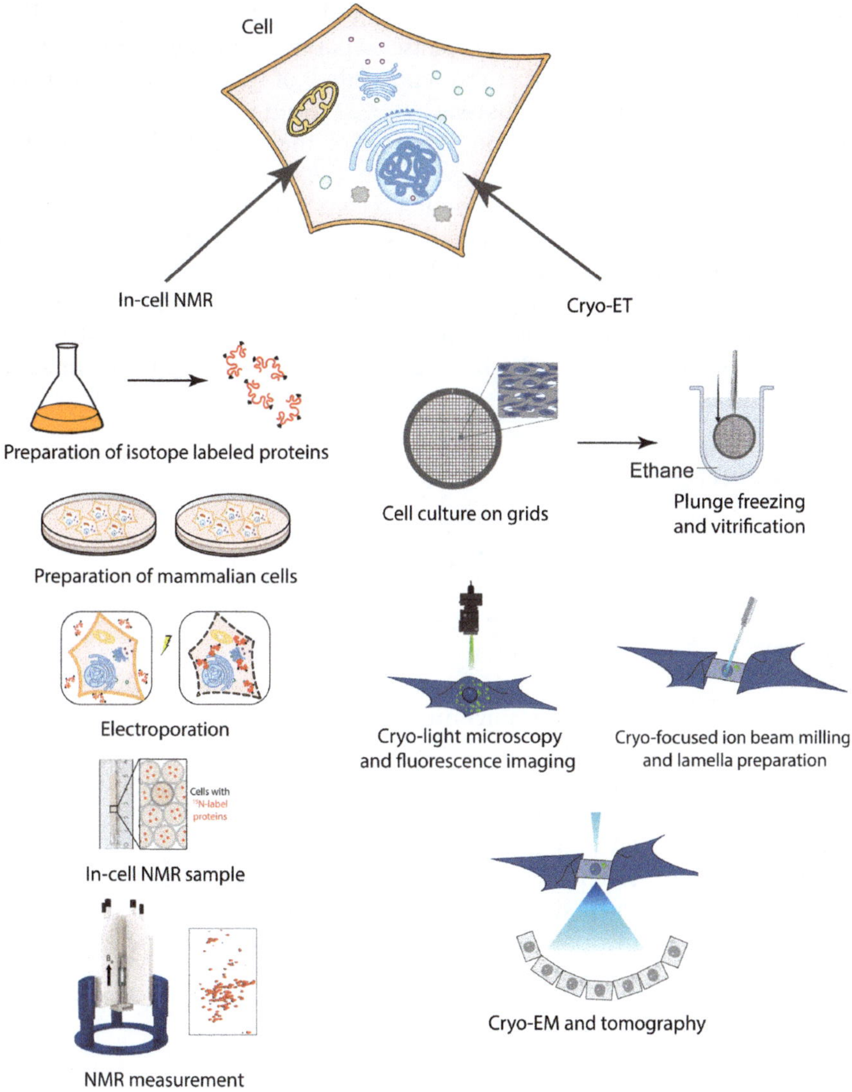

Figure 12.4 A schematic diagram of two powerful and promising approaches including in-cell NMR and cryo-ET to study the interaction between chaperones and clients in cells. The schematic diagram of cryo-ET was created from scratch with BioRender.com.

the interplay between chaperones and their client proteins in three different states including the diluted state, the dynamic phase separated state, and the aggregated fibrillar state. Since the same client protein may exhibit largely distinct properties in different states, it will be of great importance to select a suitable combination of integrative methods to study the interaction between chaperones and their client proteins in different states. New methods are urgently needed to investigate the structural details of chaperone–client interaction in both the dynamic phase separated and aggregated states. Moreover, most of our structural knowledge of the cross-talk between chaperones and client proteins was derived from the simplified *in vitro* systems. How different chaperones recognize their clients for modulating their phase separation and aggregation in the context of the cellular environment remains poorly understood. A recent in-cell NMR study provides a good example to explore how chaperones maintain their client protein from aggregation in cells (see Figure 12.4).[47] New techniques such as cryo-electron tomography (cryo-ET) (see Figure 12.4), super-resolution microscopy, and single-molecule imaging may be important for providing more structural information about the interplay between chaperones and the different states of client proteins in cells.

Abbreviations

AD	Alzheimer's disease
ALS	Amyotrophic lateral sclerosis
BS^3	Bissulfosuccinimidylsuberate
CL–MS	Cross-linking mass spectrometry
Cryo-ET	Cryo-electron tomography
CTD	C-terminal domains
DSBU	Disuccinimidyl dibutyricurea
DSS	Disuccinimidyl suberate
EDC	1-Ethyl-3-(3-dimethylaminopropyl) carbodiimide
FRAP	Fluorescence recovery after photobleaching
FTD	Frontotemporal dementia
HSQC	Heteronuclear single quantum coherence
IDPs	Intrinsically disordered proteins
IDR	Intrinsic disorder region
Kapβ2	Karyopherin-β2
LC	Low complexity
LLPS	Liquid–liquid phase separation
lncRNA	Long noncoding RNA
MD	Molecular dynamics
MLOs	Membraneless organelles
MTSL	*S*-(1-Oxyl-2,2,5,5 tetramethyl-2,5-dihydro-1*H*-pyrrol-3-yl) methyl methanesulfonothioate
ND	Neurodegenerative diseases
NTD	N-terminal domain
OMP	Outer membrane protein
PD	Parkinson's disease
PRE	Paramagnetic relaxation enhancement
pTau	Phosphorylated Tau
PY-NLS	Proline-tyrosine nuclear localization signal
RGG	Arg-Gly-Gly

RRM	RNA recognition motif
SAXS	Small-angle X-ray scattering
SGs	Stress granules
sHSPs	Small heat shock proteins
smFRET	Single molecular FRET
SPR	Surface plasmon resonance
Sulfo-NHS	Sulfo-*N*-hydroxysulfosuccinimide
TROSY	Transverse relaxation-optimized spectroscopy
VCP	Valosin-containing protein

Acknowledgements

We thank Jinge Gu, Yunpeng Sun, and Chuchu Wang for their help in making figures. We thank the funding support by the National Natural Science Foundation (NSF) of China (82188101, 32171236, and 31872716 to C. L.; 32170683 to D. L.), the Major State Basic Research Development Program (2019YFE0120600 to C. L.), the Science and Technology Commission of Shanghai Municipality (STCSM) (Grant No. 20XD1425000 and 2019SHZDZX02 to C. L.), CAS project for Young Scientists in Basic research (Grant No. YSBR-009 to S. Z.), the Shanghai Pilot Program for Basic Research – Chinese Academy of Science, Shanghai Branch (Grant No. CYJ-SHFY-2022-0xx to C. L.).

References

1. D. Balchin, M. Hayer-Hartl and F. U. Hartl, *Science*, 2016, **353**, aac4354.
2. S. F. Banani, H. O. Lee, A. A. Hyman and M. K. Rosen, *Nat. Rev. Mol. Cell Biol.*, 2017, **18**, 285–298.
3. H. Zhang, X. Ji, P. Li, C. Liu, J. Lou, Z. Wang, W. Wen, Y. Xiao, M. Zhang and X. Zhu, *Sci. China: Life Sci.*, 2020, **63**, 953–985.
4. K. You, Q. Huang, C. Yu, B. Shen, C. Sevilla, M. Shi, H. Hermjakob, Y. Chen and T. Li, *Nucleic Acids Res.*, 2020, **48**, D354–D359.
5. Y. Shin and C. P. Brangwynne, *Science*, 2017, **357**, eaaf4382.
6. S. Markmiller, S. Soltanieh, K. L. Server, R. Mak, W. Jin, M. Y. Fang, E. C. Luo, F. Krach, D. Yang, A. Sen, A. Fulzele, J. M. Wozniak, D. J. Gonzalez, M. W. Kankel, F. B. Gao, E. J. Bennett, E. Lecuyer and G. W. Yeo, *Cell*, 2018, **172**, 590–604.e513.
7. S. Jain, J. R. Wheeler, R. W. Walters, A. Agrawal, A. Barsic and R. Parker, *Cell*, 2016, **164**, 487–498.
8. J. Gu, C. Wang, R. Hu, Y. Li, S. Zhang, Y. Sun, Q. Wang, D. Li, Y. Fang and C. Liu, *Cell Res.*, 2021, **31**, 1024–1027.
9. H. Yu, S. Lu, K. Gasior, D. Singh, S. Vazquez-Sanchez, O. Tapia, D. Toprani, M. S. Beccari, J. R. Yates 3rd, S. Da Cruz, J. M. Newby, M. Lafarga, A. S. Gladfelter, E. Villa and D. W. Cleveland, *Science*, 2021, **371**, eabb4309.
10. T. Yoshizawa, R. Ali, J. Jiou, H. Y. J. Fung, K. A. Burke, S. J. Kim, Y. Lin, W. B. Peeples, D. Saltzberg, M. Soniat, J. M. Baumhardt, R. Oldenbourg,

A. Sali, N. L. Fawzi, M. K. Rosen and Y. M. Chook, *Cell*, 2018, **173**, 693–705.e622.
11. S. Qamar, G. Wang, S. J. Randle, F. S. Ruggeri, J. A. Varela, J. Q. Lin, E. C. Phillips, A. Miyashita, D. Williams, F. Strohl, W. Meadows, R. Ferry, V. J. Dardov, G. G. Tartaglia, L. A. Farrer, G. S. Kaminski Schierle, C. F. Kaminski, C. E. Holt, P. E. Fraser, G. Schmitt-Ulms, D. Klenerman, T. Knowles, M. Vendruscolo and P. S. George-Hyslop, *Cell*, 2018, **173**, 720–734.e715.
12. M. Hofweber, S. Hutten, B. Bourgeois, E. Spreitzer, A. Niedner-Boblenz, M. Schifferer, M. D. Ruepp, M. Simons, D. Niessing, T. Madl and D. Dormann, *Cell*, 2018, **173**, 706–719.e713.
13. L. Guo, H. J. Kim, H. Wang, J. Monaghan, F. Freyermuth, J. C. Sung, K. O'Donovan, C. M. Fare, Z. Diaz, N. Singh, Z. C. Zhang, M. Coughlin, E. A. Sweeny, M. E. DeSantis, M. E. Jackrel, C. B. Rodell, J. A. Burdick, O. D. King, A. D. Gitler, C. Lagier-Tourenne, U. B. Pandey, Y. M. Chook, J. P. Taylor and J. Shorter, *Cell*, 2018, **173**, 677–692.e620.
14. Z. C. Zhang and Y. M. Chook, *Proc. Natl. Acad. Sci. U. S. A.*, 2012, **109**, 12017–12021.
15. T. Murakami, S. Qamar, J. Q. Lin, G. S. Schierle, E. Rees, A. Miyashita, A. R. Costa, R. B. Dodd, F. T. Chan, C. H. Michel, D. Kronenberg-Versteeg, Y. Li, S. P. Yang, Y. Wakutani, W. Meadows, R. R. Ferry, L. Dong, G. G. Tartaglia, G. Favrin, W. L. Lin, D. W. Dickson, M. Zhen, D. Ron, G. Schmitt-Ulms, P. E. Fraser, N. A. Shneider, C. Holt, M. Vendruscolo, C. F. Kaminski and P. S. George-Hyslop, *Neuron*, 2015, **88**, 678–690.
16. Z. Liu, S. Zhang, J. Gu, Y. Tong, Y. Li, X. Gui, H. Long, C. Wang, C. Zhao, J. Lu, L. He, Y. Li, Z. Liu, D. Li and C. Liu, *Nat. Struct. Mol. Biol.*, 2020, **27**, 363–372.
17. J. Gu, Z. Liu, S. Zhang, Y. Li, W. Xia, C. Wang, H. Xiang, Z. Liu, L. Tan, Y. Fang, C. Liu and D. Li, *Proc. Natl. Acad. Sci. U. S. A.*, 2020, **117**, 31123–31133.
18. E. E. Boczek, J. Fursch, M. L. Niedermeier, L. Jawerth, M. Jahnel, M. Ruer-Gruss, K. M. Kammer, P. Heid, L. Mediani, J. Wang, X. Yan, A. Pozniakovski, I. Poser, D. Mateju, L. Hubatsch, S. Carra, S. Alberti, A. A. Hyman and F. Stengel, *Elife*, 2021, **10**, e69377.
19. A. Khong, T. Matheny, S. Jain, S. F. Mitchell, J. R. Wheeler and R. Parker, *Mol. Cell*, 2017, **68**, 808–820.e805.
20. D. Stanek and A. H. Fox, *Curr. Opin. Cell Biol.*, 2017, **46**, 94–101.
21. A. H. Fox, S. Nakagawa, T. Hirose and C. S. Bond, *Trends Biochem. Sci.*, 2018, **43**, 124–135.
22. M. Wu, G. Xu, C. Han, P. F. Luan, Y. H. Xing, F. Nan, L. Z. Yang, Y. Huang, Z. H. Yang, L. Shan, L. Yang, J. Liu and L. L. Chen, *Science*, 2021, **373**, 547–555.
23. C. Wang, Y. Duan, G. Duan, Q. Wang, K. Zhang, X. Deng, B. Qian, J. Gu, Z. Ma, S. Zhang, L. Guo, C. Liu and Y. Fang, *Mol. Cell*, 2020, **79**, 443–458.e447.
24. D. Li and C. Liu, *Trends Cell Biol.*, 2022, **32**, 1–3.

25. Q. Guo, X. Shi and X. Wang, *Non-coding RNA Res.*, 2021, **6**, 92–99.
26. V. M. Lee, M. Goedert and J. Q. Trojanowski, *Annu. Rev. Neurosci.*, 2001, **24**, 1121–1159.
27. D. Eisenberg and M. Jucker, *Cell*, 2012, **148**, 1188–1203.
28. D. Li and C. Liu, *Nat. Chem. Biol.*, 2021, **17**, 237–245.
29. C. M. Dobson, T. P. J. Knowles and M. Vendruscolo, *Cold Spring Harbor Perspect. Biol.*, 2020, **12**, a033878.
30. J. M. Nussbaum, S. Schilling, H. Cynis, A. Silva, E. Swanson, T. Wangsanut, K. Tayler, B. Wiltgen, A. Hatami, R. Ronicke, K. Reymann, B. Hutter-Paier, A. Alexandru, W. Jagla, S. Graubner, C. G. Glabe, H. U. Demuth and G. S. Bloom, *Nature*, 2012, **485**, 651–655.
31. A. M. Cuervo, L. Stefanis, R. Fredenburg, P. T. Lansbury and D. Sulzer, *Science*, 2004, **305**, 1292–1295.
32. A. Wentink, C. Nussbaum-Krammer and B. Bukau, *Cold Spring Harbor Perspect. Biol.*, 2019, **11**, a033969.
33. A. E. Renton, A. Chio and B. J. Traynor, *Nat. Neurosci.*, 2014, **17**, 17–23.
34. S. Capponi, T. Geuens, A. Geroldi, P. Origone, S. Verdiani, E. Cichero, E. Adriaenssens, V. De Winter, M. Bandettini di Poggio, M. Barberis, A. Chio, P. Fossa, P. Mandich, E. Bellone and V. Timmerman, *Hum. Mutat.*, 2016, **37**, 1202–1208.
35. L. B. Shelton, J. D. Baker, D. Zheng, L. E. Sullivan, P. K. Solanki, J. M. Webster, Z. Sun, J. J. Sabbagh, B. A. Nordhues, J. Koren 3rd, S. Ghosh, B. S. J. Blagg, L. J. Blair and C. A. Dickey, *Proc. Natl. Acad. Sci. U. S. A.*, 2017, **114**, 9707–9712.
36. C. Jia, X. Ma, Z. Liu, J. Gu, X. Zhang, D. Li and S. Zhang, *Front. Neurosci.*, 2019, **13**, 1124.
37. X. Ma, Y. Zhu, J. Lu, J. Xie, C. Li, W. S. Shin, J. Qiang, J. Liu, S. Dou, Y. Xiao, C. Wang, C. Jia, H. Long, J. Yang, Y. Fang, L. Jiang, Y. Zhang, S. Zhang, R. G. Zhai, C. Liu and D. Li, *eLife*, 2020, **9**, e51859.
38. L. Huang, T. Agrawal, G. Zhu, S. Yu, L. Tao, J. Lin, R. Marmorstein, J. Shorter and X. Yang, *Nature*, 2021, **597**, 132–137.
39. N. Chilukoti, T. B. Sil, B. Sahoo, S. Deepa, S. Cherakara, M. Maddheshiya and K. Garai, *Biophys. J.*, 2021, **120**, 476–488.
40. E. G. Rikhvanov, N. V. Romanova and Y. O. Chernoff, *Prion*, 2007, **1**, 217–222.
41. D. Cox and H. Ecroyd, *Cell Stress Chaperones*, 2017, **22**, 589–600.
42. R. Freilich, M. Betegon, E. Tse, S. A. Mok, O. Julien, D. A. Agard, D. R. Southworth, K. Takeuchi and J. E. Gestwicki, *Nat. Commun.*, 2018, **9**, 4563.
43. Z. Liu, C. Wang, Y. Li, C. Zhao, T. Li, D. Li, S. Zhang and C. Liu, *J. Biol. Chem.*, 2018, **293**, 14880–14890.
44. M. F. Ahmad, B. Raman, T. Ramakrishna and C. M. Rao, *J. Mol. Biol.*, 2008, **375**, 1040–1051.
45. J. A. Aquilina, S. Shrestha, A. M. Morris and H. Ecroyd, *J. Biol. Chem.*, 2013, **288**, 13602–13609.

46. R. Bakthisaran, K. K. Akula, R. Tangirala and C. M. Rao, *Biochim. Biophys. Acta*, 2016, **1860**, 167–182.
47. B. M. Burmann, J. A. Gerez, I. Matecko-Burmann, S. Campioni, P. Kumari, D. Ghosh, A. Mazur, E. E. Aspholm, D. Sulskis, M. Wawrzyniuk, T. Bock, A. Schmidt, S. G. D. Rudiger, R. Riek and S. Hiller, *Nature*, 2020, **577**, 127–132.
48. H. E. R. Baughman, T. T. Pham, C. S. Adams, A. Nath and R. E. Klevit, *Proc. Natl. Acad. Sci. U. S. A.*, 2020, **117**, 2923–2929.
49. Z. Hou, P. M. Wydorski, V. A. Perez, A. Mendoza-Oliva, B. D. Ryder, H. Mirbaha, O. Kashmer and L. A. Joachimiak, *Nat. Commun.*, 2021, **12**, 5338.
50. M. Sarkar, J. Kuret and G. Lee, *J. Neurosci. Res.*, 2008, **86**, 2763–2773.
51. G. E. Karagoz, A. M. Duarte, E. Akoury, H. Ippel, J. Biernat, T. Moran Luengo, M. Radli, T. Didenko, B. A. Nordhues, D. B. Veprintsev, C. A. Dickey, E. Mandelkow, M. Zweckstetter, R. Boelens, T. Madl and S. G. Rudiger, *Cell*, 2014, **156**, 963–974.
52. M. E. DeSantis, E. H. Leung, E. A. Sweeny, M. E. Jackrel, M. Cushman-Nick, A. Neuhaus-Follini, S. Vashist, M. A. Sochor, M. N. Knight and J. Shorter, *Cell*, 2012, **151**, 778–793.
53. Y. O. Chernoff, S. L. Lindquist, B. Ono, S. G. Inge-Vechtomov and S. W. Liebman, *Science*, 1995, **268**, 880–884.
54. M. E. Jackrel, M. E. DeSantis, B. A. Martinez, L. M. Castellano, R. M. Stewart, K. A. Caldwell, G. A. Caldwell and J. Shorter, *Cell*, 2014, **156**, 170–182.
55. S. N. Gates, A. L. Yokom, J. Lin, M. E. Jackrel, A. N. Rizo, N. M. Kendsersky, C. E. Buell, E. A. Sweeny, K. L. Mack, E. Chuang, M. P. Torrente, M. Su, J. Shorter and D. R. Southworth, *Science*, 2017, **357**, 273–279.
56. A. S. Wentink, N. B. Nillegoda, J. Feufel, G. Ubartaite, C. P. Schneider, P. De Los Rios, J. Hennig, A. Barducci and B. Bukau, *Nature*, 2020, **587**, 483–488.
57. M. M. Schneider, S. Gautam, T. W. Herling, E. Andrzejewska, G. Krainer, A. M. Miller, V. A. Trinkaus, Q. A. E. Peter, F. S. Ruggeri, M. Vendruscolo, A. Bracher, C. M. Dobson, F. U. Hartl and T. P. J. Knowles, *Nat. Commun.*, 2021, **12**, 5999.
58. L. J. Blair, B. A. Nordhues, S. E. Hill, K. M. Scaglione, J. C. O'Leary 3rd, S. N. Fontaine, L. Breydo, B. Zhang, P. Li, L. Wang, C. Cotman, H. L. Paulson, M. Muschol, V. N. Uversky, T. Klengel, E. B. Binder, R. Kayed, T. E. Golde, N. Berchtold and C. A. Dickey, *J. Clin. Invest.*, 2013, **123**, 4158–4169.
59. S. Alberti and A. A. Hyman, *Nat. Rev. Mol. Cell Biol.*, 2021, **22**, 196–213.
60. C. W. Pak, M. Kosno, A. S. Holehouse, S. B. Padrick, A. Mittal, R. Ali, A. A. Yunus, D. R. Liu, R. V. Pappu and M. K. Rosen, *Mol. Cell*, 2016, **63**, 72–85.
61. K. A. Burke, A. M. Janke, C. L. Rhine and N. L. Fawzi, *Mol. Cell*, 2015, **60**, 231–241.

62. A. E. Conicella, G. H. Zerze, J. Mittal and N. L. Fawzi, *Structure*, 2016, **24**, 1537–1549.
63. C. Fernandez and G. Wider, *Curr. Opin. Struct. Biol.*, 2003, **13**, 570–580.
64. Y. Jiang, P. Rossi and C. G. Kalodimos, *Science*, 2019, **365**, 1313–1319.
65. C. Wang, J. Tu, S. Zhang, B. Cai, Z. Liu, S. Hou, Q. Zhong, X. Hu, W. Liu, G. Li, Z. Liu, L. He, J. Diao, Z. J. Zhu, D. Li and C. Liu, *Nat. Commun.*, 2020, **11**, 1531.
66. F. X. Theillet, A. Binolfi, B. Bekei, A. Martorana, H. M. Rose, M. Stuiver, S. Verzini, D. Lorenz, M. van Rossum, D. Goldfarb and P. Selenko, *Nature*, 2016, **530**, 45–50.
67. L. Barbieri, E. Luchinat and L. Banci, *Nat. Protoc.*, 2016, **11**, 1101–1111.
68. G. M. Clore and J. Iwahara, *Chem. Rev.*, 2009, **109**, 4108–4139.
69. M. K. Janowska and J. Baum, *Methods Mol. Biol.*, 2016, **1345**, 45–53.
70. J. Oroz, B. J. Chang, P. Wysoczanski, C. T. Lee, A. Perez-Lara, P. Chakraborty, R. V. Hofele, J. D. Baker, L. J. Blair, J. Biernat, H. Urlaub, E. Mandelkow, C. A. Dickey and M. Zweckstetter, *Nat. Commun.*, 2018, **9**, 4532.
71. A. N. Calabrese, B. Schiffrin, M. Watson, T. K. Karamanos, M. Walko, J. R. Humes, J. E. Horne, P. White, A. J. Wilson, A. C. Kalli, R. Tuma, A. E. Ashcroft, D. J. Brockwell and S. E. Radford, *Nat. Commun.*, 2020, **11**, 2155.
72. M. Liao, E. Cao, D. Julius and Y. Cheng, *Nature*, 2013, **504**, 107–112.
73. X. Li, P. Mooney, S. Zheng, C. R. Booth, M. B. Braunfeld, S. Gubbens, D. A. Agard and Y. Cheng, *Nat. Methods*, 2013, **10**, 584–590.
74. A. W. Fitzpatrick and H. R. Saibil, *Curr. Opin. Struct. Biol.*, 2019, **58**, 34–42.
75. K. A. Verba, R. Y. Wang, A. Arakawa, Y. Liu, M. Shirouzu, S. Yokoyama and D. A. Agard, *Science*, 2016, **352**, 1542–1547.
76. R. Y. Wang, C. M. Noddings, E. Kirschke, A. G. Myasnikov, J. L. Johnson and D. A. Agard, *Nature*, 2022, DOI: 10.1038/s41586-021-04252-1.
77. V. H. Ryan, G. L. Dignon, G. H. Zerze, C. V. Chabata, R. Silva, A. E. Conicella, J. Amaya, K. A. Burke, J. Mittal and N. L. Fawzi, *Mol. Cell*, 2018, **69**, 465–479.e467.
78. X. Gui, F. Luo, Y. Li, H. Zhou, Z. Qin, Z. Liu, J. Gu, M. Xie, K. Zhao, B. Dai, W. S. Shin, J. He, L. He, L. Jiang, M. Zhao, B. Sun, X. Li, C. Liu and D. Li, *Nat. Commun.*, 2019, **10**, 2006.
79. J. Wen, L. Hong, G. Krainer, Q. Q. Yao, T. P. J. Knowles, S. Wu and S. Perrett, *J. Am. Chem. Soc.*, 2021, **143**, 13056–13064.
80. M. T. Wei, S. Elbaum-Garfinkle, A. S. Holehouse, C. C. Chen, M. Feric, C. B. Arnold, R. D. Priestley, R. V. Pappu and C. P. Brangwynne, *Nat. Chem.*, 2017, **9**, 1118–1125.
81. N. L. Fawzi, S. H. Parekh and J. Mittal, *Curr. Opin. Struct. Biol.*, 2021, **70**, 78–86.
82. D. Li and C. Liu, *Biochemistry*, 2020, **59**, 639–646.
83. Y. Li, C. Zhao, F. Luo, Z. Liu, X. Gui, Z. Luo, X. Zhang, D. Li, C. Liu and X. Li, *Cell Res.*, 2018, **28**, 897–903.
84. Y. Sun, S. Hou, K. Zhao, H. Long, Z. Liu, J. Gao, Y. Zhang, X. D. Su, D. Li and C. Liu, *Cell Res.*, 2020, **30**, 360–362.

85. D. T. Murray, M. Kato, Y. Lin, K. R. Thurber, I. Hung, S. L. McKnight and R. Tycko, *Cell*, 2017, **171**, 615–627.e616.
86. L. Q. Xu, S. Wu, A. K. Buell, S. I. Cohen, L. J. Chen, W. H. Hu, S. A. Cusack, L. S. Itzhaki, H. Zhang, T. P. Knowles, C. M. Dobson, M. E. Welland, G. W. Jones and S. Perrett, *Philos. Trans. R. Soc. London, Ser. B*, 2013, **368**, 20110410.
87. P. Arosio, T. Muller, L. Rajah, E. V. Yates, F. A. Aprile, Y. Zhang, S. I. Cohen, D. A. White, T. W. Herling, E. J. De Genst, S. Linse, M. Vendruscolo, C. M. Dobson and T. P. Knowles, *ACS Nano*, 2016, **10**, 333–341.
88. T. Scheidt, J. A. Carozza, C. C. Kolbe, F. A. Aprile, O. Tkachenko, M. M. J. Bellaiche, G. Meisl, Q. A. E. Peter, T. W. Herling, S. Ness, M. Castellana-Cruz, J. L. P. Benesch, M. Vendruscolo, C. M. Dobson, P. Arosio and T. P. J. Knowles, *Proc. Natl. Acad. Sci. U. S. A.*, 2021, **118**, e2108790118.
89. R. Tycko, *Cold Spring Harbor Perspect. Med.*, 2016, **6**, a024083.

CHAPTER 13

Structural Biology in Cells by In-cell NMR

Y. FLEISCHMANN, J. GEREZ* AND R. RIEK*

Laboratory of Physical Chemistry, ETH Zürich, Vladimir-Prelog-Weg 2, CH-8093 Zürich, Switzerland
*Emails: juan.gerez@phys.chem.ethz.ch; roland.riek@phys.chem.ethz.ch

13.1 Introduction

Albeit its success using X-ray crystallography, cryo-electron microscopy, or nuclear magnetic resonance spectroscopy (NMR), currently structural biology is mostly based on *in vitro* studies but a detailed understanding of the numerous functions and actions of biomolecules requests the establishment of dynamics and structural biology in cells. It cannot be overemphasized that the action of the vast majority of biomolecules is within cells. Moreover, the requested actions may change with time and the cellular milieu has a high protein density comprising a plethora of intracellular partners. The cell is a dynamic structurally complex and organized entity with many subcompartments including organelles and highly conserved proteinaceous structures comprising liquid phase-separated entities. The alive nature of a cell is grounded by the dynamics and multi-dimensional structural landscape of biomolecules. Attempts towards this aim include the use of a free-electron laser for which for example a structure from a protein crystal inside bacteria was determined[1] or cryo-electron microscopy and tomography which make much progress in this direction.[2] As an alternative, solid-state and in particular solution-state NMR in cells, both in prokaryotic and eukaryotic cells, is established.[3–8] NMR has the benefit of being a non-invasive method able to measure time-resolved, in liquid and at physiological temperatures during

New Developments in NMR No. 29
Biophysics of Molecular Chaperones: Function, Mechanisms and Client Protein Interactions
Edited by Sebastian Hiller, Maili Liu and Lichun He

Published by the Royal Society of Chemistry, www.rsc.org

live monitoring a protein of interest in cells, while the other methods mentioned need to freeze the cell under investigation. However, it is our view that in-cell NMR has not yet fulfilled the expectations made when it was developed more than 20 years ago[9,10] as only a relatively low number of proteins could be studied by in-cell NMR until recently,[6] as shown in Table 13.2. This chapter focuses on the in-cell NMR methodology including qualitative as well as quantitative aspects and a comparison between methodologies that yield significant methodological advances described. It also includes comments on one of its applications to chaperones which is the subject of the book the chapter is written in. For other detailed descriptions of applications, we refer to excellent reviews[3–6] including, in particular, the recent thorough review by Luchinat *et al.*,[11] which covers almost all interesting applications of in-cell NMR to date in addition to a very detailed description of the methodologies.

13.2 Historical Aspects of In-cell NMR

The first proper demonstration of NMR was reported in 1946 on a sample of water.[12,13] In the following two decades, NMR spectroscopy established itself as a reliable tool in analytical chemistry, which was mainly used to identify and study the structures of small to intermediate molecules.[14]

Only the continuous improvement of commercial spectrometers, from 60 MHz in 1961 to over 1.2 GHz today, allowed larger and more complex systems such as proteins to be studied by NMR. The development of two-dimensional techniques for NMR presented the tool that was able to bring NMR from analytical chemistry into structural biology. In 1984, the structure of a protein solved by NMR was presented for the first time.[15] Today, X-ray crystallography, cryo-electron microscopy, and NMR spectroscopy are key methods in providing accurate protein structures. From this trio, only NMR is able to study a sample in liquid form, which enables NMR to study proteins dynamically in solution while being active and interacting with their environment. To fully exploit this advantage, the environment exposed to the protein ought to resemble native-like conditions in order to prevent drawing false conclusions from protein structures not commonly present in living organisms. Early on in biological NMR, it became clear that macromolecular crowding, as it is present in cells, has a crucial effect on the structure and function of proteins.[16] However, buffers and molecular crowding agents used to mimic a densely packed cellular environment cannot imitate the complexity of the cell interior with numerous unknown potential interaction partners. Indeed, nonspecific interactions with various intracellular partners were found to be crucial to the in-cell behavior of proteins as macromolecular crowding alone fails to explain the difference observed compared to *in vitro* conditions.[17] Alternatively, cell lysates are continued to be used today and show altered thermodynamic stability for proteins compared to conventional *in vitro* studies.[18] However, in the end, only the direct observation of active proteins in the living cell can account for all phenomena existing in cells. Here, NMR spectroscopy puts forward a unique opportunity using its noninvasive character and ability, to specifically observe an isotopically labeled species within an unlabeled environment. In the following part, we will highlight a few key steps in the ongoing history of in-cell NMR.

Primary experiments that could be termed in-cell NMR consist of simply collecting intact bacterial, yeast, or mammalian cells into the NMR test tube and measuring one-dimensional experiments using natural abundance of ^{1}H, ^{13}C, ^{31}P, and others. Individual peaks could be attributed to known metabolites and used to study the metabolic processes of the living cell. As an example of this, the work by Norton is highlighted, in which yeast and mammalian cells with natural abundance ^{13}C NMR as well as with cells provided with ^{13}C labeled glucose were studied.[19]

Later on, bioreactor systems that supply fresh media to the collected cells were attached to the NMR instrument. This allows studying cells for a longer period by keeping them healthy and alive. It also allows switching media containing drugs during the ongoing observation in the spectrometer. The in-cell applications of these systems in the late 20th century were still focused on studying metabolites in natural abundance NMR.[20,21] However, bioreactor systems can be used for all kinds of in-cell applications and are not limited to metabolic studies.

Serber *et al.* were the first to demonstrate that in-cell NMR can be used to detect proteins in living cells, by expressing a protein of interest in bacteria.[10] To do so, they transformed *E. coli* cells with a plasmid containing the protein of interest (NmerA in this particular study) alongside the lac-operon, which can selectively be induced using isopropyl β-D-1-thiogalactopyranoside (IPTG). The cells were grown in lysogeny broth (LB) but transferred to a ^{15}N labeled minimal medium just before induction. This successfully resulted in a ^{15}N-labeled protein of interest while keeping the cellular background mostly unlabeled. They used ^{1}H–^{15}N heteronuclear single quantum correlation experiments (*i.e.* [^{1}H,^{15}N]-HSQC) to detect and identify the target protein. This method together with ^{1}H–^{15}N heteronuclear multiple quantum correlation experiments (*i.e.* [^{1}H,^{15}N]-HMQC) became the method of choice in in-cell NMR for many years after and is still regularly used today. Both methods deliver a "fingerprint" of the protein structure, where all amino acid residues (with the exception of proline residues and the N-terminus) contribute one peak to the spectrum, which allows chemical shift perturbations, together with peak intensity alterations to be assigned to the amino acid residues affected (please note: ^{15}N labeling allows NMR experiments which are more sensitive and significantly cheaper compared to ^{13}C labeling).

While being very effective, the overexpression technique was initially limited to bacterial cells, as there were no isotopically labeled media for mammalian cells available. In an attempt to overcome this obstacle, Selenko *et al.* and Sakai *et al.* in quick succession presented their work, in which they used ^{15}N-labeled purified proteins (produced in heterologous systems) and injected them into *Xenopus laevis* oocytes.[22,23] This marked the first-ever in-cell NMR with eukaryotic cells.

The same approach was used two years later to investigate the intrinsically unfolded protein (IDP) tau in oocytes.[24] Microinjection is a well-established method of protein delivery; however, it has the drawback that it can only be used with relatively large cells in the upper micrometer to millimeter range, as every cell needs to be injected one after another. Even if technically possible with smaller cells, the workload required to get a large enough sample is challenging.

More recent delivery methods, namely cell-penetrating peptide (CPP) delivery and streptolysin-O (SLO) delivery, explained in detail below, can be applied to a much larger variety of eukaryotic cells. The CPP delivery was first implemented for in-cell NMR by Inomata *et al.* and makes use of short positively charged peptide sequences fused to the protein of interest, which enables the internalization of the protein into the cell.[7] The peptide tag can be removed after internalization. CPP was the first transexpression (see below) technique used in mammalian cells for in-cell NMR. SLO delivery, on the other hand, relies on the bacterial toxin streptolysin-O which incorporates itself into the plasma membrane and generates pores with a diameter of 35 nm that can be used to introduce any protein of choice smaller than 150 kDa.[25] The pore formation can be reversed by addition of calcium ions to the cells.

In both publications mentioned before, a new fast multidimensional NMR method called SOFAST-HMQC (or SOFAST-HSQC) was brought into action for in-cell NMR allowing for much faster acquisition, shortening the overall measurement time.[26] This is of great value for measurements with living cells, as more time-consuming applications often struggle with a significant decrease in cell viability.

Not only new methods for the delivery or acquisition are being invented, but also the utilization of in-cell NMR is becoming ever broader. Xie *et al.* introduced small molecule interactor library in-cell NMR (SMILI-NMR) the first protein detection screening method in living cells opening a new chapter for untargeted drug research.[27]

Another milestone was achieved by Sakakibara *et al.*, who reported the first solved protein structure of ^{15}N, ^{13}C double-labeled TTHA1718 in *Escherichia coli* cells using 3D nuclear Overhauser enhancement spectroscopy (NOESY) NMR.[28] Nevertheless, it needs to be said that this was possible because the expressed protein reached artificially high protein concentrations of 3–4 mM in the cells which is only achievable using overexpression.

Conveniently, with the introduction of reliable overexpression techniques applied in in-cell NMR for yeasts,[29] insect cells,[30] and human cells,[31] overexpression was not limited to bacterial cells anymore.

The ability of in-cell NMR to study posttranslational modifications in real-time is worth mentioning and it was done by Amata *et al.*[32] Although, not the first work looking into phosphorylation patterns in mammalian cells by NMR, this work highlights the suitability of in-cell NMR to study modification sites and the kinases and phosphatases involved in the process.

The latest transexpression method being introduced and applied in in-cell NMR by the Selenko group in 2016 was electroporation.[33] This method relies on short electric pulses that cause a short-lived permeability of the plasma membrane that allows proteins from outside to enter.

Electroporation was subsequently used by Leeb and co-workers who showed how to use longitudinal and transverse relaxation rates acquired in human A2780 cells to study the dynamic behavior of several proteins in cells.[34]

Last but not least, we would like to call attention to Müntener *et al.*, who showed in 2016 that 2D NMR experiments are sufficient to obtain in-cell protein structures by simultaneously measuring protein pseudo-contact shifts (PCSs) and residual dipolar couplings (RDCs) of a protein being labeled with a paramagnetic lanthanide tag.[35] An illustration showing selected highlights of in-cell NMR history can be seen in Figure 13.1.

In-cell NMR has not yet fully lived up to its potential; however, the last two decades of in-cell NMR history have shown vast improvements and an increasingly long list of possible applications in structural biology and drug discovery. While there are still many obstacles to overcome, especially for larger proteins, to make in-cell NMR more widely applicable, it seems certain that in-cell NMR can provide biological insight otherwise hardly obtainable. In the following, methods of in-cell NMR sample preparation using overexpression or transexpression are explained in more detail.

13.3 In-cell NMR Methods Using Overexpression

In-cell NMR of mammalian systems requires that the protein of interest is stable-isotopically labeled inside living cells using ^{15}N, ^{13}C, or ^{19}F labeling. The isotopically labeled protein can either be expressed by the same cells that the NMR experiments are recorded on[36] or provided externally.[7] The advantages of the endogenous overexpression of the isotopically labeled protein are obvious since the protein is expressed, localized, and eventually post-translational modified natively. However, it suffers from the background of NMR signals due to the uncontrollable synthesis of other isotopically labeled biomolecules expressed as well with isotopes added to the media.

Overexpression is usually the method of choice in prokaryotic in-cell NMR and is rather straightforward since overexpression in a stable isotope medium is well established.[37] The medium is changed to the labeled medium 30 min to 1 hour before induction by gentle centrifugation at room temperature of the cells in order to suppress background labeling. The level of the expressed protein might be very high (*i.e.* mM concentration) challenging the biological relevance of the results obtained while the protein of interest is well above the background without further ado. In-cell NMR using overexpression of the protein of interest in yeasts *Pichia pastoris* and *Saccharomyces cerevisiae* is also rather straightforward[29,38] as protein overexpression and labeling strategies are well established. It is however rarely applied (Table 13.2). In *Pichia pastoris,* the gene encoding the protein of interest is cloned under the control of the alcohol oxidase promoter P_{AOX1}, which is strongly repressed in glucose, glycerol, or ethanol media, while induced upon change of the culture medium to ^{13}C-labeled methanol for labeling the protein of interest for the NMR investigation

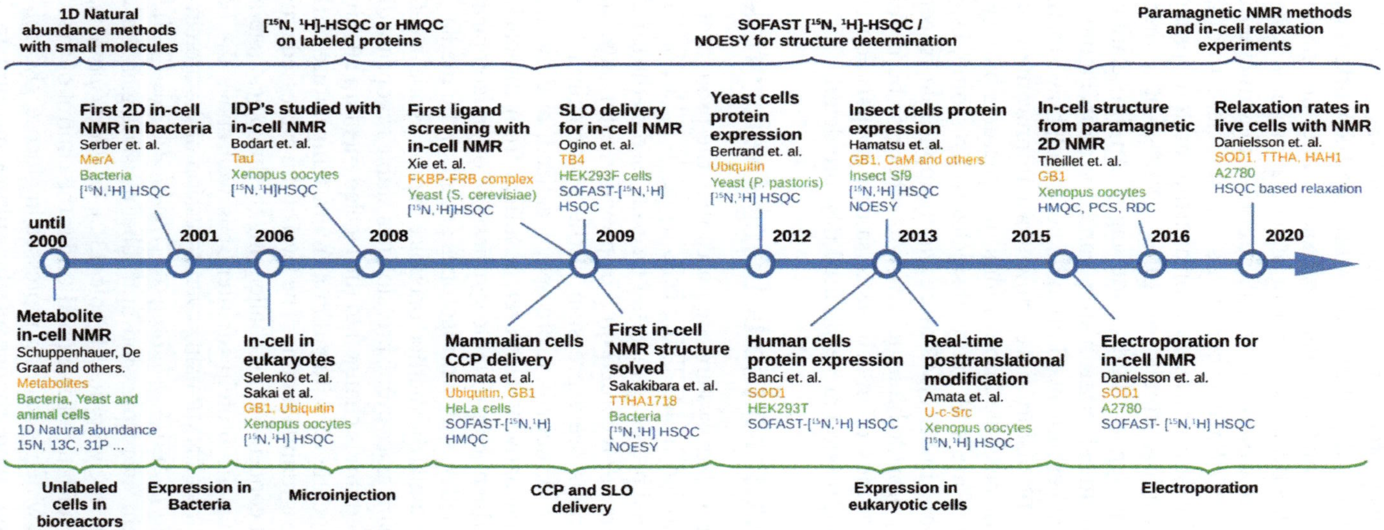

Figure 13.1 Illustration depicting selected highlights of in-cell NMR development history in a timeline format. There is an emphasis on the first appearances of new techniques and delivery methods as well as the types of cells and pulse sequences used preferably over periods of time.

(along other proteins expressed simultaneously). The existence of the cell wall poses difficulties for methods, where the protein is introduced from external sources. Due to this, overexpression is regarded as the method of choice in yeast cells. Graphical illustrations for sample preparation using overexpression are shown in Figure 13.2. Upon the establishment of commercial availability of isotope-labeled media for insect cells, baculovirus-mediated insect cell overexpression for in-cell NMR became feasible and was successfully demonstrated first by Shirakawa and Ito.[30] In their work, four different proteins were studied in sf9 cells (Table 13.2). The expression was performed for 48 h with addition of the isotope-labeled medium at 24 h in order to reduce the background. The background was further suppressed by subtracting the recorded NMR spectrum of interest from a corresponding NMR spectrum of insect cells infected with an empty vector. The later structure determination of the protein TTHA1718 in insect cells by the same group using nuclear Overhauser enhancement (NOE) indicates the principle power of in-cell NMR with insect cells.[39]

Mammalian in-cell NMR using protein overexpression was established by Banci *et al.*[31] and has since been demonstrated for a couple of proteins (Table 13.2). In these works, overexpression-based in-cell NMR in mammalian cells required commercially available uniformly labeled media for mammalian cells and a pHLsec plasmid. The original pHLsec is derived from the pLEXm plasmid having a CAG promoter, one of the strongest synthetic promoters for high-level constitutive secreted protein expression.

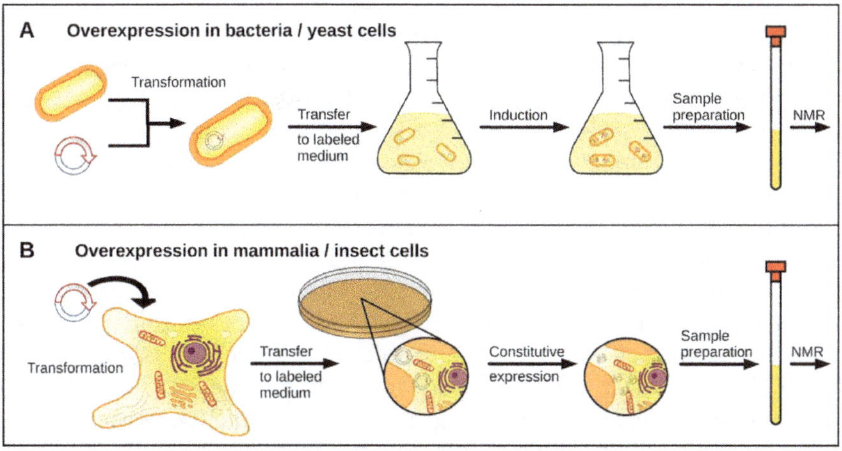

Figure 13.2 Schematic drawing of in-cell NMR sample preparation by means of overexpression. (A) Direct expression of the target protein in bacterial or yeast cells by transforming them with an inducible expression vector and transferring them to a labeled minimal medium. Several hours after induction, the cells are harvested and transferred to the NMR tube. (B) Mammalian and insect cells can be transfected with a constitutive expression vector and grown in an appropriately labeled medium. After a sufficient amount of time, the cells are harvested and transferred to the NMR tube.

However, expression of the protein of interest by pHLsec renders an intracellular protein as it is reverted to a cytoplasmic expression vector by removing the N-terminal secretion signal sequence. Upon transfection and change of the medium to the labeled medium, protein expression occurs for 48 h followed by cell collection and NMR measurements. As for insect cells, the spectral background arising from the nonspecific labeling of cellular components is strongly reduced by subtracting the NMR spectrum from a corresponding spectrum of human cells transfected with an empty vector.[40]

Alternatively, the isotopically labeled and NMR-visible protein can be obtained from an external source and then introduced into mammalian cells by microinjection (*e.g.* as in the case of *Xenopus laevis* oocytes),[22,23] electroporation,[41] cell-penetrating peptides,[7] or pore-forming toxins.[25] As the mammalian cells are not exposed to isotopically labeled precursors, the main advantage of these methods is virtually no background of the NMR signals. Additionally, they allow monitoring of immediate structural changes occurring upon the addition of the protein. For this purpose, the introduced protein can be modified according to the aim of the experiment or can be introduced unmodified.[41–43] Regardless of the method, having optimal amounts of the isotopically labeled protein inside the cells is critical for obtaining good quality spectra by in-cell NMR.

13.4 Transexpression in Mammalian Cells†

As briefly described above one of the methods to deliver isotope-labeled proteins for NMR acquisition is the introduction of exogenous molecules into mammalian cells, which constitutes a very common experimental strategy in the field of cell and molecular biology. Introducing these molecules into cells has been instrumental in the analysis and comprehensive understanding of many biological processes. For example, nucleic acids such as RNA and DNA have been delivered intracellularly at different amounts by standardized methods such as transfection or viral transduction over the last decades. In contrast to RNA and DNA, the efficient delivery of exogenous proteins into mammalian cells has not been studied in detail neither mechanistically nor quantitatively[44] and hence no unifying method has been established albeit its use in in-cell NMR as well as cell differentiation.[45] Furthermore, no terminology has been provided for this technique until recently. In recent work, our group defined "transexpression" as the process of delivering exogenous proteins into mammalian cells.[46]

As we defined it, transexpression refers therefore to the introduction of an "exogenous" protein into mammalian cells by any of the different experimental methods for such aim (*e.g.* electroporation, cell-penetrating peptides, and cell membrane permeabilization by toxins or lipoparticles). For "exogenous protein" it must be stated that the delivered protein necessarily needs to be obtained from an external source, which in most cases is a heterologous system such as bacteria. Upon intracellular delivery, the exogenous protein is now "transexpressed" in the recipient cell, forming part of its own proteome. As recipient cells do not express the exogenous protein but instead it is exogenously acquired, the transexpression term fulfills the semantic requirements for such an experimental method.[44]

13.5 Transexpression Methods for In-cell NMR

A bottleneck of current in-cell NMR protocols is the difficulty of expressing or introducing the NMR visible (isotopically labeled) protein under study into mammalian cells at the desired concentration and suitable for NMR analysis. The most common methods for transexpression are electroporation, membrane cell-penetrating peptides (CPP), and bacterial toxin-based models, all of them relying on a step where the plasma membrane is reversibly permeabilized for the uptake of exogenously added molecules. The three methods to be discussed in detail in the following have been proven to be non-invasive techniques of protein delivery and successfully applied to in-cell NMR. It is noted that there are other methods for protein delivery into mammalian cells such as the use of lipoparticles such as polycationic liposomes,[47,48] but they have not been applied yet to in-cell NMR. Schematic drawings of all the discussed transexpression methods can be seen in Figure 13.3.

13.5.1 Plasma Membrane Cell-penetrating Peptides or CPP

The first transexpression technique of interest is based on CPP. The method consists in fusing a short hydrophobic peptide with plasma membrane-permeabilizing properties to the protein of interest. Fusion of the CPP can be either in the N- or C-termini to the protein of interest. However, fusing the CPP to the N- or C-terminus of the protein might affect not only the cell-penetrating properties of CPP but also may affect the structure, dynamics, and function of the protein requesting experimental testing. In this regard, a good practice is to compare the two possible fusion proteins with either N- or C-terminus tagged with CPP *in vitro* by NMR. Once an efficient cell penetration is achieved (using for example the transexpression assay described below), the fusion protein is produced in heterologous systems as ^{15}N-labeled protein and purified. In a typical in-cell NMR experiment, several milligrams of the fusion protein are requested. The purified protein is then added directly to the culture media of mammalian cells. Administration can be performed in one single step or in multiple steps for example in one-hour intervals. It is expected that the protein is translocated through the plasma membrane by a still not fully understood mechanism. Several CPP have been reported to be suitable for in-cell NMR studies to date. For instance, the TAT HIV cell-penetrating peptide has been used multiple times in in-cell NMR experiments (Table 13.2). The aminoacidic sequence corresponding to this peptide is YGRKKRRQRRR, which ensures an efficient plasma membrane

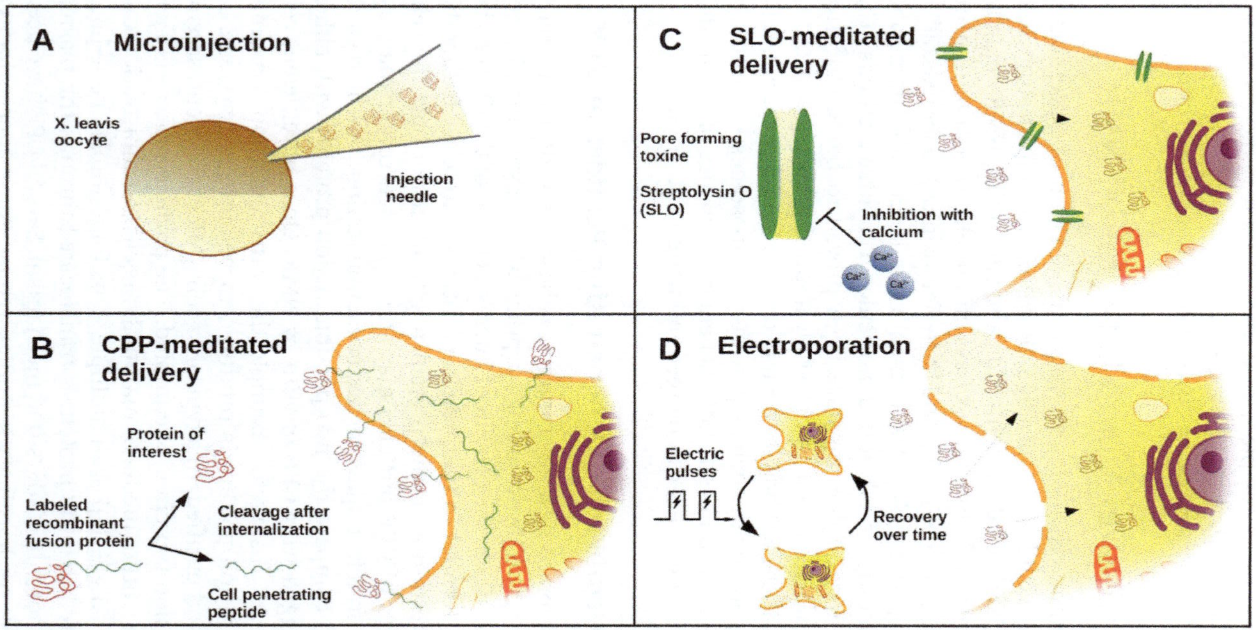

Figure 13.3 Schematic drawing of in-cell NMR sample preparation by means of transexpression of recombinantly expressed and purified labeled protein. (A) The protein of interest can be delivered *via* microinjection into *X. leavis* oocyte or other large mammalian cells. (B) Alternatively, the protein can be expressed as a fusion protein with a cell-penetrating peptide (CPP). The peptide facilitates the internalization into the cell and gets cleaved off afterward. (C) Pore-forming toxins such as streptolysin O (SLO) can be used to make the plasma membrane permeable for the target protein. The pores are closed by adding calcium ions. (D) The plasma membrane can be made permeable by applying short electric pulses (electroporation). The cells will recover and restore an intact membrane after a short period of time.

translocation of the target protein.[49,50] Parameters that affect the cell-penetrating properties of CPP which are critical for in-cell NMR experiments are amongst others the amount of protein used for exogenous administration. Removing the CPP from the mature protein might be critical for certain proteins as this tag might interfere with proper protein folding and promote or interfere with interactions with intracellular partners. Efficient removal has been achieved by adding a recognition signal for an intracellular protease or deubiquitinating enzyme.[7] As detailed below, this method requires substantially higher amounts of ^{15}N-labeled protein compared to electroporation. Moreover, protein processing can occur as the time allowed for protein delivery is in most cases >24 h.

13.5.2 Plasma Membrane Permeabilization by Bacterial Toxins

Exogenous proteins can also be transexpressed upon plasma membrane permeabilization by bacterial toxins such as Streptolysin-O (SLO). SLO binds to cell membranes containing high amounts of soluble cholesterol. It binds as a monomeric entity but the membrane-associated monomers diffuse laterally and polymerize to form oligomers that form early forms of the transmembrane pore called pre-pore. These protein–lipid complexes then undergo structural changes that result in the formation of a transmembrane β-barrel pore of approximately 35 nm in diameter localized in the cell membrane, facilitating osmotic lysis of the target cell.[51] Sublytical doses of SLO can be used for transiently permeabilizing the plasma membrane of mammalian cells, which then needs to be rapidly resealed in order to avoid the strong cytotoxic effects of SLO. To carry out in-cell NMR experiments, SLO is applied to cultured cells for a very short time immediately followed by the addition to the culture media of the ^{15}N-labeled protein of interest (produced in heterologous systems). Administration of the latter is fast usually taking minutes. After protein administration, the plasma membrane is re-sealed by a calcium treatment to the cells. Several studies (Table 13.2) show that the method has been proven to be suitable for in-cell NMR.

Membrane permeabilization by SLO has the advantage that no peptidic tags are fused to the protein of interest. The main disadvantage is, however, the secondary effects related to the intrinsic toxicity of the bacterial toxin on mammalian cells. It is known that certain cells are highly susceptible to this toxin, while others can cope with low doses applied for a very short time. Thus, the efficiency of SLO might depend on the cell type used and caution has to be taken when susceptible cells used as cellular responses to SLO might be triggered. Normally, high cell recovery is obtained by incubating the cells under normal growth conditions for 2–4 h after SLO treatment.

13.5.3 Plasma Membrane Permeabilization by Electroporation

Transexpression by electroporation consists in applying a short electric pulse to the cells to promote membrane destabilization by which extracellular molecules can be efficiently internalized. The short-lived and transient nature of electroporation makes this technique innocuous for most cell types, and thereby no further step for plasma membrane resealing is needed. Originally developed for delivering small molecules and nucleic acids not only to mammalian cells but also to bacteria and other prokaryotes and eukaryotes, electroporation has been used over the last decades in experiments of cellular and molecular biology and alike. Its extensive use over the last four decades delivering nucleic acids into mammalian cells warranties its low and tolerable side effects. However, it is important to mention that the experimental conditions used to successfully deliver small molecules and nucleic acids into mammalian cells cannot be extrapolated to proteins, which in almost all cases require completely different setups for efficient transexpression. Therefore, for in-cell NMR it is essential to screen the best conditions of electroporation that allow the efficient delivery of properly folded and functional proteins.

Similar to the CPP and SLO-based methods, electroporation has advantages and disadvantages. For better or worse, electroporation has more adjustable variables for efficient transexpression when compared to CPP and SLO. For instance, the composition of the buffer used in the electroporation step and the electric pulse itself need to be experimentally tested and systematically back-to-back compared in order to identify the best conditions of transexpression. As for the buffer used in the electroporation step, there are no limitations in terms of buffer composition provided it is innocuous for the cells. Even simple buffers such as phosphate-buffered saline (PBS) can be used, but they might not yield the highest efficiency. Some manufacturers of the devices used for electroporation offer buffers that work well for protein electroporation. However, as they are commercialized and patent-protected, the composition of these commercially available solutions is unknown *a priori*. We provide in Table 13.1 a small selection of handmade electroporation buffers that can be used. Optimization of the buffer composition for a given cell type and protein of interest is recommendable.

In addition to the electroporation buffer, the electric pulse has a strong influence not only on the transexpression efficiency but also on the cell viability and the quality of the delivered protein, as discussed in detail below. The electric pulse can be changed by shape, duration, and intensity, amongst other secondary parameters. Some electroporation devices have the possibility of freely changing these three parameters, while in others all of them are fixed and in some cases cannot be known. We have tested several pulse shapes for transexpression using several types of electroporation devices from different manufacturers. Amongst the devices that show the electric pulse parameters used, we have found that one of the most efficient pulse shapes for Hek293 and A2780 cells resulted to be a square pulse. But this is not a universal law and other cells might require different pulse shapes for the highest transexpression efficiency. Then, the duration (normally microseconds) and intensity (normally 100–1700 V) are adjusted as a function of transexpression efficiency but also of the cell death that results

Table 13.1 Recipes for handmade electroporation buffer, which have been used in our lab. The suggested recipes can lead to varying degrees of success for different cells and proteins. If handmade buffers are being used, it is advised to try several different compositions to find an ideal buffer composition for your system.

Recipe 1	Recipe 2	Recipe 3	Recipe 4	Recipe 5
Phosphate buffer (120 mM, pH 7.2)	Phosphate buffer (150 mM, pH 7.2)	Phosphate buffer (40 mM, pH 7.2)	HEPES (20 mM)	Phosphate buffer (100 mM, pH 6.5)
KCl (5 mM)	HEPES (15 mM)	KCl (5 mM)	Tris/HCl (75 mM, pH 7.2)	$MgCl_2$ (2 mM)
$MgCl_2$ (15 mM)	KCl (5 mM)	$MgCl_2$ (15 mM)	KCl (5 mM)	Mannitol (25 mM)
NaCl (50 mM)[a]	$MgCl_2$ (15 mM)	NaCl (90 mM)	$MgCl_2$ (15 mM)	Sodium succinate (25 mM)
	NaCl (50 mM)[a]	Glucose (10 mM)	NaCl (90 mM)	
		$Ca(NO_3)_2$ (0.4 mM)	Glucose (10 mM)	
			$Ca(NO_3)_2$ (0.4 mM)	

[a] NaCl in solution 1 or 2 can be replaced by 50 mM of either mannitol, sodium succinate, or sodium lactobionate.

Table 13.2 Summary of proteins or protein domains being studied by in-cell NMR listed alphabetically and including protein mass, cell type used in the study, the method used for protein delivery or overexpression, and the NMR method used. Not included are in-cell NMR studies with nucleic acids and peptides. We do not claim that the list is all-inclusive.

Protein	Type	Mass	Cell type	Delivery	NMR method	Source
ADK		40.5 kDa	*E. coli*	Overexpression	$[^{15}N,^{1}H]$-HMQC (TROSY)	94
	Fully ^{2}D labeled				$[^{15}N,^{1}H]$-CRINEPT-HMQC-TROSY	97
AK1		21.6 kDa	HeLa	Electroporation	$[^{15}N,^{1}H]$-SOFAST-HMQC	98
α-Syn		14.4 kDa	*E. coli*	Overexpression	$[^{15}N,^{1}H]$-HSQC	99 and 100
					$[^{15}N,^{1}H]$-SOFAST HMQC	95 and 101
					$[^{13}C,^{15}N]$-CON	
					$^{15}NH/D$ SOLEXY	102
			Xenopus oocyte	Microinjection	^{15}N XSTE diffusion	103
					$[^{15}N,^{1}H]$-HSQC	104
					^{19}F NMR	
			CHO-K1	CPP delivery	^{19}F NMR	105
			A2780, B65 RCSN-3 HeLa, SK-N-SH	Electroporation	$[^{15}N,^{1}H]$-HSQC PRE NMR	33 and 41
			A2780, Cos7, HeLa, HEK 293T	Electroporation	$[^{15}N,^{1}H]$-HMQC	44
			SH-SY5Y, A2780	Electroporation	$[^{15}N,^{1}H]$-SOFAST HMQC	42
Atx1 (chaperone)		8.1 kDa	*E. coli*	Overexpression	$[^{13}C,^{15}N]$-CON	95
					$[^{15}N,^{1}H]$-HSQC	
Bcl-XL	-cutloop	26 kDa	*E. coli*	Overexpression	$[^{15}N,^{1}H]$-TROSY	106
c-Src	SH3 domain	7 kDa	*E. coli*	Overexpression	$[^{15}N,^{1}H]$-HSQC	107 and 108
			Xenopus oocyte	Microinjection	^{19}F NMR	109
	Usrc (1–85)	8.4 kDa	*Xenopus* oocyte	Microinjection	$[^{15}N,^{1}H]$-SOFAST-HMQC ^{19}F NMR	32

CA1		28.8 kDa	HEK 293T	Overexpression	^{1}H NMR	110
			Red blood cells	19F covalent tag	^{19}F EXSY NMR	111
CA2		29.2 kDa	HEK 293T	Overexpression	[^{15}N,^{1}H]-SOFAST HMQC	110 and 112
Calbindin D9k		9 kDa	HeLa	CPP delivery	[^{15}N,^{1}H]-SOFAST HMQC	113
Calmodulin		16.7 kDa	*E. coli*	Overexpression	[^{15}N,^{1}H]-TROSY	106
					[^{13}C,^{1}H]-HSQC	114
			Xenopus oocyte	Microinjection	[^{15}N,^{1}H]-HSQC	23 and 104
					^{19}F NMR	104 and 115
			sf9 insect cells	Overexpression	[^{15}N,^{1}H]-HSQC	30 and 39
					[^{13}C,^{1}H]-HSQC	39
CAP-Gly1	Domain (CG1)	9 kDa	HeLa S3 cells	SLO delivery	[^{13}C,^{1}H]-methyl-TROSY	116
CCS (chaperone)	(Only the C-terminal part was detected)	29 kDa	HEK 293T	Overexpression	[^{15}N,^{1}H]-SOFAST HMQC	40
Cox17 (chaperone)		6.9 kDa	*E. coli*	Overexpression	[^{15}N,^{1}H]-HSQC	95
			HEK 293T	Overexpression	[^{15}N,^{1}H]-SOFAST HMQC [^{13}C,^{15}N]-CON	40 and 117
Cytochrome *c*		11.7 kDa	*E. coli*	Overexpression	[^{15}N,^{1}H]-HSQC	62
DJ-1 (chaperone)	(PARK7)	19.9 kDa	HEK 293T	Overexpression	[^{15}N,^{1}H]-SOFAST HMQC	118
FKBP12 FKBP		12 kDa	*E. coli*	Overexpression	[^{15}N,^{1}H]-HSQC	27
					[^{15}N,^{1}H]-HMQC (TROSY)	94
					^{13}C CP-MAS-NMR (solid-state)	117
					[^{13}C,^{15}N]-NCO-DCP-MAS-NMR (solid-state)	119
			HeLa	CPP delivery	[^{15}N,^{1}H]-HSQC	7

Table 13.2 (*Continued*)

Protein	Type	Mass	Cell type	Delivery	NMR method	Source
GB1		7 kDa	*E. coli*	Overexpression	$[^{15}N,^{1}H]$-TROSY	96 and 106
					^{19}F NMR	120
					^{19}F T_1, T_2 relaxation	121
					$[^{13}C,^{15}N]$-NOESY	122
			Xenopus oocyte	Microinjection	$[^{15}N,^{1}H]$-HSQC	22 and 104
					^{19}F NMR	104 and 123
					PCS, RDC	35 and 124
			sf9 insect cells	Overexpression	$[^{15}N,^{1}H]$-HSQC	30
					$[^{13}C,^{15}N]$-NOESY	39
			A2780, Cos7, HeLa, HEK 293T	Electroporation	$[^{15}N,^{1}H]$-HMQC	44
					$[^{13}C,^{15}N]$-NOESY	44
GB3		6.7 kDa	A2780	Electroporation	$[^{15}N,^{1}H]$-HMQC	44
HAH1 (chaperone)	(or ATOX1)	7.4 kDa	*E. coli*	Overexpression	$[^{15}N,^{1}H]$-HMQC	125
			sf9 insect cells	Overexpression	$[^{15}N,^{1}H]$-HSQC	30
			HEK 293T	Overexpression	$[^{15}N,^{1}H]$-SOFAST HMQC	40
			A2780	Electroporation	T_1, T_2 relaxation	34
HRAS	GTPase domain	19 kDa	sf9 insect cells	Overexpression	$[^{15}N,^{1}H]$-HSQC	39
					$[^{13}C,^{1}H]$-HSQC	39
			HeLa S3 cells	SLO delivery	$[^{13}C,^{1}H]$-SOFAST HMQC	64
HXK1	From yeast	53.7 kDa	*S. cerevisiae*	Overexpression	^{19}F NMR	65
MerA	Only the N-term binding domain	7 kDa	Bacterial cells (Tn501)	Overexpression	$[^{15}N,^{1}H]$-HSQC	10 and 126
			E. coli	Overexpression	$[^{15}N,^{1}H]$-HSQC	96
					$[^{15}N,^{1}H]$-TROSY	96

Mia40 (chaperone)	(or CHCHD4)	16 kDa	HEK 293T	Overexpression	$[^{15}N,^{1}H]$-SOFAST HMQC	40, 58 and 117
Nsp1	Many different FG repeats	13–30 kDa	*E. coli*	Overexpression	$[^{15}N,^{1}H]$-HSQC	127
					T_1, T_2 relaxation	127
			S. cerevisiae	Overexpression	$[^{15}N,^{1}H]$-HSQC	127
					T_1, T_2 relaxation	127
			Xenopus oocyte	Microinjection	$[^{15}N,^{1}H]$-HSQC	127
					T_1, T_2 relaxation	127
	FSFG-K (274–397)	13 kDa	*S. cerevisiae*	Overexpression	$[^{15}N,^{1}H]$-HSQC	38
PGK1	From yeast	44.7 kDa	*S. cerevisiae*	Overexpression	^{19}F NMR	65
			Neuro2a cells	Electroporation	$[^{15}N,^{1}H]$-HMQC	44
Pin1	WW domain PPI domain	18.2 kDa 4 kDa 12.5 kDa	*Xenopus* oocyte	Microinjection	$[^{15}N,^{1}H]$-SOFAST HMQC	68
Profilin 1		15 kDa	*E. coli*	Overexpression	$[^{15}N,^{1}H]$-SOFAST HMQC	128
			HEK 293T	Overexpression	$[^{15}N,^{1}H]$-SOFAST HMQC	40 and 128
ProTα		12.2 kDa	HEK 293T	Overexpression	^{1}H NMR	129
			A2780, Cos7, HeLa, HEK 293T	Electroporation	$[^{15}N,^{1}H]$-HMQC	44
Protein disulfide isomerase		54.1 kDa	*Xenopus* oocyte	Microinjection	$[^{15}N,^{1}H]$-HSQC ^{19}F NMR	104
ProtL	Only IgG binding domain	7 kDa	*E. coli*	Overexpression	$[^{15}N,^{1}H]$-HSQC ^{19}F NMR	17
PTP1E	PDZ2 domain	10 kDa	A2780, HEK 293T	Electroporation	$[^{15}N,^{1}H]$-HMQC	44
Pup		6.9 kDa	*E. coli*	Overexpression	$[^{15}N,^{1}H]$-HSQC	130 and 131
					$[^{15}N,^{1}H]$-SOFAST HMQC	132

Table 13.2 (*Continued*)

Protein	Type	Mass	Cell type	Delivery	NMR method	Source
PYK1	From yeast	54.5 kDa	*S. cerevisiae*	Overexpression	^{19}F NMR	65
SOD1	No loop 4 and 7	16 kDa (wt)	*E. coli*	Overexpression	[^{15}N,^{1}H]-SOFAST-HMQC	125, 133 and 134
			HeLa	CPP delivery	[^{15}N,^{1}H]-HMQC	135
			A2780	Electroporation	[^{15}N,^{1}H]-SOFAST HMQC	134
					T_1, T_2 relaxation	34
			HEK293T	Overexpression	[^{15}N,^{1}H]-SOFAST-HMQC	31, 40, 136 and 137
Tau	Tau40/k19	45 kDa/ 12 kDa	*Xenopus* oocyte	Microinjection	[^{15}N,^{1}H]-HSQC	24
			HEK 293T	Electroporation	[^{15}N,^{1}H]-HSQC	43
	k18	14 kDa	A2780, Cos7, HeLa, HEK 293T	Electroporation	[^{15}N,^{1}H]-HMQC	44
Thioredoxin (chaperone)		12 kDa	*E. coli*	Overexpression	[^{15}N,^{1}H]-HMQC (TROSY)	94
					[^{13}C,^{15}N]-NCO-DCP-MAS-NMR (solid-state)	119
					[^{15}N,^{1}H]-CRINEPT-HMQC-TROSY	138
			HeLa S3 cells	SLO delivery	[^{13}C,^{1}H]-methyl-TROSY	57

TTHA1718		7 kDa	*E. coli*	Overexpression	$[^{15}N,^{1}H]$-HSQC	28 and 125
					$[^{13}C,^{15}N]$-NOESY	122
					^{15}N XSTE diffusion	103
			sf9 insect cells	Overexpression	$[^{15}N,^{1}H]$-HSQC	30
					^{15}N NOESY	39
			A2780	Electroporation	T_1, T_2 relaxation	34
Tymosine β4		4.9 kDa	HEK 293F	SLO delivery	$[^{15}N,^{1}H]$-HSQC	25
Ubiquitin		8.6 kDa	*E. coli*	Overexpression	$[^{15}N,^{1}H]$-HSQC	139–141
					$[^{15}N,^{1}H]$-TROSY	106
	Version 3A		*P. pastoris*	Overexpression	$[^{15}N,^{1}H]$-HSQC	29
			sf9 insect cells	Overexpression	^{15}N NOESY	39
			Xenopus oocyte	Microinjection	$[^{15}N,^{1}H]$-HSQC	23 and 104
					^{19}F NMR	104
	Version 3A		HeLa	CPP delivery	$[^{15}N,^{1}H]$-HSQC (H–D) exchange	7
			HeLa	Electroporation	$[^{15}N,^{1}H]$-HMQC (TROSY)	94
			HEK 293T	Electroporation	$[^{15}N,^{1}H]$-HMQC	44

from the combinatorial effect of changing these two parameters. Cell death is easy to determine using for instance the trypan blue test. Another parameter that can guide protein electroporation experiments with different durations and intensities of the electric pulse is cell morphology. It is a good practice for those cells that have a well-defined morphology such as HeLa, Cos7, and U2OS and also those with fibroblast-like morphology to evaluate morphological changes that result from electroporation. In order to carry out these morphological analyses, immediately after electroporation, the cells need to be re-plated and grown for at least 2 h under normal growth conditions (37 °C and 5% CO_2). Ideally, the electroporation parameters of choice should not affect cell morphology at all assuring therefore that no cellular responses are triggered by the applied electric pulse. The number of pulses applied to the cells can vary and therefore must be also experimentally tested following the same criteria used for testing pulse duration and intensity.

Electroporation has many advantages over CPP and SLO. First of all, it does not require the addition of peptidic tags that might affect the protein structure and function. Also, it is devoid of bacterial toxins that often trigger cell responses aimed to counteract cytotoxicity. In addition, when electroporation is applied the intracellular delivery of the protein of interest is immediate. It takes place shortly after the electric pulse is applied to the cells. In this way, the protein can be analyzed right after electroporation allowing early structural analyses of the delivered protein, provided the complete removal of the non-internalized protein is assured. The latter can be easily carried out by successive washing steps of the cells with culture media or physiological buffers such as PBS. This property of electroporation allows studying for instance early events that occur on the protein of interest once it is introduced into the cells. These could be early interactions with intracellular molecules and initial posttranslational modifications that could trigger secondary and more stable and long-lived changes observed at late time points. As an example of this, it has been shown that the neuronal protein α-synuclein (α-Syn) is rapidly N-terminal acetylated upon cell take up.[41] Another advantage of electroporation is the possibility of recovering the protein under study after the electric pulse is applied. This is noted because regardless of the method used for transexpression, the delivery of proteins into mammalian cells is not efficient and thus only a small fraction of it is internalized by the cells requesting large quantities of protein (milligrams scale). In CPP and SLO-based methods, the protein that is not internalized is usually disposed of as recovering it from the culture media is inefficient due to the high quantities of BSA and other serum proteins. Electroporation allows the recycling of the isotopically labeled protein because the protein is in most cases resuspended in the electroporation buffer, then mixed with cell pellets devoid of culture media for a very short time, subjected to the electric pulse, and then separated from the cells as the cells need to be free from the non-internalized protein. It is important to mention here that the recovered protein needs, however, to be re-analyzed before it can be used in subsequent in-cell NMR experiments. The analysis of proteins that have been in contact with cells and subjected to an electric pulse is required since during electroporation certain intracellular molecules might escape from the cells and interact with the recovered protein. This can be readily performed by NMR. One can also learn from this control experiment how the protein of interest behaves with soluble molecules of the cells that escape the cells upon electroporation, something that could be considered an advantage if identifying bona fide in-cell effects is desired. Also, it is desirable to control the absence of side-effects on the protein due to the electric pulse applied.

13.6 Development of a Reporter System to Monitor Transexpression into Eukaryotic Cells

As mentioned earlier, all efforts made at present to make in-cell NMR a more powerful tool for structural biology rely on cell culture and spectroscopic advances, and very little has been done to improve the delivery of NMR-visible proteins into mammalian cells. Because this represents a bottleneck for future advances, we have contributed to this topic. In particular, we developed tools for optimizing transexpression.[46] These tools consist of reporter systems that make use of transexpression in a similar fashion as in-cell NMR does. In brief, these reporters rely on transexpressing a recombinant protein into mammalian cells that were engineered to respond to the presence of this exogenous protein (see Figure 13.4). More in detail, the reporter consists in expressing a suitable transcription factor (TF) in heterologous systems such as bacteria. The resulting recombinant protein is then purified from cell lysates using standard chromatographic techniques (Figure 13.4(A)). The second component of these

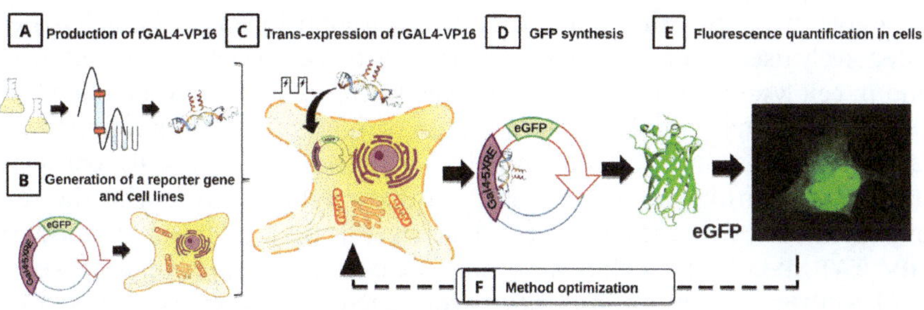

Figure 13.4 Illustration of a reporter system used to optimize and evaluate transexpression methods. (A) Recombinant (rGAL4-VP16) is expressed and purified. (B) Mammalian cells are transfected with a plasmid containing the artificial reporter gene Gal4-5XRE-eGFP, which is recognized by (rGAL4-VP16). (C) Gal4-VP16 is transexpressed into the cells by electroporation or other means of protein delivery. (D) eGFP is expressed by the transexpressed cells relative to the transexpression efficiency of the method chosen. (E) The efficiency can be quantified by measuring the fluorescence signal of the GFP generated. (F) The method of choice can be optimized to increase the amount of transexpressed protein.

reporter systems is the generation of genetically modified mammalian cells (Figure 13.4(B)). These are cells stably transfected with a reporter gene such as GFP that is transcriptionally induced by TF. The two parts come together when the recombinant TF is next transexpressed into the genetically modified mammalian cells. The introduction of TF into the cells is carried out by electroporation, CPP, or SLO-based methods (Figure 13.4(C)). The reporter system works because the transexpressed TF induces the expression of the reporter gene in the mammalian cells (Figure 13.4(D)). Thus, the more the TF delivered into the cells, the more the signal of the reporter gene (*e.g.* GFP) is obtained. Choosing reporter genes that can be easily and reliably quantified, such as the green fluorescent protein GFP or luciferase (Figure 13.4(E)), allows a fast, reliable, and inexpensive quantitative determination of the transexpression efficiency. The suitability of the reporter's systems allows therefore an easy optimization of the conditions used for protein delivery, as multiple parameters can be changed and compared accruing to the signal obtained and taken as reference values (Figure 13.4(F)). As an example below, the suitability of the reporter to compare the transexpression of TF using CPP, SLO, and electroporation is described in the following.[44] Although we discuss the suitability of the reporter system using TF as an example, any protein with detectable enzymatic activity can replace TF in this context.

The three transexpression methods described above using mammalian Cos7 cells and the reporter gene GFP as responders of the transexpressed TF are compared.[44] The TF of choice, in this case, was the fusion protein rGAL4-VP16, which consists of a 6xHIS-tagged GAL4-VP16 fusion protein containing the DNA binding domain of the yeast Gal4 protein (amino acids 1–93) fused to the C-terminal 78 amino acids of the activation domain of the VP16 protein of the herpes simplex virus.[52] This chimeric transcription factor has several favorable properties that make it an ideal protein for our aim: it has been extensively used in mammalian cells, its activity can be easily assayed *in vivo* and in cell lysates, it has a molecular weight suitable for solution-state NMR ($\sim$19 kDa), it has a globular 3D structure similar to most mammalian proteins, and it is absent in mammals precluding evolutionarily conserved interactions within the mammalian cell that might interfere with the read-outs used to evaluate the delivery efficiency. For the CPP-based method, the HIV TAT membrane cell-penetrating peptide was C-terminally fused to rGAL4-VP16, produced and purified from bacteria, and transexpressed into Cos7 cells using standard parameters. Live-cell experiments were carried out to determine the GFP signal intensity, and confocal microscopy was used in parallel on transexpressed cells to assess the cell morphology. It is found that amongst the three methods used, electroporation was the most efficient using standardized parameters.[44] Electroporation not only yielded the highest signal intensity, but also required less amount of TF compared to the CPP and SLO-based methods. Though it was a low-scale experiment, it indicates that $>$6-fold protein amounts are needed to reach the same transexpression efficiency as with the SLO method. Even lower efficiencies were obtained with the TAT CPP, where $>$15 fold protein was needed. Thus, electroporation seems to be a more efficient transexpression method when compared to CPP and SLO.

13.7 In-cell NMR Advantages and Limitations

NMR is one of the principal techniques in structural biology in addition to cryo-electron microscopy and X-ray crystallography. With solution-state NMR, the 3D structure of biomolecules can be determined at atomic resolution under physiological buffer conditions and even multi-state structures of proteins can be determined towards elucidating the structural landscape of a protein.[53] In addition, dynamics of biomolecules covering from picoseconds to milliseconds and beyond can be determined including transient interactions with client molecules. There are however two bottlenecks, NMR is a very insensitive method requesting usually high concentrations of the protein under study (usually high micromolar, but lower micromolar is possible (see Müntener *et al.*[35]), and there is a size limitation (usually <50 kDa, but see ref. 54) since with increasing tumbling time of the molecule, the NMR signals relax away. For in-cell NMR, for which ideally the protein concentration should be close to physiological levels to be biologically relevant, the consequences are: (i) an in-cell protein concentration of a few micromolar is requested limiting in-cell NMR to abundant entities typically true for chaperons of interest here. (ii) The protein should tumble, is not allowed to bind strongly, and long-lived to large entities.[55] (iii) Structure determination is challenging. For a few small systems, the 3D structure was determined by using either NOE restraint collection as demonstrated for ubiquitin or the protein G B1 domain (GB1) in sf9 insect cells[39] or *Xenopus laevis* oocytes using pseudo contact shifts.[35] However, the [^{15}N,^{1}H]-correlation spectrum, which can be usually measured within a couple of hours is a fingerprint of the structure and thus presents the structural integrity of a protein's structure, and as such in-cell NMR allows for structural studies. It is furthermore very sensitive to structural changes of the protein and transient interactions with clients and ligands. The dynamics of the system is measured usually by ^{15}N relaxation experiments covering the nanosecond–microsecond time range at residue resolution. The [^{15}N,^{1}H]-correlation spectrum measuring the chemical shifts of both ^{15}N and ^{1}H moieties is also sensitive to the buffer environment (such as pH, salt concentration, or redox potential) as well as post-translational modifications such as N-terminal acetylation,[41] oxidation,[33] and phosphorylation.[56] (iv) In-cell NMR is an ensemble-based approach in which the entire sample content is studied as a whole and as such has no spatial resolution inside the cell unless localization sequences are fused to the protein of interest or the NMR analysis is correlated to cellular imaging. In-cell NMR does therefore allow the study of the 3D structure of the protein in cells, its dynamics, its transient interactions with other biomolecules, the identification of the interaction side, and chemical modifications. To be more concrete published applications to chaperons are summarized in the following paragraphs. For other applications including drug screening by in-cell NMR, post-translational modification, redox potential, and pH sensing we refer to an excellent review by Luchinat *et al.* (2022)[11] and to Table 13.2, which lists the protein entities of which in-cell NMR investigations were published. Table 13.2 is summarized next in short.

13.8 Current In-cell NMR Investigations

To the best of our knowledge, a total of 39 different proteins were studied by in-cell NMR to date (Table 13.2). 21 of them were studied in bacteria (20 in *E. coli*), 5 in fungi, 6 in insect cells, 9 in *Xenopus laevis* oocytes, and 25 in mammalian cell lines. In the latter systems, the methods of choice for in-cell NMR were approximately equally often applied with 10 overexpression, 5 CPP, 12 electroporation, and 4 SLO delivery, respectively. The molecular weight of the systems ranges from 4 to 54 kDa. Both folded and intrinsically disordered proteins were studied.

13.9 An Overview of Previous and Future Chaperone Studies by In-cell NMR

Out of the 39 proteins studied so far by in-cell NMR and listed in Table 13.2, there are about a handful of studies on chaperones, which will be discussed in the following. Furthermore, we point towards several possible future studies. The interaction studies between HSC70 and the members of the HSP90 family with α-Syn are discussed in more detail in a separate paragraph below. The first to be mentioned are copper chaperones including the copper chaperone CCS for superoxide dismutase 1 (SOD1) associated amongst others with amyotrophic lateral sclerosis, the copper chaperone for cytochrome *c* oxidase (Cox17), which is a 7 kDa copper-binding protein, which facilitates the incorporation of copper ions into the Cu(A) site of cytochrome *c* oxidase, and the copper chaperone Antioxidant Protein1 (Atx1 or its human homologue HAH1), as well as DJ-1 also known under the name Parkin, which is a conserved, ubiquitous protein associated to a large number of intracellular processes and linked to several pathologies, including hereditary forms of Parkinson's disease, cancer, and amyotrophic lateral sclerosis. It was hypothesized to be a copper chaperone as well, before the in-cell NMR studies. The focus on copper chaperones is attributed to the interest in metalloenzymes of the Banci group which is highly active also in in-cell NMR. In their recent work, Luchinat *et al.* (2022)[11] summarize very well the action and mechanisms of these copper chaperones and their effect on superoxide dismutase 1 folding, zinc and copper-binding, and dimerization. With the use of these chaperone entities and SOD1 or thioredoxin,[57] they also review the in-cell NMR-based measurements on the redox properties of cellular environments which are strictly controlled and critical to many functional processes. The Banci group also investigated the mitochondrial intermembrane space import and assembly protein 40 (Mia40) using in-cell NMR.[40,58] For its protein folding activity in the intermembrane space of the mitochondria Mia40 can be considered to be a mitochondrial chaperone.[59,60] In the following, we want to address the proteins previously studied by in-cell NMR with known chaperone interaction partners. However, cytochrome *c* was reported to interact with histone chaperones,[61] but was regarding in-cell NMR so far only investigated in bacterial cells.[62] HRAS, which interacts with the chaperone UNC119A,[63] was the subject of two in-cell NMR studies.[39,64] Yet, both studies did not focus on the interaction with chaperones. The same is the case for the two in-cell NMR studies on PGK1[44,65] which do not explore the known interaction with Hsp90.[66] Finally, we should also address proteins whose cellular behavior is known to be indirectly affected by chaperones. In the case of Pin1, its cellular distribution was shown to be Hsp90 mediated in *Arabidopsis*.[67] The only previous in-cell NMR study in *X. laevis* oocyte did not explore this.[68] For the tau protein increased levels of Hsp70 and Hsp90 were shown to promote tau solubility in cells.[69] Again, all previous studies did not address chaperones.[24,43,44] These examples might help to illustrate that there are many future opportunities for chaperone-related studies by in-cell NMR. All concerning proteins with the feasibility of using in-cell NMR was already demonstrated.

13.10 Transient α-Synuclein Chaperone Interactions in Cells by In-cell NMR[‡]

Synucleinopathies are neurodegenerative diseases with the three major ones being Parkinson's disease, dementia with Lewy bodies, and multiple system atrophy (MSA) associated with the accumulation and aggregation of α-Syn.[71–73] The disease-relevant role of α-Syn is highlighted by mutations in the α-Syn gene (SNCA) and the duplication or triplication of the SNCA leading to early-onset Parkinson's disease in affected families.[74,75] α-Syn comprising a positively charged N-terminal domain (residues 1–60), the so-called hydrophobic NAC domain (non-amyloid-β component, residues 61–95), and the C-terminal negatively charged segment (residues 96–140) is an abundant protein in neuronal cells possessing considerable structural plasticity: (i) *in vitro* and in cells, α-Syn is an intrinsically disordered protein,[41,76] which is somewhat transiently compacted through the interaction of the positively charged/aromatic residue containing the N-terminus with its negatively charged/aromatic residue containing the C-terminus.[77] (ii) α-Syn binds in a helical state to membranes,[78] is also able to form (iii) HDL (high-density lipoprotein)-like particles,[79,80] and (iv) during pathological aggregation, α-Syn forms oligomeric species of different kinds[81] and ultimately forms mature amyloid fibrils[82–85] of polymorphic nature as revealed by cryo-electron microscopy (EM).[86–88] Chaperones are able to interact with α-Syn transiently *in vitro* as revealed for example by Burmann *et al.* studying six molecular chaperones.[70] The array included the human chaperones Hsc70 and Hsp90β, as well as the bacterial holdase chaperones SecB, Skp, SurA, and trigger factor, which feature diverse architectures and have a simple, ATP-independent functional mechanism,

‡Reproduced from ref. 70 with permission from the author, Copyright 2019.

which excludes the presence of specific features when interacting with α-Syn. Titration NMR experiments of the chaperones in the unlabeled form to uniformly ^{15}N-labeled α-Syn monitored by 2D ^{15}N–^{1}H-correlation NMR spectra resulted in signal intensity decrease of the first dozen N-terminal residues as well as *ca.* six residues around Tyr39, which are predicted to be an Hsp70-chaperone binding site, indicating a direct transient interaction of chaperones *via* these segments identifying these parts as the canonical chaperone-interaction motif of α-Syn. The time window of interaction is in the microsecond to millisecond time range. Transient interactions of α-Syn with Hsp70 identified the same canonical motif,[89] while in the same study, additional interactions of α-Syn with the co-chaperone DNAJB1 *via* the α-Syn C-terminal segment comprising residues 123–129 have also been documented.[89] Interestingly, the NMR intensity profile of α-Syn inside living mammalian cells first described by Theillet and coworkers[41] and shown in Figure 13.5(B) resembles the transient interaction between α-Syn and chaperones *in vitro* indicating that in mammalian cells α-Syn is dominantly in contact with chaperones and that the interaction is mediated *via* the amino-terminus of α-Syn. As a verification of this hypothesis, the interactome of the α-Syn amino-terminus by chemical cross-linking followed by immune-precipitation and mass-spectrometric analysis of mammalian cells, comparing the cross-linked proteomes for α-Syn relative to Δ*N*-α-Syn, which lacks residues 2–11 has been undertaken (Figure 13.5(A)). This interactome consists of a large number of molecular chaperones in the range of 30–75% abundance, including several Hsp90-variants, six Hsp70-variants, and seven different subunits of the TRiC/CCT–chaperonin. In addition, four different variants of the 14-3-3 proteins were contained in the interactome of the α-Syn amino-terminus. To further validate the stated hypothesis of transient α-Syn chaperone interactions in cells, Hsc70 was specifically knocked down by a short hairpin RNA and Hsp90β was inhibited by the addition of a mixture of geldanamycin and radicicol followed by in-cell NMR of transexpressed α-Syn (Figure 13.5(B) and (C)). Knockdown of Hsc70 resulted in no significant changes in the intensity profile. However, upon inhibiting both chaperones simultaneously effects on α-Syn interactions were observed in a time-dependent manner. Four hours after treatment with the Hsp90β inhibitors in addition to the Hsc70 knockdown, the in-cell NMR intensity profile had regained the signals at the amino-terminus and around Tyr39, indicating that the amount of free intracellular α-Syn was increased (Figure 13.5(D)). The spectral changes were even more dramatic after 24 hours of drug treatment when a global signal reduction was observed for residues 1–90, which is attributed to a transient membrane interaction. Based on these data, the combined inhibition of the two chaperones thus leads to membrane interaction of α-Syn, which was absent in the cellular ground state. In parallel, the formation of macroscopic aggregates containing α-Syn as evidenced by high-molecular weight species in cellular isolates and localization of α-Syn to the mitochondria were observed (Figure 13.5(E)).[70]

A comparison between the in-cell NMR data with the corresponding *in vitro* data allows also for the estimation of occupancy. Under the estimated

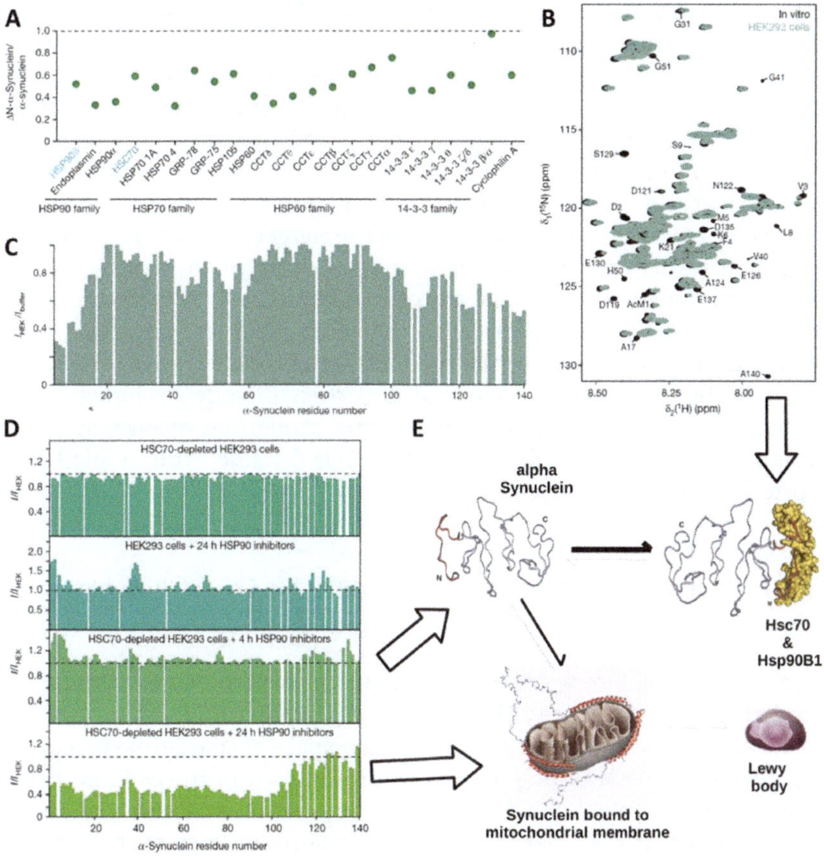

Figure 13.5 Interaction study of α-Syn and chaperones in cells. (A) Quantitative mass spectrometry analysis determining the abundance ratio of chaperones bound to Δ*N*-α-Syn relative to wild type α-Syn. Δ*N*-α-Syn is a variant of α-Syn with a deleted N-terminus. (B) Overlay of two-dimensional $[^{15}N,^{1}H]$-NMR spectra of ^{15}N-α-Syn in NMR buffer (black) and inside living HEK293 cells (blue-green). (C) Residue-resolved backbone amide NMR signal attenuation (I_{HEK}/I_{buffer}) of α-Syn in mammalian cells. (D) NMR signal attenuation in treated cells, relative to untreated cells (I/I_{HEK}). Different combinations of HSC70 depletion and HSP90 inhibition were applied, as indicated. (E) Graphical illustration of α-Syn in a chaperone bound (top right) and free (top left) form. Depletion of chaperones leads to free α-Syn which can bind to mitochondrial membranes (bottom left) from where it can aggregate and eventually form Lewy bodies (bottom right). Reproduced from ref. 70 with permission from Springer Nature, Copyright 2019.

cellular concentrations of α-Syn within the neuronal synapses of ~50 μM and by a combined concentration of Hsp70/Hsp90 chaperones of ~70 μM in the same tissue,[90] about 90% of the cellular α-Syn is in the chaperone-bound form transiently exchanging with its free intrinsically disordered state.

This finding strongly suggests a regulation mechanism of chaperones on the activity of α-Syn which might be also the case for other intrinsically disordered protein (IDP) systems. In this regard, the interference of the transient chaperone–α-Syn interaction *via* methionine oxidation of residues Met1 and Met5 in the canonical motif and tyrosine phosphorylation including Tyr39 also part of the canonical binding motif are noted.

In summary, in-cell NMR identified a generalized mechanism of transient chaperone–α-Syn interactions. A model emerges, in which chaperone-interacting forms are the dominant species of α-Syn in healthy cells and where chaperones are a master regulator of the levels of different cellular states of α-Syn. This model predicts that changes in the cellular levels of either chaperones or α-Syn or the modulation of their interaction will result in a disturbed homeostatic balance and then directly or indirectly lead to fibrillization and changes in membrane properties, eventually causing Parkinson's disease. Notably, this model is in line with a plethora of reported experimental observations.

13.11 Outlook (Focus on In-cell NMR of Mammalian Systems)

In cell structural biology is stated to be the cell biology of the future. Knowledge of the structure and dynamics of biomolecules with time-resolved and sub-cellular resolution in the coordinate space of space-time is requested for a detailed understanding of the complex biology of the cell and beyond. It is important to note that most intracellular proteins are multivalent entities that play different roles depending on their surroundings such as environment, physical interactors, or subcellular localization. Thus, upon environmental changes, a particular protein may be found in a different location within the cell and may, therefore, interact with different distinct partners resulting in structural changes. As such it is expected that subtle structural differences arising from different environments and distinctive interactions with intracellular partners could yield different structural states within the same cell. This can be exemplified by the recent findings on liquid–liquid phase separation such as stress granules which are determined by the dynamics and structure of the involved species embedded in a given environment.[91] While cryo-electron microscopy and tomography (and eventually low-temperature polarized solid-state NMR) may be able to resolve the structure of single molecules localized in a cell, solution-state NMR will elucidate its dynamics, transient interactions, and structure time-resolved while NMR is an ensemble-based method. Thus, an integration of both single-molecule cryo-EM/tomography and ensemble-based NMR is requested finally.

But to reach this aim a long avenue for in-cell NMR awaits us requesting also further methodological advances. In our view, an important step is a protein-sample-amount and cell-amount sensitive general protocol for the fast realization of in-cell NMR of a protein of interest, which we believe to have developed. This allows the study of many proteins by in-cell NMR straightforwardly and as such will enlarge the table and data sets of proteins making it possible to establish an NMR database of protein conformations in cells. This will include co-existing multiple structural states of a protein residing inside a mammalian cell. In this line, recent work by our group shows that structural multi-states can be visualized by in-cell NMR in mammalian cells as resonance splitting. Aiming at understanding the structural multivalency of proteins inside mammalian cells, Kadavath *et al.*[92] find that single NMR resonances observed *in vitro* split into daughter peaks each one representing a particular structural state of the protein under study. This hypothesis states that each particular structural state might result from the protein of interest being present in a particular subcellular localization and structural state. Recent data obtained by the use of a mitochondrial localization sequence is in agreement with this notion. It is evident, however, that further method developments to resolve the spatial resolution within a cell are key to overcoming the ensemble-based nature of in-cell NMR.

It is also important to note that in-cell structural biology (whether cryo-EM/tomography or NMR) relies on *in vitro* reference data. For example, all protein structures solved by NMR to date either in bacteria or in eukaryotic cells rely on a reference sample in buffer. This implies that structures found in cells must mirror the structures obtained *in vitro*. This requisite, although valuable for initial proof-of-concept experiments, makes structure determination in cells less attractive as both structures should be the same. But it is our belief that the field will profit from the possibility of solving protein structures in cells exclusively by NMR. It is known that there are proteins whose structures are significantly different *in vitro* and in cells. Indeed, one example of this prediction was already shown by the group of Pielak, who showed that the C-terminus of the IDP anti-sigma factor FlgM acquires structure in cells.[93] For proteins whose structure differs *in vitro* and in cells, in-cell NMR is one of the few existing methods for structure determination in cells.

It took a few years since its creation for in-cell NMR to contribute to a better understanding of certain biological processes and therefore be considered a useful tool for such studies. Amongst them in mammalian systems, in-cell NMR was used for the detailed structural characterization of protein maturation[31] redox[76] posttranslational modifications,[32] chaperones,[40,70,94–96] drug screening,[27] *etc.* The next frontier is now a comprehensive understanding of in-cell NMR and its implications for structural biology. The reader is invited to participate in it.

References

1. M. R. Sawaya, D. Cascio, M. Gingery, J. Rodriguez, L. Goldschmidt, J.-P. Colletier, M. M. Messerschmidt, S. Boutet, J. E. Koglin, G. J. Williams, A. S. Brewster, K. Nass, J. Hattne, S. Botha, R. B. Doak,

R. L. Shoeman, D. P. DePonte, H.-W. Park, B. A. Federici, N. K. Sauter, I. Schlichting and D. S. Eisenberg, *Proc. Natl. Acad. Sci. U. S. A.*, 2014, **111**, 12769–12774.
2. C. T. Ng and L. Gan, *Mol. Biol. Cell*, 2020, **31**, 87–100.
3. S. Narasimhan, G. E. Folkers and M. Baldus, *ChemPlusChem*, 2020, **85**, 760–768.
4. D. I. Freedberg and P. Selenko, *Annu. Rev. Biophys.*, 2014, **43**, 171–192.
5. E. Luchinat and L. Banci, *IUCrJ*, 2017, **4**, 108–118.
6. E. Luchinat and L. Banci, *J. Biol. Chem.*, 2016, **291**, 3776–3784.
7. K. Inomata, A. Ohno, H. Tochio, S. Isogai, T. Tenno, I. Nakase, T. Takeuchi, S. Futaki, Y. Ito, H. Hiroaki and M. Shirakawa, *Nature*, 2009, **458**, 106–109.
8. J. A. Gerez, N. C. Prymaczok and R. Riek, in *Intrinsically Disordered Proteins: Methods and Protocols*, ed. B. B. Kragelund and K. Skriver, Springer US, New York, NY, 2020, pp. 873–893.
9. Z. Serber and V. Dötsch, *Biochemistry*, 2001, **40**, 14317–14323.
10. Z. Serber, A. T. Keatinge-Clay, R. Ledwidge, A. E. Kelly, S. M. Miller and V. Dötsch, *J. Am. Chem. Soc.*, 2001, **123**, 2446–2447.
11. E. Luchinat, M. Cremonini and L. Banci, *Chem. Rev.*, 2022, **122**, 9267–9306.
12. F. Bloch, W. W. Hansen and M. Packard, *Phys. Rev.*, 1946, **70**, 474–485.
13. F. Bloch, W. W. Hansen and M. Packard, *Phys. Rev.*, 1946, **69**, 127.
14. E. D. Becker, *Anal. Chem.*, 1993, **65**, 295A–302A.
15. E. R. P. Zuiderweg, M. Billeter, R. Boelens, R. M. Scheek, K. Wüthi and R. Kaptein, *FEBS Lett.*, 1984, **174**, 243–247.
16. A. P. Minton, *Biopolymers*, 1981, **20**, 2093–2120.
17. A. P. Schlesinger, Y. Wang, X. Tadeo, O. Millet and G. J. Pielak, *J. Am. Chem. Soc.*, 2011, **133**, 8082–8085.
18. H. Welte and M. Kovermann, *ChemBioChem*, 2020, **21**, 3575–3579.
19. R. S. Norton, *Bull. Magn. Reson.*, 1981, **3**, 29–48.
20. D. L. Foxall, J. S. Cohen and J. B. Mitchell, *Exp. Cell Res.*, 1984, **154**, 521–529.
21. A. A. de Graaf, R. M. Wittig, U. Probst, J. Strohhaecker, S. M. Schoberth and H. Sahm, *J. Magn. Reson.*, 1992, **98**, 654–659.
22. P. Selenko, Z. Serber, B. Gadea, J. Ruderman and G. Wagner, *Proc. Natl. Acad. Sci. U. S. A.*, 2006, **103**, 11904–11909.
23. T. Sakai, H. Tochio, T. Tenno, Y. Ito, T. Kokubo, H. Hiroaki and M. Shirakawa, *J. Biomol. NMR*, 2006, **36**, 179–188.
24. J.-F. Bodart, J.-M. Wieruszeski, L. Amniai, A. Leroy, I. Landrieu, A. Rousseau-Lescuyer, J.-P. Vilain and G. Lippens, *J. Magn. Reson.*, 2008, **192**, 252–257.
25. S. Ogino, S. Kubo, R. Umemoto, S. Huang, N. Nishida and I. Shimada, *J. Am. Chem. Soc.*, 2009, **131**, 10834–10835.
26. P. Schanda, Ē. Kupče and B. Brutscher, *J. Biomol. NMR*, 2005, **33**, 199–211.
27. J. Xie, R. Thapa, S. Reverdatto, D. S. Burz and A. Shekhtman, *J. Med. Chem.*, 2009, **52**, 3516–3522.

28. D. Sakakibara, A. Sasaki, T. Ikeya, J. Hamatsu, T. Hanashima, M. Mishima, M. Yoshimasu, N. Hayashi, T. Mikawa, M. Wälchli, B. O. Smith, M. Shirakawa, P. Güntert and Y. Ito, *Nature*, 2009, **458**, 102–105.
29. K. Bertrand, S. Reverdatto, D. S. Burz, R. Zitomer and A. Shekhtman, *J. Am. Chem. Soc.*, 2012, **134**, 12798–12806.
30. J. Hamatsu, D. O'Donovan, T. Tanaka, T. Shirai, Y. Hourai, T. Mikawa, T. Ikeya, M. Mishima, W. Boucher, B. O. Smith, E. D. Laue, M. Shirakawa and Y. Ito, *J. Am. Chem. Soc.*, 2013, **135**, 1688–1691.
31. L. Banci, L. Barbieri, I. Bertini, E. Luchinat, E. Secci, Y. Zhao and A. R. Aricescu, *Nat. Chem. Biol.*, 2013, **9**, 297–299.
32. I. Amata, M. Maffei, A. Igea, M. Gay, M. Vilaseca, A. R. Nebreda and M. Pons, *ChemBioChem*, 2013, **14**, 1820–1827.
33. A. Binolfi, A. Limatola, S. Verzini, J. Kosten, F.-X. Theillet, H. May Rose, B. Bekei, M. Stuiver, M. van Rossum and P. Selenko, *Nat. Commun.*, 2016, 7, 10251.
34. S. Leeb, F. Yang, M. Oliveberg and J. Danielsson, *J. Phys. Chem. B*, 2020, **124**, 10698–10707.
35. T. Müntener, D. Häussinger, P. Selenko and F.-X. Theillet, *J. Phys. Chem. Lett.*, 2016, 7, 2821–2825.
36. E. Luchinat and L. Banci, *Acc. Chem. Res.*, 2018, **51**, 1550–1557.
37. N. K. Goto and L. E. Kay, *Curr. Opin. Struct. Biol.*, 2000, **10**, 585–592.
38. K. P. Wall and L. E. Hough, *Biophys. J.*, 2018, **115**, 1690–1695.
39. T. Tanaka, T. Ikeya, H. Kamoshida, Y. Suemoto, M. Mishima, M. Shirakawa, P. Güntert and Y. Ito, *Angew. Chem., Int. Ed.*, 2019, **58**, 7284–7288.
40. L. Barbieri, E. Luchinat and L. Banci, *Nat. Protoc.*, 2016, **11**, 1101–1111.
41. F.-X. Theillet, A. Binolfi, B. Bekei, A. Martorana, H. M. Rose, M. Stuiver, S. Verzini, D. Lorenz, M. van Rossum, D. Goldfarb and P. Selenko, *Nature*, 2016, **530**, 45–50.
42. A. Limatola, C. Eichmann, R. S. Jacob, G. Ben-Nissan, M. Sharon, A. Binolfi and P. Selenko, *Proteomics*, 2018, **18**, 1800056.
43. S. Zhang, C. Wang, J. Lu, X. Ma, Z. Liu, D. Li, Z. Liu and C. Liu, *Int. J. Mol. Sci.*, 2019, **20**, 90.
44. J. A. Gerez, N. C. Prymaczok, H. Kadavath, D. Ghosh, M. Bütikofer, Y. Fleischmann, P. Güntert and R. Riek, *Commun. Biol.*, 2022, **5**, 1–14.
45. V. M. Tatard, G. D'Ippolito, S. Diabira, A. Valeyev, J. Hackman, M. McCarthy, T. Bouckenooghe, P. Menei, C. N. Montero-Menei and P. C. Schiller, *Bone*, 2007, **40**, 360–373.
46. J. A. Gerez, N. C. Prymaczok and H. Kadavath, *et al.*, Protein structure determination in human cells by in-cell NMR and a reporter system to optimize protein delivery or transexpression, *Commun. Biol.*, 2022, **5**, 1322.
47. M. A. Sells, J. Li and J. Chernoff, *Biotechniques*, 1995, **19**, 72–76, 78.
48. B. Chatin, M. Mével, J. Devallière, L. Dallet, T. Haudebourg, P. Peuziat, T. Colombani, M. Berchel, O. Lambert, A. Edelman and B. Pitard, *Mol. Ther.-Nucleic Acids*, 2015, **4**, e244.

49. A. Ziegler, P. Nervi, M. Dürrenberger and J. Seelig, *Biochemistry*, 2005, **44**, 138–148.
50. N. Nischan, H. D. Herce, F. Natale, N. Bohlke, N. Budisa, M. C. Cardoso and C. P. R. Hackenberger, *Angew. Chem., Int. Ed.*, 2015, **54**, 1950–1953.
51. R. K. Tweten, *Infect. Immun.*, 2005, **73**, 6199–6209.
52. I. Sadowski, J. Ma, S. Triezenberg and M. Ptashne, *Nature*, 1988, **335**, 563–564.
53. B. Vögeli, S. Kazemi, P. Güntert and R. Riek, *Nat. Struct. Mol. Biol.*, 2012, **19**, 1053–1057.
54. R. Riek, K. Pervushin, C. Fernández, M. Kainosho and K. Wüthrich, *J. Am. Chem. Soc.*, 2001, **123**, 658–664.
55. D. S. Burz, L. Breindel and A. Shekhtman, in *Methods in Enzymology*, ed. A. K. Shukla, Academic Press, 2019, vol. 621, pp. 305–328.
56. P. Selenko, D. P. Frueh, S. J. Elsaesser, W. Haas, S. P. Gygi and G. Wagner, *Nat. Struct. Mol. Biol.*, 2008, **15**, 321–329.
57. A. Mochizuki, A. Saso, Q. Zhao, S. Kubo, N. Nishida and I. Shimada, *J. Am. Chem. Soc.*, 2018, **140**, 3784–3790.
58. L. Banci, L. Barbieri, E. Luchinat and E. Secci, *Chem. Biol.*, 2013, **20**, 747–752.
59. D. Weckbecker, S. Longen, J. Riemer and J. M. Herrmann, *EMBO J.*, 2012, **31**, 4348–4358.
60. A. Chacinska, S. Pfannschmidt, N. Wiedemann, V. Kozjak, L. K. S. Szklarz, A. Schulze-Specking, K. N. Truscott, B. Guiard, C. Meisinger and N. Pfanner, *EMBO J.*, 2004, **23**, 3735–3746.
61. K. González-Arzola, A. Guerra-Castellano, F. Rivero-Rodríguez, M. Á. Casado-Combreras, G. Pérez-Mejías, A. Díaz-Quintana, I. Díaz-Moreno and M. A. De la Rosa, *FEBS Open Bio*, 2021, **11**, 2418–2440.
62. P. B. Crowley, E. Chow and T. Papkovskaia, *ChemBioChem*, 2011, **12**, 1043–1048.
63. T. Shimizu, T. Nakamura, H. Inaba, H. Iwasa, J. Maruyama, K. Arimoto-Matsuzaki, T. Nakata, H. Nishina and Y. Hata, *J. Biol. Chem.*, 2020, **295**, 11214–11230.
64. Q. Zhao, R. Fujimiya, S. Kubo, C. B. Marshall, M. Ikura, I. Shimada and N. Nishida, *Cell Rep.*, 2020, **32**, 108074.
65. S. P. Williams, P. M. Haggie and K. M. Brindle, *Biophys. J.*, 1997, 72, 490–498.
66. X. Chen, C. Zhao, X. Li, T. Wang, Y. Li, C. Cao, Y. Ding, M. Dong, L. Finci, J. Wang, X. Li and L. Liu, *Nat. Chem. Biol.*, 2015, **11**, 19–25.
67. D. Samakovli, L. Roka, A. Dimopoulou, P. K. Plitsi, A. Žukauskaitė, P. Georgopoulou, O. Novák, D. Milioni and P. Hatzopoulos, *New Phytol.*, 2021, **231**, 1814–1831.
68. L. M. Luh, R. Hänsel, F. Löhr, D. K. Kirchner, K. Krauskopf, S. Pitzius, B. Schäfer, P. Tufar, I. Corbeski, P. Güntert and V. Dötsch, *J. Am. Chem. Soc.*, 2013, **135**, 13796–13803.
69. F. Dou, W. J. Netzer, K. Tanemura, F. Li, F. U. Hartl, A. Takashima, G. K. Gouras, P. Greengard and H. Xu, *Proc. Natl. Acad. Sci. U. S. A.*, 2003, **100**, 721–726.

70. B. M. Burmann, J. A. Gerez, I. Matečko-Burmann, S. Campioni, P. Kumari, D. Ghosh, A. Mazur, E. E. Aspholm, D. Šulskis, M. Wawrzyniuk, T. Bock, A. Schmidt, S. G. D. Rüdiger, R. Riek and S. Hiller, *Nature*, 2020, **577**, 127–132.
71. M. G. Spillantini, M. L. Schmidt, V. M.-Y. Lee, J. Q. Trojanowski, R. Jakes and M. Goedert, *Nature*, 1997, **388**, 839–840.
72. H. A. Lashuel, C. R. Overk, A. Oueslati and E. Masliah, *Nat. Rev. Neurosci.*, 2013, **14**, 38–48.
73. N. C. Prymaczok, R. Riek and J. Gerez, *Front. Hum. Neurosci.*, 2016, **10**, 608.
74. R. Krüger, W. Kuhn, T. Müller, D. Woitalla, M. Graeber, S. Kösel, H. Przuntek, J. T. Epplen, L. Schols and O. Riess, *Nat. Genet.*, 1998, **18**, 106–108.
75. P. Ibáñez, A.-M. Bonnet, B. Débarges, E. Lohmann, F. Tison, Y. Agid, A. Dürr, A. Brice and P. Pollak, *Lancet*, 2004, **364**, 1169–1171.
76. D. Eliezer, E. Kutluay, R. Bussell and G. Browne, *J. Mol. Biol.*, 2001, **307**, 1061–1073.
77. L. Salmon, G. Nodet, V. Ozenne, G. Yin, M. R. Jensen, M. Zweckstetter and M. Blackledge, *J. Am. Chem. Soc.*, 2010, **132**, 8407–8418.
78. T. S. Ulmer, A. Bax, N. B. Cole and R. L. Nussbaum, *J. Biol. Chem.*, 2005, **280**, 9595–9603.
79. J. Varkey, N. Mizuno, B. G. Hegde, N. Cheng, A. C. Steven and R. Langen, *J. Biol. Chem.*, 2013, **288**, 17620–17630.
80. C. Eichmann, P. Kumari and R. Riek, *FEBS Lett.*, 2017, **591**, 304–311.
81. G. Fusco, S. W. Chen, P. T. F. Williamson, R. Cascella, M. Perni, J. A. Jarvis, C. Cecchi, M. Vendruscolo, F. Chiti, N. Cremades, L. Ying, C. M. Dobson and A. D. Simone, *Science*, 2017, **358**, 1440–1443.
82. M. D. Tuttle, G. Comellas, A. J. Nieuwkoop, D. J. Covell, D. A. Berthold, K. D. Kloepper, J. M. Courtney, J. K. Kim, A. M. Barclay, A. Kendall, W. Wan, G. Stubbs, C. D. Schwieters, V. M. Y. Lee, J. M. George and C. M. Rienstra, *Nat. Struct. Mol. Biol.*, 2016, **23**, 409–415.
83. M. Vilar, H.-T. Chou, T. Lührs, S. K. Maji, D. Riek-Loher, R. Verel, G. Manning, H. Stahlberg and R. Riek, *Proc. Natl. Acad. Sci. U. S. A.*, 2008, **105**, 8637–8642.
84. H. Heise, W. Hoyer, S. Becker, O. C. Andronesi, D. Riedel and M. Baldus, *Proc. Natl. Acad. Sci. U. S. A.*, 2005, **102**, 15871–15876.
85. L. Bousset, L. Pieri, G. Ruiz-Arlandis, J. Gath, P. H. Jensen, B. Habenstein, K. Madiona, V. Olieric, A. Böckmann, B. H. Meier and R. Melki, *Nat. Commun.*, 2013, **4**, 2575.
86. B. Li, P. Ge, K. A. Murray, P. Sheth, M. Zhang, G. Nair, M. R. Sawaya, W. S. Shin, D. R. Boyer, S. Ye, D. S. Eisenberg, Z. H. Zhou and L. Jiang, *Nat. Commun.*, 2018, **9**, 3609.
87. R. Guerrero-Ferreira, N. M. Taylor, D. Mona, P. Ringler, M. E. Lauer, R. Riek, M. Britschgi and H. Stahlberg, *eLife*, 2018, **7**, e36402.
88. R. Guerrero-Ferreira, N. M. Taylor, A.-A. Arteni, P. Kumari, D. Mona, P. Ringler, M. Britschgi, M. E. Lauer, A. Makky, J. Verasdonck, R. Riek,

R. Melki, B. H. Meier, A. Böckmann, L. Bousset and H. Stahlberg, *eLife*, 2019, **8**, e48907.

89. A. S. Wentink, N. B. Nillegoda, J. Feufel, G. Ubartaitė, C. P. Schneider, P. De Los Rios, J. Hennig, A. Barducci and B. Bukau, *Nature*, 2020, **587**, 483–488.
90. C. Guardia-Laguarta, E. Area-Gomez, E. A. Schon and S. Przedborski, *Front. Neuroanat.*, 2015, **9**, 17.
91. J. A. Ditlev, L. B. Case and M. K. Rosen, *J. Mol. Biol.*, 2018, **430**, 4666–4684.
92. H. Kadavath, N. Cecilia Prymaczok, C. Eichmann, R. Riek and J. A. Gerez, *Angew. Chem., Int. Ed.*, 2023, **62**, e202213976.
93. M. M. Dedmon, C. N. Patel, G. B. Young and G. J. Pielak, *Proc. Natl. Acad. Sci. U. S. A.*, 2002, **99**, 12681–12684.
94. S. Majumder, J. Xue, C. M. DeMott, S. Reverdatto, D. S. Burz and A. Shekhtman, *Biochemistry*, 2015, **54**, 2727–2738.
95. I. Bertini, I. C. Felli, L. Gonnelli, V. M. Kumar and R. Pierattelli, *Angew. Chem., Int. Ed.*, 2011, **50**, 2339–2341.
96. Q. Wang, A. Zhuravleva and L. M. Gierasch, *Biochemistry*, 2011, **50**, 9225–9236.
97. C. M. DeMott, S. Majumder, D. S. Burz, S. Reverdatto and A. Shekhtman, *Biochemistry*, 2017, **56**, 4117–4126.
98. K. Inomata, H. Kamoshida, M. Ikari, Y. Ito and T. Kigawa, *Chem. Commun.*, 2017, **53**, 11245–11248.
99. C. Li, L. M. Charlton, A. Lakkavaram, C. Seagle, G. Wang, G. B. Young, J. M. Macdonald and G. J. Pielak, *J. Am. Chem. Soc.*, 2008, **130**, 6310–6311.
100. N. G. Sharaf, C. O. Barnes, L. M. Charlton, G. B. Young and G. J. Pielak, *J. Magn. Reson.*, 2010, **202**, 140–146.
101. A. Binolfi, F.-X. Theillet and P. Selenko, *Biochem. Soc. Trans.*, 2012, **40**, 950–954.
102. A. E. Smith, L. Z. Zhou and G. J. Pielak, *Protein Sci.*, 2015, **24**, 706–713.
103. C. A. Waudby, M. D. Mantle, L. D. Cabrita, L. F. Gladden, C. M. Dobson and J. Christodoulou, *J. Am. Chem. Soc.*, 2012, **134**, 11312–11315.
104. Y. Ye, X. Liu, Y. Chen, G. Xu, Q. Wu, Z. Zhang, C. Yao, M. Liu and C. Li, *Chem. - Eur. J.*, 2015, **21**, 8686–8690.
105. I. G. Zigoneanu and G. J. Pielak, *Mol. Pharmacol.*, 2012, **9**, 1024–1029.
106. G. Xu, Y. Ye, X. Liu, S. Cao, Q. Wu, K. Cheng, M. Liu, G. J. Pielak and C. Li, *Biochemistry*, 2014, **53**, 1971–1981.
107. A. E. Smith, L. Z. Zhou, A. H. Gorensek, M. Senske and G. J. Pielak, *Proc. Natl. Acad. Sci. U. S. A.*, 2016, **113**, 1725–1730.
108. S. S. Stadmiller, A. H. Gorensek-Benitez, A. J. Guseman and G. J. Pielak, *J. Mol. Biol.*, 2017, **429**, 1155–1161.
109. J. F. Thole, T. C. Fadero, J. P. Bonin, S. S. Stadmiller, J. A. Giudice and G. J. Pielak, *Biochemistry*, 2021, **60**, 451–459.
110. E. Luchinat, L. Barbieri, M. Cremonini, A. Nocentini, C. T. Supuran and L. Banci, *Angew. Chem., Int. Ed.*, 2020, **59**, 6535–6539.
111. Y. Takaoka, Y. Kioi, A. Morito, J. Otani, K. Arita, E. Ashihara, M. Ariyoshi, H. Tochio, M. Shirakawa and I. Hamachi, *Chem. Commun.*, 2013, **49**, 2801–2803.

112. E. Luchinat, L. Barbieri, M. Cremonini, M. Pennestri, A. Nocentini, C. T. Supuran and L. Banci, *Acta Crystallogr., Sect. D: Struct. Biol.*, 2021, **77**, 1270–1281.
113. D. S. S. Hembram, T. Haremaki, J. Hamatsu, J. Inoue, H. Kamoshida, T. Ikeya, M. Mishima, T. Mikawa, N. Hayashi, M. Shirakawa and Y. Ito, *Biochem. Biophys. Res. Commun.*, 2013, **438**, 653–659.
114. R. Hänsel, S. Foldynová-Trantírková, F. Löhr, J. Buck, E. Bongartz, E. Bamberg, H. Schwalbe, V. Dötsch and L. Trantírek, *J. Am. Chem. Soc.*, 2009, **131**, 15761–15768.
115. Y. Ye, X. Liu, G. Xu, M. Liu and C. Li, *Angew. Chem., Int. Ed.*, 2015, **54**, 5328–5330.
116. S. Kubo, N. Nishida, Y. Udagawa, O. Takarada, S. Ogino and I. Shimada, *Angew. Chem., Int. Ed.*, 2013, **52**, 1208–1211.
117. E. Mercatelli, L. Barbieri, E. Luchinat and L. Banci, *Biochim. Biophys. Acta, Mol. Cell Res.*, 2016, **1863**, 198–204.
118. L. Barbieri, E. Luchinat and L. Banci, *JBIC, J. Biol. Inorg. Chem.*, 2018, **23**, 61–69.
119. S. Reckel, J. J. Lopez, F. Löhr, C. Glaubitz and V. Dötsch, *ChemBioChem*, 2012, **13**, 534–537.
120. C. Kyne and P. B. Crowley, *Biochemistry*, 2017, **56**, 5026–5032.
121. Y. Ye, Q. Wu, W. Zheng, B. Jiang, G. J. Pielak, M. Liu and C. Li, *Anal. Bioanal. Chem.*, 2018, **410**, 869–874.
122. T. Ikeya, T. Hanashima, S. Hosoya, M. Shimazaki, S. Ikeda, M. Mishima, P. Güntert and Y. Ito, *Sci. Rep.*, 2016, **6**, 38312.
123. S. L. Speer, W. Zheng, X. Jiang, I.-T. Chu, A. J. Guseman, M. Liu, G. J. Pielak and C. Li, *Proc. Natl. Acad. Sci. U. S. A.*, 2021, **118**, e2019918118.
124. B.-B. Pan, F. Yang, Y. Ye, Q. Wu, C. Li, T. Huber and X.-C. Su, *Chem. Commun.*, 2016, **52**, 10237–10240.
125. X. Mu, S. Choi, L. Lang, D. Mowray, N. V. Dokholyan, J. Danielsson and M. Oliveberg, *Proc. Natl. Acad. Sci. U. S. A.*, 2017, **114**, E4556–E4563.
126. Z. Serber, R. Ledwidge, S. M. Miller and V. Dötsch, *J. Am. Chem. Soc.*, 2001, **123**, 8895–8901.
127. L. E. Hough, K. Dutta, S. Sparks, D. B. Temel, A. Kamal, J. Tetenbaum-Novatt, M. P. Rout and D. Cowburn, *eLife*, 2015, **4**, e10027.
128. L. Barbieri, E. Luchinat and L. Banci, *Sci. Rep.*, 2015, **5**, 14456.
129. C. Malicet, V. Giroux, S. Vasseur, J. C. Dagorn, J. L. Neira and J. L. Iovanna, *Proc. Natl. Acad. Sci. U. S. A.*, 2006, **103**, 2671–2676.
130. J. D. Cobbert, C. DeMott, S. Majumder, E. A. Smith, S. Reverdatto, D. S. Burz, K. A. McDonough and A. Shekhtman, *Sci. Rep.*, 2015, **5**, 9402.
131. A. Y. Maldonado, D. S. Burz, S. Reverdatto and A. Shekhtman, *PLoS One*, 2013, **8**, e74576.
132. L. Breindel, D. S. Burz and A. Shekhtman, *Commun. Biol.*, 2020, **3**, 1–9.
133. L. Banci, L. Barbieri, I. Bertini, F. Cantini and E. Luchinat, *PLoS One*, 2011, **6**, e23561.

134. J. Danielsson, X. Mu, L. Lang, H. Wang, A. Binolfi, F.-X. Theillet, B. Bekei, D. T. Logan, P. Selenko, H. Wennerström and M. Oliveberg, *Proc. Natl. Acad. Sci. U. S. A.*, 2015, **112**, 12402–12407.
135. J. Danielsson, K. Inomata, S. Murayama, H. Tochio, L. Lang, M. Shirakawa and M. Oliveberg, *J. Am. Chem. Soc.*, 2013, **135**, 10266–10269.
136. E. Luchinat, L. Barbieri, J. T. Rubino, T. Kozyreva, F. Cantini and L. Banci, *Nat. Commun.*, 2014, **5**, 5502.
137. P. Polykretis, F. Cencetti, C. Donati, E. Luchinat and L. Banci, *Redox Biol.*, 2019, **21**, 101102.
138. L. Breindel, C. DeMott, D. S. Burz and A. Shekhtman, *Biochemistry*, 2018, **57**, 540–546.
139. D. S. Burz, K. Dutta, D. Cowburn and A. Shekhtman, *Nat. Methods*, 2006, **3**, 91–93.
140. D. S. Burz, K. Dutta, D. Cowburn and A. Shekhtman, *Nat. Protoc.*, 2006, **1**, 146–152.
141. D. S. Burz and A. Shekhtman, *PLoS One*, 2008, **3**, e2571.

Subject Index